Microbiology: An Integrated Approach

Microbiology: An Integrated Approach

Editor: Jake Burns

R CALLISTO REFERENCE

www.callistoreference.com

Callisto Reference,
118-35 Queens Blvd., Suite 400,
Forest Hills, NY 11375, USA

Visit us on the World Wide Web at:
www.callistoreference.com

ISBN: 978-1-64116-548-8 (Hardback)

Cataloging-in-Publication Data

Microbiology : an integrated approach / edited by Jake Burns.
    p. cm.
Includes bibliographical references and index.
ISBN 978-1-64116-548-8
1. Microbiology. 2. Microorganisms. 3. Medical microbiology. I. Burns, Jake.
QR41.2 .M53 2022
579--dc23

Table of Contents

Preface

The purpose of the book is to provide a glimpse into the dynamics and to present opinions and studies of some of the scientists engaged in the development of new ideas in the field from very different standpoints. This book will prove useful to students and researchers owing to its high content quality.

Microbiology is the study of microbial life. This discipline encompasses the study of acellular, unicellular and multicellular organisms, such as viruses, bacteria, archaea, fungi, algae, etc. The study of microbiology is significant due to the extensive use of microorganisms in industrial fermentation processes, bioremediation of wastes and for the production of a variety of biopolymers, such as polyesters, polysaccharides and polyamides. There is immense scope of microbial use in medical science, notable among which are antibiotic production, production of beneficial vitamins and amino acids, suppression of pathogenic microbes and development of biopolymers with applications in tissue engineering and drug delivery. Research that explores microbes as therapeutic agents against cancer is also being pursued. This book brings forth some of the most innovative concepts and elucidates the unexplored aspects of microbiology. The various advancements in this area of study are glanced at and their applications as well as ramifications are looked at in detail. With state-of-the-art inputs by acclaimed experts of this field, this book targets students and professionals.

At the end, I would like to appreciate all the efforts made by the authors in completing their chapters professionally. I express my deepest gratitude to all of them for contributing to this book by sharing their valuable works. A special thanks to my family and friends for their constant support in this journey.

Editor

An extensive endoplasmic reticulum-localised glycoprotein family in trypanosomatids

Harriet Allison[1], Amanda J. O'Reilly[1], Jeremy Sternberg[2] and Mark C. Field[1],*

[1] Division of Biological Chemistry and Drug Discovery, University of Dundee, Dundee, Scotland, DD1 5EH.
[2] School of Biological Sciences, University of Aberdeen, Aberdeen, AB24 2TZ, UK.
* Corresponding Author: Mark C. Field, E-mail: mfield@mac.com

ABSTRACT **African trypanosomes are evolutionarily highly divergent parasitic protozoa, and as a consequence the vast majority of trypanosome membrane proteins remain uncharacterised in terms of location, trafficking or function. Here we describe a novel family of type I membrane proteins which we designate 'invariant glycoproteins' (IGPs). IGPs are trypanosome-restricted, with extensive, lineage-specific paralogous expansions in related taxa. In *T. brucei* three IGP subfamilies, IGP34, IGP40 and IGP48 are recognised; all possess a putative C-type lectin ectodomain and are ER-localised, despite lacking a classical ER-retention motif. IGPs exhibit highest expression in stumpy stage cells, suggesting roles in developmental progression, but gene silencing in mammalian infective forms suggests that each IGP subfamily is also required for normal proliferation. Detailed analysis of the IGP48 subfamily indicates a role in maintaining ER morphology, while the ER lumenal domain is necessary and sufficient for formation of both oligomeric complexes and ER retention. IGP48 is detected by antibodies from *T. b. rhodesiense* infected humans. We propose that the IGPs represent a trypanosomatid-specific family of ER-localised glycoproteins, with potential contributions to life cycle progression and immunity, and utilise oligomerisation as an ER retention mechanism.**

Keywords: Trypanosoma brucei, protein sorting, exocytosis, variant surface glycoprotein, endoplasmic reticulum, evolution, trypanosome.

Abbreviations:
ER - endoplasmic reticulum,
ERGIC - ER-Golgi intermediate compartment,
IGPs - invariant glycoproteins,
ISG - invariant surface glycoprotein,
PAD1/2 - protein associated with differentiation 1/2,
TMD - trans-membrane domain,
VSG - variant surface glycoprotein.

INTRODUCTION

Trypanosoma brucei is the causative agent of both Human African Trypanosomiasis and 'nagana' in livestock and game animals, important neglected tropical diseases afflicting much of sub-Saharan Africa [1]. *T. brucei* has a complex life cycle, replicating within the mammalian bloodstream/lymphatic systems and central nervous system, as well as in various organs and tissues of the tsetse fly vector [2]. During infection of the mammalian host, trypanosomes proliferate rapidly as morphologically slender forms. When a tsetse fly ingests trypanosomes, long slender forms are rapidly killed, whereas short stumpy forms are resistant and more efficiently establish infection [3]. However, at the peak of each parasitaemic wave these forms mature into $G_{1/0}$-arrested stumpy forms in response to stumpy induction factor, and partially differentiate for survival within the fly [4,5]. This process may limit parasite populations, promoting chronic infection and preventing premature host death to augment transmission [3,6]. Arrest of stumpy cells in $G_{1/0}$ ensures that most developmental changes accompanying transmission into the tsetse fly coordinate with cell cycle re-entry.

Trypanosomes express high-abundance GPI-anchored proteins at the surface in both mammalian and insect hosts. In the mammalian form there are ~1×10^7 copies per cell of the variant surface glycoprotein (VSG), which represents ~10% of total protein. Switching between VSG variants permits escape from immune responses directed towards the previously expressed VSG, enabling chronic infection [7,8]. VSG forms a dense, ~15 nm-thick surface coat, believed to protect underlying invariant epitopes from immune recognition. Due to comparatively low abundance compared to VSG, it has been challenging to identify additional membrane proteins biochemically. Evolutionary divergence adds additional complexity with defining surface or intracellular *trans*-membrane domain (TMD) proteins and membrane protein targeting signals. Knowledge of turnover mechanisms and functions is poor, despite the importance for maintaining the composition of the surface and intracellular organelles [9,10]. Several large invariant TMD glycoprotein families, ISG60, ISG65 and ISG75, expressed exclusively in bloodstream form parasites, are known [11,12] and are termed invariant as they do not exhibit antigenic variation and are expressed at the surface and endosomal system by all strains of *T. brucei*. ISG65 and ISG75 contain large N-terminal extracellular domains, a single TMD and a small C-terminal cytoplasmic tail, but lack clear sequence homology with other proteins, beyond distant structural similarity to VSG [13]. ISG75 mediates uptake of the trypanocide suramin and is likely a major protein degraded by the bloodstream form endosomal system [14,15]. Other surface proteins have been described as have some components of the endomembrane system [16,17], but many surface or organellar proteomic datasets remain to be fully validated [18,19].

Targeting signals, recognised by multiple protein complexes within the endocytic and exocytic pathways, ensure accurate protein sorting. In animals and yeasts classic endocytic signals include tyrosine-based NPXY/YXXØ and dileucine-based [DE]XXXL[LI] or DXXLL signals, which are recognised by adaptin complexes[20]. Evidence for dileucine-dependent trafficking in *T. brucei* emerged through analysis of p67, the major trypanosomatid lysosomal glycoprotein; p67 contains two [DE]XXXL[LI]-type motifs, necessary and sufficient for lysosomal targeting, and likely mediated by AP-1 [21,22]. C-terminal KDEL signals (MDDL/KQDL in trypanosomes [23,24,25] retain lumenal chaperones in the ER [26,27], while dibasic residues near C- or N-termini (KKXX or KXKXXX) interact with COP-I, to facilitate retrieval from the Golgi apparatus [28]. It is unclear if these latter signals operate in trypanosomes. Additionally, post-translational modification contributes to targeting. For example, ubiquitin (Ub) is involved in membrane trafficking and membrane internalisation [29], and the cytoplasmic domains of ISG65 and ISG75 are ubiquitylated, enabling sorting against a >100-fold VSG excess [14,30]. Although a systematic analysis has not been reported, the number of GPI-anchors per complex has also been proposed to influence targeting [31,32,33].

With the exception of these examples, the signals required for targeting most *T. brucei* TMD proteins have not been defined, and the locations of the vast majority of TMD proteins are unknown. There is clear potential for divergence in both targeting signal primary structure (e.g., MDDL v KDEL) and in the molecular mechanisms underpinning trafficking, e.g., trypanosomatid-specific clathrin-associated proteins acting in endocytosis [34]. To address this gap in our understanding, we used genome scanning to identify new families of type I TMD proteins. One, which we describe here, we termed invariant glycoproteins (IGPs) due to their expression from core genome regions distinct from the VSG antigenic variation telomeric expression sites and possessing similar size to ISGs. We found that IGPs are ER localised and retained by oligomerisation, independent of specific signals, within the cytoplasmic domain. Expression is upregulated in stumpy stage trypanosomes and IGPs are recognised by antibodies from infected humans, which suggests roles in both developmental transition and host immune response. Potential analogies to the ERGIC-53 ER lectins of mammalian cells are also discussed.

RESULTS

Identification of an extensive new type I TMD protein family

We have previously described the sorting itinerary and involvement in drug uptake of the invariant surface glycoprotein (ISG) family, and which also feature as important antigens both to the immune response and as diagnostics [14,15,30,35,36]. As the vast majority of the predicted proteins within the genome with this topology have neither defined functions nor location, we set out to extend our knowledge of the signals required for sorting type I membrane proteins in the trypanosome cell by the identification of additional type I topology proteins.

To identify predicted polypeptides bearing an N-terminal ER targeting signal plus a predicted TMD, a series of computational filters were applied to the predicted proteome of the genome reference strain, *T. brucei* TREU 927 [37] (Fig. S1). This procedure identified 208 open reading frames (ORFs) (accession numbers in Table S2), which were analysed by alignment and clustering by neighbour-joining (Fig. S2A). Our searches clearly under-sample the surface/endosomal proteome as only 33 of ~80 adenylate cyclases were recovered. Nevertheless, the screen was successful in identifying all known ISG/ISG-like ORFs. Significantly, a family of twenty hypothetical proteins, forming a distinct cluster, were identified and selected for further investigation.

Evolution of the IGP family

We named this extensive protein family invariant glycoproteins (IGPs), due to similarity to ISGs in overall predicted molecular weight, architecture, predicted topology, the presence of at least one N-glycosylation site and expression from housekeeping regions of the genome and not telomeric sites associated with antigenic variation. Twenty open reading frames encoding IGPs were identified in the *T. brucei* 927 genome (the reference genome strain), divided into three subfamilies based on sequence similarity (Fig. 1A). We designated these IGP34, IGP40 and IGP48 on account of the predicted molecular weights of their core polypeptides. All of the IGP open reading frames were recovered by the screen as additional BLASTp searches of the *T. brucei* genome failed to identify additional IGP sequences.

All twenty IGPs contain a predicted N-terminal signal and C-type lectin (CLEC) domain (Pfam PF00059), albeit with relatively weak similarity to the archetypal CLEC domain (1.24 x e^{-4}) (Fig. 1B). Although sequence identity within each subfamily is high (85 to 100%), identity and similarity between IGP subfamilies is significantly lower, specifically IGP48 against IGP40 (25-30%) and IGP48 against IGP34 (50-60%). The most conserved region within the IGP family spans the predicted CLEC domain. Residues 247-336 of the IGP48 subfamily are similar to the trimerisation domain of the LpxA-like enzyme superfamily that occurs within many bacterial transferases, including UDP-N-acetylglucosamine acyltransferase (Pfam PF00132). This region is composed of multiple DENTTV repeats, which contain an N-glycosylation

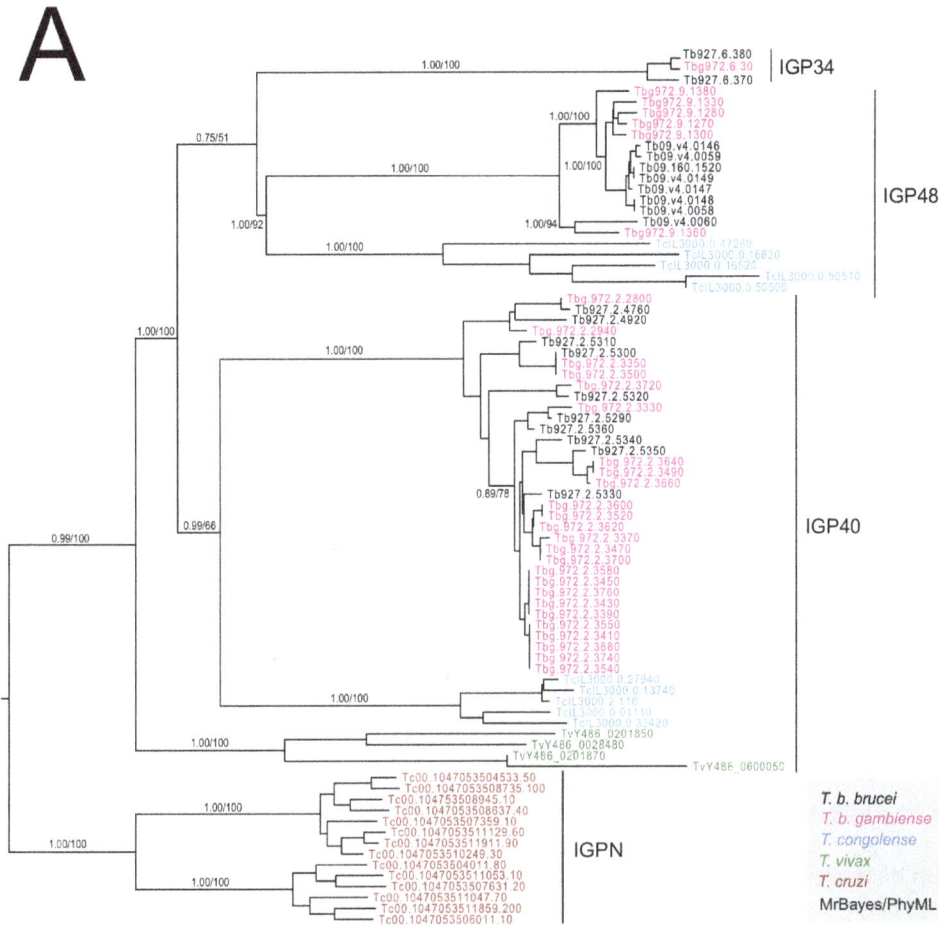

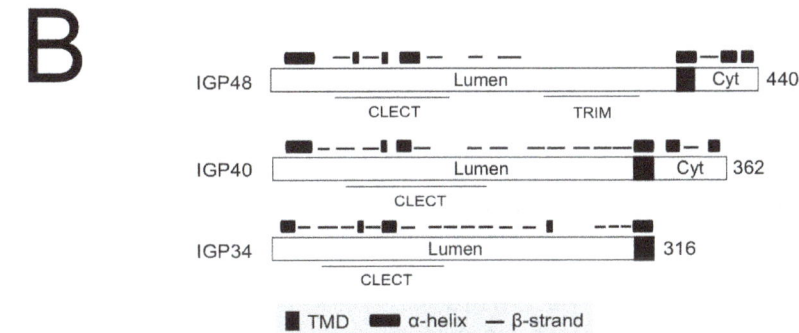

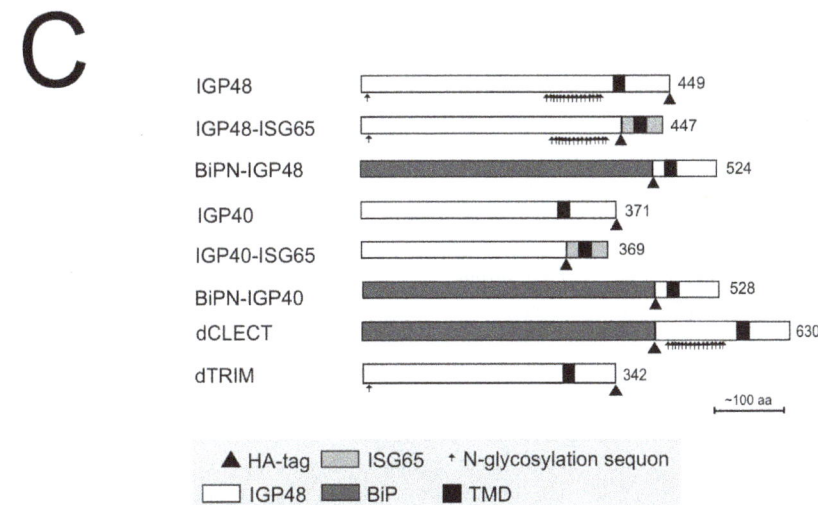

FIGURE 1: The invariant glycoprotein (IGP) family. (A) Phylogenetic reconstruction of the IGP family. The tree shown is the best Bayesian topology with branch support for important nodes indicated from both Bayesian and PhyML calculations. Clades are indicated by vertical bars. **(B)** Schematic diagram showing the secondary structure prediction and domain architecture of the IGP family. Lumen indicates the portion of the molecule predicted to be within the ER lumen and Cyt designates the predicted cytoplasmic domain. CLECT indicates C-type lectin domain. **(C)** Schematic structures of epitope-tagged and domain-swap IGP constructs used in this study. All constructs contain an HA tag as indicated by small triangle below the bar.

Abbreviations: IGP48, full length IGP48; IGP48-ISG65, IGP48 ectodomain fused to ISG65 TMD and cytoplasmic domain; BiPN-IGP48, IGP48 ectodomain replaced with BiPN, retaining IGP48 TMD and cytoplasmic domain; IGP40, full length IGP40; IGP40-ISG65, IGP40 ectodomain fused to ISG65 TMD and cytoplasmic domain; BiPN-IGP40, IGP40 ectodomain replaced with BiPN, retaining IGP40 TMD and cytoplasmic domain; dTRIM, IGP48 with predicted trimerisation domain deleted.

sequon and a potential β-turn, the latter suggesting that they may be occupied by an N-glycan.

IGP genes are present in all salivarian trypanosome genomes, plus *Leishmania*, *Leptomonas*, *Phytomonas* and the American trypanosome *T. cruzi* (Fig. S2B and S2C). Many of the IGP proteins are encoded within tandem gene arrays, and hence independent array expansion and contraction is a likely mechanism for the apparent plasticity between taxa of IGP copy number. Two IGP orthologues are also present in the recently sequenced *T. grayi* genome [38]. Significantly, no evidence for IGP-related genes was found in non-trypanosomatid Excavata genomes, specifically *Naegleria gruberi*, *Trichomonas vaginalis* and *Giardia intestinalis*, or non-excavate taxa (data not shown).

Phylogenetic analysis established the evolutionary history of the IGP family across the kinetoplastida (Fig. S2A, S2B, S3A and S3B). The IGP family possesses no obvious homology to the ISGs, and phylogenetic reconstruction of all ISG, ISG-like and IGP predicted protein sequences in *T. brucei* robustly reconstructs these as independent clades (1.0/98 Mr Bayes posterior probability/PhyML bootstrap support). Further, a scan of all eukaryotic genomes only identified IGP-related sequences in kinetoplastids, indicating the presence of a trypanosomatid-specific defining domain, which has been assigned a unique pfam ID: Pfam PF16825 (DUF5075). Expansions are common within trypanosomes and *Leishmania*, but interestingly these expansions are specific to each lineage (Fig. S2C). Moreover, the number of IGP paralogs tends to be lower in the *Leishmanias*, and the phylogeny indicates that the origin of several IGP paralogs predates speciation of the *Leishmania* and *Phytomonas* lineages. Significantly, *Angomonas deanei* and *Strigomonas culicis*, both monogenous parasites (as opposed to the digenous *Leishmanias* and trypanosomes) have only two IGP paralogs each, and *Leptomonas pyrrhocoris*, also monogenous, has just three IGP paralogs. These data suggest that the IGP family may have expanded in response to the greater burdens of infecting multiple hosts, and, moreover, that this process occurred independently on several occasions. Overall, these data suggest that the IGP gene family arose during the origin of the Kinetoplastida, and clearly predating the ISGs which are restricted to salivarian trypanosomes.

IGPs are developmentally regulated

To gain insights into the functions of the IGP proteins we initially examined their expression in the major accessible life stages of *T. brucei*. Long slender and PCF stage material was obtained from *in vitro* culture and short stumpy mRNA and protein lysates were from infected mice (kind gift of Keith Matthews, University of Edinburgh). The mRNA abundance of each IGP subfamily was assessed using primers designed to amplify all copies within a subfamily (Fig. 2A). All three IGP subfamilies are significantly up-regulated in the short stumpy bloodstream stage, most noticeably IGP34 (~10-fold compared to long slender BSF and PCF), whereas mRNA levels in the long slender BSF and PCF are relatively similar to each other.

An affinity purified rabbit anti-IGP48 polyclonal antibody was used to probe whole-cell lysates and detected a single band at 75 kDa, significantly greater than the predicted core polypeptide molecular weight of 48 kDa, likely due to post-translational modification. Probing whole-cell lysates from these same life cycle stages with anti-IGP48 confirmed that IGP48 is indeed up-regulated in the short stumpy stage by ~4-fold at the protein level compared to long slender and PCF cells (Fig. 2A). As a control, expression of PAD1 (protein associated with differentiation 1) was also monitored using a polyclonal antibody (again, kind gift of Keith Matthews). PAD1 immunoreactivity was only observed in the short stumpy bloodstream cell lysates.

IGP40 and IGP48 are N-glycosylated and locate to the ER

To further understand IGP function, we chose to determine the location of the IGPs. C-terminally HA epitope-tagged IGP48, IGP40 and IGP34 constructs were expressed in BSF 427 trypanosomes (Fig. 2B). One representative of each of the IGP48 (Tb09.v4.0147), IGP40 (Tb927.2.5330) and IGP34 (Tb927.6.380) families were selected for further study on account of demonstrating high identity with the majority of the remaining IGP paralogs in that subfamily, and hence most likely to report on the location of the remaining family members. Positive transformants were co-stained with antibody to TbBiP, an ER-resident protein [41] and HA (Fig. 2B). Fluorescence for IGP40, IGP48 and TbBiP exhibited extensive overlap, indicating that IGP40 and IGP48 have a substantial presence at the ER. Despite multiple attempts, no signal could be detected by either immunofluorescence or Western blotting for IGP34. To determine if IGP48 localised to other compartments within the cell, BSF cells expressing IGP48-HA were co-stained with TbRabX2, a Golgi marker [42], and p67, a lysosomal marker [43]. Results indicated no obvious presence in the lysosome, but potentially some presence at the Golgi complex at steady state, but which is comparatively minor compared to the extensive ER population.

Western blots of cell lysates expressing IGP40-HA or IGP48-HA ectopic constructs suggested that IGP48 migrates with an observed molecular weight ~25 kDa greater than the predicted core polypeptide, whereas the observed molecular weight of IGP40 was similar to its predicted polypeptide core, suggesting that IGP48 undergoes extensive post-translational modification. Given fifteen predicted N-glycosylation sites in the IGP48 TRIM domain, it is probable that IGP48 bears the additional weight of multiple N-glycans (Fig. 1C). To assess this possibility, cells were cultured in the presence of tunicamycin or whole cell lysates were treated with peptide N-glycanase (PNGase F) or endoglycosidase H (Endo H) prior to analysis by Western blotting (Fig. 2C). Treatment with tunicamycin and PNGase F resulted in increased mobility of ~15 kDa, whereas Endo H digestion led to a mobility shift of only ~10 kDa. This suggests that IGP48 contains a mixture of Endo H-sensitive oligomannose glycans and Endo H-resistant paucimannose and/or complex N-glycans, the latter a result of processing in the Golgi complex. Therefore, the presence of the vast majority of IGP48 in the ER at steady state suggests the

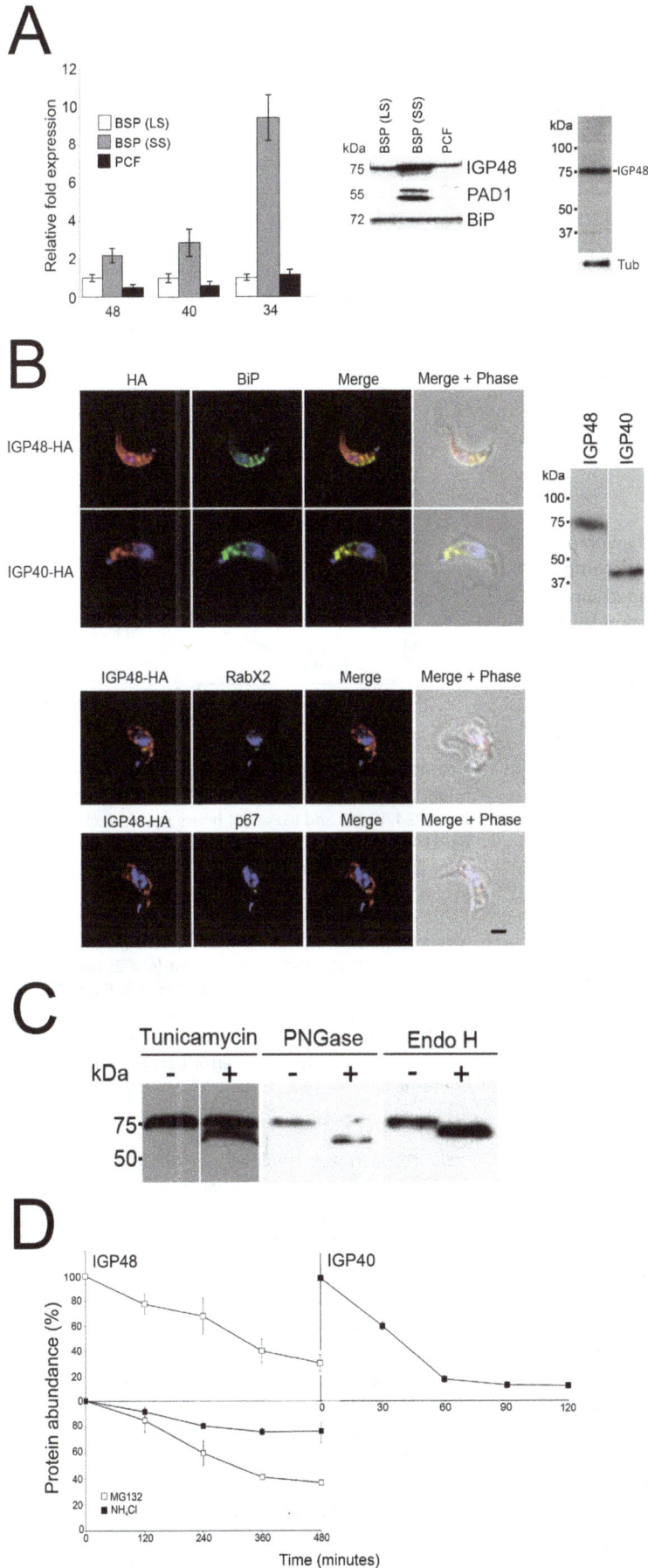

FIGURE 2: The IGP family are developmentally regulated ER proteins. (A) Copy numbers of IGP48 [39], IGP40 [40] and IGP34 [34] mRNAs measured by qRT-PCR, in different life cycle stages, normalised to long slender BSF mRNA levels at 1.0. Error bars denote standard errors of the mean from triplicate measurements on independent RNA samples. Western blot of trypanosome whole cell lysates using anti-IGP48 affinity-purified antisera raised against *E. coli*-expressed recombinant protein at 1:100 dilution. The blot was re-probed for PAD1 (protein associated with differentiation 1), and which is specifically unregulated in the stumpy bloodstream form, to validate the short stumpy lysate. Rightmost; whole cell lysate probed with anti-IGP48 antisera to validate specificity. **Abbreviations:** BSF (LS), long slender bloodstream form; BSF (SS), short stumpy bloodstream form and PCF, procyclic culture form. (B) Intracellular localisation of IGP48-HA and IGP40-HA in BSF cells under permeabilised conditions, and detected with anti-HA antibody (red). **Top panel:** co-staining with anti-TbBiP (green). **Lower panels:** co-straining with anti-TbRabX2 and anti-p67 (green) using confocal microscopy. Bar = 2μm. Inset: expression of IGP48-HA and IGP40-HA in 427 BSF cells detected by Western blotting using anti-HA antibody. (C) Digestion of IGP48 with PNGase F or Endo H, or treatment with tunicamycin results in a large molecular weight shift. IGP48 was detected in fractionated lysates using anti-HA antibody. (D) Turnover of IGP48 and IGP40. Quantification of anti-HA reactivity in lysates of cells expressing IGP48-HA and IGP40-HA following inhibition of protein synthesis with cycloheximide. Error bars represent the standard deviation and values were normalised against a loading control, BiP (n = 3).

possibility of cycling through the Golgi complex to generate Endo H-resistant glycans. The remaining ~10 kDa modification may be due to the presence of other modifications, for example O-glycans or PNGase F-resistant N-glycans, aberrant migration on SDS-PAGE or a result of oligomerisation, as investigated below. Interestingly, IGP48 also binds tomato lectin (TL), which is specific for N-linked glycans containing three or more linear repeats of N-acetyllactosamine, although recent evidence suggests that TL can also bind paucimannose structures; either structural class would reside in the Endo H-resistant N-glycan fraction [38,40]. Removal of the trimerisation domain, in which all predicted N-glycosylation sites are located, ablated TL binding, indicating that the TRIM domain harbours these N-glycans (Fig. S4). Furthermore, these data confirm the predicted topology of IGP48 as a type I membrane protein, with the N-terminus accessible to the glycosylation machinery and hence located within the ER lumen.

IGP48 is degraded in a low pH compartment

The presence of IGP40 and IGP48 predominantly within the ER, but also likely trafficking through the Golgi complex, suggested that either these proteins are degraded directly from the ER by an ER-associated degradation-related mechanism [44] or that they progress to post-Golgi compartments and are degraded in terminal endosomal compartments. Turnover of IGP48-HA and IGP40-HA was monitored following treatment of cells with cycloheximide and Western blotting with anti-HA antibody (Fig. 2D). IGP48-HA and IGP40-HA differed significantly in their stability, with a half-life of ~6 hours for IGP48-HA and ~30 minutes for IGP40-HA. Turnover of IGP48-HA was further analysed by inhibition of lysosomal functions with the weak base ammonium chloride or a proteasome inhibitor, MG132, to inhibit ER-associated degradation. Ammonium chloride clearly delayed IGP48-HA degradation with ~80% of protein remaining after 4 hours compared to ~40% in untreated cells, but no significant effect was seen with the proteasomal inhibitor. Together with the absence of a clear lysosomal pool of IGP48 (Fig. 2B), these data suggest that IGP48 is delivered to the lysosome but rapidly degraded once attaining the terminal endosomal compartment. Overall, despite a predominant presence within the ER, IGP48 may exit the ER, consistent with both the presence of Endo H-resistant N-glycans and ammonium chloride-sensitive turnover.

IGP proteins are required for normal proliferation

RNAi knockdown was used to investigate the putative roles and importance to viability of the IGP gene products, using constructs designed to suppress all members of each subfamily. Knockdown was subfamily specific, with no cross-suppression against other subfamilies detected by qRT-PCR (data not shown).

Specific down-regulation at the mRNA level of the three IGP subfamilies was confirmed by qRT-PCR after 24 hour induction (Fig. 3A) and indicated that mRNA levels for the IGP34, IGP40 and IGP48 induced cell lines were reduced by ~50%, 65% and 50% compared to uninduced

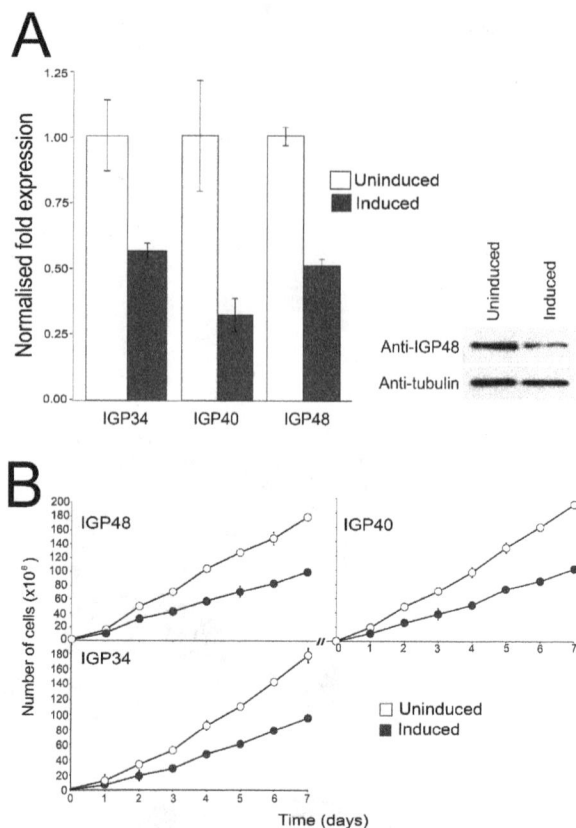

FIGURE 3: IGP40 and IGP48 are required for normal cell proliferation. (A) qRT-PCR of control and tetracycline-induced RNAi lines for IGP34, IGP40 and IGP48 24 hours post-induction (n = 3). **Right panel** shows a Western blot of IGP48 following RNAi induction after 24 hours with β-tubulin as loading control. (B) Growth curves of control and tetra-cycline-induced RNAi lines for IGP34, IGP40 and IGP48, showing a growth defect within 24 hours post induction. Representative results for one of two clonal cell lines for each construct studied are shown (n = 2) (all subsequent experiments were performed using this cell line). Cultures were diluted daily to maintain cell densities between 10^5 and 2×10^6 cells/ml, and a cumulative pseudo-growth curve is shown. Counts were carried out in triplicate, error bars represent standard error of the mean.

cells, respectively. Depletion of IGP48 at the protein level by ~50% was confirmed by Western blotting of whole cell lysates from induced cells 24 hours post-induction and probed with anti-IGP48 antibody (Fig. 3A, right panel).

Proliferation defects were observed following 24 hours of RNAi induction and continued for several days (Fig. 3B). These data suggest that each IGP subfamily is required for normal cell growth and proliferation, and that despite their similarities, these gene families have non-redundant functions. In addition, after two to three days, cells with multiple nuclei and kinetoplasts were observed (Fig. S5). This phenotype emerges following knockdown of a range of resident ER proteins, as well as many other genes in *T. brucei* [45]. However, this minor cytokinesis defect, as demonstrated by an increase in cells with two or more nuclei, is most likely a secondary defect as it occurs follow-

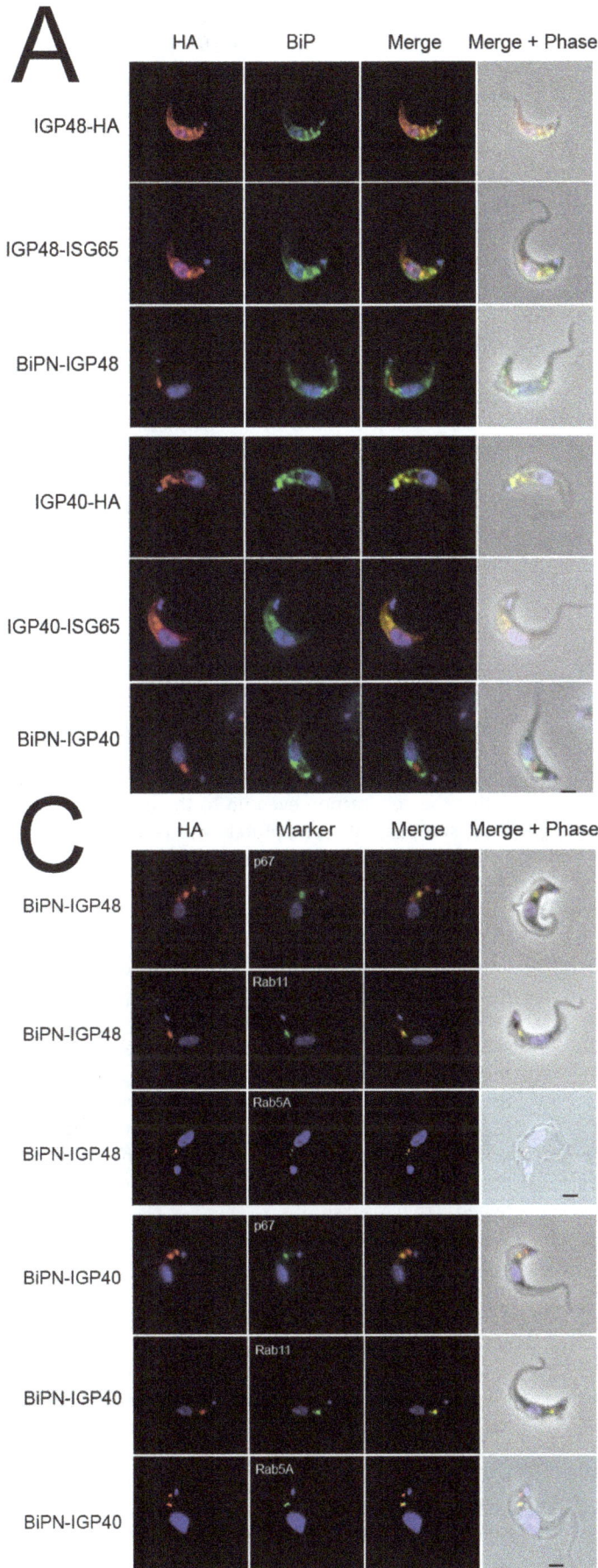

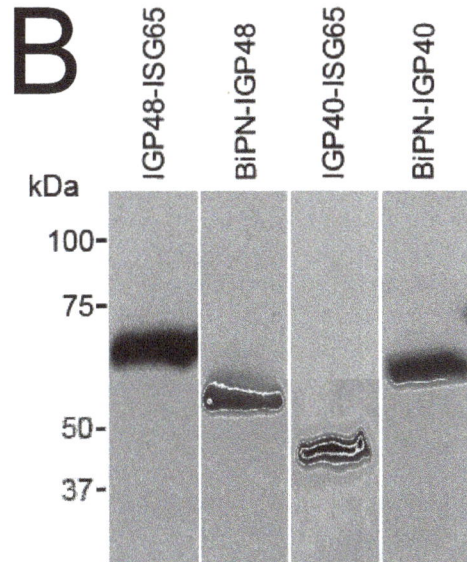

FIGURE 4: Localisation of IGP sorting signals. (A) Immunofluorescence demonstrating the locations of IGP constructs in BSF trypanosomes. HA-tagged constructs were detected in permeabilised cells with anti-HA antibody (red). The ER was stained with anti-TbBiP (green) and DNA stained with DAPI (blue). BiPN constructs, in which the lumenal/ecto-domain is replaced by the BiP ATPase domain, no longer co-localise with TbBiP and so are not retained within the ER. **(B)** Expression of IGP40-HA and IGP48-HA in 427 BSF cells detected by Western blotting using anti-HA antibody. Note that the BiPN chimeras migrate slower than predicted from their molecular weight, an observation that is consistent with the behaviour of BiPN-ISG65 chimeras reported previously [30]. **(C)** Location of BiPN constructs was determined by co-localisation with p67, Rab5A or Rab11, markers for the lysosome, early or recycling endosomes respectively (green). BiPN-IGP40 and BiPN-IGP48 both demonstrate significant overlap with all three intracellular markers. Scale bar 2 μm.

ing a significant period of proliferative impact. Therefore, each IGP subfamily is independently required for normal cellular physiology.

The IGP48 lumenal/ecto-domain is required for ER retention

To determine which IGP protein regions are required for targeting and/or ER retention a panel of deletion and domain-swap constructs was created (Fig. 1C). We chose to focus on IGP40 and IGP48 due to our inability to localise IGP34, and hence lack of information on the location of members of this subfamily. We produced constructs to investigate the presence of targeting information in the C-terminal domains of IGP40 and IGP48 by replacing these with the equivalent portion of ISG65, which in the native context, or when fused to the N-terminal domain of BiP (BiPN), support efficient trafficking to the cell surface and endosomal targeting (IGP40-ISG65 and IGP48-ISG65). To investigate lumenal/ecto-domain targeting contributions, BiPN was fused to the IGP40 and IGP48 TMD, including a short spacer region from the lumenal/ecto-domain to ensure preservation of correct membrane topology (BiPN-IGP40 and BiPN-IGP48). BiPN contains essentially no targeting information and is normally rapidly secreted [35], and has only a very modest impact on the targeting or turnover of ISG65 and ISG75 fusion constructs, where the cytoplasmic signals contain the major targeting determinants [14,30,35].

IGP40-ISG65 and IGP48-ISG65 both co-localised with BiP, suggesting an ER location indistinguishable from full-length IGP40 and IGP48 proteins (Fig. 4A), and indicating that the IGP TMD and cytoplasmic domains do not possess essential ER retention signals. By contrast, replacement of the IGP40 or IGP48 lumenal/ecto-domain with BiPN (BiPN-IGP48 and BiPN-IGP40) resulted in loss of co-localisation with BiP, suggesting that these constructs now exit the ER. A Western blot of cell lysates expressing constructs confirmed correct incorporation of the various tags and chimeras (Fig. 4B). BiPN-IGP40 and BiPN-IGP48 were localised to distinct puncta between the nucleus and kinetoplast; co-localisation with p67, Rab11 and Rab5A demonstrated that these correspond to several endosomal compartments, including the lysosome, recycling endosomes/exocytic carriers and early endosomes, indicating inefficient retention by the ER and trafficking onto the degradative arm of the endosomal system (Fig. 4C). Together, these data indicate that the basis for ER retention resides within the IGP lumenal/ecto-domain, and not the TMD or cytoplasmic domain.

The IGP40/IGP48 lumenal/ecto-domain is required for retention by the cell

The presence of BiPN-IGP fusion proteins in Rab11-positive compartments suggests that these chimeras may be exported to the cell surface and/or be delivered into endosomes. To investigate the fate of post-ER IGP chimeras, cells expressing BiPN-IGP40, BiPN-IGP48 or full-length IGP48 were surface-derivatized with biotin using a membrane-impermeant biotinylation reagent [30] and lysates fractionated into biotinylated (surface-accessible) and non-biotinylated (internal) fractions using streptavidin agarose. The relative levels of the constructs recovered in each fraction were revealed by Western blotting with anti-HA antibody (Fig. 5A). IGP48-HA was recovered entirely within the underivatized pool, suggesting that the protein was inaccessible to biotinylation, whereas a small fraction of both BiPN-IGP constructs was biotin-accessible. The distribution of native ISG75, which has a presence both at the surface and in intracellular organelles, was faithfully reflected by the biotinylation analysis [14], while RabX1, a cytoplasmic protein, was not detected in the biotinylated fraction, indicating that the cells remained intact during the procedure. To confirm that BiPN-IGP constructs are present on the cell surface, optical sections of non-permeabilised cells were taken using confocal microscopy (Fig. 5A, lower panel). In agreement with previous data, no evidence for IGP40 or IGP48 on the cell surface was obtained, but by contrast, both BiPN-IGP constructs are clearly seen at the cell periphery in central confocal sections, consistent with a surface location.

To determine whether BiPN-IGP chimeras are secreted from the cell when they reach the surface, radio-immunoprecipitations were performed on cells and culture media for BiPN-IGP48, IGP48 and BiPN [35]. Cells were pulse-radiolabeled with ^{35}S-methionine and chased, after which labelled proteins were recovered by immunoprecipitation with anti-HA antibody (Fig. 5B). IGP48 remained within the cell fraction even up to three hours, indicating that essentially all of the protein was retained by the cell. By contrast, after three hours a ~50 kDa proteolytic fragment (P) of BiPN-IGP48 was detectable in the medium, which clearly retains the HA-epitope (Fig. 5B). Essentially all BiPN was secreted as expected. Very small quantities of intact BiPN-IGP48 are shed into the medium up to 30 minutes post-chase, but the 50 kDa fragment was observed in increasing quantities up to three hours, likely reflecting a true secretion event. The molecular weight of this fragment suggests that cleavage occurs at a point immediately N-terminal to the *trans*-membrane domain of IGP48, and is consistent with BiPN-IGP48 attaining the cell surface. However, both the pulse-chase and biotinylation data indicate that the vast majority of the protein is retained within the cell.

The half-lives of BiPN-IGP48 and BiPN-IGP40 were determined to be approximately two hours and one hour respectively (Fig. 5C). This significantly fast rate of degradation and the presence of some of the BiPN-IGP constructs within the lysosome (Fig. 4B) suggests that these constructs are trafficked to this organelle. Turnover of BiPN-IGP48 was also analysed in the presence of MG132 or NH_4Cl, with both significantly increasing the half-life of this construct, so that after four hours ~70% of protein remained, compared to ~30% in untreated cells. This suggests that BiPN-IGP48 is degraded by both a lysosomal and proteasome-dependant mechanism, distinct from IGP48 itself where the proteasome has no apparent role.

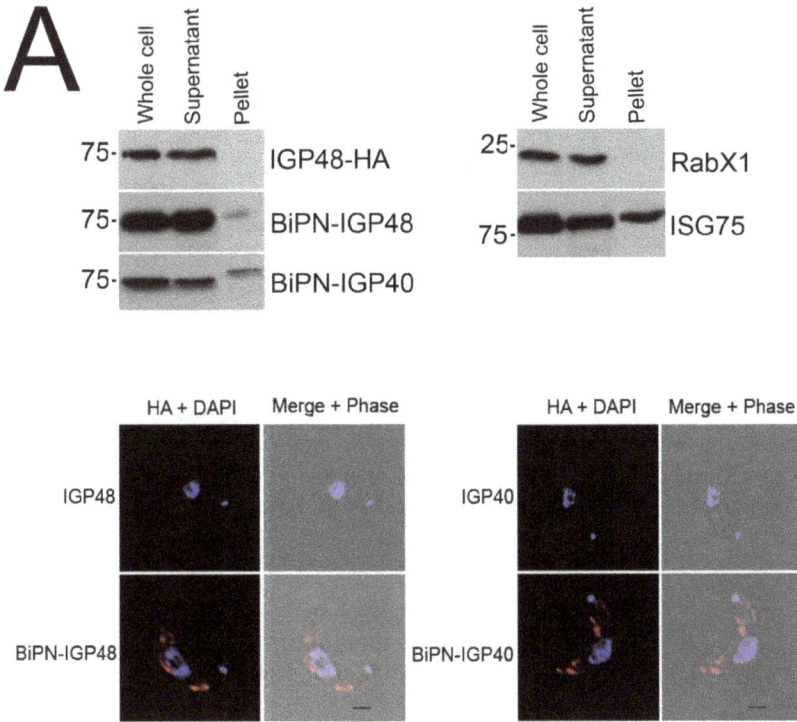

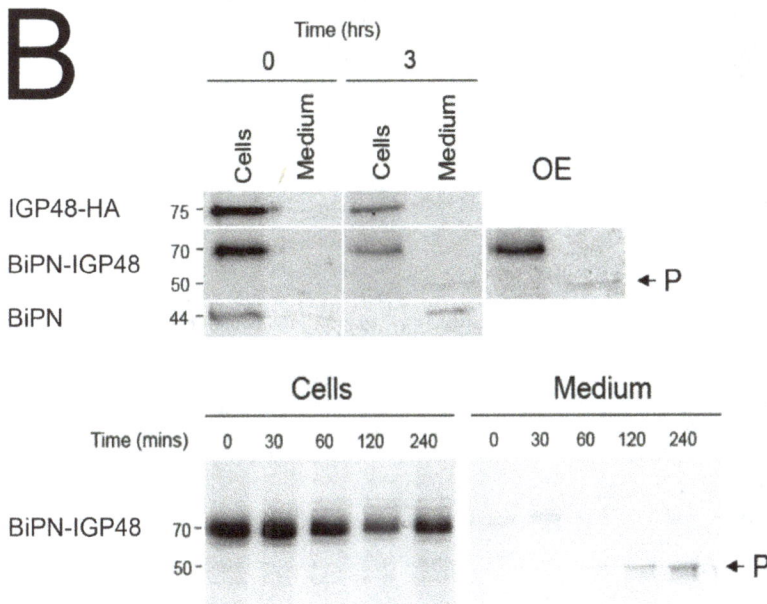

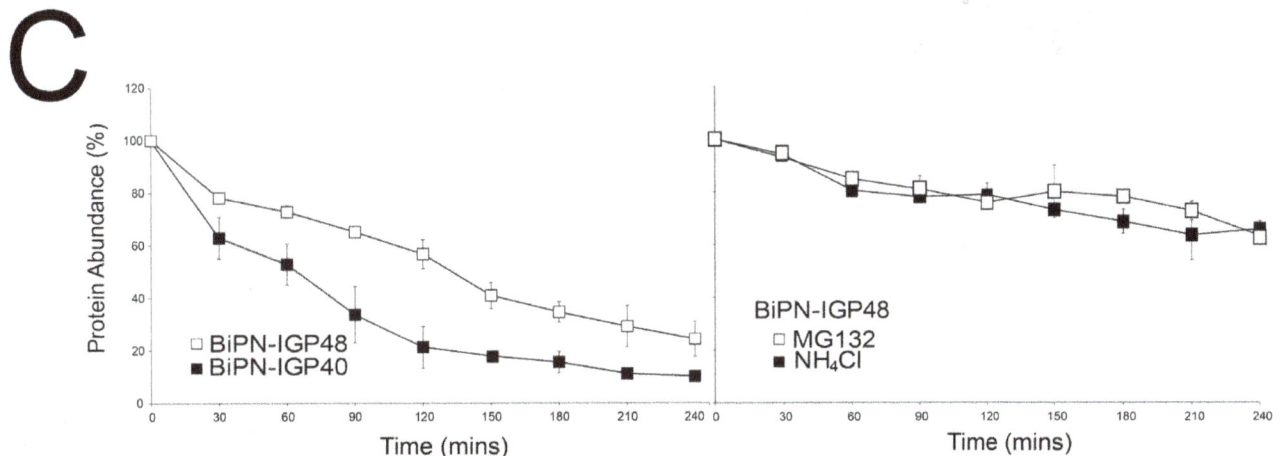

FIGURE 5: The IGP ectodomain is required for retention. (A) Western blot analysis of surface-biotinylated (Pellet) and non-biotinylated (Supernatant) cells to detect surface-exposed IGP proteins and chimeras. Blots were also probed for an intracellular (TbRabX1) and surface (ISG75) control, to demonstrate that cells are intact and that surface components are successfully biotinylated. Note that the Pellet lanes have been moved in Photoshop simply for clarity and no other manipulation has taken place. Surface presence of BiPN-IGP48 and BiPN-IGP40 was further demonstrated by confocal microscopy. Non-permeabilised cells were stained with anti-HA antibodies (red) and for DNA (blue). Scale bar = 1 μm. **(B)** Kinetics of protein secretion. BSF trypanosome cells expressing IGP48-HA, BiPN-IGP48 and BiPN were pulse-labeled with ^{35}S-Met/Cys for 15 minutes and then chased for 3 hours. At 0 and 3 hours cultures were separated into cell and medium fractions. Labeled proteins were immunoprecipitated with anti-HA and separated by SDS-PAGE. OE, OverExposure to reveal a 70 kDa proteolytic fragment (P) cleaved from BiPN-IGP48. **Lower panel:** Detailed kinetics of BiPN-IGP48 secretion. Cells were pulse-labeled with ^{35}S-Met/Cys for 15 minutes and at the indicated chase times, aliquots were treated as described above. The proteolytic BiPN-IGP48 50 kDa fragment (P) appears in the medium after 1 hour. **(C) Left panel:** Turnover kinetics of BiPN-IGP48 (open symbols) and BiPN-IGP40 (closed symbols) was determined by blocking protein synthesis with cycloheximide and detecting residual protein with anti-HA antibodies. **Right panel:** Turnover is sensitive to inhibition by 10 μM MG-132 (open symbols) or 20 mM NH$_4$Cl (closed symbols). Results were normalised to 100% at t = 0. The graph represents the mean of two independent experiments, with the standard error of the mean indicated. Numbers to the left of some panels indicate the positions of co-migrated molecular weight standards and are in kDa.

To determine the contribution the CLEC or TRIM domain makes to ER retention, the locations of dCLEC and dTRIM constructs were analysed by immunofluorescence (Fig. 6A). Both constructs co-localise with BiP and RabX2, and so appear retained in the ER with a minor presence in the Golgi complex. No co-localisation was observed with Rab11, suggesting that the proteins are unable to enter endocytic or late exocytic systems; the absence of surface staining indicates that these constructs do not reach the plasma membrane. Therefore, both the trimerisation domain and CLEC domain are independently sufficient for IGP48 ER retention, whereas removal of both releases the IGP48 *trans*-membrane and cytoplasmic domain from the ER. The observed molecular weights of both dCLEC (~65 kDa) and dTRIM (~36 kDa) (Fig. 6A, right panel) are in agreement with those predicted for these constructs, and therefore, it appears that neither construct is modified

post-translationally to the same extent as IGP48 itself. Further, the half-life of both constructs was significantly reduced compared to intact IGP48, at two hours for dCLEC and one hour for dTRIM (Fig. 6B). Therefore, the absence of either the lectin or trimerisation domain likely destabilises the IGP48 protein.

IGP48 forms higher order oligomeric complexes
The finding that retention by the ER required the lumenal domain of IGP40 or IGP48 suggested that a simple sequence-based signal may not be responsible for targeting, and we considered that assembly into a higher order complex could contribute. To investigate this, lysates of cells expressing full-length IGP48-HA, IGP48-ISG65 or BiPN-IGP48 were analysed by blue-native PAGE and Western blotted using anti-HA antibody (Fig. 7, left panel). Each protein was detected as multiple bands, and significantly

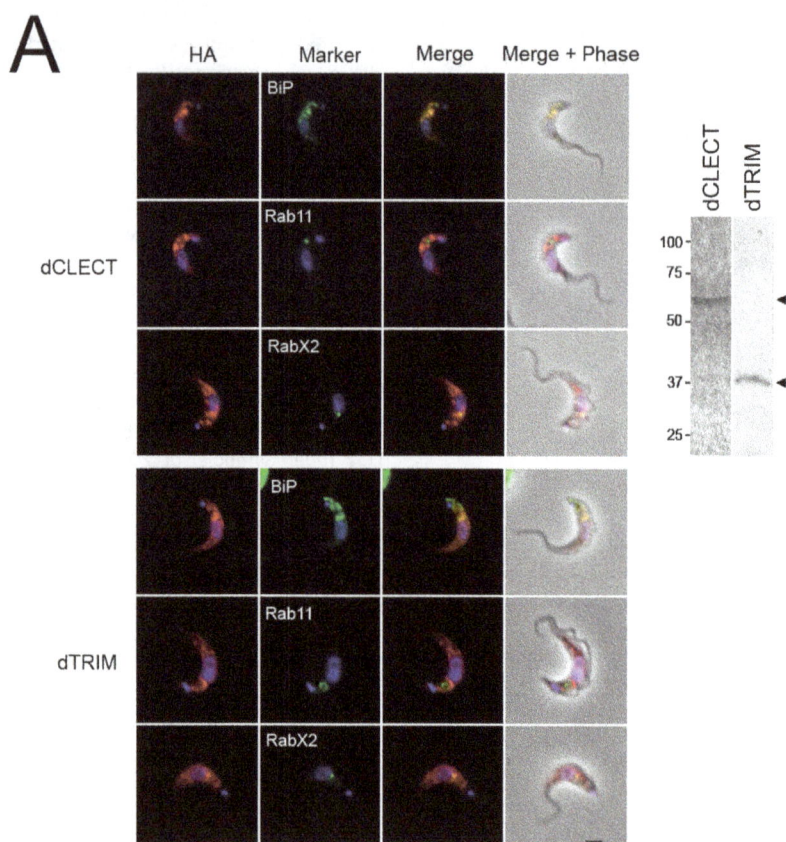

FIGURE 6: Subcellular localisation of BiPN-IGP48 chimeras is not dependent on the lumenal domain. (A) The IGP48 CLEC (dCLEC) or trimerisation lumenal domain (dTRIM) was replaced with the BiP ATPase domain and the location determined by immunofluorescence. In each case the BiPN chimera is in red and a marker protein visualised using polyclonal antibodies is in green. Scale bar = 1 μm. Verification of protein expression was carried out by Western blot (inset at right). Due to low expression, blots have been deliberately overexposed and relevant reactivity is indicated with arrows. (B) Turnover of dCLEC and dTRIM constructs. Protein degradation following cycloheximide treatment was monitored as described in Figure 2. Experiments were done in duplicate and error bars indicate standard error of the mean.

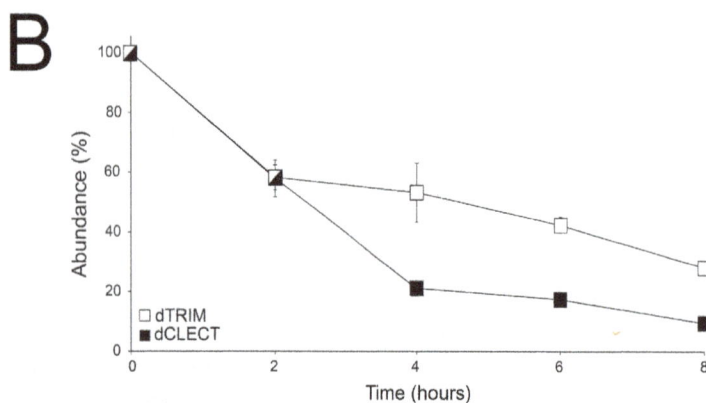

for IGP48 and IGP48-ISG65, the patterns were quite similar, with an intense band at approximately 200 kDa and a less intense doublet at ~450 kDa. This behaviour suggests incorporation into complexes, with multiple isoforms. Significantly, when the IGP48 ectodomain was replaced with BiPN, the slower migrating forms were lost and bands of only ~146 kDa and ~200 kDa forms were observed, suggesting that higher order complex formation required the IGP48 ectodomain.

To determine if these oligomeric complexes represent homomeric interactions, co-immunoprecipitations from cell lines harbouring both HA and FLAG epitope-tagged IGP constructs was performed (Fig. 7, right panel). Immunoprecipitation with anti-FLAG followed by Western blotting with anti-HA antibody revealed a band corresponding to IGP48 in the cell line containing both IGP48-HA and IGP48-FLAG, indicating that IGP48 can form homomeric interactions. Incubation of whole cell lysates of the double tag-containing cell line (IGP48-HA and IGP48-FLAG) with pro-

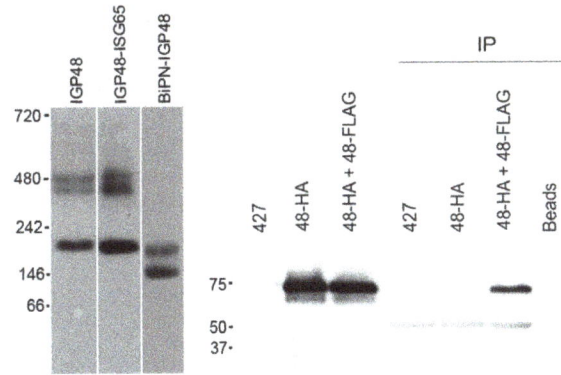

FIGURE 7: IGP48 forms a complex *in vivo*. Lysates from cells expressing various IGP48 chimeras were subjected to native PAGE, followed by detection by Western blotting. **Left:** Replacement of the IGP48 ectodomain with the BiP ATPase domain (BiPN-IGP48) results in loss of the high molecular weight (450 – 500 kDa) complexes. **Right:** Cells expressing IGP48-HA or IGP48-HA (48-HA) plus IGP48-FLAG (48-FLAG) were immunoprecipitated with anti-FLAG antibody, followed by Western blotting with anti-HA antibody. Whole cell lysates are shown to the left, and wild-type (427) and single transfected cells, as well as a bead plus lysate with no antibody IP control (Beads).

tein A beads but no anti-FLAG antibody shows that protein is not binding to the beads non-specifically, while no bands were detected from non-transfected cells or a cell line containing only IGP48-HA. Therefore, the IGP ectodomain is both necessary and sufficient for protein-protein interactions within the ER, for homo-oligomerisation and is required for ER retention.

IGP48 is up-regulated in *in vitro* surrogates of bloodstream to procyclic stage differentiation

Alterations in both copy number and protein location between trypanosome life stages are well known [46,47]. We asked if cell stage-specific expression level of IGP48 also led to changes in location. Laboratory-adapted monomorphic slender trypanosomes, including Lister 427 MITat1.2 strain, generated by long-term passage [48], have lost the ability to develop into short stumpy cells *in vivo*. However, these strains are able to differentiate into partially-differentiated cells in response to cold shock ($\Delta T > 15°C$), where the expression of PCF surface antigens is induced as well as rendering cells 1000-fold more sensitive to *cis*-aconitate-induced differentiation [49]. Differentiation can also be mimicked *in vitro* by treatment with a membrane-permeable cAMP derivative, 8-(4-chlorophenylthio)-cAMP (pCPT-cAMP) [3,50].

Whole cell lysates were generated from parasites incubated at 37°C, 20°C or with 1 mM pCPT-cAMP for twelve hours, with equivalent cell numbers used per lane (Fig. S6). Protein was separated by SDS-PAGE and analysed by Western blot using rabbit anti-IGP48 polyclonal antibody. Blots were stripped and re-probed for ISG75, to show that changes to IGP48 protein levels were specific and that there was not a global change to protein expression levels.

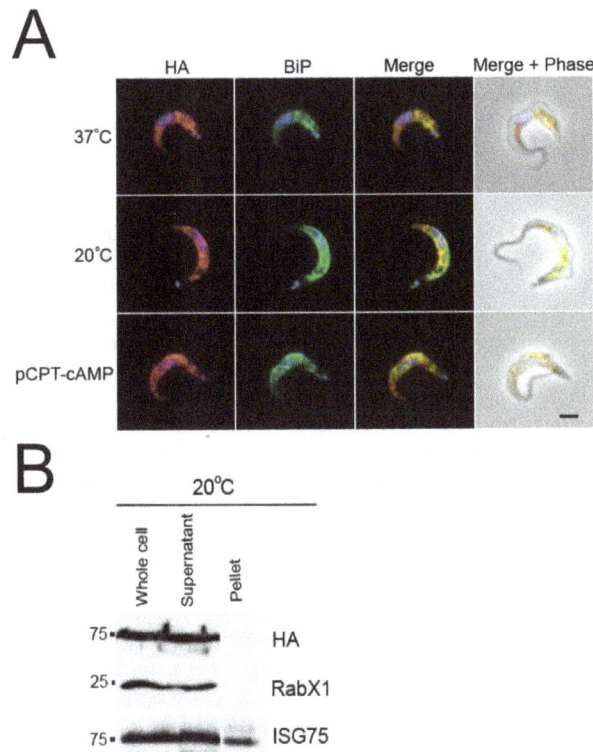

FIGURE 8: IGP48 is retained in the ER in short stumpy-like cells. (A) BSF cells expressing IGP48-HA were incubated at 37°C, 20°C (cold-shock) or with pCPT-cAMP for 12 hours. IGP48-HA was visualised with anti-HA antibody and co-stained with anti-BiP antibody. IGP48-HA remains in the ER in short stumpy-induced cells. DNA was visualised using DAPI. All images are captured at the same magnification, scale bar 2 μm. (B) Surface biotinylation was performed to determine if IGP48-HA reaches the cell surface in short stumpy-like cells. Cells were cultured *in vitro* at 37°C or 20°C for 12 hours and the biotinylation assay was carried out as described previously. IGP48-HA was detected by Western blot with anti-HA antibody. Blots were stripped and re-probed for an intracellular marker, RabX1 (localises to the ER) and a surface marker, ISG75, which localises to both the surface and endosomal compartments.

Blots were also probed for lysosomal protein, p67, which is up-regulated in cells treated with pCPT-cAMP [51] and for PAD2 which is also under thermoregulation and serves as a cold-shock [46]. Both proteins were up-regulated under conditions which promote differentiation. IGP48 protein levels increase by ~1.5-fold in cold-shock and pCPT-cAMP treated cells, showing that protein levels do increase under these conditions, albeit less dramatically than for true stumpy cells (cf. Fig. 2).

To determine whether full-length IGP48 undergoes developmentally regulated routing in cells incubated at 20°C or with 1 mM pCPT-cAMP compared to cells incubated at

37°C, immunofluorescence analysis was used (Fig. 8). Cells co-stained with anti-HA and anti-BiP antibody showed significant co-localisation, suggesting that in these differentiation models the majority of IGP48 remained in the ER. To establish if IGP48 reached the plasma membrane, surface biotinylation was carried out on cells cultured at 20°C for twelve hours (Fig. 8). IGP48-HA did not appear to reach the cell surface in these cold-shocked cells. Optical sections were taken using confocal microscopy of fixed, nonpermeabilised cells expressing N-terminal HA-tagged IGP48 stained with anti-HA antibody so that only surface proteins will be stained if present (Fig. S7). No IGP48 protein was

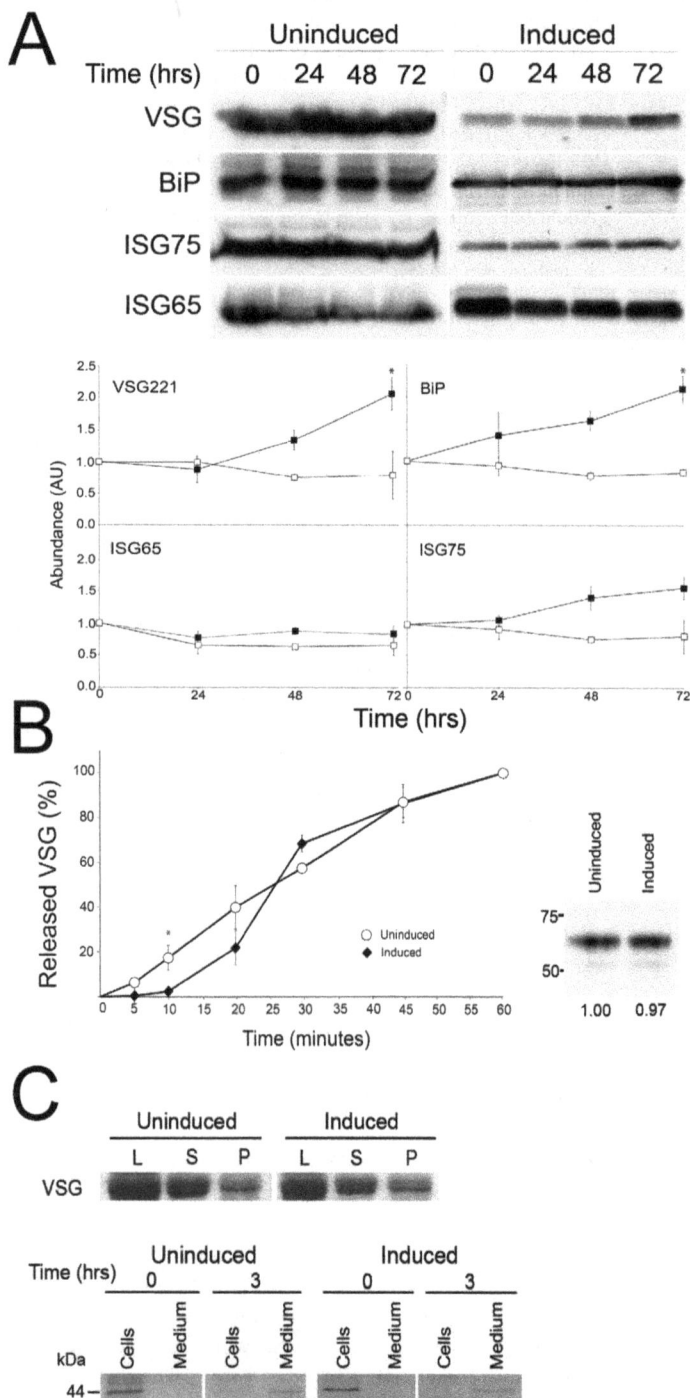

FIGURE 9: Effects of IGP knockdown on major surface protein copy number and exocytosis. (A) Cells were sampled from either induced or uninduced IGP48 RNAi cell lines at the times indicated. Membranes were probed with anti-VSG221, TbBiP, ISG65, or ISG75 and relative protein abundance was determined by densitometry. Experiments were done in duplicate and bars indicate standard error of the mean. Data were normalised to 100% at t = 0. **(B)** Export of newly synthesised VSG in induced and uninduced IGP48 RNAi cells. Surface accessible VSG was hydrolysed by GPI-PLC after hypotonic lysis of the cells. Soluble and membrane-form VSG was recovered by incubation with ConA-sepharose. Data represent the kinetics of newly synthesised VSG transported from the endomembrane system to the cell surface, shown as percent of VSG at the cell surface. Data were taken from two independent experiments and standard error of the mean is shown. Student's t-test showed statistically significant difference between induced and uninduced cells at the time point indicated with an asterisk (p < 0.05). **Right panel:** Metabolic labelling of newly synthesised VSG following 24 hour RNAi induction. Newly synthesised VSG was labelled with ^{35}S-methionine and detected by autoradiography. **(C) Top:** Hypotonic lysis, followed by separation of surface (supernatant, S) and intracellular (pellet, P) VSG 221 was visualised by SDS-PAGE and Commassie staining and levels of VSG compared to that in whole cell lysates (L). No significant difference in VSG distribution is seen between induced and uninduced cells. **Lower:** Levels of the BiPN reporter following labelling of cells with ^{35}S-methionine were detected in induced and uninduced cells, following a 3 hour chase. No significant change is seen between export of BiPN from the cell in induced compared to uninduced trypanosomes.

visible at the cell periphery in the central confocal stacks for cells incubated at 20°C, in agreement with biotinylation analysis. Therefore, it appears that IGP48 remains within the ER on the receipt of differentiation signals/mimics, and up-regulation of this protein was confirmed for both *in vitro* differentiation and *in vivo* stumpy cells.

IGP48 is not required for maintaining global glycosylation status

Since both IGP40 and IGP48 reside in the ER and possess a putative lectin-like domain, we suspected a role in quality control and/or folding of newly synthesised proteins. The impact on glycosylation status was examined by separating whole cell lysates from IGP48 RNAi induced and uninduced cells by SDS-PAGE and probing Western blots with either *Erythrina cristagalli* (EC) or *Ricinus communis* (RC) lectins conjugated to biotin (Fig. S8) to recognise N-acetyllactosamine (LacNAc) or terminal β-galactose respectively. No significant differences were found, indicating that glycosylation is not generally impacted by IGP40 or IGP48 depletion. Further, Western blotting indicated that ISG75 and ISG65 copy numbers did not significantly change between induced and uninduced cell lines, although levels of VSG221 and BiP did significantly increase (p < 0.05) (Fig. 9). A similar phenotype was present in cells silenced for proteins involved in ER quality control [45].

To analyse the internal and surface presence of BiP and VSG221 in these RNAi lines, cells were analysed by immu-nofluorescence (Fig. S9). Confocal microscopy was used to analyse central sections of permeabilised cells stained with anti-BiP at 24 and 48 hours post-induction. After 24 hours little difference is seen between induced and uninduced cells, whereas after 48 hours of induction cells with disrupted ER structures are evident. Central optical sections of permeabilised cells stained with anti-VSG221 again showed little difference between induced and uninduced cells after 24 hours, although ~ 5% of cells induced for 48 hours appeared to accumulate VSG in the ER, suggesting that by depleting IGP48, export of protein from the ER was affected. Immunofluorescence analysis of surface VSG in non-permeabilised cells showed little significant difference between the induced and uninduced cells.

To address the contribution of IGP48 in exocytosis, the export of VSG was monitored as described previously [25]. VSG export was monitored for up to 60 minutes in cells induced for IGP48 RNAi for one day (Fig. 9B, left panel). A very minor defect in VSG export to the surface was seen in induced cells at 10 minutes (p<0.05), whereas no significant difference is seen between induced and uninduced cells at later time points. Metabolic labelling indicates no apparent change to VSG synthesis levels or the distribution between internal and surface pools (Fig. 9, right panel), again similar to previous observations from silencing of ER chaperones [45]. However, previous results (Fig. 9) suggest there is accumulation in total VSG221 protein levels after 72 hours of RNAi. A possible explanation for this is that

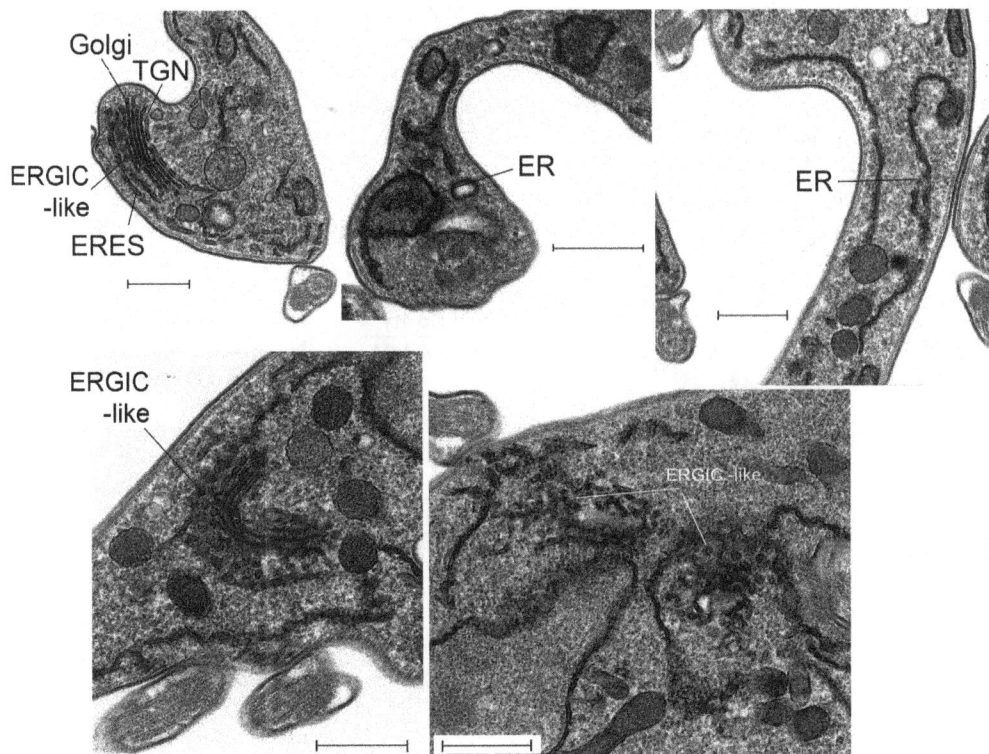

FIGURE 10: Ultrastructural analysis of IGP48 RNAi cells reveals defects in the ER. Transmission electron micrographs of IGP48 RNAi cell lines. **Top left:** Representative uninduced IGP48 RNAi cell with major secretory pathway organelles indicated. Other panels are induced cells after 24 hours induction. **Top centre:** Distorted ER with apparent lumenal inclusion. **Top right:** ER tubules with apparent normal morphology. **Lower left:** Extensive vesicles associated with the Golgi complex. Based on observations that the Golgi complex is concave towards the *trans*-face (see top left panel), these vesicles are likely ER to Golgi transport intermediates corresponding to a structure similar to the ERGIC, i.e. ER-GIC-like. **Lower right:** Examples of extensive clusters of vesicles in close association with ER tubules. Scale bars are 500 nm. **Abbreviations:** ER, endoplasmic reticulum; ERGIC, ER-Golgi intermediate compartment; TGN, *trans*-Golgi network.

VSG221 protein is being turned over less rapidly than in uninduced cells rather than increased synthesis, after prolonged IGP48 depletion. In addition, an export assay using a BiPN reporter construct showed no obvious differences between levels of exported BiPN in induced and uninduced cells after 3 hours (Fig. 9, bottom right). Overall, these data suggest that while IGP48 is located at the ER, the impact of silencing of this protein on exocytosis or the global N-glycosylation state is somewhat minimal.

IGP48 is required to maintain normal ER morphology

We used electron microscopy to assess the impact of IGP48 knockdown on the morphology of intracellular organelles in more detail (Fig. 10). There was no evidence for ER hyperplasia or large autophagosomes, which had been observed previously following knockdown of ERAP32 and ERAP18, two ER resident proteins. However, although the Golgi apparatus in induced cells appeared normal, a large number of vesicles were present in many (approximately 25%) of the cells in which the Golgi was visible, located between the ER and cisternal face of the Golgi (Fig. 10, bottom). These vesicles may represent accumulations of structures analogous to the ER-Golgi intermediate compartment (ERGIC) in mammalian cells and/or transport vesicles and possibly reflect a defect in the transport of cargos between the ER and Golgi complex. This is also consistent with an intracellular accumulation of VSG in a subset of cells (Fig. S9), and suggests that the disruption of IGP48 expression also has an influence on the production and/or consumption of ER-derived transport vesicles.

The accumulation of what appears to be vesicles at an ERGIC-like site, positioned between ER exit sites and the cis-Golgi compartments and coincident with the accumulation of VSG within the ER and other subcellular compartments, and the increase to levels of the major ER chaperone BiP, suggested the possible induction of ER stress in response to loss of IGP48. ER stress can lead to an autophagy response, and to examine this issue IGP48 knockdown

cell lines were transfected with YFP-ATG8.2, a marker for autophagosomes [48]. Following 48 hours of induced RNAi, immunofluorescence analysis was used to determine the relative number of ATG8-positive autophagosomes per cell compared to uninduced cells (Fig. S10). There was no significant difference between the relative number of ATG8-positive autophagosomes in induced compared to uninduced cells, indicating that IGP48 depletion does not activate an ATG8-dependant autophagy pathway, indicating that a major ER stress response was not present.

IGP48 contributes to the immune response in humans

The increased expression of IGPs in stumpy forms, which in infected hosts are lysed at high frequency, prompted us to ask if IGP proteins were detected by the immune system. We analysed sera selected from a panel assembled from infected and non-infected matched controls gathered from endemic areas of Africa. Significantly, an immune response against any member of the IGP family has not been previously reported.

In the *T. b. rhodesiense* patient plasmas, three types of response were observed to bacterially expressed IGP48, namely IgG plus IgM, IgG only and no detectable response (as exemplified by L16, L3 and L12 respectively, Fig. 11). Additionally, in plasma from control individuals (endemic region, uninfected and European non-endemic/exposed), no responses were detected (Table S3). No significant relationships between patient immunoglobulin responses to rIGP48 and a range of demographic, pathobiological parameters, including stage of infection, ethnicity, gender, anaemia, plasma IgG and IgM concentrations and plasma cytokine (IFN-γ, TNF-α, IL-6 and IL-10) concentrations were detected (data not shown), and nor was there a relationship to total serum IgG and IgM concentrations. It was observed that cases showing no IgG response to rIGP48 were of significantly lower thick film parasitaemia than those who responded (median counts per 10 fields 0.125 versus 20 $p<0.05$ Mann-Whitney U-test). More data are clearly

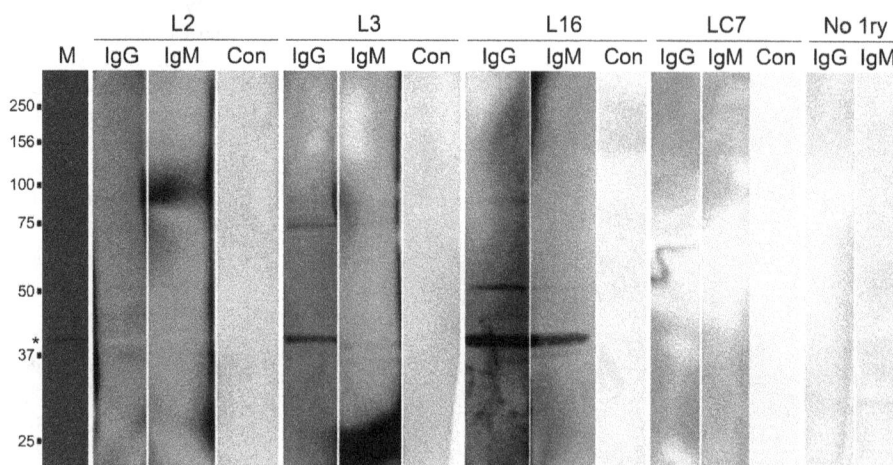

FIGURE 11: IGP48 elicits variable IgG and IgM responses in human *T. b. rhodesiense* infections. Recombinant IGP48 was resolved on a 10% SDS-PAGE gel. After Western Blotting, PVDF membranes were probed with plasma from *T. b. rhodesiense* patients and controls. **Lane M:** molecular weight markers. **Lane 1 (from left):** Ponceau red staining of IGP48 (*). **Lanes 2, 3 and 4:** membrane probed with plasma L2 with anti-IgG, anti-IgM and no secondary antibody control respectively. **Lanes 5, 6 and 7:** membrane probed with plasma L3 with anti-IgG, anti-IgM and no secondary antibody control respectively. **Lanes 8, 9 and 10:** membrane probed with plasma L16 with anti-IgG, anti-IgM and no secondary antibody control respectively. **Lanes 11, 12 and 13:** membrane probed with endemic control plasma LC7 with anti-IgG, anti-IgM and no secondary antibody control respectively. **Lanes 15 and 16:** anti-IgG and IgM controls with no primary antibody.

required to substantiate the possibility that responses to IGP are more significant in those with higher parasitaemia.

DISCUSSION

We identified a new family of glycoproteins in African trypanosomes by searching the *T. brucei* genome for membrane proteins with a predicted type I topology. We designated this new family as invariant glycoproteins, or IGPs. Phylogenetic analysis and comparative genomics indicates that IGPs are an early feature of trypanosomatid evolution, and the family exhibits frequent and lineage-specific gene expansions, even between closely related trypanosome species. This complex evolutionary history suggests that IGPs have evolved rapidly with the differing life styles between trypanosomatids. This is supported by the importance of each specific IGP subfamily to normal proliferation. Further, the absence of IGPs from *B. saltans* suggests that IGPs could have arisen coincident with acquisition of a parasitic lifestyle, and may indicate specific roles in the evolution of close association with a host and the challenges that this provides. The surface proteomes of the *Leishmanias* (including Phytomonads), African and American trypanosomes are highly distinct and may necessitate modifications to the ER environment to accommodate these differences, and are examples of well established host adaptations, critical for the transitions between mammalian and insect vectors.

The ER quality control system as described in mammals includes calnexin/calreticulin, BiP, several protein disulphide isomerases and a group of ER-degradation-enhancing α-mannosidase (EDEM) proteins. Expression levels of many of these proteins are modulated via transcriptional responses [51], which are largely absent from trypanosomatids [52,49]. However, *T. brucei*, together with other trypanosomes, does possess orthologs of hsp/dnaj chaperones, PDIs, calreticulin and EDEM pathway factors [52], while several trypanosome-specific proteins have been assigned to the ER. For example, two ER-associated proteins, ERAP18 and ERAP32, were described recently; both influence VSG surface expression and lead to ER hypertrophy [25]. Therefore IGPs and ERAPs represent distinct cohorts of trypanosome-specific ER proteins. The complexity of the IGP gene family likely compromised our ability to detect major trafficking blockades, such that while the subfamilies are individually important, some redundancy may also be present in terms of functionality during knockdown, despite our use of pan-subfamily constructs.

While the presence of the lectin domain suggests a role in glycosylation or glycan recognition, we were unable to detect a major defect in the N-glycans for IGP48 knockdown, as probed by a small panel of lectins. While this finding indicates that IGP48 subfamily members are not major effectors of N-glycan levels, it is possible that only a subset of specific glycoproteins are affected, as described for mammalian ERGIC-53 [53], or that an unrelated function for IGP48 within the ER is important.

Significantly, the structure of the IGPs and the oligomerisation behaviour of IGP48 suggests that these proteins have at least architectural similarities to ERGIC-53, which itself consists of a C-type lectin domain and a stalk, followed by a TMD and cytoplasmic domain. Further, ERGIC-53 forms homohexamers and also cycles between the ER and Golgi complex, which is suggested for IGP48, based on localisation and analysis of the N-glycan structural classes. However, there is no sequence relationship between the IGPs and ERGIC-53, and if ERGIC-53 and IGP are derived from a common ancestor, we would expect them to have similar cargo interactions. As the known cargo for ERGIC-53 are specific to higher eukaryotes, this is difficult to test, but exploration of the interactions between IGPs and cargo proteins is currently being attempted by immunoaffinity purification. Again, the absence of bulk effects is one more significant similarity in function between IGPs and ERGIC-53, as ERGIC-53 mutations appear to affect only a small subset of secretory proteins, while up-regulation of ERGIC-53 following tunicamycin-mediated stress suggests responsiveness to changing conditions within the ER [54], possibly mirroring the slender to stumpy transition, as suggested here for the IGPs.

The absence of an obvious role for the cytoplasm-oriented portion of IGP40 or IGP48 was unexpected, as this region of the protein is expected to interact with coatomer protein systems. However, we mapped the region responsible for ER-retention to the N-terminal lumenal/ectodomain. For IGP48 we could demonstrate that this portion of the protein allows homo-oligomerisation and the formation of high molecular weight complexes, which we speculate also form *in vivo*. Both the CLEC and trimerisation domains of IGP48 appear sufficient to retain the protein within the ER. As wild type proteins remain within the parasite, this may suggest that retention is due to truncated IGPs retaining an ability to associate with wild type IGP48 proteins, but it is unclear if IGPs can form heterocomplexes, either between IGP subfamilies or with unrelated proteins. We noted the presence of an extensive region of VDENTT heptad repeats in IGP48, which is potentially an oligomerisation signal. Direct characterisation of IGP oligomers is clearly required to address this issue. Removal of the IGP40 and IGP48 ectodomain released these proteins from the ER and into endosomes, the cell surface and into proteolysed secreted fragments. This failure to be specifically targeted to the plasma membrane, or to a specific intracellular compartment, suggests that additional potent targeting signals are unlikely to be present in the TMD or cytoplasmic regions of either IGP40 or 48 subfamilies, and represents a relatively nonspecific localisation, resulting from the absence of any post-ER targeting information.

The IGP family is up-regulated in the stumpy life cycle stage at the mRNA level and, at least for IGP48, the protein level. Unlike PAD2, the localisation of IGP48 is unaltered in stumpy cells and remains within the ER, making it unlikely that the IGP family are developmentally routed during differentiation, but that they remain within the ER where they may act in remodelling of the cell surface during developmental transition via altering of the ER environment. Increased expression in stumpy forms may be coupled to recognition by the human immune system. Specifically, as stumpy cells die, intracellular IGP will be released and sampled by the immune system. Immune recognition of IGP48 was confirmed using *T. b. rhodesiense* sleeping sickness patient sera from Uganda. While endemic controls showed no antibody response to IGP48, infected individuals showed a range of responses. The lack of any relationship to total IgG and IgM is an evidence that this is a specific response, rather than a result of polyclonal Ig activation [55], and is consistent with variable serum IgG responses to invariant trypanosome antigens, including ISG64, ISG65, and ISG75 [36]. The relationship of parasitaemia to detectable IgG responses to IGP48, while based on a very small sample, was significant, and this requires further investigation to determine its immunological impact and whether the monitoring of IGP antibody is of utility in assessing disease progression.

In conclusion, the IGP family represents an extensive, rapidly evolving family of *trans*-membrane domain proteins. Some IGP subfamily members were localised to the ER, and in African trypanosomes all three subfamilies appear to be essential for normal cell proliferation. An unusual mechanism of retention by oligomerisation is suggested by a combination of complex formation and requirement for the lumenal/ecto-domain for both aspects, and taken together there is considerable evidence that the behaviour of the IGPs is similar to ERGIC-53. We speculate that IGP oligomers may form a specialised matrix for folding and/or trafficking of VSG or of surface macromolecules during the developmental switch between VSG and procyclin, but clearly requires direct analysis of IGP interactions to move beyond speculation. Significantly, increased expression in short stumpy forms suggests a specific role in conditioning the ER during transition between hosts, and the detection of IGP48 by human sera provides evidence that antigens released from stumpy cells during necrosis can be sampled by the immune system.

MATERIALS AND METHODS
Informatics
A series of *in silico* filters for identification of predicted type I *trans*-membrane domain proteins were used (Fig. S1); type I topology has a single TMD with the N-terminus lumenal/extracellular and the C-terminus within the cytosol. The TREU927 predicted proteome (http://tritrypdb.org/tritrypdb/) was scanned using SignalP HMM for the presence of a putative N-terminal ER-targeting signal (http://www.cbs.dtu.dk/services/SignalP-2.0/). Sequences predicted to contain signal peptides and/or signal anchors were retained. As ER-targeting signals can be confused with mitochondrial N-terminal signals, the resulting sequence cohort was scanned with Mitoprot (http://ihg.gsf.de/ihg/mitoprot.html), Predotar (http://www.hsls.pitt.edu/obrc/index.php?page=URL1043959 648) and TargetP (http://www.cbs.dtu.dk/services/TargetP/). Predicted mitochondrial proteins were removed. Using SignalP HMM predictions, mature protein sequences were generated

for all retained proteins in order to negate recognition of the signal sequence as a TMD. These sequences were then entered into TMHMM (http://www.cbs.dtu.dk/services/TMHMM/) to select sequences containing a single TMD. Finally, GPI-SOM (http://gpi.unibe.ch) and big-PI (http://mendel.imp.ac.at/gpi/gpi_server.html) were used to remove proteins containing a predicted C-terminal GPI-anchor signal. A neighbour joining tree was generated from the remaining sequences using ClustalW2 (Fig. S2A).

To find any further members of the IGP family across the eukaryotic lineage, the twenty *T. brucei* IGP sequences were used as BLASTp [56] queries against a panel of eukaryotic predicted proteomes. A ClustalW [57] neighbour-joining tree was generated from a ClustalW alignment of all hits and a trypanosome-specific cluster, containing the twenty IGP queries, was recovered. These cluster-derived sequences were aligned with Muscle [58]. Poorly-aligned sequences in the N-terminal and the C-terminal regions were removed in Jalview [59]. This alignment/domain query was then used to search a panel of eukaryotic predicted proteomes with PSI-BLAST [60] as described [61], until the search converged. A ClustalW neighbour joining tree was generated from a ClustalW alignment of all matches. IGP paralogs were found in most trypanosomatid genomes (see Fig. S2C). Poorly-aligned sequences in the N-terminal and the C-terminal regions were again removed in Jalview. The domain was submitted to pfam and subsequently classed as pfam family PF16825, domain DUF5075.

For phylogenetic reconstructions, all protein sequence alignments were generated by ClustalW2 [57] and manually edited to remove poorly aligned regions. For phylogenetic reconstructions PhyML [62] and MrBayes [63] were used with default parameters. Protein domain predictions were performed at Pfam (http://pfam.xfam.org/) and Superfamily (http://supfam.cs.bris.ac.uk/SUPERFAMILY/) (Fig. S2B).

Cell culture
Long slender bloodstream form (BSF) *T. brucei* were maintained as described [64]. Briefly, BSF Molteno Institute trypanosomal antigen type (MITat) 1.2 cells, derived from Lister 427 and expressing VSG 221, were cultured in HMI-9 complete medium [65] at 37°C with 5% CO_2 in a humid atmosphere in non-adherent culture flasks with vented caps. Cells were maintained at densities between 1 x 10^5 and 2 x 10^6 cells/ml. For RNAi, the single marker T7$^{RNAP/TETR}$ BSF cells (SMB) line was used [34]. Expression of integrated plasmid constructs was maintained in MITat and SMB cells using G418 antibiotic selection at 2.5 μg/ml and G418 plus hygromycin selection at 2 μg/ml, respectively. Induction of RNAi was carried out with tetracycline (Sigma) at 1.0 μg/ml. Procyclic culture form cells were grown exactly as described using SDM79 medium [34]. BSF stumpy stage mRNA and protein extracts were obtained from cells propagated in mice, a kind gift from Keith Matthews (University of Edinburgh).

In vitro short stumpy differentiation/cold shock
Cells were incubated with 1 mM 8-(4-chlorophenylthio)-cAMP (pCPT-cAMP) [66], or incubated at 20°C [47] for 12 hours to induce states with resemblance to differentiating cells in monomorphic 427 BSF cells.

Plasmid constructs and transfections

Constructs for ectopic expression were designed so that an HA- or FLAG-epitope tag was incorporated into the gene sequence. For RNAi, silencing fragments were selected using RNAit [67]. Primers used for PCR amplification are given in Table S1. PCR products were cloned into the BSF expression vector pXS519 [68] or p2T7 [69] vector for RNAi. All constructs were verified by standard sequencing methods (Geneservice Ltd.). YFP-ATG8.2::GL2166, an expression plasmid encoding a fusion protein between YFP and one *T. brucei* ATG8 paralog was a kind gift from Jeremy Mottram, University of Glasgow. Prior to introduction into trypanosomes, about 15 µg of pXS5 and p2T7 constructs were linearised with XhoI or NotI (NEB) respectively. Transgenic BSF lines were generated by electroporation using an Amaxa Nucleofector®II, incubated for ~6 hours and selected in the presence of the appropriate drug. Viable cells were taken from plates where less than 50% of wells contained transformants (considered likely clonal) and further expanded in the presence of antibiotic(s).

IGP48 protein expression and antibody production

The IGP48 ectodomain sequence was amplified and cloned into the pQE30 His$_6$-tag bacterial expression vector (Qiagen) as described [42]. Protein expression was induced at OD_{600nm} = 0.6 by addition of 0.1 mM IPTG. Cells were incubated for 2 hours at 37°C, pelleted and lysed in lysis buffer (1 x PBS, 100 µg/ml lysozyme, mini-complete EDTA-free protein inhibitor tablets, Sigma). Cells were crushed with a cell homogenisor (One Shot Model, Constant Systems Ltd.), Triton X-100 added to 1% and rotated at 4°C for 30 minutes. The lysate was centrifuged at 10 000 g for 30 minutes at 4°C to pellet cell debris. His$_6$-IGP48 was affinity purified from the soluble fraction using the QIAexpressionist system (Qiagen). To further purify the recombinant protein, the bound fraction was subjected to SDS-PAGE and the resulting band corresponding to His$_6$-IGP48 excised, washed with PBS and used to immunise rabbits (CovalAb, France).

To affinity purify anti-IGP48 antibodies from rabbit serum, the His$_6$-IGP48 ectodomain was coupled to CNBr-sepharose-4B following the manufacturer's instructions (GE Healthcare). Serum and beads were then applied to a Poly-Prep Chromatography Column (BioRad) and the column washed five times with PBS. Anti-IGP48 antibody was eluted with 100 mM glycine pH 2.5 and neutralised immediately with 1 M Tris pH 8.0. Eluted antibody was washed four times in PBS using Vivaspin 20 columns (Sartorius Stedim Biotechnology), resuspended in storage buffer (PBS plus 30% glycerol and 0.01% sodium azide) and stored at -20°C.

Quantitative RT-PCR (qRT-PCR)

1 x 10^8 cells were harvested at 800 g for 10 minutes at 4°C and washed in ice-cold PBS and quick-frozen in dry ice for 1 minute. RNA was purified using an RNeasy minikit (Qiagen) according to the manufacturer's instructions. The RNA concentration was quantified using an ND-1000 spectrophotometer (Nanodrop Technologies). qRT-PCR was performed using iQ-SYBR green Supermix on a MiniOpticon real-time PCR (RT-PCR) detection system (Bio-Rad), and quantified using OPTICON3 software (Bio-Rad) [52]. The following primers were used: IGP48-RTF (CTGCAGGCTGCCAGCTCTG), IGP48-RTR (TTTAATCTCCCGTACGCAGG), IGP40-RTF (CTG-CATGTGACTGCTGCT), IGP40-RTR (TGAAAGGGTATA-

CAACTGACC), IGP34-RTF (ATTGCGTCTACCGATGGAAC), IGP34-RTR (TAGACTCCTCATCTGAATGC). Data were normalised against TERT (telomerase reverse transcriptase) (TERT-RTF (GAGCGTGTGACTTCCGAAGG) and TERT-RTR (AG-GAACTGTCACGGAGTTTGC).

Western blot analysis

Protein samples were typically resolved on 10% SDS-PAGE after solubilisation in SDS sample buffer [70] and then transferred to polyvinylidene fluoride (PVDF) membranes (Millipore). Non-specific binding sites on the membrane were blocked, and Western blotting was carried out following standard procedures.

Immunofluorescence microscopy

Immunofluorescence analysis (IFA) was as described [64]. Antibodies for IFA were used at the following dilutions: mouse anti-HA-epitope IgG (Santa Cruz Biotechnology Inc.) at 1:1000, rabbit anti-Rab11 at 1:200 [44], rabbit anti-Rab5a at 1:200 [68], mouse anti-p67 at 1:1000 [43], mouse anti-BiP at 1:10 000 [41], rabbit anti-VSG221 at 1:1000 [68], rabbit anti-RabX2 at 1:50 [71], rabbit anti-IGP48 (this study) at 1:50. Secondary antibodies were used at the following dilutions: anti-mouse Oregon Green (Molecular Probes) at 1:1000 and anti-rabbit Cy3 (Sigma) at 1:1000. Cells were examined on a Nikon Eclipse E600 epifluorescence microscope fitted with optically matched filter blocks and a Hamamatsu ORCA charge-coupled-device camera. Digital images were captured using Metamorph software (Universal Imaging Corp.) on a computer running the Windows XP operating system (Microsoft Inc.) and the raw images processed using Photoshop CS6 software (Adobe Systems Inc.). Confocal z-sections were acquired using a Leica DMIRE2 microscope and deconvolved using Huygens Professional software.

N-glycosylation analysis

Bloodstream form trypanosomes were cultured in the presence of tunicamycin (Sigma) added to complete media at 1 µg/ml, which prevents further cell proliferation [52]. For treatment with PNGase F, 1 x 10^8 cells were washed in PBS and incubated with lysis buffer (1% NP-40, 100 mM NaPO$_4$, pH 7.5) plus protease inhibitor cocktail (Roche) and then heated to 95°C for 15 minutes. Samples were treated with 10 mU PNGase F (NEB) overnight at 37°C. A second aliquot of PNGase F was added and the reaction continued for 2 hours prior to analysis. For treatment with Endoglycosidase H, 1 x 10^8 cells were washed and lysed by incubation with 1 ml of 5 mM EDTA and protease inhibitor cocktail (Roche), followed by two cycles of freeze-thawing. Samples were centrifuged to separate crude membranes and the pellet dissolved in 10 mM Tris-HCl, pH 8.0, 1 mM EDTA, 1% SDS. 5 mU/ml of Endoglycosidase H (Calbiochem) was added to each sample, and then 5 mU/ml again 12 hours prior to analysis.

Protein turnover

Protein synthesis was blocked by the addition of cycloheximide (35 µM) and IGP copy number estimated by Western blotting [64]. Antibodies were used at the following dilutions: mouse anti-HA at 1:8000, mouse anti-β-tubulin at 1:20 000, rabbit anti-BiP at 1:10 000 and horseradish peroxidase-conjugated anti-(rabbit IgG) and anti-(mouse IgG) (Sigma) at 1:10 000. Antigen copy number was estimated by densitometry using ImageJ on scanned films derived from low exposure blots (National Institutes of Health).

Biotinylation

Mid-log phase cells (1×10^7) were collected and washed three times in vPBS. Biotinylation and separation of biotinylated and non-biotinylated proteins were all carried out as described [14]. Samples were incubated in SDS sample buffer at 95°C, and both fractions subjected to SDS-PAGE and Western blotting. Antibodies were used at the following dilutions: mouse anti-HA at 1:8,000, rabbit anti-RabX1 (1:1 000) or rabbit anti-ISG75 (1:8 000) and blot signals quantified using ImageJ software.

Radioimmunoprecipitation

1×10^7 cells were washed in PBS and resuspended in 500 µl of Met/Cys-free RPMI 1640 (Sigma) medium supplemented with 10% dialyzed fetal bovine serum (FBS), followed by incubation at 37°C for 1 hour. Cells were pulse-labeled for 1 hour with EasyTag EXPRESS^{35}S protein labeling mix (Perkin Elmer) at a specific activity of 200 µCi/ml and chased by addition of HMI-9 medium for 3 hours. Cells were washed in ice-cold PBS, lysed in 100 µl RIPA buffer (25 mM Tris-Cl, pH 7.5, 150 mM NaCl, 1% NP40, 0.5% sodium deoxycholate, 0.1% SDS) for 15 minutes on ice, and incubated for 5 minutes at 95°C in RIPA buffer containing 1% SDS. Lysates were pre-cleared for 1 hour with Pansorbin (Calbiochem) and incubated overnight with mouse anti-HA (1:100) at 4°C. Immunocomplexes were isolated by incubation with protein A-sepharose (Sigma) for 1 hour and subjected to SDS-PAGE. Gels were dried on Whatman 3MM paper and exposed to autoradiographic films for up to 1 week.

Immunoprecipitation

Cells were grown to logarithmic phase and harvested by centrifugation (800 g, 10 minutes, 4°C) and washed twice in PBS. Cells were lysed in 500 µl of NP40 buffer (1% NP40, 150 mMNaCl, 50 mM Tris HCl pH 7.5, 1 mM EDTA, protease inhibitor cocktail) and incubated on ice for 10 minutes. Lysates were centrifuged at 16 000 g for 15 minutes at 4°C to remove nuclei and cell debris, and the supernatant transferred to a fresh tube. Five microliters of rat anti-HA or mouse anti-FLAG was added to each sample, and the mixture incubated overnight at 4°C. Immune complexes were isolated by addition of 50 µl of protein A-sepharose (Generon) or Dynabeads®Protein G (Invitrogen) and incubated at 4°C with rolling for 2 hours. Beads were recovered by centrifugation at 3000 g for 5 minutes (protein A-sepharose) or by magnetisation (Dynabeads®) and washed extensively with NP40 buffer. The supernatant was discarded and bound proteins eluted by addition of 1 x SDS sample buffer and incubation at 95°C for 10 minutes. Samples were subjected to protein electrophoresis and Western blotting for detection. Antibodies were used at 1:8000 rat anti-HA and 1:800 mouse anti-FLAG.

Ultrastructural analysis

Tetracycline-induced and non-induced RNAi cell lines were grown to a density of 1×10^6 cells/ml and rapidly fixed in culture by the addition of isothermal glutaraldehyde to the culture flask, to a final concentration of 2.5% (w/v), as described previously [72]. The culture flask was rocked gently for 10 minutes at 37°C, followed by centrifugation at 800 g for 5 minutes and harvested cells resuspended in 2.5% glutaraldehyde in PBS for another 30 minutes at room temperature. Fixed cells were post-fixed and embedded as described previously [34]. Ultra-thin (70 nm) sections were viewed on a Philips CM100 electron microscope (FEI-Philips).

VSG and BiPN transport

VSG export was monitoredas described [64] with a few modifications. Parasites were labeled with EasyTag EXPRESS ^{35}S protein labeling mix (PerkinElmer) at a specific activity of 200 µCi/ml and incubated for 7 minutes at 37°C. Following separation of membrane-form and soluble VSG by hypotonic lysis and binding to ConA, fractions were washed and resuspended in sample buffer and loaded onto 10% SDS-polyacrylamide gels at 1×10^6 cell equivalents per lane. Gels were fixed, stained and exposed to X-ray film. VSG band intensity was quantified using ImageJ. BiPN export was monitored as described previously [73]. Transformed IGP48 RNAi BSF cells expressing BiPHA9 were pulse-labeled for 1 hour at 37°C and then chased for 1 hour, taking samples at regular time points. Aliquots were separated into cells and medium by centrifugation, and labeled polypeptides in both fractions immunoprecipitated with anti-HA and analysed as for VSG export.

Trypanosomiasis plasma samples and Western blotting

Plasma was prepared from heparinised blood samples from patients diagnosed with *T. b. rhodesiense* infection and non-infected local controls in Tororo, Uganda in 2003. Details of the recruitment and diagnostic methods for these samples have been described previously [54] as well as measurement of serum IgG and IgM concentrations [74]. Sample collection and subsequent analyses were approved by ethical review by the Uganda Ministry of Health and the Grampian Research Ethics Committee, and were subject to informed consent. All plasma was stored at -80°C after collection. For immunoprobing, PVDF membranes were wetted in 50% methanol (v/v), washed with PBS-T (PBS, 0.05% Tween-20) and blocked for 1 hour in PBS-T plus 5% BSA (w/v). After washing in PBS-T, filters were probed with human plasma samples diluted 1:1000 in PBS-T, 5% BSA for 3 hours with gentle rocking. Filters were then washed four times with PBS-T and probed with either no secondary antibody, peroxidase-conjugated goat anti-human IgG (Invitrogen 62-7520) or peroxidase-conjugated goat anti-human IgM (Invitrogen 62-7120) diluted 1:1000 in PBS, 5% BSA. After 1 hour incubation, filters were washed five times in PBS-T. Filters were flooded with a solution comprising equal volumes of freshly mixed 0.02% H_2O_2 in 20 mM Tris-HCl pH 8.5 and 2.5 mM Luminol, 0.4 mM p-Coumaric acid in 20 mM Tris-HCl, pH 8.5. After excess ECL reagent was removed, luminescence was visualised using a Fusion SL imaging system (Peqlab).

ACKNOWLEDGMENTS

This work was supported by a program grant from the Wellcome Trust (082813 to MCF), and a studentship from the BBSRC (to HA), both of which are gratefully acknowledged. We are also grateful to James Bangs (University of Buffalo) for anti-BiP antibody and Jeremy Mottram (University of Glasgow) for the ATG8-tagged cell line, Catarina Gadelha (University of Nottingham) for advice with ultrastructural analysis, Mike Ferguson (University of Dundee) for suggestions on N-glycosylation studies and finally Keith Matthews (University of

Edinburgh) for stumpy cell RNA and protein lysates and antibodies to PAD1 and PAD2.

REFERENCES

1. M.P. Barrett, R.J. Burchmore, A. Stich, J.O. Lazzari, A.C. Frasch, J.J. Cazzulo, and S. Krishna (**2003**). The trypanosomiases. **The Lancet**, vol. 362, pp. 1469-1480.

2. K.R. Matthews (**2005**). The developmental cell biology of Trypanosoma brucei. **Journal of Cell Science**, vol. 118, pp. 283-290.

3. D.P. Nolan, S. Rolin, J.R. Rodriguez, J. Van Den Abbeele, and E. Pays (**2000**). Slender and stumpy bloodstream forms of Trypanosoma brucei display a differential response to extracellular acidic and proteolytic stress. **European Journal of Biochemistry**, vol. 267, pp. 18-27.

4. E. Vassella, B. Reuner, B. Yutzy, and M. Boshart (**1997**). Differentiation of African trypanosomes is controlled by a density sensing mechanism which signals cell cycle arrest via the cAMP pathway. **Journal of cell science**, 110 (Pt 21):2661-71.

5. J.R. Seed, and J.B. Sechelski (**1989**). Mechanism of Long Slender (LS) to Short Stumpy (SS) Transformation in the African Trypanosomes. **The Journal of Protozoology**, vol. 36, pp. 572-577.

6. K.A. Lythgoe, L.J. Morrison, A.F. Read, and J.D. Barry (**2007**). Parasite-intrinsic factors can explain ordered progression of trypanosome antigenic variation. **Proceedings of the National Academy of Sciences**, vol. 104, pp. 8095-8100.

7. C.R. Turner (**1997**). The rate of antigenic variation in fly-transmitted and syringe-passaged infections of Trypanosoma brucei. **FEMS Microbiology Letters**, vol. 153, pp. 227-231.

8. E. Pays (**2005**). Regulation of antigen gene expression in. **Trends in Parasitology**, vol. 21, pp. 517-520.

9. A.P. Jackson, H.C. Allison, J.D. Barry, M.C. Field, C. Hertz-Fowler, and M. Berriman (**2013**). A Cell-surface Phylome for African Trypanosomes. **PLoS Neglected Tropical Diseases**, vol. 7, pp. e2121.

10. M.C. Field, and M. Carrington (**2009**). The trypanosome flagellar pocket. **Nat Rev Micro**, vol. 7, pp. 775-786.

11. K. Ziegelbauer, and P. Overath (**1993**). Organization of two invariant surface glycoproteins in the surface coat of Trypanosoma brucei. **Infection and immunity**, 61(11):4540-5.

12. D.G. Jackson, H.J. Windle, and H.P. Voorheis (**1993**). The identification, purification, and characterization of two invariant surface glycoproteins located beneath the surface coat barrier of bloodstream forms of Trypanosoma brucei. **The Journal of biological chemistry**, 268(11):8085-95.

13. M. Carrington, and J. Boothroyd (**1996**). Implications of conserved structural motifs in disparate trypanosome surface proteins. **Molecular and Biochemical Parasitology**, vol. 81, pp. 119-126.

14. K.F. Leung, F.S. Riley, M. Carrington, and M.C. Field (**2011**). Ubiquitylation and Developmental Regulation of Invariant Surface Protein Expression in Trypanosomes. **Eukaryotic Cell**, vol. 10, pp. 916-931.

15. S. Alsford, S. Eckert, N. Baker, L. Glover, A. Sanchez-Flores, K.F. Leung, D.J. Turner, M.C. Field, M. Berriman, and D. Horn (**2012**). High-throughput decoding of antitrypanosomal drug efficacy and resistance. **Nature**, 482(7384):232-6.

16. S. Alexandre, P. Paindavoine, P. Tebabi, A. Pays, S. Halleux, M. Steinert, and E. Pays (**1990**). Differential expression of a family of putative adenylate/guanylate cyclase genes in Trypanosoma brucei. **Molecular and Biochemical Parasitology**, vol. 43, pp. 279-288.

17. D. Steverding (**2000**). The transferrin receptor of Trypanosoma brucei. **Parasitology International**, vol. 48, pp. 191-198.

18. D.J. Bridges, A.R. Pitt, O. Hanrahan, K. Brennan, H.P. Voorheis, P. Herzyk, H.P. de Koning, and R.J.S. Burchmore (**2008**). Characterisation of the plasma membrane subproteome of bloodstream form Trypanosoma brucei. **Proteomics**, vol. 8, pp. 83-99.

19. M.L.S. Güther, M.D. Urbaniak, A. Tavendale, A. Prescott, and M.A.J. Ferguson (**2014**). High-Confidence Glycosome Proteome for Procyclic Form Trypanosoma brucei by Epitope-Tag Organelle Enrichment and SILAC Proteomics. **J. Proteome Res.**, vol. 13, pp. 2796-2806.

20. J. Hirst, L. D. Barlow, G.C. Francisco, D.A. Sahlender, M.N.J. Seaman, J.B. Dacks, and M.S. Robinson (**2011**). The Fifth Adaptor Protein Complex. **PLoS Biol**, vol. 9, pp. e1001170.

21. C.L. Allen, D. Liao, W. Chung, and M.C. Field (**2007**). Dileucine signal-dependent and AP-1-independent targeting of a lysosomal glycoprotein in Trypanosoma brucei. **Molecular and Biochemical Parasitology**, vol. 156, pp. 175-190.

22. N.N. Tazeh, and J.D. Bangs (**2007**). Multiple Motifs Regulate Trafficking of the LAMP-Like Protein p67 in the Ancient Eukaryote Trypanosoma brucei. **Traffic**, vol. 8, pp. 1007-1017.

23. J.D. Bangs (**1986**). Posttranslational modification and intracellular transport of a trypanosome variant surface glycoprotein. **The Journal of Cell Biology**, vol. 103, pp. 255-263.

24. M.P. Hsu, M.L. Muhich, and J.C. Boothroyd (**1989**). A developmentally regulated gene of trypanosomes encodes a homolog of rat protein-disulfide isomerase and phosphoinositol-phospholipase C. **Biochemistry**, vol. 28, pp. 6440-6446.

25. Y. Wang, M. Wang, and M.C. Field (**2010**). Trypanosoma brucei: Trypanosome-specific endoplasmic reticulum proteins involved in variant surface glycoprotein expression. **Experimental Parasitology**, vol. 125, pp. 208-221.

26. S. Munro, and H.R. Pelham (**1987**). A C-terminal signal prevents secretion of luminal ER proteins. **Cell**, vol. 48, pp. 899-907.

27. H.R. Pelham (**1988**). Evidence that luminal ER proteins are sorted from secreted proteins in a post-ER compartment. **The EMBO journal**, 7(4):913-8.

28. L. Ellgaard, and A. Helenius (**2003**). Quality control in the endoplasmic reticulum. **Nature Reviews Molecular Cell Biology**, vol. 4, pp. 181-191.

29. R.C. Aguilar, and B. Wendland (**2003**). Ubiquitin: not just for proteasomes anymore. **Current Opinion in Cell Biology**, vol. 15, pp. 184-190.

30. W. Chung, M. Carrington, and M.C. Field (**2004**). Cytoplasmic Targeting Signals in Transmembrane Invariant Surface Glycoproteins of Trypanosomes. **Journal of Biological Chemistry**, vol. 279, pp. 54887-54895.

31. K.J. Schwartz (**2005**). GPI valence and the fate of secretory membrane proteins in African trypanosomes. **Journal of Cell Science**, vol. 118, pp. 5499-5511.

32. S. Mayor, and H. Riezman (**2004**). Sorting GPI-anchored proteins. **Nature Reviews Molecular Cell Biology**, vol. 5, pp. 110-120.

33. P.T. Manna, C. Boehm, K.F. Leung, S.K. Natesan, and M.C. Field (**2014**). Life and times: synthesis, trafficking, and evolution of VSG. **Trends in Parasitology**, vol. 30, pp. 251-258.

34. V.O. Adung'a, C. Gadelha, and M.C. Field (**2013**). Proteomic Analysis of Clathrin Interactions in Trypanosomes Reveals Dynamic Evolution of Endocytosis. **Traffic**, vol. 14, pp. 440-457.

CONFLICT OF INTEREST
The authors declare no conflict of interest.

35. W. Chung, K.F. Leung, M. Carrington, and M.C. Field (**2008**). Ubiquitylation is Required for Degradation of Transmembrane Surface Proteins in Trypanosomes. **Traffic**, vol. 9, pp. 1681-1697.

36. L. Sullivan, S.J. Wall, M. Carrington, and M.A.J. Ferguson (**2013**). Proteomic Selection of Immunodiagnostic Antigens for Human African Trypanosomiasis and Generation of a Prototype Lateral Flow Immunodiagnostic Device. **PLoS Neglected Tropical Diseases**, vol. 7, pp. e2087.

37. M. Berriman (**2005**). The Genome of the African Trypanosome Trypanosoma brucei. **Science**, vol. 309, pp. 416-422.

38. S. Kelly, A. Ivens, P.T. Manna, W. Gibson, and M.C. Field (**2014**). A draft genome for the African crocodilian trypanosome Trypanosoma grayi. **Scientific Data**, vol. 1.

39. S. Qin, N. Kawasaki, D. Hu, H. Tozawa, N. Matsumoto, and K. Yamamoto (**2012**). Subcellular localization of ERGIC-53 under endoplasmic reticulum stress condition. **Glycobiology**, vol. 22, pp. 1709-1720.

40. K.J. Schwartz, R.F. Peck, and J.D. Bangs (**2013**). Intracellular Trafficking and Glycobiology of TbPDI2, a Stage-Specific Protein Disulfide Isomerase in Trypanosoma brucei. **Eukaryotic Cell**, vol. 12, pp. 132-141.

41. J.D. Bangs, L. Uyetake, M.J. Brickman, A.E. Balber, and J.C. Boothroyd (**1993**). Molecular cloning and cellular localization of a BiP homologue in Trypanosoma brucei. Divergent ER retention signals in a lower eukaryote. **Journal of cell science**, 105 (Pt 4):1101-13.

42. H. Field, M. Farjah, A. Pal, K. Gull, and M.C. Field (**1998**). Complexity of Trypanosomatid Endocytosis Pathways Revealed by Rab4 and Rab5 Isoforms in Trypanosoma brucei. **Journal of Biological Chemistry**, vol. 273, pp. 32102-32110.

43. R.J. Kelley, D.L. Alexander, C. Cowan, A.E. Balber, and J.D. Bangs (**1999**). Molecular cloning of p67, a lysosomal membrane glycoprotein from Trypanosoma brucei. **Molecular and Biochemical Parasitology**, vol. 98, pp. 17-28.

44. T.R. Jeffries, G.W. Morgan, and M.C. Field (**2001**). A developmentally regulated rab11 homologue in Trypanosoma brucei is involved in recycling processes. **Journal of cell science**, 114(Pt 14):2617-26.

45. M.C. Field, T. Sergeenko, Y. Wang, S. Böhm, and M. Carrington (**2010**). Chaperone Requirements for Biosynthesis of the Trypanosome Variant Surface Glycoprotein. **PLoS ONE**, vol. 5, pp. e8468.

46. S. Dean, R. Marchetti, K. Kirk, and K.R. Matthews (**2009**). A surface transporter family conveys the trypanosome differentiation signal. **Nature**, vol. 459, pp. 213-217.

47. M. Engstler (**2004**). Cold shock and regulation of surface protein trafficking convey sensitization to inducers of stage differentiation in Trypanosoma brucei. **Genes & Development**, vol. 18, pp. 2798-2811.

48. R.A. Williams, K.L. Woods, L. Juliano, J.C. Mottram, and G.H. Coombs (**2009**). Characterization of unusual families of ATG8-like proteins and ATG12 in the protozoan parasite Leishmania major. **Autophagy**, vol. 5, pp. 159-172.

49. M. Engstler (**2004**). Kinetics of endocytosis and recycling of the GPI-anchored variant surface glycoprotein in Trypanosoma brucei. **Journal of Cell Science**, vol. 117, pp. 1105-1115.

50. E. Wirtz, S. Leal, C. Ochatt, and G. Cross (**1999**). A tightly regulated inducible expression system for conditional gene knock-outs and dominant-negative genetics in Trypanosoma brucei. **Molecular and Biochemical Parasitology**, vol. 99, pp. 89-101.

51. C. Patil, and P. Walter (**2001**). Intracellular signaling from the endoplasmic reticulum to the nucleus: the unfolded protein response in yeast and mammals. **Current Opinion in Cell Biology**, vol. 13, pp. 349-355.

52. V.L. Koumandou, S. Natesan, T. Sergeenko, and M.C. Field (**2008**). The trypanosome transcriptome is remodelled during differentiation but displays limited responsiveness within life stages. **BMC Genomics**, vol. 9, pp. 298.

53. H. Hauri, C. Appenzeller, H. Andersson, and F. Kappeler (**1999**). **Nature Cell Biology**, vol. 1, pp. 330-334.

54. L. MacLean, M. Odiit, A. MacLeod, L. Morrison, L. Sweeney, A. Cooper, P. Kennedy, and J. Sternberg (**2007**). Spatially and Genetically Distinct African Trypanosome Virulence Variants Defined by Host Interferon-γ Response. **The Journal of Infectious Diseases**, vol. 196, pp. 1620-1628.

55. B. Greenwood, and H. Whittle (**1980**). The pathogenesis of sleeping sickness. **Transactions of the Royal Society of Tropical Medicine and Hygiene**, vol. 74, pp. 716-725.

56. S.F. Altschul, W. Gish, W. Miller, E.W. Myers, and D.J. Lipman (**1990**). Basic local alignment search tool. **Journal of Molecular Biology**, vol. 215, pp. 403-410.

57. M. Larkin, G. Blackshields, N. Brown, R. Chenna, P. McGettigan, H. McWilliam, F. Valentin, I. Wallace, A. Wilm, R. Lopez, J. Thompson, T. Gibson, and D. Higgins (**2007**). Clustal W and Clustal X version 2.0. **Bioinformatics**, vol. 23, pp. 2947-2948.

58. R.C. Edgar (**2004**). MUSCLE: multiple sequence alignment with high accuracy and high throughput. **Nucleic Acids Research**, vol. 32, pp. 1792-1797.

59. A.M. Waterhouse, J.B. Procter, D.M.A. Martin, M. Clamp, and G.J. Barton (2009). Jalview Version 2-a multiple sequence alignment editor and analysis workbench. **Bioinformatics**, vol. 25, pp. 1189-1191.

60. D.P. Nolan, M. Geuskens, and E. Pays (**1999**). N-linked glycans containing linear poly-N-acetyllactosamine as sorting signals in endocytosis in Trypanosoma brucei. **Current Biology**, vol. 9, pp. 1169-S1.

61. A.J. O'Reilly, J.B. Dacks, and M.C. Field (**2011**). Evolution of the Karyopherin-β Family of Nucleocytoplasmic Transport Factors; Ancient Origins and Continued Specialization. **PLoS ONE**, vol. 6, pp. e19308.

62. S. Guindon, and O. Gascuel (**2003**). A Simple, Fast, and Accurate Algorithm to Estimate Large Phylogenies by Maximum Likelihood. **Systematic Biology**, vol. 52, pp. 696-704.

63. F. Ronquist, and J.P. Huelsenbeck (**2003**). MrBayes 3: Bayesian phylogenetic inference under mixed models. **Bioinformatics**, vol. 19, pp. 1572-1574.

64. K.F. Leung, J.B. Dacks, and M.C. Field (**2008**). Evolution of the Multivesicular Body ESCRT Machinery; Retention Across the Eukaryotic Lineage. **Traffic**, vol. 9, pp. 1698-1716.

65. H. Hirumi, and K. Hirumi (**1989**). Continuous Cultivation of Trypanosoma brucei Blood Stream Forms in a Medium Containing a Low Concentration of Serum Protein without Feeder Cell Layers. **The Journal of Parasitology**, vol. 75, pp. 985.

66. T. Breidbach, E. Ngazoa, and D. Steverding (**2002**). Trypanosoma brucei: in vitro slender-to-stumpy differentiation of culture-adapted, monomorphic bloodstream forms. **Experimental Parasitology**, vol. 101, pp. 223-230.

67. S. Redmond, J. Vadivelu, and M.C. Field (**2003**). RNAit: an automated web-based tool for the selection of RNAi targets in Trypanosoma brucei. **Molecular and Biochemical Parasitology**, vol. 128, pp. 115-118.

68.　A. Pal, B.S. Hall, D.N. Nesbeth, H.I. Field, and M.C. Field (**2001**). Differential endocytic functions of Trypanosoma brucei Rab5 isoforms reveal a glycosylphosphatidylinositol-specific endosomal pathway. **The Journal of biological chemistry**, 277(11):9529-39.

69.　V.P. Alibu, L. Storm, S. Haile, C. Clayton, and D. Horn (**2005**). A doubly inducible system for RNA interference and rapid RNAi plasmid construction in Trypanosoma brucei. **Molecular and Biochemical Parasitology**, vol. 139, pp. 75-82.

70.　U.K. Laemmli (**1970**). Cleavage of Structural Proteins during the Assembly of the Head of Bacteriophage T4. **Nature**, vol. 227, pp. 680-685.

71.　S.K.A. Natesan, L. Peacock, K.F. Leung, K.R. Matthews, W. Gibson, and M.C. Field (**2009**). The Trypanosome Rab-Related Proteins RabX1 and RabX2 Play No Role in IntraCellular Trafficking but May Be Involved in Fly Infectivity. **PLoS ONE**, vol. 4, pp. e7217.

72.　C. Gadelha, S. Rothery, M. Morphew, J.R. McIntosh, N.J. Severs, and K. Gull, (**2009**). Membrane domains and flagellar pocket boundaries are influenced by the cytoskeleton in African trypanosomes. **Proceedings of the National Academy of Sciences**, vol. 106, pp. 17425-17430.

73.　J.D. Bangs, E.M. Brouch, D.M. Ransom, and J.L. Roggy (**1996**). A Soluble Secretory Reporter System in Trypanosoma brucei: Studies On Endoplasmic Reticulum Targeting. **Journal of Biological Chemistry**, vol. 271, pp. 18387-18393.

74.　MacLean L, Reiber H, Kennedy PGE, Sternberg JM (**2012**) Stage Progression and Neurological Symptoms in Trypanosoma brucei rhodesiense Sleeping Sickness: Role of the CNS Inflammatory Response. **PLoS Neglected Tropical Diseases** 6(10): e1857.

Functional analysis of lipid metabolism genes in wine yeasts during alcoholic fermentation at low temperature

María López-Malo[1,2], Estéfani García-Ríos[1], Rosana Chiva[1] and José Manuel Guillamon[1,]*

[1] Departamento de Biotecnología de los alimentos, Instituto de Agroquímica y Tecnología de los Alimentos (CSIC), Avda, Agustín Escardino, 7, E-46980-Paterna, Valencia, Spain.

[2] Biotecnologia Enològica. Departament de Bioquímica i Biotecnologia, Facultat d'Enologia, Universitat Rovira i Virgili, Marcel·li Domingo s/n, 43007, Tarragona, Spain.

* Corresponding Author: José Manuel Guillamón, E-mail: guillamon@iata.csic.es

ABSTRACT Wine produced by low-temperature fermentation is mostly considered to have improved sensory qualities. However few commercial wine strains available on the market are well-adapted to ferment at low temperature (10 – 15°C). The lipid metabolism of *Saccharomyces cerevisiae* plays a central role in low temperature adaptation. One strategy to modify lipid composition is to alter transcriptional activity by deleting or overexpressing the key genes of lipid metabolism. In a previous study, we identified the genes of the phospholipid, sterol and sphingolipid pathways, which impacted on growth capacity at low temperature. In the present study, we aimed to determine the influence of these genes on fermentation performance and growth during low-temperature wine fermentations. We analyzed the phenotype during fermentation at the low and optimal temperature of the lipid mutant and overexpressing strains in the background of a derivative commercial wine strain. The increase in the gene dosage of some of these lipid genes, e.g., *PSD1, LCB3, DPL1* and *OLE1,* improved fermentation activity during low-temperature fermentations, thus confirming their positive role during wine yeast adaptation to cold. Genes whose overexpression improved fermentation activity at 12°C were overexpressed by chromosomal integration into commercial wine yeast QA23. Fermentations in synthetic and natural grape must were carried out by this new set of overexpressing strains. The strains overexpressing *OLE1* and *DPL1* were able to finish fermentation before commercial wine yeast QA23. Only the *OLE1* gene overexpression produced a specific aroma profile in the wines produced with natural grape must.

Keywords: wine, industrial yeast, cold adaptation, lipids, mutant, stable overexpression, fermentation.

Abbreviations:
DHS - dihydrosphingosine,
FA - fatty acids,
GT - generation time,
NM - natural must,
PC - phosphatidylcholine,
PE - phosphatidylethanolamine,
PS - phosphatidylserine,
PHS - phytosphingosine,
SM - synthetic must.

INTRODUCTION

Temperature is one of the most important parameters to affect the length and rate of alcoholic fermentation and final wine quality. Many winemakers prefer low-temperature fermentation (10 - 15°C) for the production of white and rosé wine because it improves taste and aroma characteristics. This improved quality can be attributed not only to the prevention of volatilization of primary aromas, but also to the increased synthesis of secondary aromas. Thus the final wine possesses greater terpenes retention, reduced higher alcohols and an increased proportion of ethyl and acetate esters in the total volatile compounds [1–4]. Another positive aspect is that low temperatures reduce the growth of acetic and lactic bacteria, thus making it easier to control alcoholic fermentation.

Despite low-temperature fermentations offering interesting improvements, this practice also has its disadvantages. The optimal growth and fermentation temperature for *Saccharomyces cerevisiae* is 25 - 28°C. Restrictive low temperature increases the lag phase and lowers the growth rate, leading to sluggish and stuck fermentations [5]. Therefore, the quality of those wines produced at low temperature depends on the yeast's ability to adapt to cold.

The importance of lipid composition in the yeast adaptive response at low temperature is well-known [1,4,6,7]. A drop in temperature leads to diminished membrane fluidity [8]. To counteract this membrane rigidity, yeasts were able to develop several mechanisms to maintain appropriate fluidity. The most commonly studied involves increased unsaturation and reduced average chain length of fatty acids (FA) [1,4]. Recently, [7] also reported new common changes in the lipid composition of different industrial species and strains of *Saccharomyces* after growth at low temperature. Despite specific strain-/species-dependent responses, the results showed that the medium chain FA and triacylglyceride content increased at low temperatures, whereas phosphatidic acid content and the phosphatidylcholine/phosphatidylethanolamine (PC/PE) ratio decreased. In this way, cells can also be influenced by the environment during wine fermentation because yeast can incorporate fatty acids from the medium into its own phospholipids [1,9]. In grapes, unsaturated fatty acids represent the major component of total lipids. The most abundant is linoleic acid (C18:2), followed by oleic (C18:1), linolenic (C18:3) and palmitoleic acid (C16:1) [10].

In *S. cerevisiae*, these metabolic changes are primarily governed by the regulation of the transcriptional activity of those genes involved in the lipid biosynthesis pathway. Tai *et al.* [11] compared different genome-wide transcriptional studies of *S. cerevisiae* grown at low temperature. They concluded that the lipid metabolism genes were the only ones whose activity was clearly regulated by low tempera-

ture. In a recent work, we also demonstrated that the main differences between the metabolic profiling of *S. cerevisiae* growing at 12°C and 28°C were related to lipid metabolism [12]. In another study by our group, we also screened the importance of most of the genes belonging to the phospholipid, sterol and sphingolipid pathways in adaptation to low temperature by analyzing the effect on growth in a laboratory and an industrial strain [13]. From this previous study, the genes whose deletion and overexpression showed the greatest effect on growth were the following: *PSD1*, *CHO2* and *OPI3*, of the phospholipid metabolism; *ERG3*, *ERG6* and *IDI1*, of the ergosterol pathway; *LCB3*, *LCB4* and *DPL1*, belonging to the sphingolipid pathway; and *OLE1*, the only desaturase of *S. cerevisiae*. The aim of the present study was to conduct an in-depth study of these selected genes in a context that mimicked wine fermentation conditions. Firstly, we analyzed the gene activity of the selected genes in several low-temperature fermentations of synthetic grape must in wild-type and overexpressing strains. We then characterized the effect of the mutations and overexpressions in a derivative wine strain on growth and fermentation activity in wine fermentations at low and

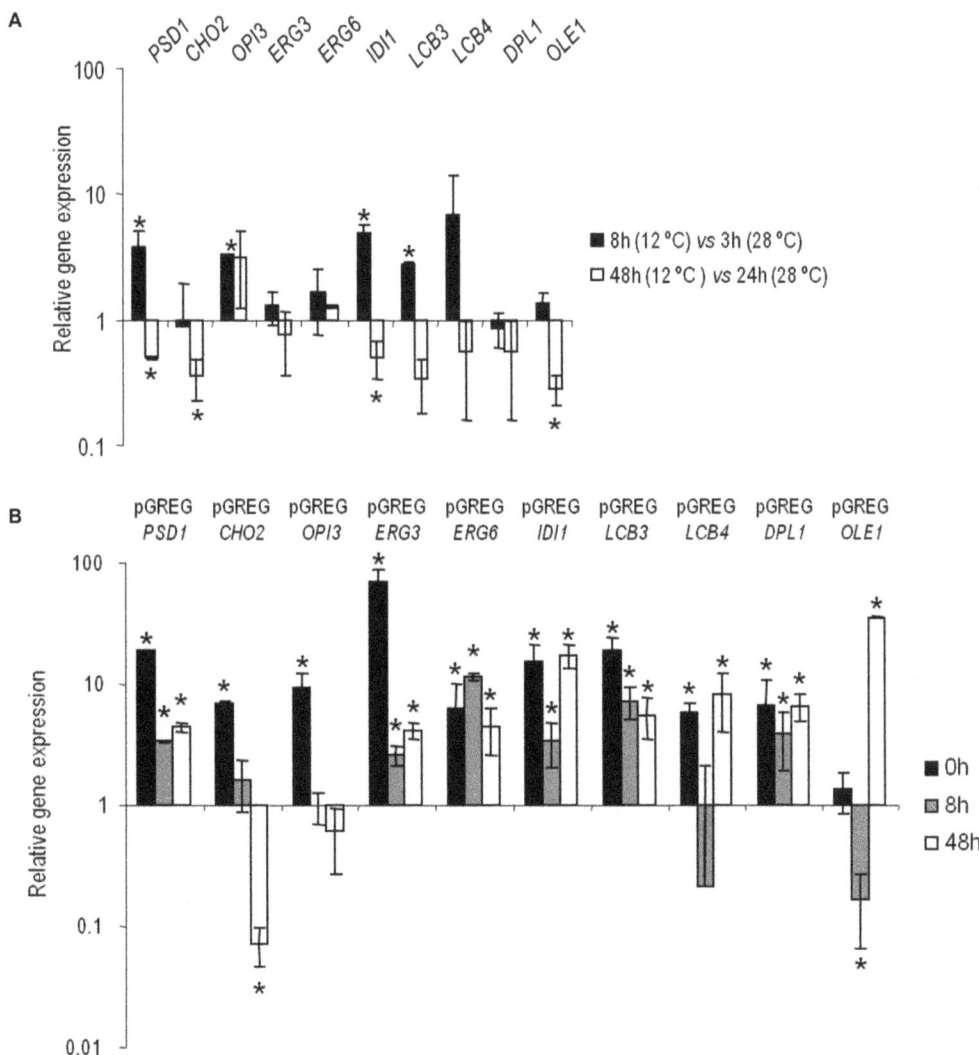

FIGURE 1: (A) Relative expression of the selected genes in the haploid strain *ho*QA23 at different stages of alcoholic fermentation. The gene expression in the lag and exponential phases of fermentation at 12°C are shown as a relative value in comparison to the control fermentation at 28°C. Values over 1 indicate higher gene expression at 12°C, whereas those under 1 indicate higher gene expression at 28°C. **(B)** Relative expression of the overexpressed genes. The differences in gene expression in the selected overexpressing strains are shown in relation to control *ho*QA23-pGREG (set as value 1). Values over 1 indicate higher gene expression than the control, whereas values under 1 indicate lower gene expression in comparison to the control. *The results with statistically significant differences (P-value ≤ 0.05).

optimum temperature. The increase in the gene dosage of some of these lipid genes improved both growth and fermentation activity in low-temperature fermentations, thus confirming their positive role during wine yeast adaptation at low temperature. Finally, the genes that showed an improved phenotype were overexpressed by integrating one or more copies in the delta regions of the genome of commercial wine yeast QA23 [14]. These stable overexpressing strains were retested in synthetic and natural grape must fermentations. The fermentative aroma compounds obtained in these wines were also analyzed.

RESULTS
Gene expression
Expression of the selected genes during fermentations at 12°C vs. 28°C

The changes in the gene expressions at low temperature of *PSD1*, *CHO2*, *OPI3*, *ERG3*, *ERG6*, *IDI1*, *LCB3*, *LCB4*, *DPL1* and *OLE1* were analyzed in the control *ho*QA23 strain during the first fermentation stages at 12°C and 28°C. Prior to taking samples, growth curves were analyzed to select the hours corresponding to the lag and exponential phases at both temperatures (the same OD in both curves). Thus

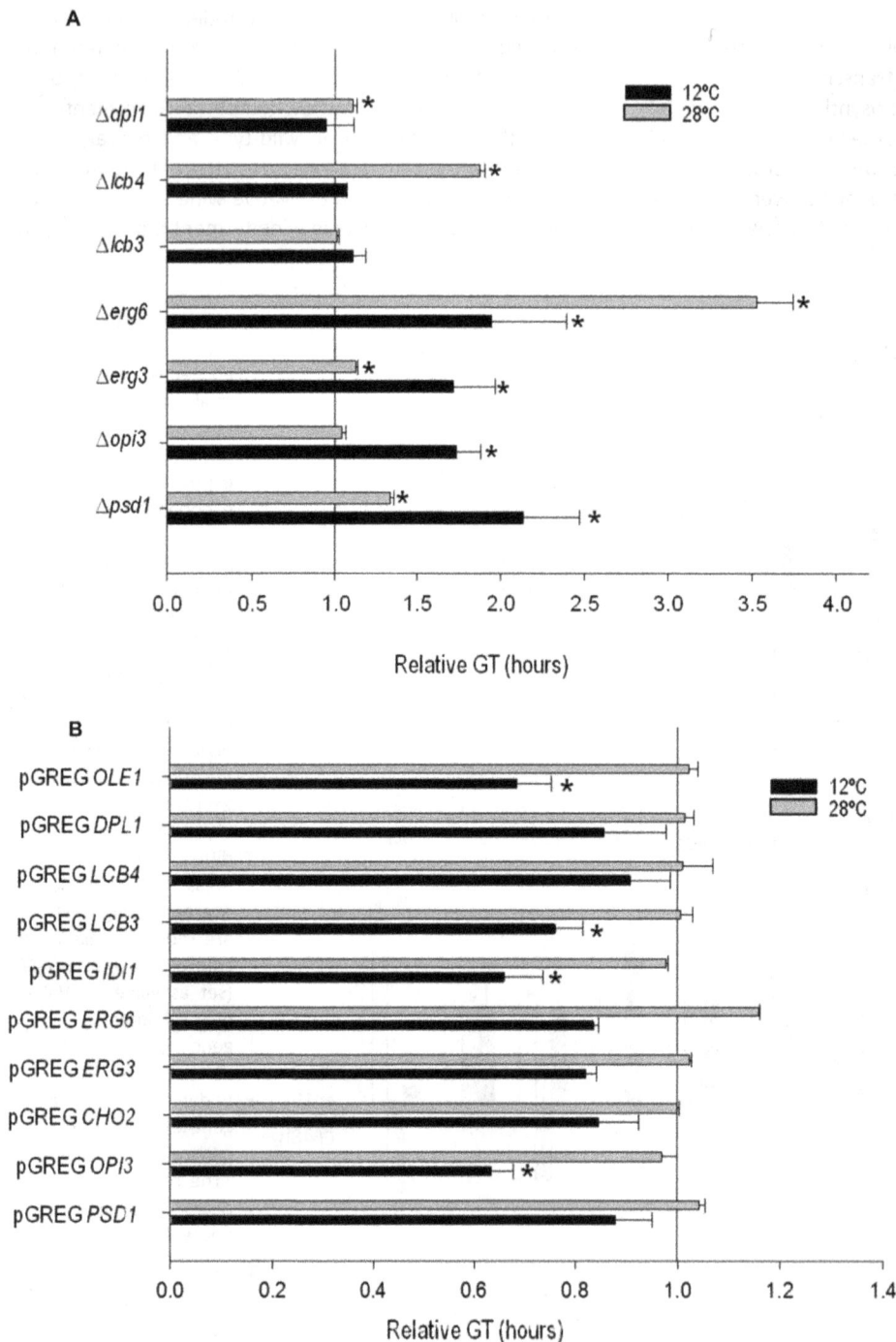

FIGURE 2: Generation time (GT) of **(A)** the mutant and **(B)** selected overexpressing strains grown at 12°C (black bars) and at 28°C (gray bars) normalized with the GT of their control strains *ho*QA23 and *ho*QA23-pGREG (normalized as value 1). The GT for the control strains was as follows: 11.59 h ± 3.12 h and 3.48 h ± 0.06 h for *ho*QA23 and 13.83 h ± 0.05 h and 3.63 h ± 0.05 h for *ho*QA23-pGREG at 12°C and 28°C, respectively. *Statistically significant differences (P-value ≤ 0.05) if compared with the control strain at the same temperature.

samples were taken in the lag (3 h at 28°C and 8 h at 12°C) and exponential phases (24 h at 28°C and 48 h at 12°C) during fermentation. The relative gene expression is shown in Figure 1A. The values higher and lower than 1 indicate a higher and lower gene expression at 12°C in comparison to 28°C. Save very few exceptions, these lipid genes showed higher activity during the lag or adaptation period at low temperature and, conversely, they were more active in the exponential phase at 28°C.

Verification of the overexpression in haploid wine strain hoQA23

Having determined gene activity at both temperatures in the key wine fermentation phases, we aimed to validate and quantify the overexpression of the constructed strains. Samples were taken before inoculation (time 0) and at the same time points (8 h and 48 h) at low temperature. The relative gene expression values of the overexpressing strains, normalized with the values of the control haploid strain (hoQA23-pGREG), are shown in Figure 1B. All the constructed strains presented increased overexpressed gene activity, which ranged from 3.5- to 68-fold more than the control strain at most of the time points analyzed. Thus we can verify that the constructed strains overexpressed the gene of interest.

Phenotypic analysis of the mutant and overexpressing strains of hoQA23

Determination of generation time (GT) during growth in a synthetic grape must (SM)

In order to determine the importance of the deletion or overexpression of the selected genes on growth at low temperature in wine fermentation, we calculated the GT of the mutant and overexpressing strains at 12°C and 28°C in SM (Fig. 2). All the phospholipid and sterol mutants showed worse growth than the control strain at 12°C, whereas no significant differences were observed for the sphingolipid mutants at this temperature. The GT of several mutants also increased at 28°C if compared with hoQA23. However, the differences were much larger at 12°C than at 28°C in Δpsd1 and Δerg3.

Likewise, the GT of the overexpressing strains was also determined (Fig. 2). Most of the overexpressing strains showed a substantially shorter GT at low temperature, although significant differences were noted only for pGREG OPI3, pGREG IDI1, pGREG LCB3 and pGREG OLE1.

Fermentation activity of the mutant and overexpressing strains of hoQA23

The fermentation kinetics of the mutant and overexpressing strains were estimated by calculating the time required to ferment 5% (T5), 50% (T50) and 100% (T100) of the sugars in the SM (Fig. 3). T5, T50 and T100 approximately match the beginning (lag phase), middle (end of the exponential phase) and end of fermentation, respectively. It should be highlighted that parental hoQA23 (control strain) and the same strain transformed with empty vector pGREG (the control strain of the overexpressing strains) showed differences in the T5, T50 and T100 (data provided in the

figure legend). These differences may be explained by the presence of geneticin in the fermentations of the overexpressing strains and their resistance to this antibiotic encoded in the plasmid.

Deletion of some genes impaired the low-temperature fermentation performance of the wine strain. This was especially remarkable for Δpsd1 and Δerg3, which were significantly delayed at the beginning of the process (T5) (more than 30 h and 60 h, respectively). The Δpsd1, Δopi3, Δerg3 and Δerg6 mutant strains also needed more time to ferment 50% of the sugars (T50) and did not finish the fermentation process at low temperature. Although not as long, a similar delay in fermentation was also observed at 28°C for the Δopi3, Δerg3 and Δerg6 strains, but not for Δpsd1. This strain was considerably affected at low temperature, but was not affected at all at 28°C. Deletion of genes Δlcb4 and Δlcb3 affected the fermentation capacity at both low and optimum temperature. The latter gene deletion produced a stuck fermentation at 28°C.

Conversely, several overexpressing strains showed quicker fermentation activity at low temperatures. The overexpressions of OLE1, DPL1 and LCB3 resulted in a shorter T5, T50 and T100. Despite pGREG PSD1 did not start fermentation before the control, this strain displayed greater fermentation activity at T50 and finished almost 2 days before the fermentation if compared with the control hoQA23-pGREG strain. However, the overexpressions of ERG3 and ERG6 resulted in a serious delay throughout the fermentation process at 12°C. pGREG ERG3 and pGREG ERG6 obtained longer T5 and T50, and were unable to finish fermentation. Interestingly, the overexpressions of PSD1 and OLE1 had no effect on fermentation length at 28°C.

Stable overexpression of the selected genes in commercial wine yeast QA23

Based on the previous results, we selected the four genes DPL1, LCB3, OLE1 and PSD1 to construct stable overexpressing strains in the genetic background of the commercial wine yeast QA23. These copies were integrated by homologous recombination into the repetitive delta elements of Ty1 and Ty2. The correct integration of one or more copies was verified by PCR with primers homologous to the δ sequences. The overexpression of these strains was verified during wine fermentation in natural "Parellada" grape must at low temperature. The relative gene expression values were normalized with the commercial wine strain QA23 values (Fig. 4.). The four strains showed an overexpression of the target genes but, in all cases, the level of overexpression was lower than in the overexpressing strains of hoQA23, constructed by transformation with centromeric plasmids.

These stable overexpressing strains were used to ferment both the SM and natural must (NM) of two different grape varieties (Albariño and Parellada). Yeast growth during fermentations was similar between the overexpressing strains and the commercial QA23 (data not shown). Minor differences were observed in the density reduction in the fermentations of both SM and NM carried out by the over-

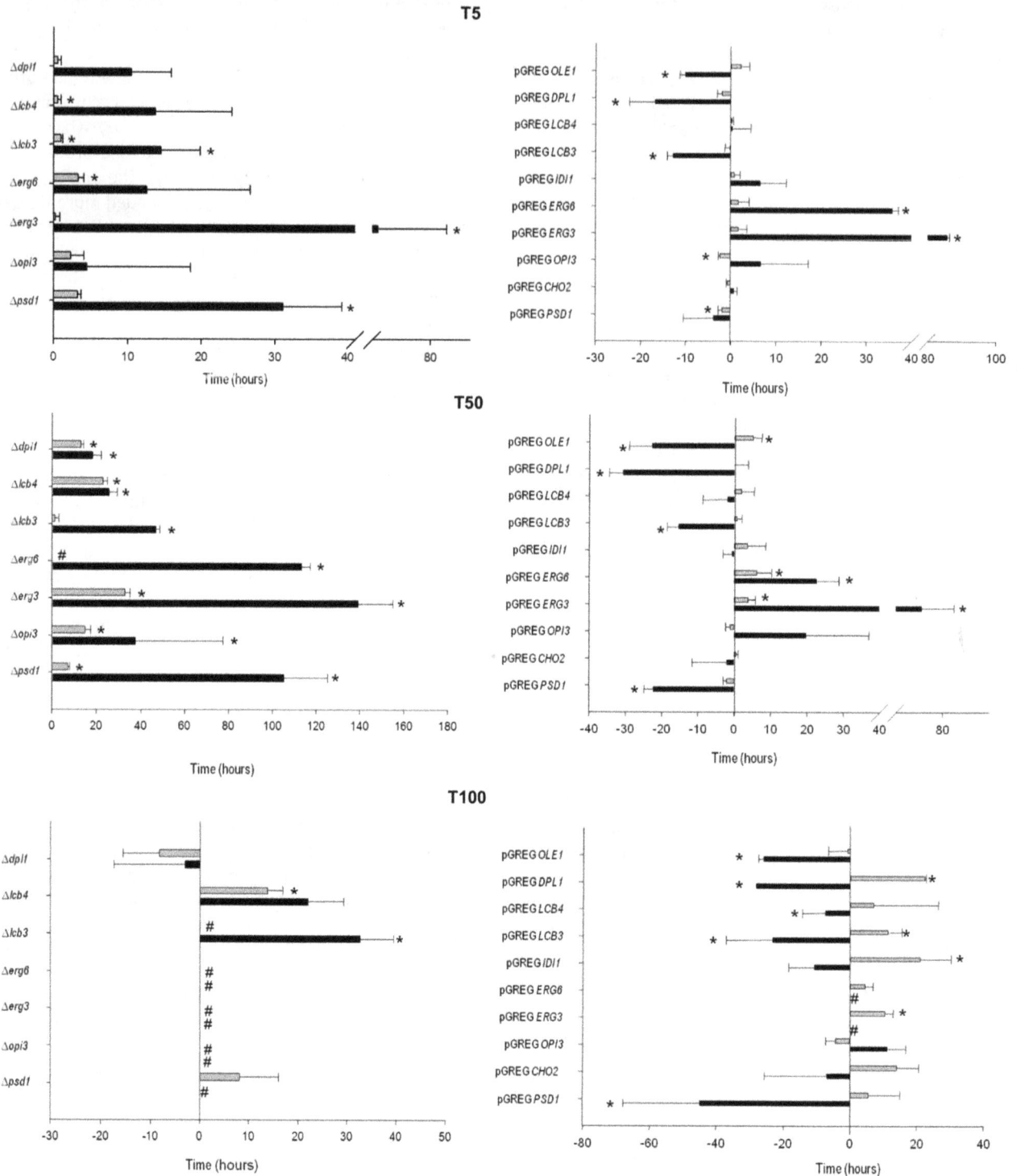

FIGURE 3: Determination of the time required by the mutant and the selected overexpressing strains to ferment 5% (T5), 50% (T50) and 100% (T100) of the initial sugar content in SM at 12°C (black bars) and 28°C (gray bars). Positive and negative values respectively represent the increases and decreases in time (hours) of the mutant and the overexpressing strains if compared to the control strains (normalized as value 0). The fermentation time of the control strains are: *ho*QA23 at 12°C T5 = 27 h ± 3.18 h, T50 = 96.19 h ± 3.97 h T100 = 251.44 h ± 10.34 h; at 28°C T5 = 6.23 h ± 0.93 h, T50 = 44.95 h ± 0.93 h, T100 = 131.14 h ± 2.32 h and *ho*QA23-pGREG at 12°C T5 = 41.63 h ± 7.16 h, T50 = 119.81 h ± 11.93 h T100 = 271.69 h ± 21.48 h; at 28°C T5 = 8.49 h ± 0.64 h, T50 = 38.40 h ± 1.56 h, T100 = 121.73 h ± 3.36 h. # Indicates stuck fermentation before T50 or T100. * Indicates statistically significant differences (P-value ≤ 0.05).

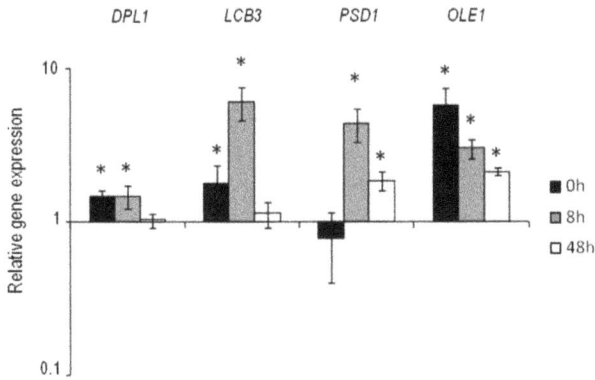

FIGURE 4: Relative expression of the overexpressed genes in the commercial wine yeast QA23. The differences in gene expression in the selected overexpressing strains are shown in relation to control QA23 (set as value 1). Values over 1 indicate higher gene expression than the control, whereas values under 1 indicate lower gene expression in comparison to the control. *The results with statistically significant differences (P-value ≤ 0.05).

expressing strains as compared to that performed by commercial wine strain QA23 (Fig. 5). The overexpression of δDPL1δ and δPSD1δ resulted in a shorter T5 in the fermentations performed in SM, but no difference was found at the end of fermentation. Only δOLE1δ was able to ferment 50% of sugars faster than the control in the "Parellada" grape fermentations. Moreover, δOLE1δ and δDPL1δ finished the fermentation process more quickly than QA23 in both "Parellada" and "Albariño" NM grape must fermentations.

We also analyzed the fermentative aroma compounds (higher alcohols, acetate esters and ethyl esters) in the

wines obtained with both the overexpressing strains and the commercial wine yeast QA23 in "Albariño" grape must. A principal component analysis was performed to explore the effect of the overexpression of these genes on aroma composition (Fig. 6). The two first components were retained and explained 90.5% of total variance. The first principal component (PC1) accounted for 55.5% of total variance and was marked by high components loadings for ethyl lactate (+0.614) and isoamyl alcohol (+0.557). The second component loading explained 35% of the variation and was marked by high positive component loadings for ethyl lactate (+0.510) and isoamyl alcohol (+0.230), and by

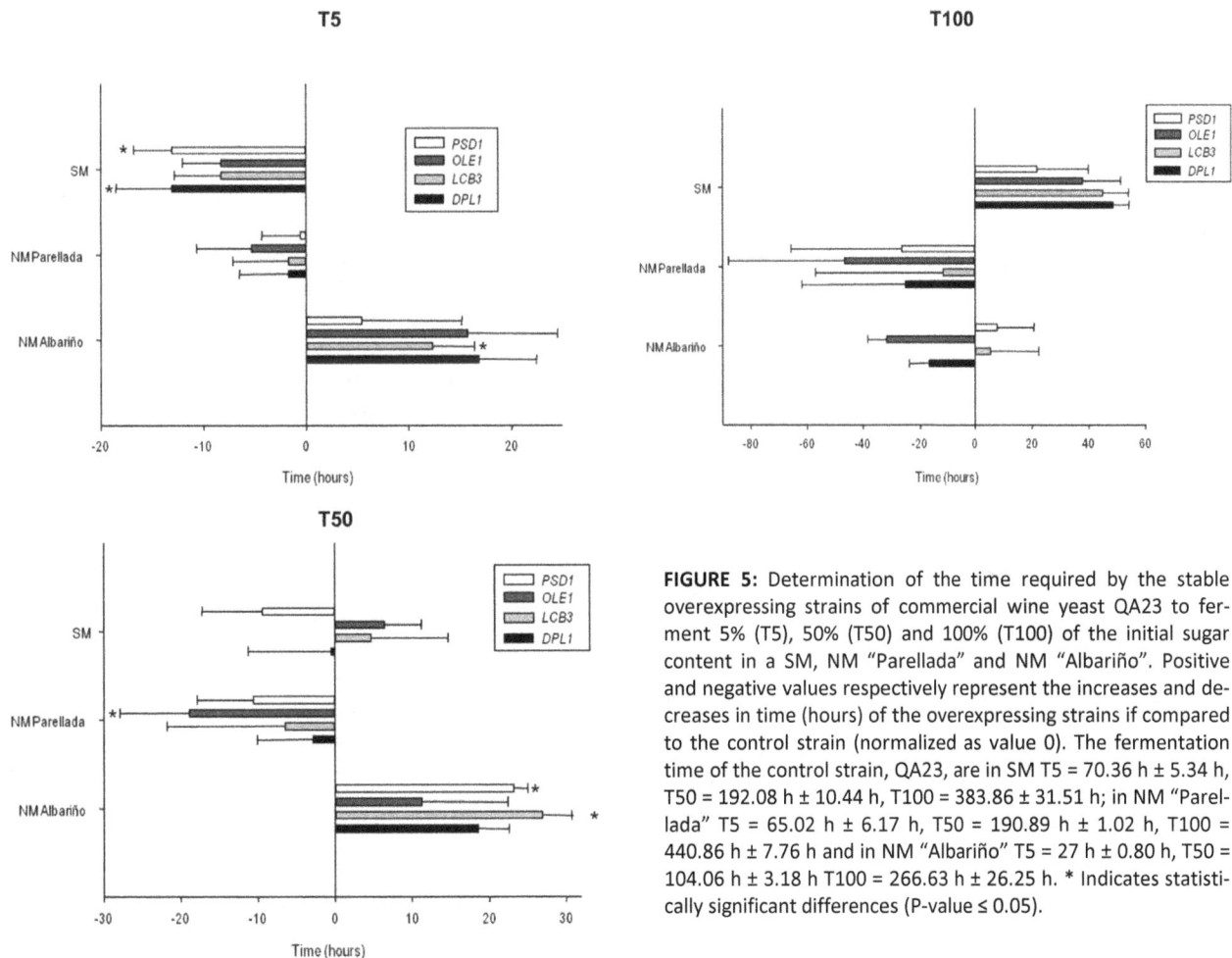

FIGURE 5: Determination of the time required by the stable overexpressing strains of commercial wine yeast QA23 to ferment 5% (T5), 50% (T50) and 100% (T100) of the initial sugar content in a SM, NM "Parellada" and NM "Albariño". Positive and negative values respectively represent the increases and decreases in time (hours) of the overexpressing strains if compared to the control strain (normalized as value 0). The fermentation time of the control strain, QA23, are in SM T5 = 70.36 h ± 5.34 h, T50 = 192.08 h ± 10.44 h, T100 = 383.86 ± 31.51 h; in NM "Parellada" T5 = 65.02 h ± 6.17 h, T50 = 190.89 h ± 1.02 h, T100 = 440.86 h ± 7.76 h and in NM "Albariño" T5 = 27 h ± 0.80 h, T50 = 104.06 h ± 3.18 h T100 = 266.63 h ± 26.25 h. * Indicates statistically significant differences (P-value ≤ 0.05).

a high negative loading for ethyl acetate (-0.823).

The δOLE1δ strain clearly separated from the other strains in the down-left quadrant, denoting the most specific aroma profile. The wine produced by this strain was the poorest in isoamyl alcohol, ethyl acetate and ethyl lactate.

DISSCUSSION

Low-temperature fermentations produce wines with greater aromatic complexity. Nonetheless, the success of these fermentations greatly depends on the adaptation of yeast cells to cold. Changes in the plasma membrane composition have been directly related with the yeast adaptive response at different environmental temperatures in many studies [1,4,6,7]. In our previous study [13], we screened most of the mutants of laboratory strain BY4742 encoding enzymes of the phospholipid, sterol and sphingolipid pathways in their growth capacity at 12°C. Those genes whose deletion showed growth impairment at low temperature were also deleted and overexpressed in the derivative haploid hoQA23. In this previous study, determination of growth parameters was carried out in minimal medium (SC) to avoid interferences of the other stresses exerted during wine fermentation (osmotic, pH, ethanol, etc.). Despite the many phenotypic differences observed between the laboratory and the commercial wine yeast strains, we detected some key lipid metabolism genes in promoting better growth at low temperature [13]. We are, however, aware that these mutant and overexpressing strains with differential phenotypes at low temperature should be tested in an environment that mimics grape must fermentation. The aim of this study was to confirm the importance of these genes in growth and fermentation activity at low temperature by using a SM.

In this study, a SM without anaerobic factors was used to avoid the incorporation of some sterols and unsaturated fatty acids from the medium [1]. If we compare the generation time of the mutants growing in SC [13] and SM (this study), phospholipid (Δpsd1 and Δopi3) and sterol mutants (Δerg3 and Δerg6) showed strongly impaired growth at low temperature, regardless of the media. However, no difference in the growth of sphingolipid mutants in SM was noted. Likewise, similar results were observed when we analyzed the growth of the overexpressing strains growing in SC and SM at low temperature. The overexpressions of OPI3, IDI1 and LCB3 produced a phenotype with better growth in both culture media at low temperature in comparison to the control strain. Unlike the results obtained in SC, the overexpression of OLE1 in SM enhanced growth at low temperature. All these results demonstrate the importance of testing growth capacity in an environment that mimics grape must fermentation.

The analysis of fermentation performance showed that the mutants with worst growth at 12°C were unable to finish low-temperature fermentation (Δpsd1, Δopi3, Δerg3 and Δerg6). We also observed stuck fermentations at 28°C in the fermentation carried out by Δerg3 and Δerg6. These genes are involved in the last steps of ergosterol biosyn-

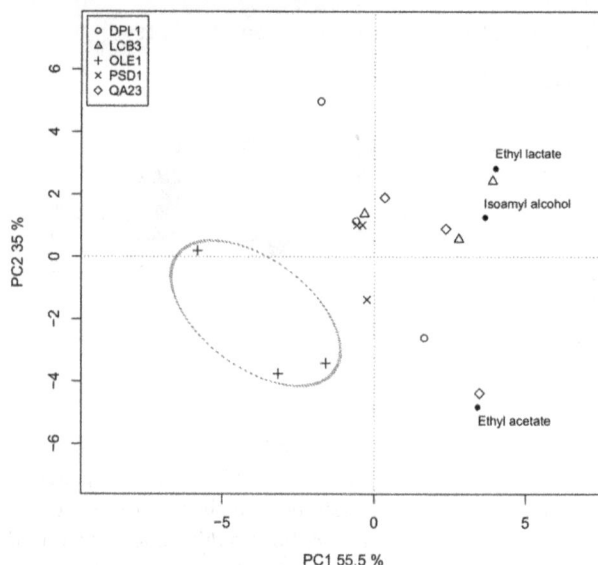

FIGURE 6: Biplot of the two first components of the principal component analysis according to the aroma composition of the selected overexpressing strains of QA23 in "Albariño" grape must at 12°C. Open symbols represent the samples and filled circles the aroma compounds with higher loadings.

thesis, and their function must be crucial for growth in SM and fermentation activity because these strains were strongly affected in both activities, with minimum influence of fermentation temperature. The most specific response to low temperature was related to PSD1. Growth and fermentation performance were barely affected in the Δpsd1 and pGREG PSD1 strains at optimum temperature. Nevertheless, these strains presented major phenotypic differences in comparison to the control performed in low temperature fermentations. PSD1 encodes a phosphatidylserine decarboxylase (Psd1p) of the mitochondrial membrane, which converts PS into PE. Recent works have related increases in PE or decreases in the PC/PE ratio as a general response to low temperature in different strains and species of Saccharomyces [7,15].

Another specific response at low temperature has also been observed in overexpressing strain pGREG OLE1, which showed improved fermentation performance and a shorter generation time than the control strain, as previously reported by [16].

The overexpression of two sphingolipid genes (LCB3 and DPL1) improved fermentation activity at 12°C. The pGREG LCB3 strain also showed a shorter GT at low temperatures in comparison to the control strain. The LCB3 gene encodes a phosphatase that is capable of dephosphorylating long-chain bases, dihydrosphingosine-1-phosphate (DHS-1-P) and phytosphingosine-1-phosphate (PHS-1-P), and the DPL1 gene encodes a lyase, which cleaves the same long-base phosphates [17]. Mandala et al. [18] demonstrated that Δlcb3 and Δdpl1 dramatically enhanced survival upon severe heat shock. Conversely, our data evidence that the overexpression of these genes improves growth and fermentation performance at low temperature.

In our opinion, although the importance of these genes in yeast cold adaptation is quite conclusive, these data were obtained in a derivative haploid of an industrial strain and using SM. In an attempt to take another step forward to approach industrial conditions, we decided to overexpress the four genes showing a specific response at low temperature in the industrial strain QA23 to subsequently test these new overexpressing strains in both synthetic and natural grape musts. When using non integrative plasmids, gene overexpression requires the cultivation of overexpressing strains in the presence of antibiotics or in a chemically-defined medium in order to maintain the plasmid by selection pressure. We recently adapted a novel, efficient method of stable gene overexpression in the industrial wine strains of *S. cerevisiae* [14]. This strategy is based on multi-copy chromosomal integration by homologous recombination with ubiquitous δ elements, which are integral parts of yeast transposons [19]. These new overexpressing strains did not show major phenotypic differences in low-temperature fermentation if compared with industrial strain QA23. The different phenotype shown by the overexpressing strains constructed by the two different methods (chromosome integration and centromeric plasmid) could be explained by the different ploidy of parental strains or the different number of new copies of the target gene, which resulted in a lower overexpression levels in the overexpressing strains constructed by chromosome integration. Despite the minor differences observed, fer-

mentation length was shorter for strains δOLE1δ and δDPL1δ if compared to commercial wine yeast QA23 both in "Parellada" and "Albariño" grape must fermentations.

As changes in the fatty acid profile can have a direct impact on aroma production [4], we analyzed the fermentative aroma compounds in the wines obtained with both the overexpressing strains and commercial wine yeast QA23. The overexpression of these genes did not lead to major modifications in the aroma profile of the final wines, except for the wines fermented by strain δOLE1δ, which achieved a poorer production of several aroma compounds (ethyl lactate, isoamyl alcohol and ethyl acetate). Saerens *et al.* [20] reported an indirect correlation between unsaturated fatty acids and ethyl acetate.

In summary, most of the results have supported the screening done in the previous study because the constructed mutants exhibited impaired growth and fermentation activity, whereas the overexpressing strains of these genes reduced the GT and fermentation length. Genes such as *DPL1, LCB3, OLE1* and *PSD1* have been seen to play a crucial role in cold adaptation, and the genetic manipulation of these genes may improve the performance of wine yeasts in low-temperature fermentations. Construction of overexpressing strains by chromosomal integration is a clean, safe method that can be used in the wine industry. Moreover we can increase the overexpression by integrating more copies of the target gene in successive rounds of transformations of the same commercial wine strain.

TABLE 1. List of the lipid genes used in this study.

Standard name	Systematic name	Molecular function	Substrate	Product
PSD1	YNL169C	Phosphatidylserine decarboxylase I	PS	PE
CHO2	YGR157W	Phosphatidylethanolamine N-Methyltransferase	PE	M-PE/ MM-PE
OPI3	YJR073C	Phospholipid methyltranferase	MM-PE	PC
ERG3	YLR056W	C-5 sterol desaturase	Episterol	5,7,24(28)-ergostatrienol
ERG6	YML008C	Sterol 24-C methyltransferase	Zymosterol	Fecosterol
IDI1	YPL117C	Isopentenyl-diphosphate delta-isomerase	Delta3-isopentenyl-PP	Dimethylallil-pyrophosphate
LCB3	YJL134W	Sphingosine-1-phosphate phosphatase	DHS-P, PHS-P	DHS, PHS
LCB4	YOR171C	D-erythro-sphingosine kinase	DHS, PHS	DHS-P, PHS-P
DPL1	YDR294C	Sphinganine-1-phosphate aldolase	DHS-P	Palmitaldehyde Phosphoryl-ethanolamine
OLE1	YGL055W	Stearoyl-CoA-desaturase	Saturated fatty acids	Unsaturated fatty acids

MATERIALS AND METHODS
Construction of mutant and overexpressing strains
Most of the deleted mutant and the selected overexpressing strains were constructed in our previous work [13] in the background of a derivative haploid of commercial wine strain QA23 (hoQA23) (Lallemand S.A., Canada) [21]. All the genes were deleted using the short flanking homology (SFH) method based on the KanMX4 deletion cassette [22] and were overexpressed by cloning into the centromeric plasmid pGREG505, as described in [23]. These genes are listed in Table 1. IDI1, OLE1 and CHO2 were overexpressed only because the deletion of the two former genes produced an unviable phenotype [24] and the deletion of CHO2 caused an auxotroph phenotype for choline in derivative wine strain hoQA23 [13]. The haploid QA23 strain transformed with empty plasmid pGREG505 (hoQA23-pGREG) was used as control of the overexpressing strains.

Moreover in this study, the stable overexpressing strains were constructed by integrating one or more copies of genes DPL1, LCB3, OLE1 and PSD1 into the genome of the commercial wine yeast strain QA23. To do this, the method proposed by [19] was followed with some modifications [14]. This genetic transformation system allows the integration of the selected gene in the δ sequences of the S. cerevisiae genome. Briefly, KanMX4 was integrated approximately 400 bp downstream of the stop codon of the gene of interest. After checking the correct integration of KanMX4, a new PCR product incorporating the gene of interest, with its own promoter and the gene of resistance to geneticin (KanMX4), was obtained. These PCR fragments were generated with primers D1-Forward and KanD2-Reverse, which contain homologous tails to the δ sequences of Ty [19]. The expression cassettes for genes DPL1, LCB3, OLE1 and PSD1 were used to transform wine yeast strain QA23. Transformants were selected by geneticin resistance and PCR was used to test the correct insertion of the cassettes into the δ sequences. The new overexpressing strains were named δDPL1δ, δLCB3δ, δOLE1δ and δPSD1δ.

Gene expression analysis by real-time quantitative PCR
Total RNA of 10^8 cell/ml was isolated as described by [25] and was resuspended in 50 μl of DEPC-treated water. Total RNA suspensions were purified using the High Pure Isolation kit (Roche Applied Science, Germany) according to the manufacturer's instructions. RNA concentrations were determined using a NanoDrop ND-1000 spectrophotometer (NanoDrop Technologies, USA), and RNA quality was verified electrophoretically on a 0.8% agarose gel. Solutions and equipment were treated so that they were RNase-free, as outlined in [26].

Total RNA was reverse-transcribed with Superscript™ II RNase H⁻ Reverse Transcriptase (Invitrogen, USA) in a GenAmp PCR System 2700 (Applied Biosystems, USA). The reaction contained 0.5 μg of Oligo (dT)₁₂₋₁₈ Primer (Invitrogen, USA) and 0.8 μg of total RNA as a template in a total reaction volume of 20 μl. Following the manufacturer's guidelines, after denaturation at 70°C for 10 min, cDNA was synthesized at 42°C for 50 min, and then the reaction was inactivated at 70°C for 15 min.

The primers were designed with the Saccharomyces Genome Database (SGD), except for housekeeping gene ACT1, which was previously described by [27]. All the amplicons were shorter than 100 bp, which ensured maximal PCR efficiency and the most precise quantification. Real-Time Quantitative PCR was performed in LightCycler® 480 SYBR Green I

Master (Roche, Germany). The SYBR PCR reactions contained 2.5 μM of each PCR primer, 5 μl of cDNA and 10 μl of SYBR Green I Master (Roche, Germany) in a 20-μl reaction.

All the PCR reactions were mixed in a LightCycler® 480 Multiwell Plate 96 (Roche, Germany) and cycled in a LightCycler® 480 Instrument II, 96-well thermal cycler (Roche, Germany) under the following conditions: 95°C for 5 min, and 45 cycles at 95°C for 10 sec, at 55°C for 10 sec and 72°C for 10 sec. Each sample had two controls that were run in the same PCR: no amplification control (sample without reverse transcriptase reaction) to avoid interference by contaminant genomic DNA and no template control (sample with no RNA template) to avoid interference by primer-dimer formation. All the samples were analyzed in triplicate with the LightCycler® 480 Software, version 1.5 (Roche, Germany) and the expression values were averaged. The gene expression levels are shown as a relative value in comparison to the control. Housekeeping gene ACT1 was used as an endogenous reference gene to normalize input amounts.

Generation time
Growth was monitored at 600 nm in a SPECTROstar Omega instrument (BMG Labtech, Offenburg, Germany) at 12°C and 28°C, as described in [13].

Growth parameters were calculated from each treatment by directly fitting OD measurements versus time to the reparametized Gompertz equation proposed by [28]:

$$y=D*exp\{-exp[((\mu_{max}*e)/D)*(\lambda-t))+1]\}$$

where $y=ln(OD_t/OD_0)$, OD_0 is the initial OD and OD_t is the OD at time t; $D=ln(OD_t/OD_0)$ is the asymptotic maximum, μ_{max} is the maximum specific growth rate (h⁻¹), and λ is the lag phase period (h) [29]. The R code (statistical software R v.2.15 (R Development Core Team, 2013)) was used to fit the results to the reparametized Gompertz equation. Generation time was calculated using the equation $t_d=ln2/\mu$. Values were normalized by dividing, with its control, the generation time of strains hoQA23 or hoQA23-pGREG. Values lower than 1 indicated a shorter generation time, whereas values higher than 1 indicated a longer generation time as compared to the control.

Fermentations
All strains were cultured in the SM (pH 3.3) described by [30], but with 200 g/L of reducing sugars (100 g/L glucose + 100 g/L fructose) and without anaerobic factors [27]. The following were utilized: organic acids, malic acid 5 g/L, citric acid 0.5 g/L and tartaric acid 3 g/L; mineral salts KH_2PO_4 750 mg/L, K_2SO_4 500 mg/L, $MgSO_4$ 250 mg/L, $CaCl_2$ 155 mg/L, NaCl 200 mg/L, $MnSO_4$ 4 mg/L, $ZnSO_4$ 4 mg/L, $CuSO_4$ 1 mg/L, KI 1 mg/L, $CoCl_2$ 0.4 mg/L, H_3BO_3 1 mg/L and $(NH4)_6Mo_7O_{24}$ 1 mg/L; vitamins myo-inositol 20 mg/L, calcium pantothenate 1.5 mg/L, nicotinic acid 2 mg/L, chlorohydrate thiamine 0.25 mg/L, chlorohydrate pyridoxine 0.25 mg/L and biotin 0.003 mg/L. The assimilable nitrogen source used was 300 mg N/L (120 mg N/L as ammonium and 180 mg N/L in the amino acid form). Geneticin was also added (200 mg /L) to the SM of the overexpressing strains to ensure plasmid stability.

The overexpressing strains constructed by chromosomal integration were also cultured in two natural grape musts: "Albariño" grape must, which contained about 200 g/L of reducing sugars (100 g/L glucose + 100 g/L fructose); "Parellada" grape must, which contained about 180 g/L of reducing sugars (90 g/L glucose + 90 g/L fructose). Prior to inoculation, the grape must was treated with 1 ml/L of Velcorin (trade name for dimethyldicarbonate; Merck, Hohenbrunn, Germany). The use of this antimicrobial agent resulted in the practical elimination of the microbiota of the NM, tested by plating the grape must on YPD plates and incubated for 72 h at 30°C.

In the fermentations performed by the mutant and overexpressing strains of hoQA23, and also in those performed in "Albariño" by the overexpressing strains of QA23, the inoculated population came from an overnight culture in YPD at 30°C. In order to avoid other stresses (osmotic, pH, etc.) to the inoculum produced by changing from YPD to grape must, the fermentations carried out by the stable overexpressing strains of QA23 in SM and NM "Parellada" were inoculated with the cells from an overnight culture at 30°C in the same fermentation media. After counting microscopically, the appropriate dilution of the overnight culture was transferred to the grape must to achieve an initial cell concentration of 2×10^6 cells/ml.

Fermentation activity of the mutant and overexpressing strains of hoQA23 were tested at 28°C and 12°C, and fermentation activity of stable overexpressing strains were analyzed only at 12°C. Fermentations were performed with continuous orbital shaking at 100 rpm. Fermentations were carried out in laboratory-scale fermenters using 100-ml bottles filled with 60 ml of media, which were fitted with closures that enabled carbon dioxide to escape and samples to be removed. Yeast cell growth was determined by absorbance at 600 nm and by plating samples at the end of fermentation on YPD agar at an adequate dilution to be incubated for 2 days at 30°C. Fermentation was monitored by measuring the density of the media (g/L) using a Densito 30 PX densitometer (Mettler Toledo, Switzerland). Fermentation was considered to have been completed when density was below 998 g/L. Residual sugars were also determined by HPLC in a Surveyor Plus Chromatograph (Thermo Fisher Scientific, Waltham, MA, USA).

Volatile aroma compounds
Higher alcohols and esters were analyzed based on a headspace solid-phase microextraction (SPME) technique using a 100 µm poly-dimetylsiloxane (PDMS) fiber (Supelco, Sigma-Aldrich, Spain). Aliquots of 1.5 ml of the sample were placed into 15 ml vials and 0.35 g of NaCl and 20 µl of 2-heptanone (0.005%) was added as an internal standard. Vials were closed with screwed caps and 13 mm silicone septa. Solutions were attired for 2 h to obtain the required headspace-liquid equilibrium. Fibers were injected through the vial septum and exposed to the headspace for 7 min to then be desorbed for 4 min in a gas chromatograph (TRACE GC Ultra, Thermo Scientific), with a flame ionization detector (FID) equipped with an HP INNOWax 30 m x 0.25 mm capillary column coated with a 0.25 m layer of cross-linked polyethylene glycol (Agilent Technologies). The carrier gas was helium (1 ml/min) and the oven temperature program utilized was: 5 min at 35°C, 2°C/min to 150°C, 20°C/min to 250 °C. The injector and detector temperatures were maintained at 220°C and 300°C respectively. A chromatographic signal was recorded by the ChromQuest program. Volatiles compounds were identified by comparing the retention time for reference compounds. Volatile compound concentrations were determined using calibration graphs of the corresponding standard volatile compounds.

Statistical data processing
All the experiments were repeated at least 3 times. Data are reported as the mean value ± SD. Significant differences among the control strain, the mutant and the overexpressing strains were determined by t-tests (SPSS 13 software package, USA). The statistical level of significance was set at P ≤ 0.05. A principal component analysis was done using the vegan package (rda function) of the statistical software R, v.2.15 [31].

ACKNOWLEDGMENTS
This work has been financially supported by grants AGL2010-22001-C02-01 and PROMETEOII/2014/042 from the Spanish government and the Generalitat Valenciana, respectively, awarded to JMG. MLM wishes to thank the Spanish government for her FPI grant. The authors also thank to Ana Cristina Adam for her assistance with the gene expression analysis.

CONFLICT OF INTEREST
The authors declare no conflict of interest.

REFERENCES
1. Beltran G, Novo M, Guillamón JM, Mas A, and Rozès N (2008). Effect of fermentation temperature and culture media on the yeast lipid composition and wine volatile compounds. Int J Food Microbiol 121(2): 169–77.

2. Llauradó J, Rozès N, Bobet R, Mas A, and Constantí M (2002). Low temperature alcoholic fermentations in high sugar concentration grape musts. J Food Sci 67(1): 268–273.

3. Beltran G, Torija MJ, Novo M, Ferrer N, Poblet M, Guillamón JM, Rozès N, and Mas A (2002). Analysis of yeast populations during alcoholic fermentation: a six year follow-up study. Syst Appl Microbiol 25(2): 287–93.

4. Torija MJ, Beltran G, Novo M, Poblet M, Guillamón JM, Mas A, and Rozès N (2003). Effects of fermentation temperature and Saccharomyces species on the cell fatty acid composition and presence of volatile compounds in wine. Int J Food Microbiol 85(1-2): 127–136.

5. Bisson LF (1999). Stuck and sluggish fermentations. Am J Enol Vitic 50 (1): 107–119.

6. Henschke PA and Rose AH (1991). Plasma membrane. In: Rose AH, J.S. H, editors. The yeasts, vol. IV: yeast Organelles. Academic Press Limited, London, UK; pp. 297–345.

7. Redón M, Guillamón JM, Mas A, and Rozès N (2011). Effect of growth temperature on yeast lipid composition and alcoholic fermentation at low temperature. Eur Food Res Technol 232(3): 517–527.

8. Russell NJ (1990). Cold adaptation of microorganisms. Philos Trans R Soc Lond B Biol Sci 326(1237): 595–608, discussion 608–11.

9. Redón M, Guillamón JM, Mas A, and Rozès N (2009). Effect of lipid supplementation upon Saccharomyces cerevisiae lipid composition and fermentation performance at low temperature. Eur Food Res Technol 228(5): 833–840.

10. Castela P, Mesias J, and Maynar J (1985). Evolution de la teneur en lipides totaux neutres et polaires dans les raisins Macabeo au cours de leur cycle végétatif. Sci Aliments 59: 587–597.

11. Tai SL, Daran-Lapujade P, Walsh MC, Pronk JT, and Daran J (2007). Acclimation of Saccharomyces cerevisiae to low temperature: a

chemostat-based transcriptome analysis. **Mol Biol Cell** 18(12): 5100–12.

12. López-Malo M, Querol A, and Guillamón JM (**2013**). Metabolomic comparison of *Saccharomyces cerevisiae* and the cryotolerant species *S. bayanus var. uvarum* and *S. kudriavzevii* during wine fermentation at low temperature. **PLoS One** 8(3): e60135.

13. López-Malo M, Chiva R, Rozes N, and Guillamón JM (**2013**). Phenotypic analysis of mutant and overexpressing strains of lipid metabolism genes in *Saccharomyces cerevisiae*: implication in growth at low temperatures. **Int J Food Microbiol** 162(1): 26–36.

14. Chiva R, López-Malo M, Salvadó Z, Mas A, and Guillamón JM (**2012**). Analysis of low temperature-induced genes (LTIG) in wine yeast during alcoholic fermentation. **FEMS Yeast Res** 12(7): 831–43.

15. Tronchoni J, Rozès N, Querol A, and Guillamón JM (**2012**). Lipid composition of wine strains of *Saccharomyces kudriavzevii* and *Saccharomyces cerevisiae* grown at low temperature. **Int J Food Microbiol** 155(3): 191–8.

16. Kajiwara S, Aritomi T, Suga K, Ohtaguchi K, and Kobayashi O (**2000**). Overexpression of the *OLE1* gene enhances ethanol fermentation by *Saccharomyces cerevisiae*. **Appl Microbiol Biotechnol** 53(5): 568–74.

17. Dickson RC and Lester RL (**1999**). Metabolism and selected functions of sphingolipids in the yeast *Saccharomyces cerevisiae*. **Biochim Biophys Acta** 1438(3): 305–21.

18. Mandala SM, Thornton R, Tu Z, Kurtz MB, Nickels J, Broach J, Menzeleev R, and Spiegel S (**1998**). Sphingoid base 1-phosphate phosphatase: a key regulator of sphingolipid metabolism and stress response. **Proc Natl Acad Sci U S A** 95(1): 150–5.

19. Guerra OG, Rubio IGS, da Silva Filho CG, Bertoni RA, Dos Santos Govea RC, and Vicente EJ (**2006**). A novel system of genetic transformation allows multiple integrations of a desired gene in *Saccharomyces cerevisiae* chromosomes. **J Microbiol Methods** 67(3): 437–45.

20. Saerens SMG, Delvaux F, Verstrepen KJ, Van Dijck P, Thevelein JM, and Delvaux FR (**2008**). Parameters affecting ethyl ester production by *Saccharomyces cerevisiae* during fermentation. **Appl Environ Microbiol** 74(2): 454–61.

21. Salvadó Z, Chiva R, Rozès N, Cordero-Otero R, and Guillamón JM (**2012**). Functional analysis to identify genes in wine yeast adaptation to low-temperature fermentation. **J Appl Microbiol** 113(1): 76–88.

22. Güldener U, Heck S, Fielder T, Beinhauer J, and Hegemann JH (**1996**). A new efficient gene disruption cassette for repeated use in budding yeast. **Nucleic Acids Res** 24(13): 2519–24.

23. Jansen G, Wu C, Schade B, Thomas DY, and Whiteway M (**2005**). Drag&Drop cloning in yeast. **Gene** 344(0): 43–51.

24. Giaever G, Chu AM, Ni L, Connelly C, Riles L, Véronneau S, Dow S, Lucau-Danila A, Anderson K, André B, Arkin AP, Astromoff A, El-Bakkoury M, Bangham R, Benito R, Brachat S, Campanaro S, Curtiss M, Davis K, Deutschbauer A, Entian K-D, Flaherty P, Foury F, Garfinkel DJ, Gerstein M, Gotte D, Güldener U, Hegemann JH, Hempel S, Herman Z, et al. (**2002**). Functional profiling of the *Saccharomyces cerevisiae* genome. **Nature** 418(6896): 387–91.

25. Sierkstra LN, Verbakel JM, and Verrips CT (**1992**). Analysis of transcription and translation of glycolytic enzymes in glucose-limited continuous cultures of *Saccharomyces cerevisiae*. **J Gen Microbiol** 138(12): 2559–66.

26. Sambrook J, Fritsch E., and Maniatis T (**1989**). Molecular cloning : a laboratory manual. New York : Cold Spring Harbor Laboratory Press.

27. Beltran G, Novo M, Rozès N, Mas A, and Guillamón JM (**2004**). Nitrogen catabolite repression in *Saccharomyces cerevisiae* during wine fermentations. **FEMS Yeast Res** 4(6): 625–32.

28. Zwietering MH, Jongenburger I, Rombouts FM, and van 't Riet K (**1990**). Modeling of the bacterial growth curve. **Appl Environ Microbiol** 56(6): 1875–81.

29. Salvadó Z, Guillamón JM, Salazar G, Querol A, and Barrio E (**2011**). Temperature adaptation markedly determines evolution within the genus *Saccharomyces*.

30. Riou C, Nicaud JM, Barre P, and Gaillardin C (**1997**). Stationary-phase gene expression in *Saccharomyces cerevisiae* during wine fermentation. **Yeast** 13(10): 903–15.

31. R Core Team (**2013**). R: A language and environment for statistical computing.

3

Cell wall dynamics modulate acetic acid-induced apoptotic cell death of *Saccharomyces cerevisiae*

António Rego[#], Ana Marta Duarte[#], Flávio Azevedo[#], Maria João Sousa, Manuela Côrte-Real and Susana R. Chaves*

Centro de Biologia Molecular e Ambiental, Departamento de Biologia, Universidade do Minho, Braga, Portugal.
[#] António Rego, Ana Marta Duarte and Flávio Azevedo contributed equally to this work.
* Corresponding Author: Susana R. Chaves, University of Minho, Department of Biology, Campus de Gualtar; 4710-057 Braga, Portugal; E-mail: suchaves@bio.uminho.pt

ABSTRACT Acetic acid triggers apoptotic cell death in *Saccharomyces cerevisiae*, similar to mammalian apoptosis. To uncover novel regulators of this process, we analyzed whether impairing MAPK signaling affected acetic acid-induced apoptosis and found the mating-pheromone response and, especially, the cell wall integrity pathways were the major mediators, especially the latter, which we characterized further. Screening downstream effectors of this pathway, namely targets of the transcription factor Rlm1p, highlighted decreased cell wall remodeling as particularly important for acetic acid resistance. Modulation of cell surface dynamics therefore emerges as a powerful strategy to increase acetic acid resistance, with potential application in industrial fermentations using yeast, and in biomedicine to exploit the higher sensitivity of colorectal carcinoma cells to apoptosis induced by acetate produced by intestinal propionibacteria.

Keywords: yeast, apoptosis, acetic acid, MAPK, CWI.

Abbreviations:
PCD - programmed cell death,
CRC - colorectal carcinoma,
CWI - cell wall integrity,
HOG - High Osmolarity Glycerol,
MOMP - mitochondrial outer,
membrane permeabilization,
PKC - Protein kinase C,
MAPKKK - Mitogen Activated Protein kinase kinase kinase,
MAPKK - MAP kinase kinase,
MAPK - MAP kinase,
ROS - reactive oxygen species,
ECM - extracellular matrix,
PI - propidium iodide,
DHE - dihydroethidium,
CFU - colony-forming units.

INTRODUCTION

Saccharomyces cerevisiae is currently a well-established eukaryotic model organism used in the elucidation of molecular mechanisms of programmed cell death (PCD) pathways [1]. In particular, acetic acid-induced apoptosis is among the best-characterized yeast apoptotic pathways, due to the interest of modulating this response for applications in both biotechnology and biomedicine [2]. Indeed, there is an increasing number of studies aiming to develop improved yeast strains for use in fermentations, a process often hindered by excessive levels of acetic acid [3, 4]. On the other hand, it has been found that colorectal carcinoma (CRC) cells are particularly sensitive to short-chain fatty acids produced by propionibacteria (including acetate) that reside in the intestine, generating an interest in exploring novel probiotics as a prevention/therapeutic tool in CRC [5-

8]. In yeast, acetic acid triggers a PCD process with features similar to mammalian apoptosis, such as exposure of phosphatidylserine on the outer leaflet of the cytoplasmic membrane, chromatin condensation and DNA fragmentation [9]. Like in mammalian cells, mitochondria play a key role in this process. Indeed, different alterations in mitochondrial structure and function occur during acetic acid-induced apoptosis, including reduction in cristae number and mitochondrial swelling [10], a transient mitochondrial hyper-polarization followed by depolarization, production of reactive oxygen species (ROS), decrease in cytochrome oxidase activity and mitochondrial outer membrane permeabilization (MOMP), with concomitant release of cytochrome *c* and yeast Aif1p [11-13]. Several proteins regulating acetic acid-induced apoptosis have already been identified, such as Por1p (yeast voltage dependent anion channel), which protects cells from apoptosis triggered by acetic acid, and ADP/ATP carrier proteins, yeast orthologs of the adenine nucleotide transporter, which seem to mediate MOMP and cytochrome c release [14]. Mitochondrial proteins involved in fission/fusion, namely Fis1p, Dnm1p and Mdv1p [15], have also been implicated in the execution of the yeast apoptotic program induced by acetic acid, as has the cathepsin D homologue Pep4p, important for mitochondrial degradation in this process [16]. The Ras–cAMP–PKA pathway has also been shown to mediate acetic acid-induced apoptosis, both in *S. cerevisiae* and *C. albi-*

cans [17]. Despite the large number of proteins shown to be involved, the complexity of the networks contributing to acetic acid-induced cell death and their interrelationships are still elusive.

Mitogen Activated Protein Kinase (MAPK) cascades are important signaling pathways that allow yeast cells to adjust to changing environment conditions. These pathways regulate various important processes, from cell proliferation and differentiation to cell death. MAPK cascades normally contain three protein kinases that act in sequence: a MAP kinase kinase kinase (MAPKKK, MAP3K, MEKK or

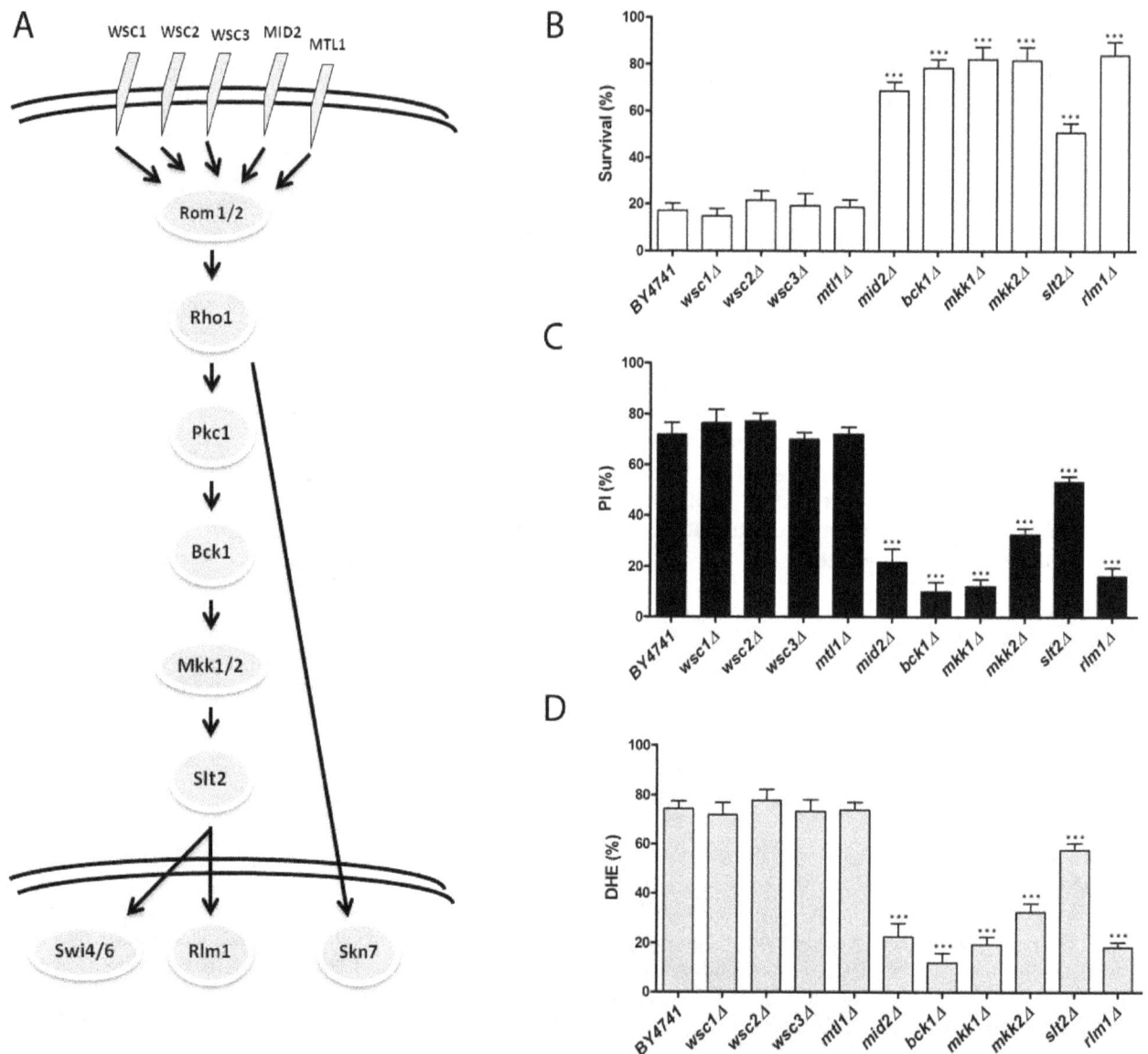

FIGURE 1: The role of the cell wall integrity (CWI) signaling pathway in acetic acid-induced apoptosis. (A) Overview of the pathway. (B) Survival of the wild type (BY4741) and indicated isogenic yeast strains exposed to 110 mM acetic acid, at pH 3.0 for 200 min. Values represent means ± SD of at least three independent experiments. (C) Percentage of cells displaying propidium iodide (PI) internalization assessed by flow cytometry after treatment with 110 mM acetic acid, at pH 3.0 for 200 min. (D) Percentage of intracellular ROS levels assessed by flow cytometry after treatment with 110 mM acetic acid, at pH 3.0 for 200 min. Values in (C) and (D) are represented as means ± SD of at least three independent experiments with at least 20000 cells counted in each time point. Asterisks represent significant statistical difference from control by One-way ANOVA test: (* represents $p < 0.05$ and *** $p < 0.001$).

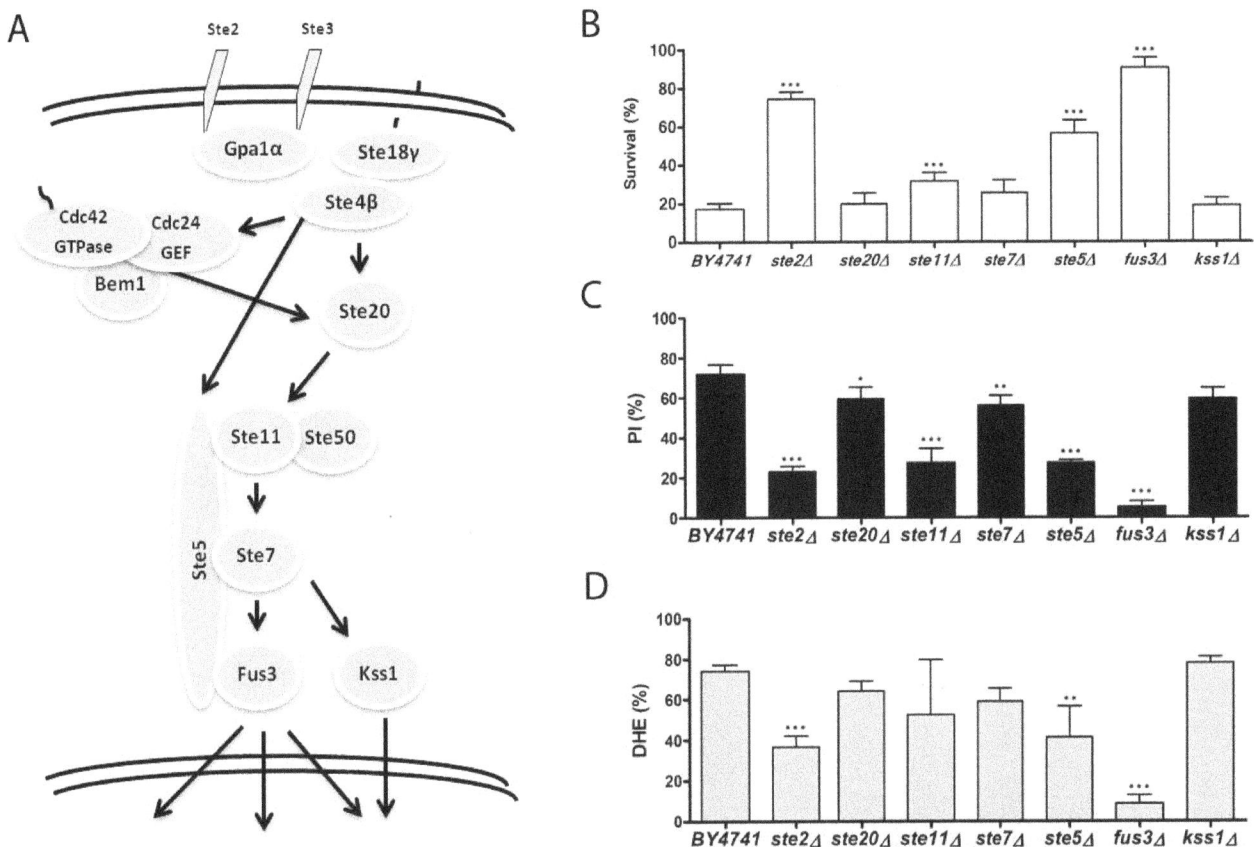

FIGURE 2: The role of the mating-pheromone response signaling pathway in acetic acid-induced apoptosis. (A-C) panels as described in Figure 1.

MKKK), a MAP kinase kinase (MAPKK, MAP2K, MEK or MKK), and a MAP kinase (MAPK). Therefore, when the cascade is activated, the MAPKKK phosphorylates the MAPKK, which in turn phosphorylates both the threonine and tyrosine residues of a conserved -Thr-X-Tyr- motif within the activation loop of the MAPK [18]. MAPKs phosphorylate a diverse set of well-characterized substrates, including transcription factors, translational regulators, MAPK-activated protein kinases (MAPKAP kinases), phosphatases, and other classes of proteins, thereby regulating metabolism, cellular morphology, cell cycle progression, and gene expression in response to a variety of extracellular stresses and molecular signals [19]. The specificity of the MAPK pathways is regulated at several levels, including kinase-kinase and kinase-substrate interactions, co-localization of kinases by scaffold proteins, and inhibition of cross-talk/output by the MAPKs themselves [20]. *S. cerevisiae* contains five MAPKs, Fus3p, Kss1p, Hog1p, Slt2p/Mpk1p and Smk1p, in five functionally distinct cascades, associated with the mating-pheromone response, invasive growth/pseudohyphal development, high osmolarity, cell wall integrity (CWI), and sporulation, respectively [21]. The five MAP kinases are controlled by four MAPKKs, Ste7p (regulating Fus3p and Kss1p), Pbs2p (regulating Hog1p) and the redundant pair

Mkk1p/Mkk2p (regulating Slt2p/Mpk1p), and by four MAPKKKs, Ste11p, the redundant pair Skk2p/Skk22p and Bck1p. The specificity of signal transduction is guaranteed by scaffold proteins [22], Ste5p for the mating-pheromone response pathway, and Pbs2p for the High Osmolarity Glycerol (HOG) pathway.

It has been reported that exposure to non-lethal concentrations of acetic acid activates the HOG pathway [23], and also leads to phosphorylation of Slt2p, a MAPKK from the CWI pathway [24]. These results suggest an intricate relation between CWI and HOG signaling in response to growth in the presence of acetic acid. In this work, we aimed to characterize the involvement of MAPK signaling pathways in cell death induced by acetic acid in *S. cerevisiae*.

RESULTS

Components of the MAPK pathways modulate acetic acid-induced cell death

In order to investigate the involvement of the different MAPK signaling pathways in acetic acid-induced cell death, we assessed whether deletion of components of these pathways affected the viability of *S. cerevisiae* cells in response to acetic acid. In Figures 1 through 4, a simplified

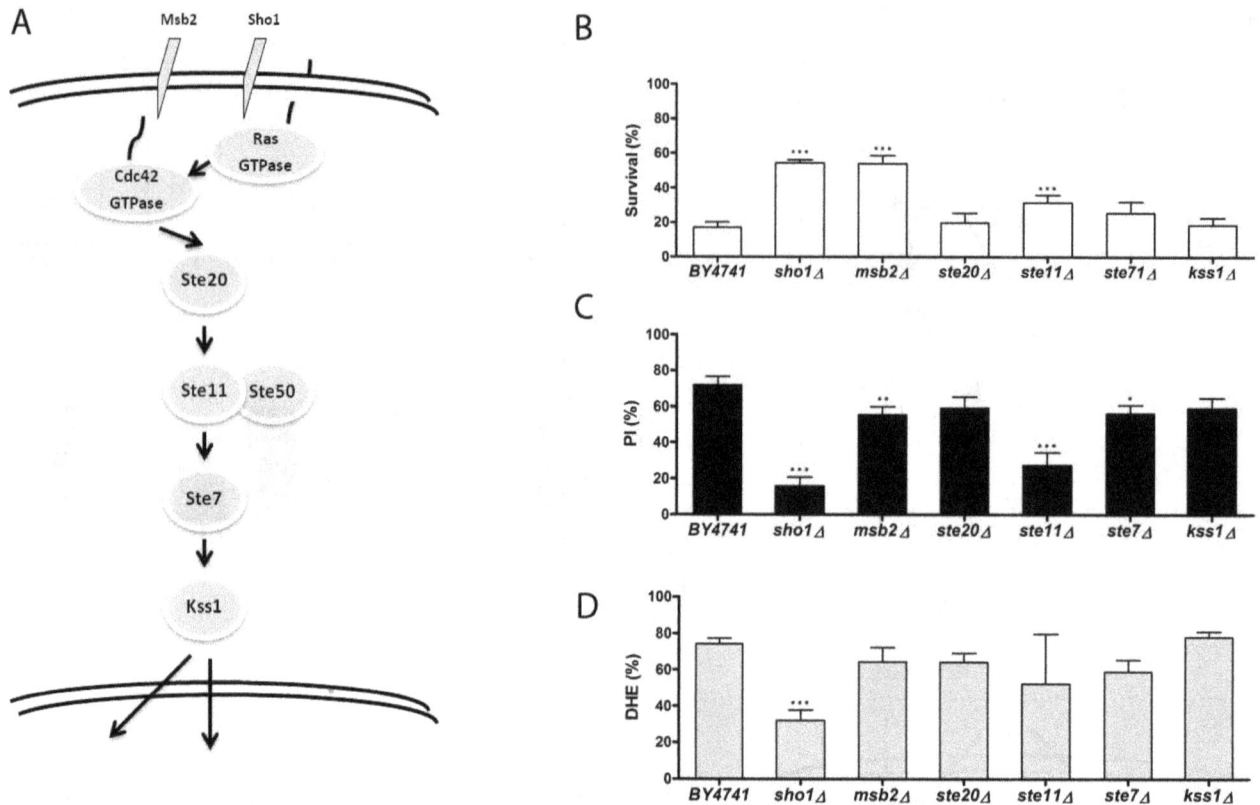

FIGURE 3: The role of the invasive growth/pseudohyphal development signaling pathway in acetic acid-induced apoptosis. (A-C) panels as described in Figure 1.

model of the MAPK pathways is represented in the (A) panels, and the viability of the different mutants is shown in the (B) panels. We found that several mutants of the MAPK components were significantly more resistant to acetic acid-induced cell death than the wild type strain. These included multiple components of the mating-pheromone response pathway (ste2Δ, ste5Δ and fus3Δ) and most components of the cell wall integrity pathway (mutants mid2Δ, bck1Δ, mkk1Δ/mkk2Δ, slt2/mpk1Δ). The mutants sho1Δ and msb2Δ, lacking the two membrane signaling proteins common to both invasive growth/pseudohyphal development and of the HOG pathway, and the mutant ssk22Δ, a member of the redundant pair of MAPKKK of the HOG pathway, were also significantly more resistant to acetic acid-induced cell death than the wild type strain.

Acetic acid induces a mitochondria-dependent apoptotic cell death in *S. cerevisiae* that displays characteristic apoptotic markers such as ROS accumulation, phosphatidylserine externalization, chromatin condensation, DNA fragmentation and mitochondrial dysfunction with release of cytochrome *c* [9, 12]. We therefore also assessed loss of plasma membrane integrity and ROS accumulation in the mutant strains exposed to acetic acid by staining cells with PI and DHE, respectively, and analyzing the fluorescence by flow cytometry. We found that, in general, mutant strains

with higher resistance to acetic acid had a lower percentage of cells displaying an accumulation of ROS and a lower percentage of cells with compromised plasma membrane integrity than the wild type strain (Fig. 1-4, C and D panels), confirming the involvement of the mating-pheromone response, HOG and CWI pathways, but not the invasive growth/pseudohyphal development pathway, in acetic acid-induced regulated cell death. In fact, though deletion mutants of some components of the latter pathway display a resistance phenotype, namely Msb2p, Sho1p and Ste11p, they are shared by other pathways, and the only MAPK of the pathway, Kss1p, does not seem to be involved.

As mentioned above, we have previously shown that acetic acid triggers a cell death program with hallmarks of mitochondria-dependent apoptosis, including MOMP and translocation of cytochrome *c* from the mitochondria into the cytosol. Since all mutants in the CWI MAPKKK/MAPKK/MAPK cascade were more resistant to acetic acid and displayed lower ROS accumulation, we next determined whether there was also decreased MOMP, to further support the involvement of mitochondria in the regulation of acetic acid-induced programmed cell death by the CWI signaling pathway. To this end, we assessed the levels of cytochrome *c* in cytosolic and mitochondrial extracts of untreated and acetic acid-treated cultures of wild type BY4741 and the CWI mutants bck1Δ and slt2Δ (dele-

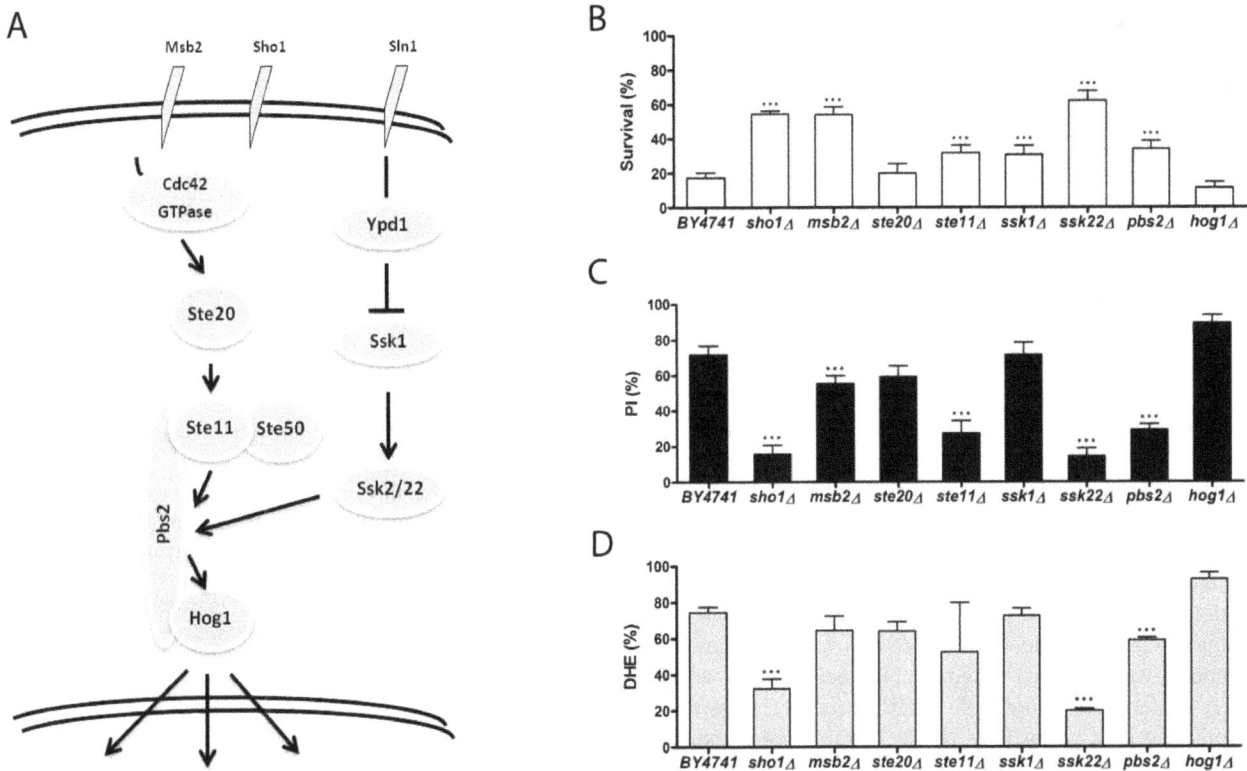

FIGURE 4: The role of the high osmolarity glycerol (HOG) signaling pathway in acetic acid-induced apoptosis. (A-C) panels as described in Figure 1.

tion mutants of the MAPKKK and the MAPK of the pathway, respectively). In agreement with our previous results [12], exposure of wild type cells to acetic acid resulted in depletion of cytochrome c from mitochondria and consequent detection in the cytosolic fraction (Fig. 5). In contrast, we did not detect any depletion of cytochrome c from mitochondria or its translocation to the cytosol in acetic acid-treated $bck1\Delta$ or $slt2\Delta$ mutant cells, indicating that the CWI pathway mediates acetic acid-induced apoptosis through a mitochondrial pathway.

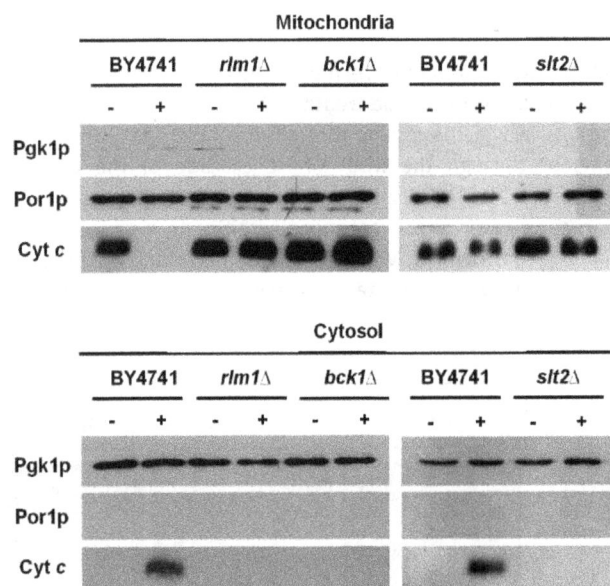

FIGURE 5: CWI mutants are defective in acetic acid-induced cytochrome c release. Western blot analysis of cytochrome c in S. cerevisiae strains BY4741, $bck1\Delta$, $slt2\Delta$ and $rlm1\Delta$ before (-) and after (+) exposure to 120 mM acetic acid, pH 3.0, for 200 min, in both mitochondrial and cytosolic fractions. Cytosolic phosphoglycerate kinase (Pgk1p) and mitochondrial porin (Por1p) levels were used as loading control of cytosolic and mitochondrial fractions, respectively. A representative experiment is shown of at least two independent experiments with similar results.

Over-activation of the CWI pathway sensitizes cells to acetic acid exposure

As shown above, impairment of the CWI pathway results in increased resistance to acetic acid-induced cell death. We therefore next sought to determine whether over-activation of this pathway would result in increased sensitivity to acetic acid. We transformed wild type cells with a plasmid expressing BCK1-20 and respective empty plasmid control [25], as it has previously been shown that over-expression of this BCK1 allele resulted in constitutive activation of the CWI pathway. We used the S. cerevisiae W303 strain as the wild type control due to the plasmid selective marker, and confirmed the bck1Δ mutant in this background also displayed resistance (not shown). Indeed, we found that over-expression of the Bck1 protein led to increased sensitivity to acetic acid (Fig. 6), providing further evidence that induction of the CWI pathway mediates acetic acid-induced cell death.

CWI pathway mutants display differential sensitivity to multiple stresses

To determine whether CWI mutants are specifically resistant to acetic acid-induced cell death or to death stimuli in general, we assessed the sensitivity of bck1Δ, slt2Δ, and rlm1Δ mutants to other cell death inducers by semi-quantitative spot assay (Fig. 7). All mutants were more resistant to acetic, propionic and butyric acid-induced cell death than the wild type strain, though to a different extent. Mutants were also slightly resistant to hydrogen peroxide-induced cell death, but not to methyl methanesulfonate-induced cell death. This indicates that the CWI pathway is particularly involved in acid-induced cell death, but is not a general stress response pathway.

CWI pathway mutants are more sensitive to zymolyase digestion after acetic acid treatment

Mutants in which signaling through the upstream components or through the MAP kinase cascade of the CWI pathway is blocked display cell wall defects with varying degrees of severity and are more sensitive to a variety of stimuli [26]. However, we determined that many of these mutants are more resistant to acetic acid-induced cell

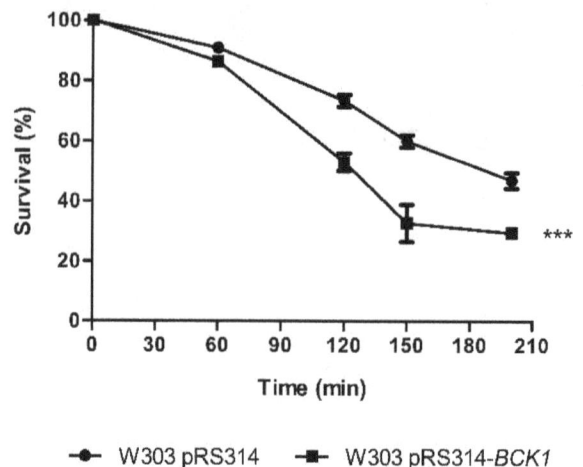

FIGURE 6: Stimulation of the CWI pathway sensitizes cells to acetic acid-induced cell death. Survival of wild type (W303) cells over-expressing Bck1p (pRS314-BCK1-20) or empty plasmid control (pRS314) exposed to 90 mM acetic acid, at pH 3.0 for 200 min. Values represent means ± SD of at least three independent experiments. Asterisks represent significant statistical difference from control by Two-way ANOVA test: (p < 0.001).

death. It has also been reported that weak-acid stress leads to cell wall remodeling, decreasing cell wall porosity [27]. We therefore assessed whether there were differences in cell wall structural integrity of CWI mutants mid2Δ, bck1Δ, mkk1Δ and mkk2Δ in comparison with wild type cells after exposure to acetic acid, using a zymolyase sensitivity assay. All the CWI mutants tested were more susceptible to digestion with zymolyase after exposure to acetic acid than the wild type strain (Fig. 8), indicating that they display a resistant phenotype despite their cell wall defect.

Rlm1p and its target genes involved in cell wall organization/biogenesis and cell wall structure modulate acetic acid-induced cell death

The final and most prominent consequence of the activa-

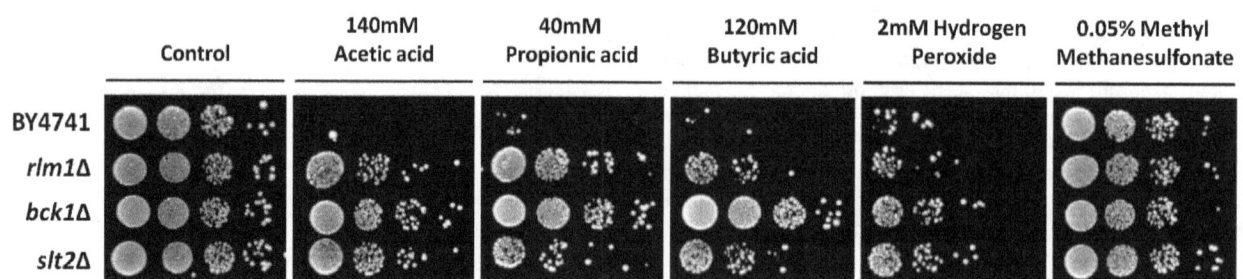

FIGURE 7: Sensitivity of CWI mutants to different stimuli. Survival of the wild type (BY4741) and indicated isogenic yeast strains exposed to 140 mM acetic acid, pH 3.0, 40 mM propionic acid, pH 3.0, 120 mM butyric acid, pH 3.0, 2 mM hydrogen peroxide, or 0.05% methyl methanesulfonate for 180 minutes at 30°C. Representative images are shown from at least 3 independent experiments.

tion of the CWI pathway by cell wall stress is the induction of an adaptive transcriptional program coordinated by Slt2p/Mpk1p and mostly mediated by the transcription factor Rlm1p [28]. Notably, we observed that deletion of *RLM1* led to resistance to acetic acid (Fig. 1B, C and D) and impaired acetic acid-induced cytochrome *c* release into the cytosol (Fig. 5). Furthermore, the *rlm1Δ* mutant was more susceptible to zymolyase digestion, in agreement with the phenotype of CWI mutants, described above. We therefore sought to determine the involvement of Rlm1p target genes in acetic acid-induced apoptosis. With the aid of bioinformatics tools, in particular the data available in the database YEASTRACT (http://www.yeastract.com/), we could identify 205 genes putatively regulated by Rlm1p, of which 29 are essential. To identify genes regulated by Rlm1p required for resistance to acetic acid induced-cell death, we screened the strains mutated in all the non-essential genes under Rlm1p control from the EUROSCARF haploid mutant deletion collection (EUROSCARF; http://web.uni-frankfurt.de/fb15/mikro/euroscarf/). The 176 mutant strains were patched onto 96-dot arrays and incubated in synthetic complete liquid medium (SC-Gal) containing 250 mM acetic acid at pH 3.0. The presence of viable cells was tested at 100, 200, 300 and 400 min and compared with that of wild type cells. Of the 176 mutants tested, 103 were more resistant to acetic acid-induced cell death and 28 were more sensitive while the other 45 mutants had a phenotype similar to that of the wild type strain (Table S1). To further validate our results, we determined the viability of 50 randomly selected mutant strains from the resistant and sensitive datasets and compared the phenotype with that obtained in the screening. The phenotype of 47 strains was confirmed, two mutant strains scored as sensitive in the 96-plate assay displayed no differences from wild type when tested individually, and one was more resistant in the 96-plate assay but also did not display any differences from wild type when tested individually (not shown).

In the dataset of resistant strains, the Biological Process most significantly enriched according to Gene Ontology classification (FUNSPEC analysis http://funspec.med.utoronto.ca/) was "fungal-type cell

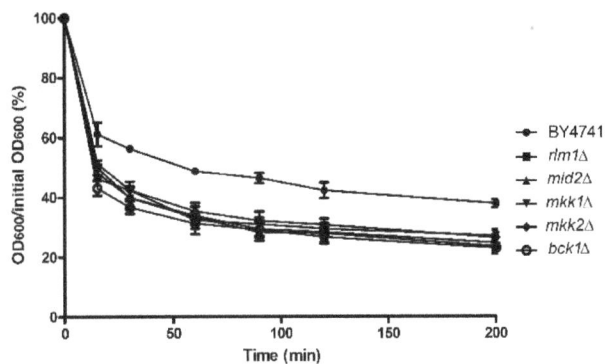

FIGURE 8: Sensitivity of CWI mutants to digestion with zymolyase. Cells were exposed to 110 mM acetic acid, at pH 3.0 for 200 min, digested with zymolyase 20T for up to 200 min, and optical density (600 nm) assessed over time. Values represent means ± SD of three independent experiments.

wall organization", enclosing genes coding for proteins involved in hydrolysis of O-glycosyl compounds (*EXG2, UTR2, CRH1, BGL2* and *EXG1*), namely glucan exo-1,3-beta-glucosidase activity (*EXG2, BGL2, EXG1*), cell wall proteins containing a putative GPI-attachment site (*PST1, YLR194C*), a putative GPI-anchored aspartic protease (*YPS6*), and cell wall mannoproteins (*CCW12, CCW14*) (Table 1). Deficiency in proteins Flc1p, Flc2p, Rim21p and Dfg5p, also involved in the "cell wall biogenesis", conferred resistance to acetic acid-induced cell death as well. These results indicate that cell wall remodeling plays a decisive role in the induction of apoptosis. Several genes with a function in polarized growth were also enriched, likely through their involvement in the modulation of Cdc42p and Rho proteins, essential for this process. These included *PXL1*, similar to metazoan paxillin, involved in adhesion, and *GIC2*, a Cdc42 effector, whose deletion conferred resistance to acetic acid. Accordingly, deletion of *BEM2*, a RhoGAP (Rho GTPase activating protein), resulted in sensitivity to acetic acid, presumably due to the increased Cdc42-GTP levels observed in this mutant [29].

TABLE 1. Categories that were significantly enriched (p-value below 0.01) based on physiological function of the genes whose deletion increases the resistance to acetic acid-induced cell death.

Category, biological process	p-value	Genes in the dataset
fungal-type cell wall organization [GO:0031505]	4.626e-07	*PST1 EXG2 UTR2 CRH1 BGL2 SLT2 SIM1 YPS6 CCW12 YLR194C EXG1 CCW14*
cellular cell wall organization [GO:0007047]	7.386e-06	*EXG2 UTR2 CRH1 BGL2 CCW12 YLR194C EXG1 DFG5 SUN4*
fungal-type cell wall biogenesis [GO:0009272]	1.567e-05	*FLC2 DFG5 RIM21 FLC1*
cell wall chitin metabolic process [GO:0006037]	0.0002317	*UTR2 CRH1*

TABLE 2. Categories that were significantly enriched (p-value below 0.01) based on physiological function of the genes whose deletion increases the susceptibility to acetic acid-induced cell death.

Category, biological process	p-value	Genes in the dataset
fungal-type cell wall organization [GO:0031505]	0.001243	HSP150 CWP1 CWP2 PIR3
response to stress [GO:0006950]	0.002342	CTT1 HSP150 TSL1 HOR7

In the data set of sensitive strains, the biological process most significantly enriched according to Gene Ontology classification was also "fungal-type cell wall organization", followed by "response to stress". The "response to stress" class included the cytosolic catalase (CTT1), the subunit of the threalose 6-phosphate synthase/phosphatase complex (TLS1), and a protein of unknown function (HOR7). The "fungal-type cell wall organization" class, in contrast with the genes represented in the dataset of resistant strains, was composed in this case of genes that code for proteins involved in the stability of the cell wall, namely O-mannosylated heat shock proteins (HSP150 and PIR3) and cell wall manoproteins (CWP1, CWP2) (Table 2). Therefore, proteins regulated by Rlm1p that ensure the stability of the cell wall protect cells from acetic acid-induced cell death.

The phenotype of the rlm1Δ mutant is the result of several responses; since deletion of some Rlm1p target genes results in resistance (those involved in cell wall remodeling), and of others in sensitivity (those involved in cell wall stability), and the overall phenotype of the rlm1Δ mutant is resistance to acetic acid-induced cell death, the more prevalent Rlm1p-mediated response to acetic acid seems to be cell wall remodeling.

DISCUSSION

In this study, we performed a comprehensive analysis of the MAPK signaling pathways involved in acetic acid-induced apoptotic cell death. Absence of Ste11p MAP kinase, shared by the mating-pheromone response, invasive growth/pseudohyphal development and HOG pathways, resulted in higher cell survival. However, since deficiency in the MAPK of the invasive growth/pseudohyphal development pathway, Kss1p, did not alter sensitivity to acetic acid, this pathway does not seem to be involved in apoptosis induced by acetic acid. On the other hand, although several components of the HOG signaling pathway, both specific and common to other MAPK pathways, have a pro-death role in this process, deletion of the MAPK of the HOG pathway tends to confer sensitivity to acetic acid. Therefore, the results support the interpretation that the HOG pathway does not play a relevant role in signaling acetic acid-induced apoptosis, or that it has a dual role. These results are in accordance with those obtained in a genome-wide screen for the identification of positive and negative regulators of acetic acid-induced cell death, where the HOG pathway was also not identified as relevant in this process [30]. In this analysis, the term "Sporulation result-

ing in formation of a cellular spore" was enriched and, consistently, we found that the mating-pheromone response signals cell death. Indeed, deficiency in several components of this pathway, and particularly in its specific MAPK, resulted in higher resistance to acetic acid-induced apoptosis. Absence of different CWI pathway components also conferred resistance to acetic acid, sustaining that this pathway is another major mediator of acetic acid-induced apoptosis. Of the mutants in CWI sensors, only mid2Δ displayed a resistant phenotype, suggesting a possible role for this sensor. Other sensors may also play a role, as their function may be redundant, and deletion of multiple genes would be required. Since a crosstalk exists between MAPK pathways, we also cannot exclude activation by intracellular signals. Also, the CWI MAPK mutant slt2Δ was slightly less resistant to acetic acid-induced cell death than the other CWI mutants. This can reflect its involvement in other processes and differential regulation, such as by PTP genes, Knr4p, Cdc37p, or the Hsp90 chaperone [31-33]. Our results also highlight the different involvement of MAPK pathways in resistance to acetic acid-induced cell death and to chronic exposure to acetic acid. Indeed, it has been previously shown that impairment of the HOG pathway results in increased sensitivity to growth in the presence of acetic acid, whereas deletion of SLT2 had no effect [34] or resulted in reduced growth in acidic pH [35, 36]. Notably, both Slt2p and Hog1p were phosphorylated in response to sub-lethal concentrations of acetic acid [23, 24], though we only observed the phosphorylation of Hog1p in response to lethal concentrations used in our assay, for the time points tested (not shown, and Figure S1). This shows that there is not always an obvious relation between protein phosphorylation and the response/phenotype of a particular pathway, as has been found in other studies (e.g., [33]). In this study, we focused on how the CWI pathway regulates acetic acid-induced apoptosis.

The yeast cell wall is a strong and rigid barrier that protects cells from extreme changes in the environment. It has four major functions: 1) stabilization of internal osmotic conditions, 2) protection against physical stress, 3) maintenance of cell shape, which is a precondition for morphogenesis, and 4) a scaffold for proteins [37]. It consists of an inner layer of load-bearing polysaccharides (glucan polymers and chitin), acting as a scaffold for a protective outer layer of mannoproteins that extend into the medium [37]. The yeast cell wall is a dynamic structure, and its composition changes in response to several stress conditions, such as heat stress, hypo-osmotic shock, cell wall stress, as well as carbon source, nutrient, or oxygen availability and in the presence of acetic acid [38, 39]. Accordingly, exposure to acetic acid renders the cell wall more resistant to lyticase digestion, reflecting an adaptation mechanism that allows cells to grow better in the presence of this weak acid [27]. Our results now show that, in contrast, a more resistant cell wall is not needed for higher resistance to acetic acid-induced cell death. Indeed, CWI mutants, known to display cell wall defects [26], were more sensitive to zymolyase digestion but more resistant to acetic acid-induced cell death. Therefore, in order to identify the relevant functions regulated by this MAPK pathway that are involved in the higher resistance to apoptosis induced by acetic acid, we screened for targets of Rlm1p, the main downstream mediator of Slt2p signaling.

Rlm1p targets comprise genes involved in a multitude of processes, which are not restricted to genes with a cell wall function. Accordingly, several classes were represented in the datasets of genes regulated by Rlm1p whose deletion resulted in altered sensitivity to acetic acid-induced cell death, including several previously implicated in this process. These classes include proteins involved in sphingolipid metabolism [40], as well as genes implicated in the oxidative stress response [41] and mitochondrial components [10-12, 14, 42]. Modulation of the CWI pathway can therefore affect multiple functions involved in acetic acid-induced cell death. However, as expected, most genes found are involved in stabilization or remodeling of the cell wall, as well as vesicle trafficking and polarized growth, all affecting cell wall structure.

The results from our screen indicate that the stabilization of the cell wall is important for the cell's ability to resist to acetic acid-induced cell death, while cell's engagement in cell wall remodeling compromises its survival. Indeed, many genes required for cell wall stability were found in the sensitive dataset. Moreover, several genes involved in the modulation of Cdc42p and Rho proteins were found, which seem to be associated with the function of these proteins in polarized growth. Since polarized growth requires re-organization of the actin cytoskeleton as well as cell wall remodeling, these processes are intimately connected. This highlights the crosstalk between the CWI and mating-pheromone response MAPK pathways we found as mainly involved in acetic acid-induced cell death, and their intricate regulation.

The results obtained in this study may impact different biotechnological processes and biomedical applications. High levels of acetic acid produced during acid catalyzed-hydrolysis of lignocelluloses, used as raw material to produce bioethanol, or formed during industrial fermentation processes, often compromise the yeast fermentative performance [4, 43]. One way to overcome the inhibition of fermentation process is to render industrial strains more resistant to this weak acid. Identifying molecular determinants of sensitivity to acetic acid, and of strategies to increase strain resistance, is therefore of utmost importance. Specifically, modulation of upstream signaling pathways is of great interest, since a number of genes and processes

are affected to produce a desirable outcome, rather than affecting specific downstream genes with limited functions, which the cells often adapt to through redundant/compensatory mechanisms. In the future, it will be interesting to determine how modulating the CWI signaling pathway impacts yeast fermentative performance, namely industrial ethanol production from lignocellulosic hydrolysates highly enriched in acetic acid.

Many of the cellular and metabolic features that constitute hallmarks of tumor cells include higher glycolytic energetic dependence, lower mitochondrial functionality, increased cell division and metabolite synthesis [44]. Notably, these same alterations result in higher sensitivity of yeast cells to acetic acid [30], consistent with the specific sensitivity of CRC cells to short chain fatty acids, including acetate and propionate, and reinforcing the exploitation of yeast as a model system to elucidate the molecular basis of this sensitivity. Therefore, despite obvious differences between the extracellular matrix (ECM) and the yeast cell wall, it would be interesting to determine whether increased ECM dynamics could also underlie the higher susceptibility of CRC cells to acetate-induced apoptosis, or whether modulation of this process or of MAPK pathways could further potentiate the sensitivity of these cells to acetate, without compromising viability of healthy adjacent cells. Indeed, modulating MAPK signaling pathways has previously been suggested as a strategy in colorectal cancer treatment, though particular molecular components to be targeted have not been identified, nor has its efficacy been evaluated [45].

In summary, our work indicates that the mating-pheromone response and CWI MAPK pathways are involved in signaling acetic acid-induced cell death, as blocking signal transduction in these pathways renders cells more resistant to programmed cell death induced by acetic acid. This resistance is achieved through regulation of several processes, of which alterations in the cell wall were particularly evident. Modulation of the CWI MAPK signaling pathway therefore emerges as a powerful strategy to increase resistance of yeast strains to acetic acid through multiple effector processes, with potential application in biotechnology as a way to avoid stuck or sluggish alcoholic fermentations. Our results also open new avenues of research into the regulation of acetate-induced apoptosis in mammals, with particular impact for the design of novel therapeutic opportunities against colorectal carcinoma based on the modulation of MAPK pathways.

MATERIALS AND METHODS
Yeast strains and growth conditions
The yeast S. cerevisiae strain BY4741 (MATa his3Δ1 leu2Δ0 met15Δ0 ura3Δ0) [46] and isogenic mutant strains were used throughout this work, except for determination of the BCK1-20 overexpression phenotype, where W303-1A was used due to auxotrophy requirements (MATa, ura3-52, trp1Δ 2, leu2-3,112, his3-11, ade2-1, can1-100). Cells were maintained in rich medium (YPD) (1% yeast extract, 2% glucose, 2% bacto-peptone, 2% agar) and grown in synthetic complete medium (SC-Gal) (0.67% Bacto-yeast nitrogen base w/o amino acids

(Difco), 2% galactose and 0.2% Dropout mix). Galactose was used as the carbon and energy source to address mitochondrial function, as this leads to higher mitochondrial mass because galactose is less effective in the repression of respiratory metabolism [47]. The sensitivity of several strains was assessed in YPD and the results were comparable (e.g., rlm1Δ, not shown).

Acetic acid treatments: quantitative c.f.u. counts
Yeast cells were grown overnight in liquid SC-Gal (or SC-Gal without tryptophan) until exponential growth-phase (OD_{600nm} = 0.5-0.6) at 30°C with agitation (200 rpm). Cells were harvested by centrifugation and suspended in fresh SC-Gal medium (pH 3.0) with 90-120 mM acetic acid, and incubated for 200 minutes at 30°C in 50 mL Erlenmeyer flasks with an air: liquid ratio of 5:1 in a mechanical shaker at 200 rpm. Samples were taken at different time points, diluted to 10^{-4} in 1:10 serial dilutions in deionized sterilized water, and 40 µL drops were spotted on YPD agar plates in replicates of seven. Colony forming units (c.f.u.) were counted after 48 h incubation at 30°C. Cell viability was calculated as percentage of c.f.u.s in relation to time zero.

Semi-quantitative spot assays:
Yeast cells were grown overnight in SC-Gal medium until exponential growth-phase (OD_{600nm} = 0.5-0.6) at 30°C at 200 rpm. Cells were harvested by centrifugation and suspended in fresh medium with 140 mM acetic acid, pH 3.0, 40 mM propionic acid, 120 mM butyric acid, 2 mM hydrogen peroxide, or 0.05% methyl methanesulfonate and incubated for 180 minutes at 30°C in 50 mL Erlenmeyer flasks with an air: liquid ratio of 5:1. Samples were taken at different time points, diluted to 10^{-4} in 1:10 serial dilutions in deionized sterilized water, and 5 µL drops of each dilution were spotted on YPD agar plates. Plates were photographed after incubation for 48 h at 30°C.

96 well plate screen
Mutant strains deleted for Rlm1p target genes were patched in ordered arrays of 96 on YPD plates and grown at 30°C for 2 days. Yeast cells were inoculated into 96-well plates containing synthetic complete, 2% galactose medium with a pin-replicator, and grown for 24 hours at 30°C. Cultures were diluted 100 fold using a multichannel pipette into SC-Gal medium at pH 3.0, containing 250 mM acetic acid (this concentration was optimized for the culture conditions used in the 96 well plate screen). At different times of incubation (100, 200, 300 and 400 minutes), cells were replicated into 96-well plates containing YPD medium, using a pin replicator, as described in [30]. After incubation at 30°C for two days, optical density (640 nm) was measured to assess cell growth reflecting the presence of viable cells in the inoculum, using a microplate reader (Molecular Devices SpectraMax Plus).

Zymolyase sensitivity assay
To monitor structural changes in the yeast cell wall, a Zymolyase (Medac; Medacshop) sensitivity assay was performed as described in [48]. Briefly, after treatment with 110 mM acetic acid for 200 min, cells were harvested by centrifugation, washed with sterile distilled water and resuspended in 0.1 mM sodium phosphate buffer (pH 7.5). After adding 60 µg/ml of Zymolyase, cell lysis was followed by measuring the decrease in the OD_{600nm} of each cell suspension.

Flow cytometry
During acetic acid treatment, samples were also taken to assess loss of plasma membrane integrity and accumulation of reactive oxygen species (ROS) by flow cytometry, using an EPICS® XL™ (Beckman COULTER®) flow cytometer equipped with an argon-ion laser emitting a 488 nm beam at 15mW. Cells were collected by centrifugation, washed in deionized water, suspended in phosphate buffered saline (PBS) and stained with 1 µg/mL propidium iodide (PI, Sigma) or 2 µM/mL dihydroethidium (DHE, Sigma) for 10 and 30 min, respectively, at room temperature, in the dark. Monoparametric detection of PI fluorescence was performed using FL-3 (488/620 nm) and detection of DHE was performed using FL-4 (488/675 nm).

Assessment of cytochrome c release
Mitochondrial and cytosolic fractions of untreated and acetic acid-treated cells were prepared as described in [14] and protein concentration determined using the Bradford method and BSA as standard [49]. Mitochondrial integrity was assessed by measuring citrate synthase activity [14]. Fractions were separated on a 12.5% SDS-polyacrylamide gel and transferred to a Hybond-P Polyvinylidene difluoride membrane (PVDF; GE Healthcare). Membranes were incubated with the primary antibodies mouse monoclonal anti-yeast phosphoglycerate kinase (Pgk1p) antibody (1:5000, Molecular Probes), mouse monoclonal anti-yeast porin (Por1p) antibody (1:10000, Molecular Probes) and rabbit polyclonal anti-yeast cytochrome c (Cyc1p) antibody (1:2000, custom-made by Millegen), followed by incubation with secondary antibodies against mouse or rabbit IgG-peroxidase (1:5000; Sigma Aldrich). Pgk1p and Por1p were used as a loading control for cytosolic and mitochondrial fractions, respectively. Immunodetection of bands was revealed by chemiluminescence (ECL, GE Healthcare).

ACKNOWLEDGMENTS
We thank Dr. Levin (Boston University) for the plasmid expressing BCK1-20. This work was supported by FCT/MEC through Portuguese funds (PIDDAC) - PEst-OE/BIA/UI4050/2014, PTDC/BIA-BCM/69448/2006, FCT-ANR/BEX-BCM/0175/2012, PTDC/AGR-ALI/102608/2008, as well as fellowships to F.A (SFRH/BD/80934/2011), A.R (SFRH/BD/79523/2011) and S.C (SFRH/ BPD/89980/2012).

CONFLICT OF INTEREST
The authors declare no conflict of interest.

REFERENCES
1. Carmona-Gutierrez D, Eisenberg T, Buttner S, Meisinger C, Kroemer G, Madeo F (**2010**). Apoptosis in yeast: triggers, pathways, subroutines. **Cell Death Differ** 17(5): 763-773.

2. Sousa MJ, Ludovico P, Rodrigues F, Leão C, Côrte-Real M, editors (**2012**). Stress and Cell Death in Yeast Induced by Acetic Acid. InTech, Rijeka.

3. Alexandre H, Charpentier C (**1998**). Biochemical aspects of stuck and sluggish fermentation in grape must. **J Ind Microbiol Biotechnol** 20(1): 20-27.

4. Vilela-Moura A, Schuller D, Mendes-Faia A, Silva RD, Chaves SR, Sousa MJ, Corte-Real M (**2011**). The impact of acetate metabolism on yeast fermentative performance and wine quality: reduction of volatile acidity of grape musts and wines. **Appl Microbiol Biotechnol** 89(2): 271-280.

5. Jan G, Belzacq AS, Haouzi D, Rouault A, Metivier D, Kroemer G, Brenner C (**2002**). Propionibacteria induce apoptosis of colorectal carcinoma cells via short-chain fatty acids acting on mitochondria. **Cell Death Differ** 9(2): 179-188.

6. Lan A, Bruneau A, Bensaada M, Philippe C, Bellaud P, Rabot S, Jan G (**2008**). Increased induction of apoptosis by Propionibacterium freudenreichii TL133 in colonic mucosal crypts of human microbiota-associated rats treated with 1,2-dimethylhydrazine. **Br J Nutr** 100(6): 1251-1259.

7. Lan A, Bruneau A, Philippe C, Rochet V, Rouault A, Herve C, Roland N, Rabot S, Jan G (**2007**). Survival and metabolic activity of selected strains of Propionibacterium freudenreichii in the gastrointestinal tract of human microbiota-associated rats. **Br J Nutr** 97(4): 714-724.

8. Marques C, Oliveira CS, Alves S, Chaves SR, Coutinho OP, Corte-Real M, Preto A (**2013**). Acetate-induced apoptosis in colorectal carcinoma cells involves lysosomal membrane permeabilization and cathepsin D release. **Cell Death Dis** 4:e507.

9. Ludovico P, Sousa MJ, Silva MT, Leao C, Corte-Real M (**2001**). *Saccharomyces cerevisiae* commits to a programmed cell death process in response to acetic acid. **Microbiology** 147(Pt 9): 2409-2415.

10. Pereira C, Silva RD, Saraiva L, Johansson B, Sousa MJ, Corte-Real M (**2008**). Mitochondria-dependent apoptosis in yeast. **Biochim Biophys Acta** 1783(7): 1286-1302.

11. Wissing S, Ludovico P, Herker E, Buttner S, Engelhardt SM, Decker T, Link A, Proksch A, Rodrigues F, Corte-Real M, Frohlich KU, Manns J, Cande C, Sigrist SJ, Kroemer G, Madeo F (**2004**). An AIF orthologue regulates apoptosis in yeast. **J Cell Biol** 166(7): 969-974.

12. Ludovico P, Rodrigues F, Almeida A, Silva MT, Barrientos A, Corte-Real M (**2002**). Cytochrome *c* release and mitochondria involvement in programmed cell death induced by acetic acid in *Saccharomyces cerevisiae*. **Mol Biol Cell** 13(8): 2598-2606.

13. Giannattasio S, Atlante A, Antonacci L, Guaragnella N, Lattanzio P, Passarella S, Marra E (**2008**). Cytochrome *c* is released from coupled mitochondria of yeast en route to acetic acid-induced programmed cell death and can work as an electron donor and a ROS scavenger. **Febs Letters** 582(10): 1519-1525.

14. Pereira C, Camougrand N, Manon S, Sousa MJ, Corte-Real M (**2007**). ADP/ATP carrier is required for mitochondrial outer membrane permeabilization and cytochrome *c* release in yeast apoptosis. **Mol Microbiol** 66(3): 571-582.

15. Fannjiang Y, Cheng WC, Lee SJ, Qi B, Pevsner J, McCaffery JM, Hill RB, Basanez G, Hardwick JM (**2004**). Mitochondrial fission proteins regulate programmed cell death in yeast. **Genes Dev** 18(22): 2785-2797.

16. Pereira C, Chaves S, Alves S, Salin B, Camougrand N, Manon S, Sousa MJ, Corte-Real M (**2010**). Mitochondrial degradation in acetic acid-induced yeast apoptosis: the role of Pep4 and the ADP/ATP carrier. **Mol Microbiol** 76(6): 1398-1410.

17. Phillips AJ, Crowe JD, Ramsdale M (**2006**). Ras pathway signaling accelerates programmed cell death in the pathogenic fungus Candida albicans. **Proc Natl Acad Sci U S A** 103(3): 726-731.

18. Marshall CJ (**1994**). MAP kinase kinase kinase, MAP kinase kinase and MAP kinase. **Curr Opin Genet Dev** 4(1): 82-89.

19. Chen RE, Thorner J (**2007**). Function and regulation in MAPK signaling pathways: lessons learned from the yeast *Saccharomyces cerevisiae*. **Biochim Biophys Acta** 1773(8): 1311-1340.

20. Burack WR, Shaw AS (**2000**). Signal transduction: hanging on a scaffold. **Curr Opin Cell Biol** 12(2): 211-216.

21. Hunter T, Plowman GD (**1997**). The protein kinases of budding yeast: six score and more. **Trends Biochem Sci** 22(1): 18-22.

22. Pawson T, Scott JD (**1997**). Signaling through scaffold, anchoring, and adaptor proteins. **Science** 278(5346): 2075-2080.

23. Mollapour M, Piper PW (**2006**). Hog1p mitogen-activated protein kinase determines acetic acid resistance in *Saccharomyces cerevisiae*. **FEMS Yeast Res** 6(8): 1274-1280.

24. Mollapour M, Shepherd A, Piper PW (**2009**). Presence of the Fps1p aquaglyceroporin channel is essential for Hog1p activation, but suppresses Slt2(Mpk1)p activation, with acetic acid stress of yeast. **Microbiology** 155(Pt 10): 3304-3311.

25. Lee KS, Levin DE (**1992**). Dominant mutations in a gene encoding a putative protein kinase (BCK1) bypass the requirement for a *Saccharomyces cerevisiae* protein kinase C homolog. **Mol Cell Biol** 12(1): 172-182.

26. Jendretzki A, Wittland J, Wilk S, Straede A, Heinisch JJ (**2011**). How do I begin? Sensing extracellular stress to maintain yeast cell wall integrity. **Eur J Cell Biol** 90(9): 740-744.

27. Simoes T, Mira NP, Fernandes AR, Sa-Correia I (**2006**). The SPI1 gene, encoding a glycosylphosphatidylinositol-anchored cell wall protein, plays a prominent role in the development of yeast resistance to lipophilic weak-acid food preservatives. **Appl Environ Microbiol** 72(11): 7168-7175.

28. Levin DE (**2005**). Cell wall integrity signaling in *Saccharomyces cerevisiae*. **Microbiol Mol Biol Rev** 69(2): 262-291.

29. Knaus M, Pelli-Gulli MP, van Drogen F, Springer S, Jaquenoud M, Peter M (**2007**). Phosphorylation of Bem2p and Bem3p may contribute to local activation of Cdc42p at bud emergence. **EMBO J** 26(21): 4501-4513.

30. Sousa M, Duarte AM, Fernandes TR, Chaves SR, Pacheco A, Leao C, Corte-Real M, Sousa MJ (**2013**). Genome-wide identification of genes involved in the positive and negative regulation of acetic acid-induced programmed cell death in *Saccharomyces cerevisiae*. **BMC Genomics** 14:838.

31. Mattison CP, Spencer SS, Kresge KA, Lee J, Ota IM (**1999**). Differential regulation of the cell wall integrity mitogen-activated protein kinase pathway in budding yeast by the protein tyrosine phosphatases Ptp2 and Ptp3. **Molecular and Cellular Biology** 19(11): 7651-7660.

32. Martin-Yken H, Dagkessamanskaia A, Basmaji F, Lagorce A, Francois J (**2003**). The interaction of Slt2 MAP kinase with Knr4 is necessary for signalling through the cell wall integrity pathway in *Saccharomyces cerevisiae*. **Molecular Microbiology** 49(1): 23-35.

33. Hawle P, Horst D, Bebelman JP, Yang XX, Siderius M, van der Vies SM (**2007**). Cdc37p is required for stress-induced high-osmolarity glycerol and protein kinase C mitogen-activated protein kinase pathway functionality by interaction with Hog1p and Slt2p (Mpk1p)v. **Eukaryot Cell** 6(3): 521-532.

34. Kawahata M, Masaki K, Fujii T, Iefuji H (**2006**). Yeast genes involved in response to lactic acid and acetic acid: acidic conditions caused by the organic acids in *Saccharomyces cerevisiae* cultures induce expression of intracellular metal metabolism genes regulated by Aft1p. **Fems Yeast Research** 6(6): 924-936.

35. de Lucena RM, Elsztein C, Simoes DA, de Morais MA (**2012**). Participation of CWI, HOG and Calcineurin pathways in the tolerance of *Saccharomyces cerevisiae* to low pH by inorganic acid. **J Appl Microbiol** 113(3): 629-640.

36. Claret S, Gatti X, Doignon F, Thoraval D, Crouzet M (**2005**). The

Rgd1p Rho GTPase-activating sensor are required at low pH protein and the Mid2p cell wall for protein kinase C pathway activation and cell survival in *Saccharomyces cerevisiae*. **Eukaryot Cell** 4(8): 1375-1386.

37. Klis FM, Boorsma A, De Groot PW (**2006**). Cell wall construction in Saccharomyces cerevisiae. **Yeast** 23(3): 185-202.

38. Lesage G, Bussey H (**2006**). Cell wall assembly in *Saccharomyces cerevisiae*. **Microbiol Mol Biol Rev** 70(2): 317-343.

39. Levin DE (**2011**). Regulation of cell wall biogenesis in *Saccharomyces cerevisiae*: the cell wall integrity signaling pathway. **Genetics** 189(4): 1145-1175.

40. Rego A, Costa M, Chaves SR, Matmati N, Pereira H, Sousa MJ, Moradas-Ferreira P, Hannun YA, Costa V, Corte-Real M (**2012**). Modulation of mitochondrial outer membrane permeabilization and apoptosis by ceramide metabolism. **PLoS One** 7(11): e48571.

41. Giannattasio S, Guaragnella N, Corte-Real M, Passarella S, Marra E (**2005**). Acid stress adaptation protects *Saccharomyces cerevisiae* from acetic acid-induced programmed cell death. **Gene** 354:93-98.

42. Buttner S, Eisenberg T, Carmona-Gutierrez D, Ruli D, Knauer H, Ruckenstuhl C, Sigrist C, Wissing S, Kollroser M, Frohlich KU, Sigrist S, Madeo F (**2007**). Endonuclease G regulates budding yeast life and death. **Mol Cell** 25(2): 233-246.

43. Palmqvist E, Hahn-Hagerdal B (**2000**). Fermentation of lignocellulosic hydrolysates. I: inhibition and detoxification. **Bioresource Technol** 74(1): 17-24.

44. Hanahan D, Weinberg RA (**2011**). Hallmarks of cancer: the next generation. **Cell** 144(5): 646-674.

45. Fang JY, Richardson BC (**2005**). The MAPK signalling pathways and colorectal cancer. **Lancet Oncol** 6(5): 322-7.

46. Brachmann CB, Davies A, Cost GJ, Caputo E, Li J, Hieter P, Boeke JD (**1998**). Designer deletion strains derived from *Saccharomyces cerevisiae* S288C: a useful set of strains and plasmids for PCR-mediated gene disruption and other applications. **Yeast** 14(2): 115-132.

47. Herrero P, Fernandez R, Moreno F (**1985**). Differential Sensitivities to Glucose and Galactose Repression of Gluconeogenic and Respiratory Enzymes from *Saccharomyces-Cerevisiae*. **Arch Microbiol** 143(3): 216-219.

48. Pacheco A, Azevedo F, Rego A, Santos J, Chaves SR, Côrte-Real M, Sousa M (2013). C2-Phytoceramide Perturbs Lipid Rafts and Cell Integrity in Saccharomyces cerevisiae in a Sterol-Dependent Manner. PLoS ONE 8(9): e74240.

49. Bradford MM (**1976**). A rapid and sensitive method for the quantitation of microgram quantities of protein utilizing the principle of protein-dye binding. **Anal Biochem** 72:248-254.

4

Loss of wobble uridine modification in tRNA anticodons interferes with TOR pathway signaling

Viktor Scheidt[1,#], André Jüdes[1,#], Christian Bär[1,2,#], Roland Klassen[1] and Raffael Schaffrath[1,*]

[1] Institut für Biologie, Abteilung Mikrobiologie, Universität Kassel, D-34132 Kassel, Germany.

[2] Present address: Molecular Oncology Program, Spanish National Cancer Centre (CNIO), Melchor Fernandez Almagro 3, Madrid, Spain.

[#] These authors contributed equally to the study.

[*] Corresponding Author: Raffael Schaffrath, Institut für Biologie, Abteilung Mikrobiologie, Universität Kassel, Heinrich-Plett-Str. 40; D-34132 Kassel, Germany; E-mail: schaffrath@uni-kassel.de

ABSTRACT Previous work in yeast has suggested that modification of tRNAs, in particular uridine bases in the anticodon wobble position (U34), is linked to TOR (target of rapamycin) signaling. Hence, U34 modification mutants were found to be hypersensitive to TOR inhibition by rapamycin. To study whether this involves inappropriate TOR signaling, we examined interaction between mutations in TOR pathway genes (*tip41Δ, sap190Δ, ppm1Δ, rrd1Δ*) and U34 modification defects (*elp3Δ, kti12Δ, urm1Δ, ncs2Δ*) and found the rapamycin hypersensitivity in the latter is epistatic to drug resistance of the former. Epistasis, however, is abolished in tandem with a *gln3Δ* deletion, which inactivates transcription factor Gln3 required for TOR-sensitive activation of NCR (nitrogen catabolite repression) genes. In line with nuclear import of Gln3 being under control of TOR and dephosphorylation by the Sit4 phosphatase, we identify novel TOR-sensitive *sit4* mutations that confer rapamycin resistance and importantly, mislocalise Gln3 when TOR is inhibited. This is similar to *gln3Δ* cells, which abolish the rapamycin hypersensitivity of U34 modification mutants, and suggests TOR deregulation due to tRNA undermodification operates through Gln3. In line with this, loss of U34 modifications (*elp3Δ, urm1Δ*) enhances nuclear import of and NCR gene activation (*MEP2, GAP1*) by Gln3 when TOR activity is low. Strikingly, this stimulatory effect onto Gln3 is suppressed by overexpression of tRNAs that usually carry the U34 modifications. Collectively, our data suggest that proper TOR signaling requires intact tRNA modifications and that loss of U34 modifications impinges on the TOR-sensitive NCR branch via Gln3 misregulation.

Keywords: Saccharomyces cerevisiae, TOR signaling, rapamycin, Gln3, NCR, Sit4, Elongator complex, tRNA anticodon modification, tRNase zymocin.

Abbreviations:
NCR - nitrogen catabolite repression,
PP2A - type 2A protein phosphatases,
PPIases - peptidyl-prolyl cis/trans-isomerases,
TOR - target of rapamycin.

INTRODUCTION

While cell growth and proliferation are typically characterized by active *de novo* protein synthesis, cell quiescence goes along with translational downregulation. So, translational activity and metabolic cycling need to be tightly regulated in response to growth signals [1]. In eukaryotes including budding yeast, this is coordinated by the nutrient-sensitive TOR (target of rapamycin) kinase pathway, which among others promotes the biogenesis of components essential for translation, i.e. ribosomal proteins, rRNAs and tRNAs [1, 2]. During maturation, tRNAs undergo many posttranscriptional modifications which appear to occupy roles in TOR-dependent processes [3-5]. In line with this

notion are TOR-indicative defects in filamentous growth and metabolic cycling as well as GAAC (general amino acid control) response signatures typical of tRNA modification mutants including those that fail to form 5-methoxycarbonylmethyl-2-thiouridine (mcm5s2U) onto anticodon wobble uridines (U34) [5-7]. The mcm5s2U34 modification depends on two pathways one of which requires the Elongator complex (Elp1-Elp6) for mcm5 side chain formation while the second one (Uba4, Urm1, Ncs2/Ncs6) provides S-transfer for s2 thiolation [8-10]. In support of a link between TOR signaling and tRNA modification, Elongator and thiolation mutants are hypersensitive to TOR inhibition by caffeine and rapamycin [10-14].

When TOR activity is low due to nitrogen-starvation (or rapamycin treatment), type 2A protein phosphatases (PP2A) including Sit4 are typically freed from TOR control and dephosphorylate TOR pathway targets including Gln3 [1, 2]. Gln3 is required for transcription of NCR (nitrogen catabolite repression) genes and its release from TOR phosphoinhibition by Sit4 dephosphorylation leads to its nuclear import and NCR gene activation [2, 15]. Consistent with its TOR-sensitive role, loss of Gln3 causes rapamycin resistance. Gln3 activation also involves genes that encode regulators of PP2A including Sit4 (*TAP42, TIP41, SAP190, PPM1/2, RRD1/2*) and that, when mutated, protect against rapamycin [2, 15-18]. Among these, the *RRD1/2* products are PPIases (peptidyl-prolyl *cis/trans*-isomerases) which bind and activate PP2A enzymes through conformational changes thought to confer substrate specificity [19-21]. Consistent with such activator role for Sit4, *rrd1/2Δ* mutants accumulate Gln3 in its phosphorylated form [22]. Yet, *sit4Δ* cells are hypersensitive to rapamycin and this trait is unaltered in tandem with *rrd1/2Δ* null-alleles, which alone cause drug resistance [17, 22, 23]. This implies that the hypersensitivity of *sit4Δ* cells is independent of Rrd1/2 function. A TOR-independent role for Sit4 is indeed known, and was shown to confer growth inhibition by zymocin [17, 24], a lethal tRNase toxin which kills cells by cleaving mcm5s2U34 modified anticodons (see above). Hence, tRNA modification defects typical of Elongator and *sit4Δ* mutants cause zymocin resistance [8, 12, 24-26]. *sit4Δ* mutants accumulate hyperphosphorylated Elongator forms demonstrating a link between Sit4 dephosphorylation and Elongator's tRNA modification function [27, 28]. This Sit4 role is independent of Rrd1/2 but requires the Sit4 partner proteins Sap185 and Sap190 [17, 23, 29], which is why a *sap185Δ190Δ* double mutant copies defects typical of *sit4Δ* and Elongator mutants (i.e. Elongator hyperphosphorylation, loss of tRNA modification, zymocin resistance and rapamycin hypersensitivity) [25-28].

Here, we show that the rapamycin hypersensitivity of Elongator and U34 thiolation mutations (*elp3Δ, urm1Δ*), which are epistatic to TOR signaling mutations (*tip41Δ, sap190Δ, ppm1Δ, rrd1Δ*), is suppressed by overexpression of tRNAs known to undergo U34 modification and entirely abolished by a *GLN3* deletion. This implies improper tRNA functioning due to loss of anticodon modifications interferes with the TOR signaling pathway through Gln3. Consistently, we can correlate mislocalisation of and upregulated NCR gene (*MEP2, GAP1*) activation by Gln3 with tRNA anticodon (U34) modification defects. This strongly suggests tRNA modifications are required for proper signaling into TOR-sensitive activation of NCR genes and regulation of Gln3.

RESULTS AND DISCUSSION
Previously, it was shown that inactivation of either Sit4 (*sit4Δ*) or all four Sap proteins (*sap4Δ155Δ185Δ190Δ*) required for Sit4 function causes rapamycin hypersensitivity and that this trait is not altered in combination with an *rrd1Δ* null-allele [17, 23]. Intriguingly, reintroduction of

SAP190 into *rrd1ΔsapΔΔΔΔ* cells reestablishes the drug resistance typical of *rrd1Δ* cells alone suggesting the trait depends on Sit4 and Sap190 [17]. This is similar to Sit4 dependent Elongator dephosphorylation, which requires Sap185 and Sap190 and promotes Elongator's tRNA modification function [25, 27, 28]. Elongator dephosphorylation can be suppressed by high dosage of *KTI12*, a gene coding for a potential regulator of Elongator and Sit4 [12, 27]. Strikingly, multicopy *KTI12* gene dosage was also found to confer rapamycin sensitivity in an *rrd1Δ* background (Supplemental Figure 1). With Kti12 being intimately linked to Elongator [8, 12, 30, 31], we observed that both Elongator and *KTI12* gene deletions (*elp3Δ, kti12Δ*) phenocopied each other and caused *rrd1Δ* cells to become rapamycin sensitive (Figure 1A). Besides Elongator function, formation of the mcm5s2U34 modification also requires components of the U34 thiolation pathway (i.e. Urm1, Uba4 and Ncs2/Ncs6) [8-10]. When we combined sulfur transfer defects (*urm1Δ, ncs2Δ*) with an *rrd1Δ* allele, the resulting drug sensitivity basically copied the above epistasis seen with Elongator and *kti12* mutations (Figure 1A). Similarly, on studying growth inhibition by caffeine, a TOR inhibitor drug distinct from rapamycin [32], we again found epistasis between U34 modification defects and an *rrd1Δ* null-allele confirming that the genetic interactions seen were specific to the TOR pathway (Supplemental Figure 2).

Next, we included TOR signaling mutants with defects in the Gln3 pathway (*sap190Δ, tip41Δ* or *ppm1Δ*) [15, 18] into our assays. We found that on their own, they copied the rapamycin resistance of *rrd1Δ* cells, but in tandem with an *elp3Δ* mutation, they also failed to counter the drug sensitivity typical of Elongator mutants (Figure 1B). Consistent with our observation that the drug sensitivity of thiolation defects was independent of *rrd1Δ* (Figure 1A), we observed that the rapamycin hypersensitivity of an *urm1Δ* mutant was not altered in tandem with a *tip41Δ* null-allele, which alone is a potent rapamycin suppressor (Figure 1B). Taken together, our data indicate that TOR-related phenotypes (rapamycin and caffeine hypersensitivity), which result from loss of U34 modifications in Elongator and thiolation mutants, are not modulated by Gln3 pathway gene mutations that cause protection against TOR inhibitor drugs.

Gln3 is a TOR-sensitive transcription factor necessary for NCR gene activation [33]. We found that the rapamycin resistance of a *gln3Δ* mutant was hardly altered in tandem with either *elp3Δ* or *urm1Δ* null-alleles (Figure 1C). So in striking contrast to the above TOR signaling mutants, the drug phenotype of *gln3Δ* cells is insensitive to tRNA modification defects indicating the rapamycin hypersensitivity of *elp3Δ* and *urm1Δ* mutants requires Gln3 function. Activity of Gln3 is largely regulated at the level of nucleo-cytoplasmic shuttling and cytosolic localization involves phosphoinhibition of Gln3 by TOR and interaction with Ure2 for cytosolic retention [34]. Consistently, inactivation of Ure2 in *ure2Δ* cells leads to nuclear import of Gln3, NCR gene activation and rapamycin hypersensitivity [34]. Having found that Gln3 is critical for the rapamycin hypersensitivity

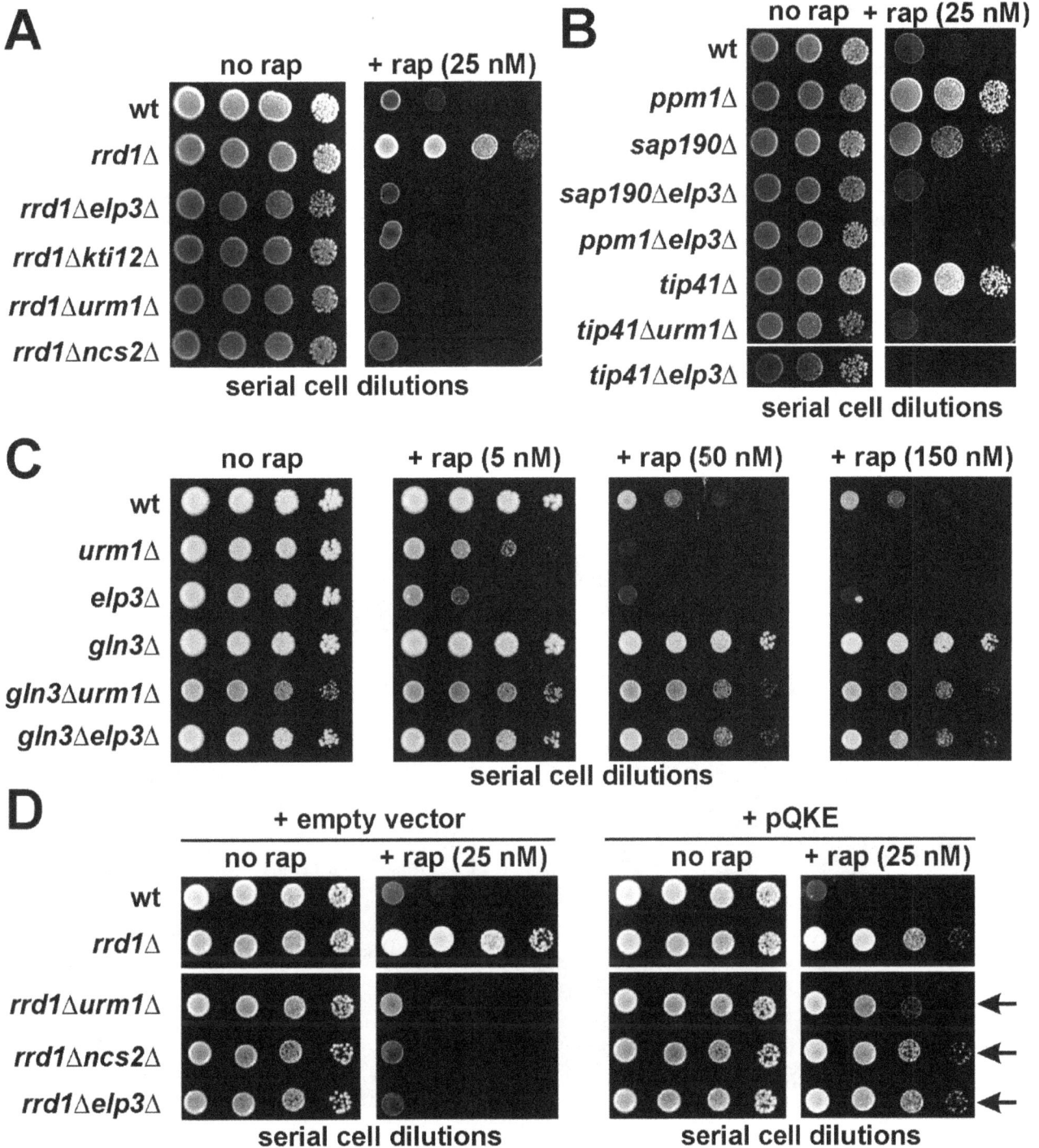

FIGURE 1: Genetic interaction between mutations in TOR signaling and tRNA modification pathways. (A) Rapamycin hypersensitivity due to loss of tRNA modification in Elongator and U34 thiolation mutants is epistatic over TOR signaling mutant *rrd1Δ*. **(B)** Other mutations in TOR pathway signaling genes cannot alter the rapamycin sensitive phenotype of Elongator or U34 thiolation mutants. **(C)** Hypersensitivity of U34 modification mutants to TOR inhibition by rapamycin requires Gln3 function. **(D)** Overexpression of tRNA species (tRNA^{Gln}, tRNA^{Lys} and tRNA^{Glu}) known to undergo U34 anticodon modification suppresses the rapamycin sensitivity of *rrd1Δ* cells carrying tRNA modification defects and reinstates drug tolerance. Shown are drug responses with vector controls (+ empty vector: left panels) or in response to tRNA overexpression from multicopy plasmid (+ pQKE: right panels) carrying tRNA^{Gln} (Q) tRNA^{Lys} (K) and tRNA^{Glu} (E) genes. Phenotypic suppression is indicated by arrows. In (A-D), ten-fold serial dilutions of yeast tester strains with genetic backgrounds as indicated were spotted onto YPD media containing no drug (no rap) or various doses (5, 20, 25, 50 and 150 nM) of rapamycin (+ rap) and grown for 3-5 days at 30°C. Lack of growth indicates rapamycin sensitive or hypersensitive responses while growth in the presence of the TOR inhibitor drug equals rapamycin resistance.

of U34 modification mutants, we asked if this trait may involve *URE2* function. On comparing rapamycin phenotypes between *ure2Δ* and *urm1Δ* single as well as *ure2Δurm1Δ* double mutants, we observed a hypersensitive response of the latter to the TOR inhibitor drug (Supplemental Figure 3). Such phenotype suggests that in combination, defects in Ure2 and U34 modification are additive and result in enhanced sensitivity to TOR inhibition by rapamycin.

Studies in yeast have shown that U34 tRNA modification defects and associated phenotypes are suppressible by overexpressing tRNAs that would normally undergo the U34 anticodon modifications. Therefore, we repeated the above assays in the absence (empty vector) and presence of a multicopy plasmid (pQKE) that allows for overexpression of tRNAs (tRNA^{Gln} [Q], tRNA^{Lys} [K] and tRNA^{Glu} [E]) known to be mcm5s2U34 modified in wild-type cells (see above). As illustrated in Figure 1D, higher-than-normal lev-

els of these tRNA species had efficient suppressor function and reconferred a rapamycin tolerant trait to *rrd1Δelp3Δ*, *rrd1Δurm1Δ* and *rrd1Δncs2Δ* double mutants that almost compares to the drug tolerance of an *rrd1Δ* single mutant alone. Similarly, we found that the caffeine sensitive phenotype typical of the *rrd1Δelp3Δ* double mutant was suppressed by tRNA overexpression (Supplemental Figure 4) to confer a drug tolerant trait resembling the *rrd1Δ* mutant alone. Taken together, these suppression data strongly suggest that the effect U34 modification defects obviously have on TOR pathway mutants including *rrd1Δ* cells can be ascribed to loss of Elongator's tRNA modification function (*elp3Δ*) and deficient thiolation (*urm1Δ*, *ncs2Δ*). So, proper tRNA anticodon (U34) modification and translation-related tRNA functions that are associated with it are apparently required for intact TOR pathway signaling.

With Gln3 localisation involving dephosphorylation by Sit4, we next asked whether TOR-dependent roles of this

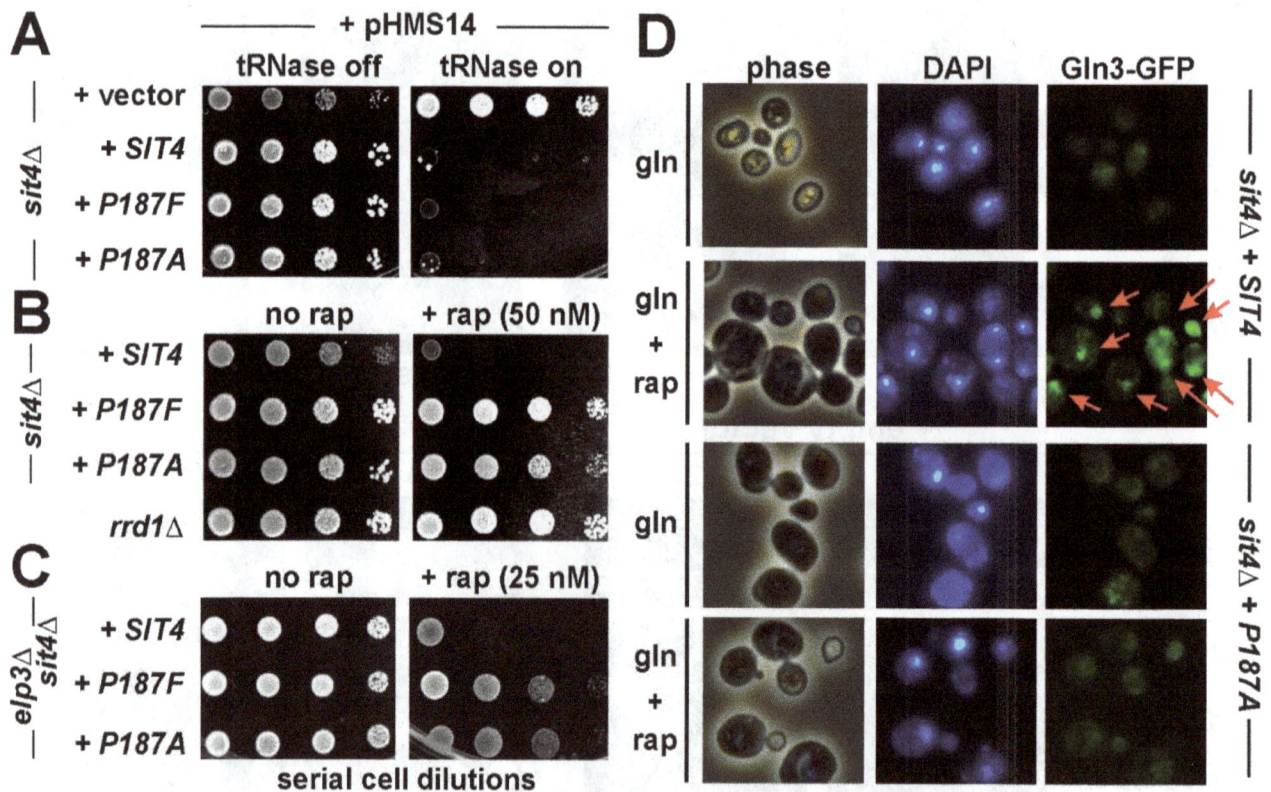

FIGURE 2: Substitutions of Sit4 proline residue 187 identify novel *sit4* mutants that separate TOR-dependent phosphatase functions from TOR-insensitive ones. (A) TOR-independent zymocin γ-toxin tRNase assay. The indicated *sit4Δ* backgrounds carrying empty vector, wild-type *SIT4* and the P187F/A substitution alleles were transformed with the *GAL1*::γ-toxin expression vector (pHMS14) [12] and spotted onto glucose repressing (tRNase off) or galactose inducing (tRNase on) media. Growth was for 3 days at 30°C. Resistance or sensitivity towards tRNase toxicity is distinguished by growth or lack of growth, respectively. (B) TOR-sensitive rapamycin phenotype. Ten-fold serial cell dilutions of a TOR signaling mutant (*rrd1Δ*) and *sit4Δ* cells with genetic backgrounds as indicated in (A) were spotted onto medium containing rapamycin (+ rap, 50 nM) or no drug (no rap). Lack of growth indicates drug sensitivity, growth equals rapamycin resistance. (C) The novel *sit4* separation of function mutations suppress the sensitivity of Elongator mutants to rapamycin. A *sit4Δelp3Δ* double mutant with genetic backgrounds as indicated in (A) was grown on plates with no drug (no rap) or containing rapamycin (+ rap, 25 nM). (D) Gln3 mislocalises as a result of the P187 substitution and fails to be imported into the nucleus under conditions of TOR inhibition. Cells carrying pRS416-GFP-Gln3 were grown in minimal medium containing glutamine (gln) as the sole N-source with or without 10 nM rapamycin (rap) and images taken in phase contrast, DAPI- and GFP-fluorescence modes (phase, DAPI, Gln3-GFP). Arrows indicate GFP signals and foci that co-localise with DAPI-stained nuclei.

multifunctional PP2A phosphatase could be distinguished from TOR-independent ones. Mammalian PP2A phosphatase activator (PTPA) and yeast Rrd1/2 are related PPIases with roles in phosphatase regulation [18-22, 35]. Isomerization of human PP2A by PTPA involves a proline residue (P190) [19] which in yeast Sit4 highly likely aligns to P187. To study the significance of P187, substitutions (P187A/F) were generated and analysed in zymocin and rapamycin assays indicative for TOR-independent and TOR-dependent Sit4 functions [12, 17, 18]. As for the zymocin assays, a *SIT4* wild-type allele, the substitutions (P187A/F) and empty vector were co-transformed into a *sit4Δ* reporter strain with pHMS14. pHMS14 allows for galactose-inducible expression of the tRNase γ-toxin subunit of zymocin, which in the presence of active Sit4 (and Elongator's intact tRNA modification function) cleaves anticodons and becomes lethal [25, 36, 37]. Upon galactose induction, *sit4Δ* cells (with empty vector) survived tRNase expression. The *SIT4* wild-type cells and both substitution (P187A/F) mutants, however, became killed (Figure 2A) indicating that P187 is dispensable for the role Sit4 plays in zymocin inhibition and Elongator's tRNA modification function. In contrast, the rapamycin assays (Figure 2B) show that, similar to *rrd1Δ* cells, the *sit4* mutants (P187A/F) are resistent against TOR inhibition by the drug. Based on these novel differential *sit4* phenotypes, the substitutions (P187A/F) thus separate TOR-independent (zymocin action) from TOR-dependent (rapamycin inhibition) Sit4 functions. Remarkably, the novel *sit4* mutations suppressed the rapamycin hypersensitivity of *elp3Δ* cells (Figure 2C). This is similar to the *gln3Δ* scenario (Figure 1C) and suggests that in the P187A/F mutants, non-functional Gln3 may be responsible for this *elp3Δ* suppressor effect. To address this issue in more detail, we found that in relation to *SIT4* wild-type cells, the P187A/F mutants indeed mislocalised a GFP-Gln3 reporter and failed to shuttle the transcription factor into the nucleus following TOR inhibition by rapamycin (Figure 2D). So, cytosolic accumulation of GFP-Gln3 under conditions of TOR inactivation is in good agreement with our data showing that the P187A/F mutants evoke rapamycin resistance (Figure 2B), a trait indicative for Gln3 inactivation and copied by loss of Gln3 function in the *gln3Δ* null-mutant (Figure 1C).

Therefore, we next asked whether the loss of tRNA modification in *elp3Δ* or *urm1Δ* mutants may interfere with TOR signaling through effects onto Gln3. Using RT-PCR and qPCR, we studied TOR-sensitive transcription of NCR genes (*GAP1, MEP2*) by Gln3 in response to alterations of tRNA anticodon modification. The strains were either wild-type or deleted for *ELP3* or *URM1* or they carried *sit4Δ* (and *sit4Δelp3Δ*) or *gln3Δ* null-alleles in which phosphorylated Gln3 should remain cytosolic due to Sit4 defects (*sit4Δ*, *sit4Δelp3Δ*) or no NCR gene activation should occur in the first place (*gln3Δ*). Also, all strains were *gat1Δ* to eliminate gene activation by a transcription factor redundant to Gln3. As Gln3-independent controls, we monitored actin *ACT1* (RT-PCR) and *ALG9* (qPCR) gene transcription [38]. Under good nitrogen conditions (glutamine) and in the absence of rapamycin, we hardly detected any *MEP2* and *GAP1* activa-

tion by Gln3 in the tester strains (Figure 3A and Supplemental Figure 5A). This is consistent with phosphoinhibition by TOR and cytosolic localisation of Gln3 [22, 33, 34]. As expected, no *MEP2* transcription was seen in *gln3Δ*, *sit4Δ* or *sit4Δelp3Δ* mutants (Figure 3A). Consistent with Gln3 being released from TOR control by rapamycin, addition of the drug triggered basal gene activation by Gln3 (Figure 3A and Supplemental Figure 5A). Strikingly, however, *MEP2* and *GAP1* gene transcription was significantly upregulated in *elp3Δ* or *urm1Δ* cells (Figure 3A and Supplemental Figure 5A) and based on validation by qPCR [38], *MEP2* transcription increased roughly eight-fold (*elp3Δ*) and six-fold (*urm1Δ*) in relation to wild-type cells with proper tRNA modifications (Figure 3B). The finding that gene activation by Gln3 is significantly enhanced in the tRNA modification mutants is in line with our data showing that rapamycin hypersensitivity of *elp3Δ* or *urm1Δ* cells does require *GLN3* function (Figure 1C). However, it is noteworthy that irrespective to TOR inhibition by rapamycin, even under conditions of TOR suppression by a poor nitrogen source (proline), constitutive Gln3-dependent transcription apparently differs between U34 modification mutants and wild-type cells (Supplemental Figure 5B).

As for conditions of good nitrogen (glutamine) supply, our RT-PCR data show that enhanced *MEP2* activation strictly depends on *SIT4* function (Figure 3A), strongly suggesting that the transcription activation effects of U34 modification defects on Gln3 operate through the Sit4 phosphatase and require Gln3 dephosphorylation for nuclear localisation. Consistent with this notion, we observed that in the absence of TOR inhibition by rapamycin and in drastic contrast to wild-type cells, GFP-tagged Gln3 was significantly mislocalised in cells lacking the *ELP3* gene and accumulated in the nucleus (Figure 3C). Equally important and consistent with our phenotypic suppression data above (Figure 1D and Supplemental Figure 4) was our finding that Gln3 mislocalisation in the Elongator mutant could be efficiently suppressed by overexpressing tRNAs (tRNA[Gln], tRNA[Lys] and tRNA[Glu]) usually carrying U34 anticodon modifications (Figure 3C). This suggests it is a tRNA-related function and/or process that, when impaired or deficient due to loss of Elongator's tRNA modification function, affects proper Gln3 localisation and subsequent transcriptional activation. Whether misregulated Gln3 localization accounts for the activated NCR response signature in the Elongator mutant is an attractive option since it entirely consists with a previous report from 2007 [39] that placed Urm1 upstream of Gln3 and also showed Gln3 mislocalisation in an *urm1Δ* mutant, which by then was solely thought to be deficient in urmylation (a ubiquitin-like protein conjugation pathway) [39]. However, with the recently advanced evidence showing that Urm1 has dual roles in protein urmylation and tRNA anticodon (U34) thiolation [8-10], Gln3 mislocalization shared between *elp3Δ* (Figure 3C) and *urm1Δ* [39] cells further supports our notion that it is loss of tRNA anticodon modification due to Elongator inactivation or U34 thiolation defects (rather than protein urmylation) which enhances NCR gene activation by Gln3, particularly under conditions of low TOR activity (Figure 3A and

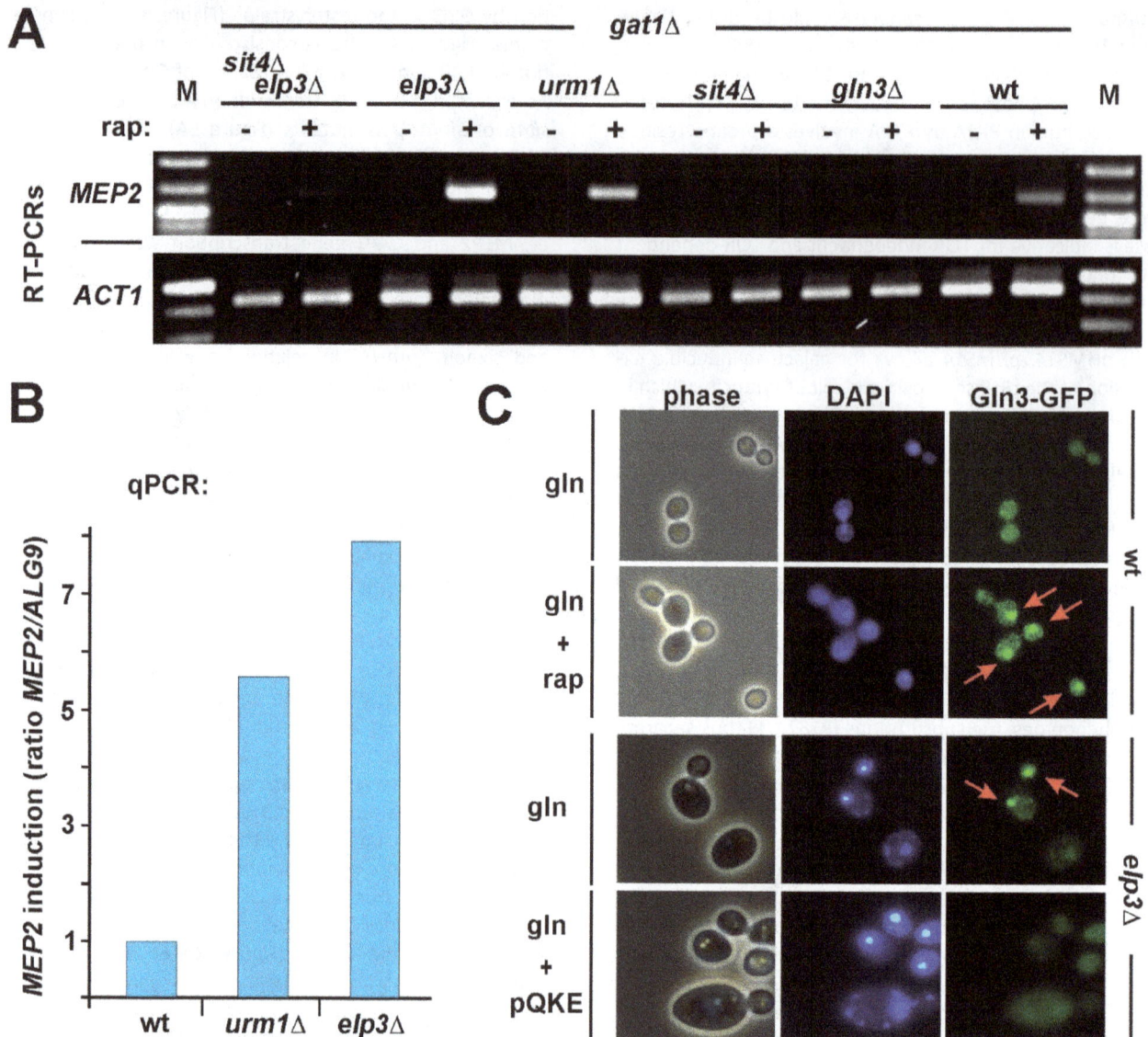

FIGURE 3: Loss of U34 anticodon modification causes nuclear Gln3 mislocalisation and enhances TOR-sensitive gene activation by Gln3. (A, B) RT-PCR (A) and qPCR (B) reveal that the TOR-sensitive *MEP2* gene transcription by Gln3 is enhanced in anticodon modification mutants that lack Elongator and U34 thiolation activities. Total RNA was isolated from the indicated strains cultivated with good nitrogen source (glutamine) supply in the absence (-) or presence (+) of 50 mM (rap) rapamycin. Following RT-PCR, the transcriptional induction of the *MEP2* gene was analysed in comparison to actin (*ACT1*) transcription (A) and quantified by qPCR in relation to *ALG9* transcription (B) (mean values of triplicates). **(C)** Gln3 mislocalises to the nucleus in an *elp3Δ* Elongator mutant, a property suppressible by tRNA overexpression. Wild-type cells carrying pRS416-GFP-Gln3 were grown in minimal medium containing glutamine (gln) as the sole N-source with or without 10 nM rapamycin (rap). For localization studies with the Elongator mutant in response to tRNA overexpression, *elp3Δ* cells carrying pRS416-GFP-Gln3 together with empty vector control or multicopy tRNA plasmid (pQKE) were used and images taken in phase contrast, DAPI- and GFP-fluorescence modes (phase, DAPI, Gln3-GFP). Arrows indicate nuclear localization of GFP-tagged Gln3.

Supplemental Figure 5A).

Our observation that improper tRNA modification deregulates transcription factor Gln3, is very reminiscent of two previous reports showing that Elongator-linked tRNA modification defects as well as KEOPS mutants with inappropriate N6-threonylcarbamoyl adenosine (t6A) modification in the anticodon stem loop can also misregulate transcription factor Gcn4 and the GAAC response associated with it [5, 6]. Intriguingly, this is in line with another study

showing that a *sap185Δ190Δ* mutant which lacks the Sit4 partner proteins, Sap185 and Sap190, caused *GCN4* induction, too [23]. This is significant since *sap185Δ190Δ* cells copy defects and phenotypes typical of *sit4Δ* and Elongator mutants including loss of tRNA modification and rapamycin hypersensitivity [23, 27-29]. Since protein dephosphorylation by Sit4 not only regulates Elongator's tRNA modification function [27-29] but also operates on the two TOR pathway branches, NCR (Gln3) and GAAC (Gcn4) [15], it will be very important to study how the loss of tRNA modifications elicits Gln3 and Gcn4 response signatures at the molecular level. In yeast, almost all Elongator-linked phenotypes studied are rescued or at least partially suppressed by overexpressing tRNAs whose anticodons would normally carry the mcm5s2U34 modifications [10, 40-42]. Consistent with this, we found that overexpressing tRNA^Gln, tRNA^Lys and tRNA^Glu is sufficient to suppress phenotypes (Figure 1D and Supplemental Figure 4) and properties that are typical of U34 modification mutants including Gln3 mislocalisation (Figure 3C). This strongly suggests that a translational defect in these tRNA modification mutants operates in deregulated TOR signaling. In support of this, a recent study has shown that U34 modifications indeed promote tRNA decoding functions during mRNA translation elongation [42].

Collectively, our data suggest that proper tRNA modifications play a positive role in the TOR signaling pathway. As a result of loss of anticodon modification, U34-minus cells undergo TOR suppression and enhance TOR-sensitive gene activation by Gln3 (NCR). Whether or not this operates upstream of TOR or outside the TOR network at a downstream level impinging on Gln3 activation, Ure2 interaction and/or nuclear import of Gln3 (Figure 4) is not entirely distinguishable from our preliminary evidence presented in this report. Alternatively, enhanced signaling in the U34 modification mutants may result from a combination of both options (Figure 4). In support of such scenario, our data show that drastic NCR gene activation by Gln3 in elp3Δ cells requires TOR inhibition by rapamycin for Gln3 mobilisation in concert with the Sit4 phosphatase, which (in addition to release of Gln3 from Ure2) further contributes to nuclear import of Gln3 and NCR gene transcription activation by Gln3. Whether the enhanced Gln3 activation (together with Gcn4 induction reported previously [5, 6, 43]) in tRNA modification mutants reflect cellular (stress) responses to overcome scenarios typically encountered upon TOR inhibition or suppression, i.e. poor nitrogen supply (NCR) or amino acid depletion (GAAC), are attractive hypotheses that need to be addressed in further studies. Given a report, however, that amino-acyl-tRNA synthetases have been found upregulated in tRNA modification mutants [5], tRNA charging defects, which may be associated with loss of U34 modifications, could potentially feed into TOR pathway signaling.

MATERIALS AND METHODS

General methods, yeast growth and strains

Routine yeast growth was in yeast extract, peptone, and dextrose (YPD) rich or synthetic complete (SC) minimal media [44]. For TOR modulation, growth media were supplemented with proline (poor N-source) or glutamine (good N-source). For testing the effect of TOR inhibitor drugs, 2.5 - 150 nM rapamycin (Calbiochem) or 2.5 - 15 mM, caffeine (Sigma) were added to YPD medium, and yeast growth was monitored for 3 - 5 days at 30°C. Yeast transformation used the PEG-lithium-acetate method [45] and tRNA overexpression involved multicopy plasmids pKQ/E carrying tRNA^Lys UUU, tRNA^Gln UUG and tRNA^Glu UUC genes [10]. To assess the response to lethal induction of zymocin's tRNase subunit (γ-toxin), strains were transformed with pHMS14 [12], and growth was followed under conditions of tRNase expression (galactose) and repression (glucose). Substitutions of Sit4 proline residue 187 to either alanine or phenylalanine were generated using the two-step fusion PCR approach [46]. Mutants generated in this study by one-step PCR mediated gene deletion [47] derived from previously described strain backgrounds and used knock-out primers specific for ELP3, KTI12, URM1, NCS2 and GLN3 [12, 13, 17].

RT-PCR and qPCR methods

Total yeast RNA isolation and RT-PCR were carried out as previously described [36] with the following oligonucleotides ACT1-FW (5'-CTT CCG GTA GAA CTA CTG GT-3'); ACT1-RV (5'-CCT TAC GGA CAT CGA CAT CA-3'); GAP1-FW (5'-TCC CGC TTC GCT ACT GAT TG-3'); GAP1-RV (5'-GCA GAG TTA CCG ACA GAT AA-3'); MEP2-FW (5'-GGT ATG TTT GCC GCA GTC AC-3') and MEP2-RV (5'-ACC ACC CAC ACC ATG GAT AG-3'). Real-time PCR

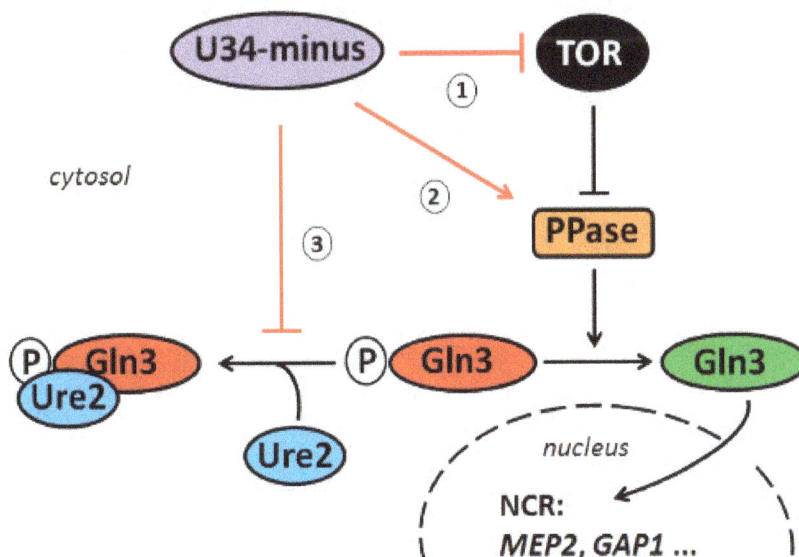

FIGURE 4: Working model for interactions between the TOR pathway and tRNA modifications. Loss of anticodon wobble uridine (U34) modifications enhances rapamycin sensitivity and NCR gene activation by Gln3 suggesting U34-minus mutants dampen TOR signaling. Such buffer function may operate upstream of TOR (1) or outside from TOR (2, 3) affecting steps that counter TOR-sensitive Gln3 inhibition. The latter (2, 3) may involve dephosporylation by a phosphatase (PPase) such as Sit4 (2) and/or Gln3 release from Ure2 (3) for mobilization and nuclear import. Alternatively, the effects of inappropriate U34 modifications on Gln3 misregulation could result from a combination of these options (1 - 3).

used a Mastercycler (Eppendorf), the SensiFAST™ SYBR® No-ROX Kit (BIOLINE) and the *ALG9* standardization protocol [38]. qPCR involved *ALG9*-qFW (5'-GTC ACG GAT AGT GGC TTT GG-3'); *ALG9*-qRV (5'-TGG CAG CAG GAA AGA ACT TG-3'); *MEP2*-qFW (5'-GTA TGT TTG CCG CAG TCA CC-3') and *MEP2*-qRV (5'-CAG ACC CAG CAT GCA ATA GG-3') oligonucleotides.

Cellular localization of Gln3-GFP

Yeast strains carrying pRS416-GFP-Gln3 [48, 49] were grown in yeast nitrogen base media with glutamine as the sole N-source, left untreated or treated with 10 nM rapamycin for 20 min. Subsequently, cells were fixed by adding 3.7% formaldehyde directly to the medium and incubated for 10 min at room temperature and washed once with water. Cells were resuspended in water containing 1 µg·ml^{-1} 4,6 diamidino-2-phenylindole (DAPI, Sigma, Germany). Following washing with water, cells were analyzed using an Olympus BX53 microscope with appropriate filters for DAPI and GFP fluorescence. Images were captured using the CellSens 1.6 software package (Olympus).

ACKNOWLEDGMENTS
We thank Drs S. Leidel and T. Cooper for tRNA overexpression and GFP-Gln3 localisation plasmids. We appreciate discussion with Dr T. Cooper and support by Zentrale Forschungsförderung (ZFF #1798), Universität Kassel, Germany, and Deutsche Forschungsgemeinschaft (SCHA750/15 & SCHA750/18), Bonn, Germany.

CONFLICT OF INTEREST
The authors declare there is no conflict of interest.

REFERENCES

1. Broach JR (**2012**). Nutritional control of growth and development in yeast. **Genetics** 192(1): 73-105.

2. Loewith R and Hall MN (**2011**). Target of rapamycin (TOR) in nutrient signaling and growth control. **Genetics** 189(4): 1177-1201.

3. El Yacoubi B, Bailly M, and de Crécy-Lagard V (**2012**). Biosynthesis and function of posttranscriptional modifications of transfer RNAs. **Annu Rev Genet** 46: 69-95.

4. Cullen PJ and Sprague GF Jr (**2012**). The regulation of filamentous growth in yeast. **Genetics** 190(1): 23-49.

5. Zinshteyn B and Gilbert WV (**2013**). Loss of a conserved tRNA anticodon modification perturbs cellular signaling. **PLoS Genet** 9(8): e1003675.

6. Daugeron MC, Lenstra TL, Frizzarin M, El Yacoubi B, Xipeng Liu X, Baudin-Baillieu A, Lijnzaad P, Decourty L, Saveanu C, Jacquier A, Holstege FCP, de Crécy-Lagard V, van Tilbeurgh H, and Libri D (**2011**). Gcn4 misregulation reveals a direct role for the evolutionary conserved EKC/KEOPS in the t6A modification of tRNAs. **Nucleic Acids Res** 39(14): 6148-6160.

7. Laxman S, Sutter BM, Wu X, Kumar S, Guo X, Trudgian DC, Mirzaei H, and Tu BP (**2013**). Sulfur amino acids regulate translational capacity and metabolic homeostasis through modulation of tRNA thiolation. **Cell** 154(2): 416-429.

8. Huang B, Johansson MJ, and Byström AS (**2005**). An early step in wobble uridine tRNA modification requires the Elongator complex. **RNA** 11(4): 424-436.

9. Noma A, Sakaguchi Y, and Suzuki T (**2009**). Mechanistic characterization of the sulfur-relay system for eukaryotic 2-thiouridine biogenesis at tRNA wobble positions. **Nucleic Acids Res** 37(4): 1335-1352.

10. Leidel S, Pedrioli PG, Bucher T, Brost R, Costanzo M, Schmidt A, Aebersold R, Boone C, Hofmann K, and Peter M (**2009**). Ubiquitin-related modifier Urm1 acts as a sulphur carrier in thiolation of eukaryotic transfer RNA. **Nature** 458(7235): 228-232.

11. Chan TF, Carvalho J, Riles L, and Zheng XF (**2000**). A chemical genomics approach toward understanding the global functions of the target of rapamycin protein (TOR). **Proc Natl Acad Sci USA** 97(24): 13227-13232.

12. Frohloff F, Fichtner L, Jablonowski D, Breunig KD, and Schaffrath R (**2001**). *Saccharomyces cerevisiae* Elongator mutations confer resistance to the *Kluyveromyces lactis* zymocin. **EMBO J** 20(8): 1993-2003.

13. Fichtner L, Jablonowski D, Schierhorn A, Kitamoto HK, Stark MJR, and Schaffrath R (**2003**). Elongator's toxin-target (TOT) function is nuclear localization sequence dependent and suppressed by post-translational modification. **Mol Microbiol** 49(5): 1297-1307.

14. Goehring AS, Rivers DM, and Sprague GF Jr (2003). Urmylation: a ubiquitin-like pathway that functions during invasive growth and budding in yeast. **Mol Biol Cell** 14(11): 4329-4341.

15. Düvel K and Broach JR (**2004**). The role of phosphatases in TOR signaling in yeast. **Curr Top Microbiol Immunol** 279: 19-38.

16. Neklesa TK and Davis RW (**2008**). Superoxide anions regulate TORC1 and its ability to bind Fpr1:rapamycin complex. **Proc Natl Acad Sci USA** 105(39): 15166-15171.

17. Jablonowski D, Täubert JE, Bär C, Stark MJ, and Schaffrath R (**2009**). Distinct subsets of Sit4 holophosphatases are required for inhibition of *Saccharomyces cerevisiae* growth by rapamycin and zymocin. **Eukaryot Cell** 8(11): 1637-1647.

18. Rempola B, Kaniak A, Migdalski A, Rytka J, Slonimski PP, and di Rago JP (**2000**). Functional analysis of *RRD1* (YIL153w) and *RRD2* (YPL152w), which encode two putative activators of the phosphotyrosyl phosphatase activity of PP2A in *Saccharomyces cerevisiae*. **Mol Gen Genet** 262(6): 1081-1092.

19. Jordens J, Janssens V, Longin S, Stevens I, Martens E, Bultynck G, Engelborghs Y, Lescrinier E, Waelkens E, Goris J, and Van Hoof C (**2006**). The protein phosphatase 2A phosphatase activator is a novel peptidyl-prolyl cis/trans-isomerase. **J Biol Chem** 281(10): 6349-6357.

20. Leulliot N, Vicentini G, Jordens J, Quevillon-Cheruel S, Schiltz M, Barford D, van Tilbeurgh H, and Goris J (**2006**). Crystal structure of the PP2A phosphatase activator: implications for its PP2A-specific PPIase activity. **Mol Cell** 23(3): 413-424.

21. Hombauer H, Weismann D, Mudrak I, Stanzel C, Fellner T, Lackner DH, and Ogris E (**2007**). Generation of active protein phosphatase 2A is coupled to holoenzyme assembly. **PLoS Biol** 5(6): e155.

22. Zheng Y and Jiang Y (**2005**). The yeast phosphotyrosyl phosphatase activator is part of the Tap42-phosphatase complexes. **Mol Biol Cell** 16(4): 2119-2127.

23. Rohde JR, Campbell S, Zurita-Martinez SA, Cutler NS, Ashe M, and Cardenas ME (**2004**). TOR controls transcriptional and translational programs via Sap-Sit4 protein phosphatase signaling effectors. **Mol Cell Biol** 24(19): 8332-8341.

24. Jablonowski D and Schaffrath R (**2007**). Zymocin, a composite chitinase and tRNase killer toxin from yeast. **Biochem Soc Trans** 35(6): 1533-1537.

25. Jablonowski D, Butler AR, Fichtner L, Gardiner D, Schaffrath R, and Stark MJ (**2001**). Sit4p protein phosphatase is required for sensitivity of *Saccharomyces cerevisiae* to *Kluyveromyces lactis* zymocin. **Genetics** 159(4): 1479-1489.

26. Huang B, Lu, J, and Byström AS (**2008**). A genome-wide screen identifies genes required for formation of the wobble nucleoside 5-methoxycarbonylmethyl-2-thiouridine in *Saccharomyces cerevisiae*. **RNA** 14(10): 2183-2194.

27. Jablonowski D, Fichtner L, Stark MJ, and Schaffrath R (**2004**). The yeast elongator histone acetylase requires Sit4-dependent dephosphorylation for toxin-target capacity. **Mol Biol Cell** 15(3): 1459-1469.

28. Mehlgarten C, Jablonowski D, Breunig KD, Stark MJ, and Schaffrath R (**2009**). Elongator function depends on antagonistic regulation by casein kinase Hrr25 and protein phosphatase Sit4. **Mol Microbiol** 73(5): 869-881.

29. Luke MM, Della Seta F, Di Como CJ, Sugimoto H, Kobayashi R, and Arndt KT (**1996**). The SAP, a new family of proteins, associate and function positively with the SIT4 phosphatase. **Mol Cell Biol** 16(6): 2744-2755.

30. Fichtner L, Frohloff F, Bürkner K, Larsen M, Breunig KD, and Schaffrath R (**2002**). Molecular analysis of *KTI12/TOT4*, a *Saccharomyces cerevisiae* gene required for *Kluyveromyces lactis* zymocin action. **Mol Microbiol** 43(3): 783-791.

31. Mehlgarten C, Jablonowski D, Wrackmeyer U, Tschitschmann S, Sondermann D, Jäger G, Gong Z, Byström AS, Schaffrath R, and Breunig KD (**2010**). Elongator function in tRNA wobble uridine modification is conserved between yeast and plants. **Mol Microbiol** 76(5): 1082-1094.

32. Reinke A, Chen JC, Aronova S, and Powers T. (**2006**). Caffeine targets TOR complex I and provides evidence for a regulatory link between the FRB and kinase domains of Tor1p. **J Biol Chem** 281(42): 31616-31626.

33. Beck T and Hall MN (**1999**). The TOR signalling pathway controls nuclear localization of nutrient-regulated transcription factors. **Nature** 402(6762): 689-692.

34. Bertram PG, Choi JH, Carvalho J, Ai W, Zeng C, Chan TF, and Zheng XF (**2000**). Tripartite regulation of Gln3 by TOR, Ure2, and phosphatases. **J Biol Chem** 275(46): 35727-35733.

35. Van Hoof C, Martens E, Longin S, Jordens J, Stevens I, Janssens V, and Goris J (**2005**). Specific interactions of PP2A and PP2A-like phosphatases with the yeast PTPA homologues, Ypa1 and Ypa2. **Biochem J** 386(1): 93-102.

36. Jablonowski D, Zink S, Mehlgarten C, Daum G, and Schaffrath R (**2006**). tRNA^Glu wobble uridine methylation by Trm9 identifies Elongator's key role for zymocin-induced cell death in yeast. **Mol Microbiol** 59(2): 677-688.

37. Lu J, Huang B, Esberg A, Johansson MJ, and Byström AS (**2005**). The *Kluyveromyces lactis* gamma-toxin targets tRNA anticodons. **RNA** 11(11): 1648-1654.

38. Teste MA, Duquenne M, François JM, and Parrou JL (**2009**). Validation of reference genes for quantitative expression analysis by real-time RT-PCR in *Saccharomyces cerevisiae*. **BMC Mol Biol** 10: 99.

39. Rubio-Texeira M (**2007**). Urmylation controls Nil1p and Gln3p-dependent expression of nitrogen-catabolite repressed genes in *Saccharomyces cerevisiae*. **FEBS Lett** 581(3): 541-550.

40. Esberg A, Huang B, Johansson MJ, and Byström AS (**2006**). Elevated levels of two tRNA species bypass the requirement for elongator complex in transcription and exocytosis. **Mol Cell** 24(1): 139-148.

41. Chen C, Huang B, Eliasson M, Rydén P, and Byström AS (**2011**). Elongator complex influences telomeric gene silencing and DNA damage response by its role in wobble uridine tRNA modification. **PLoS Genet** 7(9): e1002258.

42. Rezgui VA, Tyagi K, Ranjan N, Konevega AL, Mittelstaet J, Rodnina MV, Peter M, and Pedrioli PG (**2013**). tRNA tKUUU, tQUUG, and tEUUC wobble position modifications fine-tune protein translation by promoting ribosome A-site binding. **Proc Natl Acad Sci USA** 110(30): 12289-12294.

43. Krogan NJ and Greenblatt JF (**2001**). Characterization of a six-subunit holo-elongator complex required for the regulated expression of a group of genes in *Saccharomyces cerevisiae*. **Mol Cell Biol** 21(23): 8203-8212.

44. Sherman F (**2002**). Getting started with yeast. **Meth Enzymol** 350: 3-41.

45. Gietz D, St. Jean, A Woods, RA and Schiestl RH (**1992**). Improved method for high efficiency transformation of intact yeast cells. **Nucleic Acids Res** 20(6): 1425.

46. Hobert O (**2002**). PCR fusion-based approach to create reporter gene constructs for expression analysis in transgenic *C. elegans*. **Biotechniques** 32(4): 728-730.

47. Janke C, Magiera MM, Rathfelder N, Taxis C, Reber S, Maekawa H, Moreno-Borchart A, Doenges G, Schwob E, Schiebel E, and Knop M (**2004**). A versatile toolbox for PCR-based tagging of yeast genes: new fluorescent proteins, more markers and promoter substitution cassettes. **Yeast** 21(11): 947-962.

48. Giannattasio S, Liu Z, Thornton J, Butow RA (**2005**). Retrograde response to mitochondrial dysfunction is separable from TOR1/2 regulation of retrograde gene expression. **J Biol Chem** 280(52): 42528-42535.

49. Georis I, Tate JJ, Cooper TG, Dubois E (**2011**). Nitrogen-responsive regulation of GATA protein family activators Gln3 and Gat1 occurs by two distinct pathways, one inhibited by rapamycin and the other by methionine sulfoximine. **J Biol Chem** 286(52): 44897-44912.

Heat shock protein 90 and calcineurin pathway inhibitors enhance the efficacy of triazoles against *Scedosporium prolificans* via induction of apoptosis

Fazal Shirazi and Dimitrios P. Kontoyiannis*

Department of Infectious Diseases, Infection Control and Employee Health, The University of Texas M.D. Anderson Cancer Center, Houston, TX 77030, U.S.A.

* Corresponding Author: Dimitrios P. Kontoyiannis, MD, ScD, Department of Infectious Diseases, Infection Control and Employee Health, Unit 402, The University of Texas MD Anderson Cancer Center, 1515 Holcombe Boulevard; Houston, TX 77030, USA; Email: dkontoyi@mdanderson.org

ABSTRACT *Scedosporium prolificans* is a pathogenic mold resistant to current antifungals, and infection results in high mortality. Simultaneous targeting of both ergosterol biosynthesis and heat shock protein 90 (Hsp90) or the calcineurin pathway in *S. prolificans* may be an important strategy for enhancing the potency of antifungal agents. We hypothesized that the inactive triazoles posaconazole (PCZ) and itraconazole (ICZ) acquire fungicidal activity when combined with the calcineurin inhibitor tacrolimus (TCR) or Hsp90 inhibitor 17-demethoxy-17-(2-propenylamino) geldanamycin (17AAG). PCZ, ICZ, TCR and 17AAG alone were inactive *in vitro* against *S. prolificans* spores (MICs > 128 µg/ml). In contrast, MICs for PCZ or ICZ in combination with TCR or 17AAG (0.125-0.50 µg/ml) were much lower compared with drug alone. In addition PCZ and ICZ in combination with TCR or 17AAG became fungicidal. Because apoptosis is regulated by the calcineurin pathway in fungi and is under the control of Hsp90, we hypothesized that this synergistic fungicidal effect is mediated via apoptosis. This observed fungicidal activity was mediated by increased apoptosis of *S. prolificans* germlings, as evidenced by reactive oxygen species accumulation, decreased mitochondrial membrane potential, phosphatidylserine externalization, and DNA fragmentation. Furthermore, induction of caspase-like activity was correlated with TCR or 17AAG + PCZ/ICZ-induced cell death. In conclusion, we report for the first time that PCZ or ICZ in combination with TCR or 17AAG renders *S. prolificans* exquisitely sensitive to PCZ or ICZ via apoptosis. This finding may stimulate the development of new therapeutic strategies for patients infected with this recalcitrant fungus.

Keywords: apoptosis, 17AAG, calcineurin, itraconazole, posaconazole, reactive oxygen species.

INTRODUCTION

Scedosporium prolificans is an emerging filamentous fungus that causes severe, frequently fatal pulmonary or disseminated opportunistic infections in immunocompromised patients [1]. *S. prolificans* is inherently resistant to treatment with a wide range of antifungals, including the new generation of broad-spectrum triazoles [1-4]. Hence, new therapeutic strategies for *Scedosporium* infections are urgently needed.

In pathogenic fungi, the calcineurin pathway and heat shock protein 90 (Hsp90) play major roles in maintaining fungal homeostatic cell responses, including resistance to antifungal agents [5-10]. The calcineurin inhibitor tacrolimus (TCR) is an immunosuppressive agent widely used in solid organ and hematopoietic stem cell transplant recipients to prevent graft rejection [11]. TCR binds to the intracellular protein immunophilin FKB12 and forms a complex, thereby inhibiting activation of the calcineurin pathway. *In vitro* studies have suggested synergy between triazoles and calcineurin inhibitors against *Aspergillus* spp. and the Mucorales [12-14]. Our group recently reported that treatment with the combination of TCR and posaconazole (PCZ)

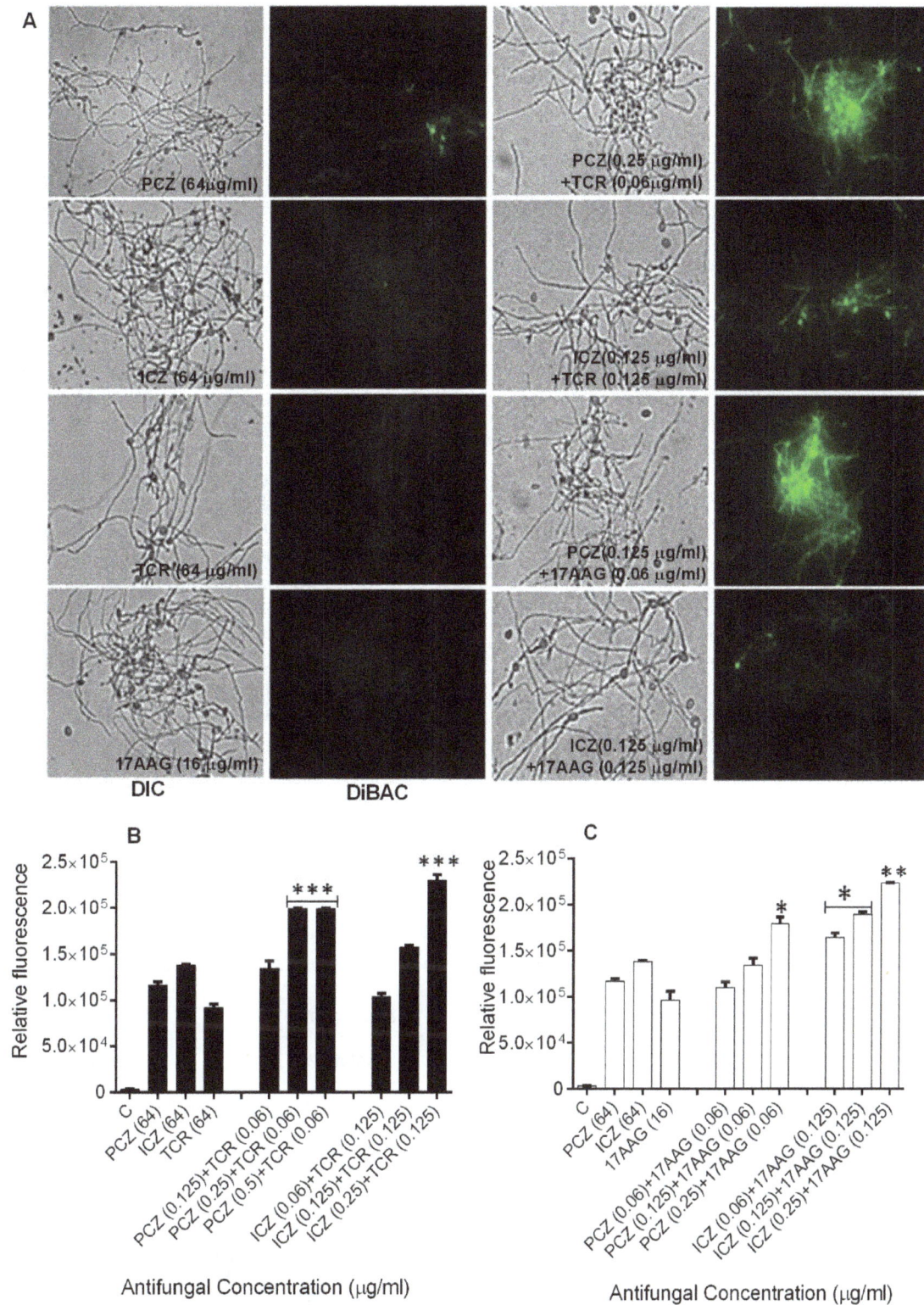

FIGURE 1: Fungicidal action of PCZ and ICZ alone and in combination with TCR and 17AAG against *S. prolificans* germlings (isolate 1). (A) Fluorescent images of *S. prolificans* germlings stained with the morbidity dye DiBAC. **(B, C)** Relative fluorescence levels in *S. prolificans* germlings treated with PCZ or ICZ plus TCR **(B)** or 17AAG **(C)** as determined using DiBAC staining. The experiments were performed in triplicate and repeated three times. *p<0.05; **p<0.001; ***p<0.0001 (compared with untreated control germlings and germlings exposed to drug alone). Error bars on graphs indicate standard deviation. DIC, differential interference contrast.

improves control of invasive, necrotizing cutaneous mucormycosis in immunosuppressed mice compared with PCZ alone [15].

Hsp90 is a molecular chaperone involved in stress responses of *Candida albicans* and *Aspergillus* spp. and plays a major role in echinocandin resistance via regulation of the calcineurin pathway [6, 16]. Specifically, pharmacological inhibition of Hsp90 by 17-demethoxy-17-(2-propenylamino) geldanamycin (17AAG) prevents azole resistance and abrogates this resistance in *C. albicans* and *A. fumigatus* in a human host [6, 16]. In addition, researchers recently suggested a role for the calcineurin pathway in regulation of apoptosis in fungi [17, 18]. However, the role of Hsp90 in apoptosis remains unclear. Therefore, simultaneous targeting of both ergosterol biosynthesis and Hsp90 or calcineurin pathways in *S. prolificans* may be an important strategy for restoring the potency of antifungal agents. Specifically, we hypothesized that TCR or 17AAG in combination with the triazoles PCZ or itraconazole (ICZ) induces apoptosis in *S. prolificans*. Thus, we examined the effects of TCR and 17AAG co-administration on PCZ and ICZ activity using several *in vitro* methods to evaluate induction of apoptosis in *S. prolificans*.

RESULTS
PCZ and ICZ are inactive when used alone against *S. prolificans*, but exhibit significant fungicidal activity when combined with TCR or 17AAG

Individually, PCZ, ICZ, TCR, and 17AAG were inactive against *S. prolificans* (isolates 1 to 3), with minimum inhibitory concentrations (MICs) ranging from 32 to128 μg/ml. In contrast, the combination of PCZ or ICZ with either TCR or 17AAG rendered *S. prolificans* exquisitely more sensitive to the triazoles than did use of the triazoles alone (Table 1). Specifically, in combination with TCR or 17AAG, PCZ and ICZ were synergistic, with a fractional inhibitory concentra-

tion index (ΣFIC) of 0.5. In addition, bis-[1,3-dibutylbarbituric acid] trimethine oxonol (DiBAC) vital staining revealed enhanced uptake of stain and plasma membrane damage in *S. prolificans* germlings (isolates 1 and 2) exposed to PCZ or ICZ in combination with TCR or 17AAG (Figure 1 A-C; Table S1). Use of PCZ or ICZ (0.125-0.25 μg/ml) in combination with TCR or 17AAG resulted in 2.0- to 2.5-fold greater plasma membrane damage than did the use of triazoles alone.

Detection of intracellular Reactive Oxygen Species (ROS) accumulation and loss of mitochondrial membrane potential ($\Delta\Psi_m$) in *S. prolificans* (isolates 1 and 2) germlings in response to treatment with PCZ or ICZ combined with TCR or 17AAG

Staining of *S. prolificans* germlings with dihydrorhodamine (DHR)-123 (red fluorescence) and rhodamine (Rh)-123 (green fluorescence) was most prominent in germlings treated with PCZ or ICZ in combination with TCR or 17AAG (Figures 2 and 3). A small percentage of control germlings and germlings treated with PCZ or ICZ alone exhibited positive staining for DHR-123 and Rh-123 (Figures 2 and 3). Staining with DHR123 and Rh-123 increased markedly when triazoles were combined with TCR or 17AAG, respectively (1.2-2.1 fold increase in fluorescence intensity), compared with triazoles alone (Figures 2 and 3 A-C). Isolate 2 in particular, had 1.0-2.1 fold and 1.3-2.1 fold increase in fluorescence for ROS accumulation and loss of mitochondrial potential, respectively, over germlings treated with triazoles alone (Table S1). Accumulation of intracellular ROS and disruption of $\Delta\Psi_m$ are important steps in mitochondria-mediated apoptosis. These data indicate that treatment with PCZ or ICZ combined with TCR or 17AAG can trigger apoptosis in *S. prolificans* due to accumulation of ROS.

TABLE 1. *In vitro* antimicrobial activity of PCZ and ICZ in combination with TCR or 17AAG against *S. prolificans* isolates.

Drugs	MIC (μg/ml)		
	Isolate 1	Isolate 2	Isolate 3
PCZ	128 (>128)*	128 (>128)	128 (>128)
ICZ	128 (>128)	128 (>128)	128 (>128)
TCR	128 (>128)	128 (>128)	128 (>128)
17AAG	32 (128)	64 (>128)	64 (>128)
PCZ + TCR (0.06 μg/ml)	0.50 (0.50)	0.25 (1.00)	0.25 (0.50)
ICZ + TCR (0.125 μg/ml)	0.25 (1.00)	0.25 (0.50)	0.25 (0.50)
PCZ + 17AAG (0.06 μg/ml)	0.25 (0.50)	0.25 (1.00)	0.25 (0.50)
ICZ + 17AAG (0.125 μg/ml)	0.125 (0.50)	0.125 (0.50)	0.125 (0.50)

*MFC is given in parenthesis.

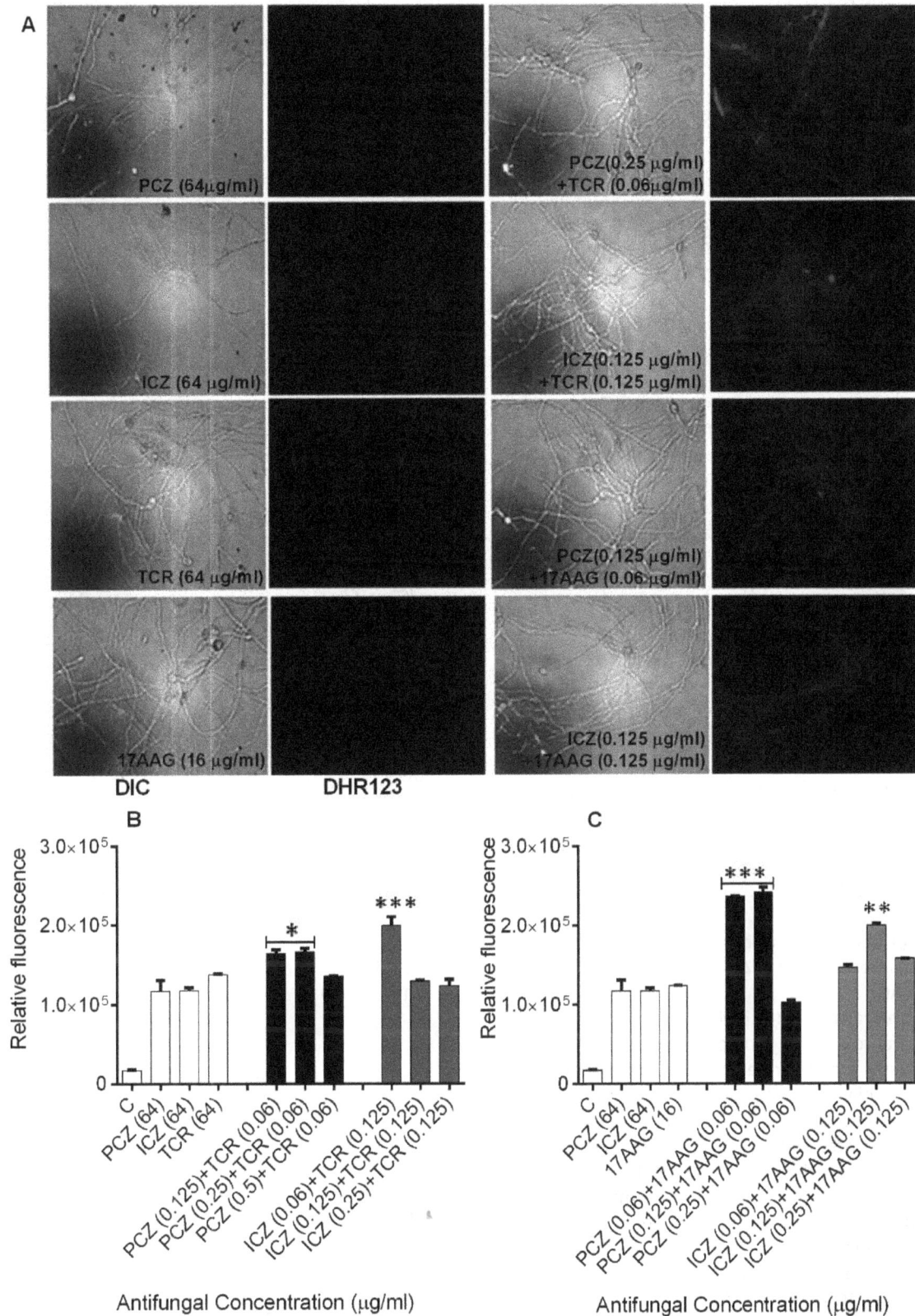

FIGURE 2: Intracellular ROS accumulation as detected by DHR-123 in *S. prolificans* isolate 1 germlings treated with PCZ or ICZ with either TCR or 17AAG, was measured using fluorescence spectrophotometry. (A) Fluorescent images of *S. prolificans* stained with DHR-123. (B) Relative fluorescence levels in *S. prolificans* germlings treated with PCZ or ICZ in combination with TCR. (C) Measurement of fluorescence of germlings treated with PCZ or ICZ in combination with 17AAG. *p<0.05; **p<0.001; ***p<0.0001. Error bars on graphs indicate standard deviation. DIC, differential interference contrast.

Evidence of apoptosis in *S. prolificans* (isolates 1 and 2) induced by treatment with PCZ or ICZ in combination with TCR or 17AAG

Because various drugs can induce both apoptosis and necrosis in mammalian cells [19], we sought to differentiate between apoptotic and necrotic *S. prolificans* protoplast using annexin V-fluorescein isothiocyanate (FITC)–propidium iodide (PI) double staining, in which apoptotic cells are stained with annexin V-FITC (green), whereas the nuclei of necrotic cells are stained with PI (red) [20-22]. Incubation of *S. prolificans* (isolate 1) protoplasts in the presence of PCZ (0.25 µg/ml) or ICZ (0.125 µg/ml) in combination with TCR (0.060-0.125 µg/ml) at 37°C for 3 h led to annexin V-FITC staining of 35-50% of the protoplasts. We found that 30-40% of protoplasts exhibited annexin V-FITC staining, when incubated with PCZ or ICZ in combination with 17AAG (0.060-0.125 µg/ml) (Table 2). In *S. prolificans* isolate 2, however, 40-65% of protoplasts were apoptotic after incubation with PCZ or ICZ in combination with TCR, and 35-70% were apoptotic after incubation with PCZ or ICZ with 17AAG (Table S1). We observed no annexin V-FITC staining in untreated protoplasts (Table 2). These results suggested that a fungicidal property of PCZ and ICZ was due to induction of apoptosis in *S. prolificans* cells, especially in combination with TCR or 17AAG.

To confirm the apoptotic features of PCZ and ICZ in *S. prolificans* germlings, we evaluated nuclear DNA fragmentation using a terminal deoxynucelotidyl transferase dUTP nick end labeling (TUNEL) assay. *S. prolificans* germlings exposed to PCZ or ICZ for 3 h at 37°C exhibited marked nuclear DNA fragmentation in a concentration-dependent manner (Table 2). The proportion of TUNEL-positive germlings was higher in both isolate 1 (50-60%) and isolate 2 (30-60%) in the presence of PCZ or ICZ (0.125 µg/ml) combined with 17AAG than when combined with TCR (isolate 1, 20-40%; isolate 2, 35-55%) (Tables 2 and S1).

Induction of caspase-like activity in *S. prolificans* (isolate 1) germlings treated with PCZ or ICZ in combination with TCR or 17AAG

Caspases are activated in the early stages of apoptosis and play a central role in the apoptotic cascade [23, 24]. Although caspases are not present in fungi, researchers have identified orthologs of mammalian caspases, called metacaspases in fungi (25). We stained *S. prolificans* germlings (isolate 1) pretreated with PCZ or ICZ in combination with TCR or 17AAG with the cell-permeable, broad-spectrum caspase inhibitor CaspACE-Z-VAD-FMK. In this staining, a green fluorescent signal is a direct measure of the amount of active caspase in a cell. *S. prolificans* germlings with activated metacaspases, treated with azoles in combination with TCR or 17 AAG were stained green, whereas germlings exposed to azoles alone remained unstained. This result indicated that treatment with PCZ or ICZ plus TCR or 17AAG triggered an apoptotic pathway in *S. prolificans* germlings via activation of metacaspases (Figure 4).

TABLE 2. Percentage of *S. prolificans* (isolate 1) cells stained with annexin V, TUNEL and PI for detection of phosphatidylseriene exposure, DNA fragmentation and cell membrane integrity respectively.

Drugs (µg/ml)	Apoptotic protoplast %		
	Annexin V	TUNEL	PI
Control	-	-	4.0±1.0
PCZ (64.0)	-	-	-
ICZ (64.0)	-	-	5.0±1.0
TCR (64.0)	-	-	3.0±1.0
17AAG (16.0)	-	-	3.0±1.0
PCZ + TCR (0.06 µg/ml)			
0.125	10.0±0.0	18.0±2.0	-
0.25	50.0±3.0	35.0±1.0	-
0.5	35.0±2.0	-	15.0±1.0
ICZ + TCR (0.125 µg/ml)			
0.060	23.0±2.0	12.0±1.0	3.0±1.0
0.125	30.0±1.0	35.0±1.0	5.0±1.0
0.25	40.0±2.0	40.0±2.0	15.0±1.0
PCZ + 17AAG (0.06 µg/ml)			
0.060	10.0±1.0	12.0±1.0	-
0.125	35.0±3.0	60.0±4.0	3.0±1.0
0.25	40.0±2.0	22.0±1.0	12.0±2.0
ICZ + 17AAG (0.125 µg/ml)			
0.060	5.0±1.0	20.0±1.0	-
0.125	30.0±1.0	50.0±4.0	5.0±0.0
0.25	30.0±2.0	10.0±1.0	15.0±0.0

-, Not detected (0% of cells showed particular apoptotic marker)

*MFC is given in parenthesis.

DISCUSSION

We hypothesized that TCR and 17AAG enhance the negligible activity of the ergosterol biosynthesis inhibitors PCZ and ICZ, to the point that they become fungicidal, and that this fungicidal activity is mediated through apoptosis in *S. prolificans*. The calcineurin pathway and Hsp90 are important for the survival of pathogenic fungi because they

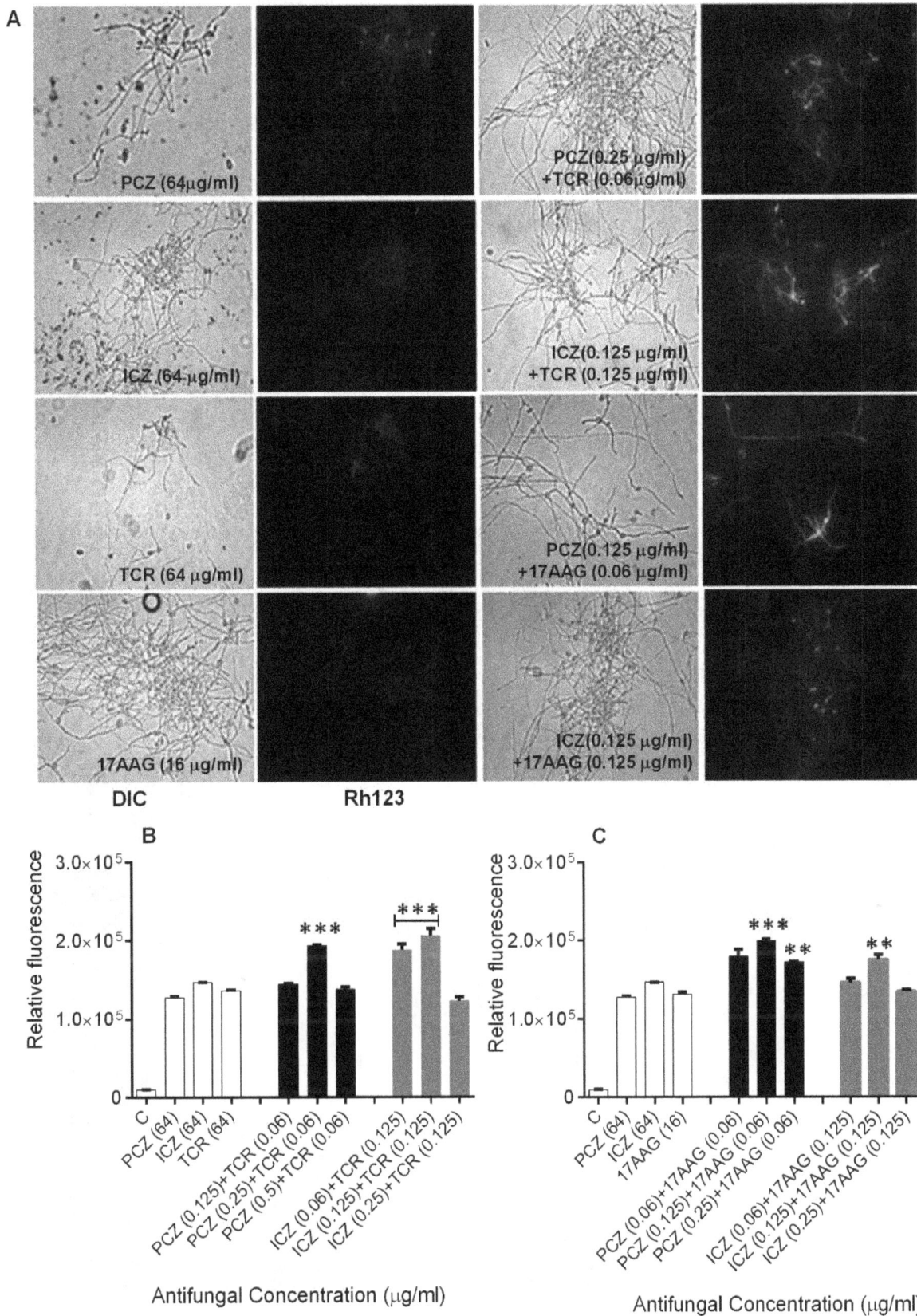

FIGURE 3: Changes in $\Delta\Psi_m$ in *S. prolificans* isolate 1 germlings triggered by treatment with PCZ or ICZ combined with TCR or 17AAG. (A) Fluorescence images of *S. prolificans* germlings stained with Rh-123. **(B, C)** Relative fluorescence levels in *S. prolificans* germlings treated with PCZ or ICZ plus TCR **(B)** and PCZ or ICZ plus 17AAG **(C)**. **p<0.001; ***p<0.0001. Error bars on graphs indicate standard deviation. DIC, differential interference contrast.

have central roles in various cellular processes, including morphogenetic transition and development of antifungal tolerance and resistance [7, 16]. Inhibition of the calcineurin pathway and Hsp90 in combination with administration of conventional antifungal agents may have broad therapeutic potential in patients with fungal infections [16, 26]. Owing to the immunosuppressive properties of calcineurin inhibitors and the role of Hsp90 in controlling the calcineurin pathway, clinical use of a combination of TCR or 17AAG with triazole for treatment of *S. prolificans* infection would ultimately require a novel antifungal agent that selectively targets fungal stress pathways without having collateral effects on human immune cells.

We found evidence of synergy of PCZ and ICZ with TCR and 17AAG in *S. prolificans in vitro*, which is consistent with data on other fungal species [12, 16, 18, 21]. In addition, we used multiple markers of cell death to show that apoptosis is a mechanism of PCZ/ICZ- and TCR/17AAG-induced cell death. We corroborated the rate of apoptosis

in *S. prolificans* germlings using assays for detection of phosphatidylserine (PS) by annexin V-FITC, ROS accumulation by DHR-123 staining and decreased mitochondrial membrane potential by Rh123, DNA damage by TUNEL staining, and activation of caspase-like activity by CaspACE FITC-VAD-FMK. In each of the assays, apoptosis was evident at PCZ, ICZ, TCR, and 17AAG concentrations (0.125-0.250 µg/ml) that were below the MIC of triazoles. Taken together, these data indicate that PCZ or ICZ combined with TCR or 17AAG at concentrations below the MIC causes apoptosis in *S. prolificans* germlings.

We found that induction of apoptosis and the fungicidal activity of PCZ and ICZ in combination with TCR or 17AAG correlated with increased plasma and mitochondrial membrane disruption, PS externalization, DNA fragmentation, and ROS accumulation in *S. prolificans* germlings (isolates 1 and 2) (Tables 1, 2 and S1, Figures 1-4). Calcineurin activity is known to contribute to the fungicidal effects of Hsp90 inhibitors. 17-AAG in particular induces

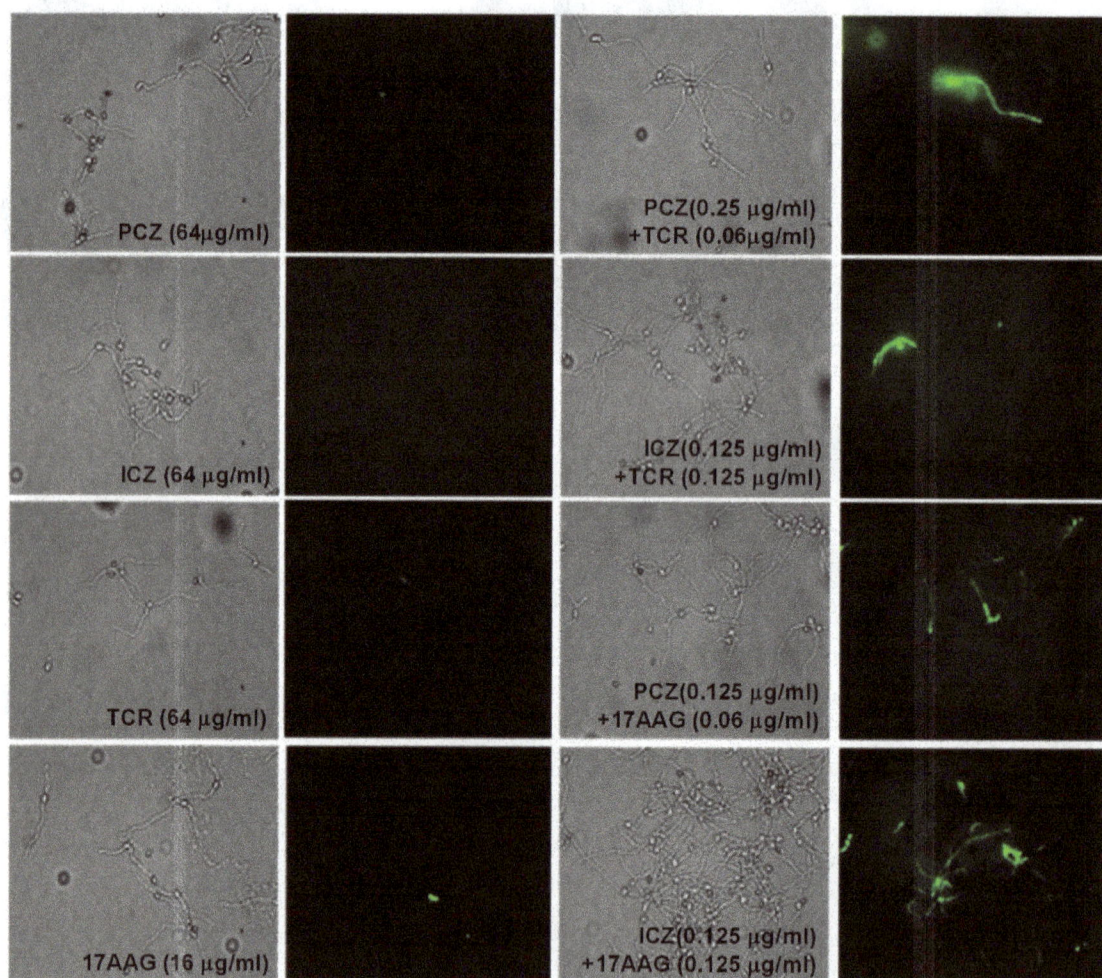

FIGURE 4: Detection of metacaspase (caspase-like) activity using CaspACE FITC-VAD-FMK probe in germlings of *S. prolificans* (isolate 1) treated with PCZ or ICZ in combination with TCR or 17AAG. Shown are fluorescent images of activated metacaspases of *S. prolificans* germlings treated with drugs alone and in combination with TCR or 17AAG.

apoptosis in colon carcinoma-derived cell lines (27), so determination of whether inhibition of Hsp90 can induce apoptotic cell death in fungi would be of interest. Dai *et al.* [28] demonstrated the role of Hsp90 in apoptosis in *C. albicans* and showed that inhibition of Hsp90 attenuated apoptosis by regulating the calcineurin pathway. Several fungi undergo apoptosis in response to antifungal treatment and various other stimuli [19]. Additional studies providing better understanding of fungal apoptotic pathways would promote the discovery of much-needed antifungal therapies.

Our results indicated that disruption of mitochondrial integrity by a combination of PCZ/ICZ with TCR or 17AAG induced apoptosis in *S. prolificans*. Our study in *S. prolificans* and studies in other fungi showed that 17AAG inhibits Hsp90, causing mitochondria-mediated apoptosis in rat histiocytomas [29]. Also, Shirazi and Kontoyiannis [18] showed that increased apoptosis after exposure to TCR was correlated with increased intracellular ROS accumulation in Mucorales. Furthermore, translocation of mitochondrial cyt *c* to the cytosol has led to binding of cyt *c* with apoptotic protease-activating factor to form a complex with caspase-9, resulting in caspase activation [23, 24, 30]. Release of cyt *c* requires an increase in mitochondrial membrane permeability during apoptosis [23]. As in Mucorales, our results in *S. prolificans* also demonstrated that, ROS formation, changes in $\Delta\Psi_m$, and cyt *c* release were associated with apoptosis [20-22]. Authors have also reported ROS-induced apoptosis in *A. nidulans*, *Fusarium oxysporum*, and *C. albicans* [31-33]. Sharon *et al.* [25] reported that apoptotic pathways in fungi seem to be mitochondrion-dependent, and can be powerful sources of superoxide radicals in cells undergoing miconazole and farnesol-induced apoptosis [34].

Authors have reported accumulating evidence that different stimuli induce different apoptotic pathways in yeasts and other fungi [35, 36]. In mammals, apoptosis is regulated by activation of caspases, which cleave specific substrates and trigger apoptotic death [37]. Now it is evident that caspase-like proteolytic activity may exist not only in multicellular organisms but also unicellular organisms, such as fungi. In the current study, we observed caspase-like activity in *S. prolificans* germlings upon exposure to PCZ or ICZ with TCR or 17AAG. Further studies are needed to demonstrate how proteases contribute to apoptotic fungal death.

In conclusion, we have shown for the first time that co-administration of inhibitors of the ergosterol biosynthesis pathways with an inhibitor of calcineurin or Hsp90 induces apoptosis in the recalcitrant fungus *S. prolificans*. This fungicidal synergistic interaction requires further study, as it may be a useful therapeutic strategy for infections caused by pathogenic fungi for which treatment options are extremely limited.

MATERIALS AND METHODS
Drugs
PCZ stock (5 mg/ml; Merck & Co., Inc.) was prepared in distilled water. ICZ (5 mg/ml; Janssen Pharmaceuticals), TCR (1 mg/ml; Medisca), and 17AAG (Sigma) stocks were prepared in ethanol, and aliquots were stored at -20°C in the dark until use.

Isolates and growth conditions
Three clinical isolates of *S. prolificans* (*S.p*-071507 [isolate 1], 071826 [isolate 2], and 674802 [isolate 3]) were grown on freshly prepared Sabouraud dextrose agar plates. After 48 h of incubation at 37°C, spores were collected and washed twice in sterile phosphate-buffered saline (PBS). The spores were then counted using a hemocytometer and stored at 4°C in PBS.

Susceptibility testing
Broth microdilution was performed according to the Clinical and Laboratory Standards Institute method [38]. Briefly, two-fold serial PCZ and ICZ dilutions were prepared in flat-bottomed 96-well microtiter plates (100 µl/well) in the presence or absence of TCR or 17AAG (0.060-0.125 µg/ml). Drug-free wells were used as controls. Each well was inoculated with 100 µl of freshly isolated *S. prolificans* spores (2-3 days old; 1×10^4 spores/ml) suspended in RPMI medium. After 48 h of incubation at 37°C, the MICs of PCZ and ICZ were determined visually as the lowest drug concentrations resulting in complete growth inhibition. To determine the minimum fungicidal concentrations (MFC) of PCZ and ICZ, an aliquot (20 µl) from each well that exhibited 100% growth inhibition was plated onto YPD agar (1% yeast extract, 2% peptone, 2% dextrose and 2% agar) plates. After 24 h of incubation at 37°C, the MFC was recorded as the lowest drug concentration at which no growth was observed.

For all of the wells of the microtiter plates that corresponded to MICs, the sum of the fractional inhibitory concentrations (ΣFIC) was calculated for each well using the equation ΣFIC = FICA + FICB = (CA/MICA) + (CB/MICB), in which MICA and MICB are the MICs of drugs A and B alone, respectively, and CA and CB are the concentrations of the drugs in combination, respectively, in all of the wells corresponding to an MIC. Synergy was defined as a ΣFIC of up to 0.5. Indifference was defined as a ΣFIC of at least 0.5 but no more than 4.0. Antagonism was defined as a ΣFIC greater than 4.0.

Viability assay
S. prolificans germlings (isolates 1 and 2) treated with TCR or 17AAG (0.060-0.125 µg/ml) along with PCZ (0.06-0.50 µg/ml) or ICZ (0.06-0.25 µg/ml) for 3 h were stained with DiBAC (Molecular Probes) as described previously [21, 22].

Annexin V-FITC–PI double staining of *S. prolificans* (isolates 1 and 2)
The apoptosis marker PS is located on the inner leaflet of the lipid bilayer of the cytoplasmic membrane and is translocated to the outer leaflet at the onset of apoptosis [39-41]. PS can be detected using staining with annexin V-FITC, which binds to it. Germlings treated with PCZ (0.06-0.50 µg/ml) or ICZ (0.06-0.25 µg/ml) in combination with TCR or 17AAG (0.060 and 0.125 µg/ml) were digested with a lysing enzyme mixture (0.25 mg/ml chitinase, 15 U of lyticase, and 20 mg/ml lysing enzyme; Sigma) for 3 h at 30°C. After digestion, *S. prolificans* protoplasts were stained with annexin V-FITC (BD Pharmingen) and PI at room temperature for 15 min and ob-

served under a fluorescence microscope to assess the externalization of PS as described previously [39].

Detection of intracellular ROS accumulation and $\Delta\Psi_m$ in S. prolificans germlings (isolates 1and 2)

ROS plays an important role as an early initiator of apoptosis in yeasts and other filamentous fungi [20-22]. The amount of ROS in S. prolificans germlings was measured using DHR-123 (Sigma) staining [20-22]. The mitochondrial membrane potential was assessed by staining with Rh-123 (Sigma), a fluorescent dye that diffuses in the matrix in response to electric potential as described [20-22]. Intracellular ROS levels and $\Delta\Psi_m$ in S. prolificans germlings were measured after treatment with PCZ (0.060-0.50 µg/ml) or ICZ (0.060-0.25 µg/ml) in combination with TCR and 17 AAG (0.060 and 0.125 µg/ml) for 3 h at 37°C using a fluorimetric assay with DHR-123 and Rh-123 staining [20-22, 42].

Measurement of DNA damage in S. prolificans (isolates 1 and 2)

DNA fragmentation, a characteristic of apoptosis, was detected in S. prolificans using a TUNEL assay. Germlings pretreated with PCZ (0.06-0.50 µg/ml) or ICZ (0.06-0.25 µg/ml) in combination with TCR or 17AAG (0.060 and 0.125 µg/ml) for 3 h at 37°C were fixed with 3.7% formaldehyde for 30 min on ice and digested using a lysing enzyme mixture. Enzyme-digested germlings were used to detect DNA fragmentation using a TUNEL assay as described by Madeo et al. [41]. The protoplasts were observed for fluorescence with excitation and emission wavelengths of 488 nm and 520 nm, respectively.

Detection of metacaspase activity using CaspACE FITC-VAD-FMK in S. prolificans germlings (isolates 1 and 2)

Active metacaspases in S. prolificans germlings were detected using CaspACE FITC-VAD-FMK (Promega) according to the manufacturer's instructions [20-22]. Briefly, germlings pretreated with PCZ (0.06-0.50 µg/ml) or ICZ (0.06-0.25 µg/ml) in combination with TCR or 17AAG (0.060 and 0.125 µg/ml) for 3 h at 37°C were collected, washed in PBS, resuspended in 10 µM FITC-VAD-FMK, and incubated again for 2 h at 30°C. Apoptosis in the S. prolificans germlings was inhibited in the presence of the caspase inhibitor z-VAD-FMK (Sigma) at final concentrations of 40 µM. After incubation, germlings were washed twice in PBS and observed microscopically for fluorescence with excitation and emission settings of 488 nm and 520 nm, respectively.

Statistical Analysis

For all assays, three independent experiments were performed in triplicate. Comparisons of multiple treatment groups were performed by using two-way analysis of variance with post-hoc paired comparisons using Dunnett's test. Calculations were made using the InStat software program (GraphPad Software). Two-tailed P values of less than 0.05 were considered statistically significant.

ACKNOWLEDGMENTS

D.P.K. acknowledges the Frances King Black Endowed Professorship for Cancer Research. This research was supported in part by the National Institutes of Health through MD Anderson's Cancer Center Support Grant P30CA016672.

CONFLICT OF INTEREST

D.P.K has received research support and honoraria from Pfizer, Astellas Pharma US, and Merck and Co. Inc, F.S reports no conflicts.

REFERENCES

1. Gosbell IB, Morris ML, Gallo JH, Weeks KA, Neville S, Rogers AH, Andrews RH, and Ellis DH (**1999**). Clinical, pathologic and epidemiologic features of infection with Scedosporium prolificans: four cases and review. **Clin Microbiol Infect** 5:672–686.

2. Cuenca-Estrella M, Ruiz-Diez B, Martinez-Suarez JV, Monzon A, and Rodriguez-Tudela JL (**1999**). Comparative in-vitro activity of voriconazole (UK-109,496) and six other antifungal agents against clinical isolates of Scedosporium prolificans and Scedosporium apiospermum. **J Antimicrob Chemother** 43:149–151.

3. Cortez KJ, Roilides E, Quiroz-Telles F, Meletiadis J, Antachopoulos C, Knudsen T, Buchanan W, Milanovich J, Sutton DA, Fothergill A, Rinaldi MG, Shea YR, Zaoutis T, Kottilil S, and Walsh TJ (**2008**). Infections caused by Scedosporium spp. **Clin Microbiol Rev** 21: 157e197.

4. Lamaris GA, Chamilos G, Lewis RE, Safdar A, Raad I, and Kontoyiannis DP (**2006**). Scedosporium infection in a tertiary care cancer center: a review of 25 cases from 1989–2006. **Clin Infect Dis** 43: 1580–1584.

5. Bader T, Schroppel K, Bentink S, Aqabian N, Kohler G, and Morschhauser J (**2006**). Role of calcineurin in stress resistance, morphogenesis, and virulence of a Candida albicans wild-type strain. **Infect Immun** 74: 4366–4369.

6. Cowen LE, and Lindquist S (**2005**). Hsp90 potentiates the rapid evolution of new traits: drug resistance in diverse fungi. **Science** 309: 2185–2189.

7. Cowen LE (**2008**). The evolution of fungal drug resistance: modulating the trajectory from genotype to phenotype. **Nat Rev Microbiol** 6: 187–198.

8. Reedy JL, Filler SG, and Heitman J (**2010**). Elucidating the Candida albicans calcineurin signaling cascade controlling stress response and virulence. **Fungal Genet Biol** 47: 107–116.

9. Steinbach WJ, Cramer RA, Perfect BZ, Asfaw YG, Sauer TC, Najvar LK, Kirkpatrick WR, Patterson TF, Benjamin DK, Heitman J, and Perfect JR (**2006**). Calcineurin controls growth, morphology, and pathogenicity in Aspergillus fumigatus. **Eukaryot Cell** 5: 1091–1103.

10. Wandinger SK, Richter K, and Buchner J (**2008**). The Hsp90 chaperone machinery. **J Biol Chem** 283: 18473–18477.

11. Kahan BD (**2003**). Timeline: Individuality: the barrier to optimal immunosuppression. **Nat Rev Immunol** 3: 831–838.

12. Dannaoui E, Afeltra J, Meis JF, and Verweij PE (**2002**). In vitro susceptibilities of zygomycetes to combinations of antimicrobial agents. **Antimicrob Agents Chemother** 46: 2708–2711.

13. Kontoyiannis DP, Lewis RE, Osherov N, Albert ND, and May GS (**2003**). Combination of caspofungin with inhibitors of the calcineurin pathway attenuates growth in vitro in Aspergillus species. **J Antimicrob Chemother** 51: 313–316.

14. Narreddy S, Manavathu E, Chandrasekar PH, Alangaden GJ, and Revankar SG (**2010**). In vitro interaction of posaconazole with calcineurin inhibitors and sirolimus against zygomycetes. **J Antimicrob Chemother** 65: 701–703.

Heat shock protein 90 and calcineurin pathway inhibitors enhance the efficacy of triazoles...

63

15. Lewis RE, Ben-Ami R, Best L, Albert N, Walsh TJ, and Kontoyiannis DP (**2012**). Tacrolimus enhances the potency of posaconazole against *Rhizopus oryzae* in vitro and in an experimental model of mucormycosis. **J Infect Dis** 5: 834–841.

16. Cowen LE, Singh SD, Kohler JR, Collins C, Zaas AK, Schell WA, Aziz H, Mylonakis E, Perfect JR, Whitesell L, and Lindquist S (**2009**). Harnessing Hsp90 function as a powerful, broadly effective therapeutic strategy for fungal infectious disease. **Proc Natl Acad Sci U S A** 106: 2818–2823.

17. Lu H, Zhu Z, Dong L, Jia X, Sun X, Yan L, Chai Y, Jiang Y, and Cao Y (**2011**). Lack of trehalose accelerates H_2O_2-induced *Candida albicans* apoptosis through regulating Ca^{2+} signaling pathway and caspase activity. **PLoS One** 6(1): e15808.

18. Shirazi F, and Kontoyiannis DP (**2013**). The calcineurin pathway inhibitor tacrolimus enhances the in vitro activity of azoles against Mucorales via apoptosis. **Eukaryot Cell** 12 (9): 1225-1234.

19. Ramsdale M (**2008**). Programmed cell death in pathogenic fungi. **Biochim Biophys Acta** 1783: 1369–1380.

20. Barbu EM, Shirazi F, McGrath DM, Albert N, Sidman RL, Pasqualini R, Arap W, and Kontoyiannis DP (**2013**). An antimicrobial peptidomimetic induces Mucorales cell death through mitochondria-mediated apoptosis. **PLoS One** 8 (10): e76981.

21. Shirazi F, and Kontoyiannis DP (**2013**). Mitochondrial respiratory pathways inhibition in *Rhizopus oryzae* potentiates activity of posaconazole and itraconazole via apoptosis. **PLoS One** 8 (5): e63393.

22. Shirazi F, Pontikos MA, Walsh TJ, Albert N, Lewis RE, and Kontoyiannis DP (**2013**). Hyperthermia sensitizes *Rhizopus oryzae* to posaconazole and itraconazole action through apoptosis. **Antimicrob Agents Chemother** 57 (9): 4360-4368.

23. Wu XZ, Chang WQ, Cheng AX, Sun LM, and Lou HX (**2010**). Plagiochin E, an antifungal active macrocyclic bis(bibenzyl), induced apoptosis in *Candida albicans* through a metacaspase-dependent apoptotic pathway. **Biochim Biophys Acta** 1800: 439–447.

24. Cho J, and Lee DG (**2011**). The antimicrobial peptide arenicin-1 promotes generation of reactive oxygen species and induction of apoptosis. **Biochim Biophys Acta** 1810: 1246–1251.

25. Sharon A, Finkelstein A, Shlezinger N, and Hatam I (**2009**). Fungal apoptosis: function, genes and gene function. **FEMS Microbiol Rev** 33: 833–854.

26. LaFayette SL, Collins C, Zaas AK, Schell WA, Betancourt-Quiroz M, Gunatilaka AA, Perfect JR, and Cowen LE (**2010**). PKC signaling regulates drug resistance of the fungal pathogen *Candida albicans* via circuitry comprised of Mkc1, calcineurin, and Hsp90. **PLoS Pathog** 6: e1001069.

27. Hostein I, Robertson D, DiStefano F, Workman P, and Clarke PA (**2001**). Inhibition of signal transduction by the Hsp90 inhibitor 17-allylamino-17-demethoxygeldanamycin results in cytostasis and apoptosis. **Cancer Res** 61: 4003–4009.

28. Dai B, Wang Y, Li D, Xu Y, Liang R, Zhao LX, CaoYB, Jia JH, and Jiang YY (**2012**). Hsp90 is involved in apoptosis of *Candida albicans* by regulating the calcineurin-caspase apoptotic pathway. **PLoS One** 7: e45109.

29. Taiyab A, Sreedhar AS, and Rao CM (**2009**). Hsp90 inhibitors, GA and 17AAG, lead to ER stress-induced apoptosis in rat histiocytoma. **Biochem Pharmacol** 78: 142-152.

30. Cho J, and Lee DG (**2011**). Oxidative stress by antimicrobial peptide pleurocidin triggers apoptosis in *Candida albicans*. **Biochemie** 93: 1873–1879.

31. Semighini CP, Hornby JM, Dumitru R, Nickerson KW, and Harris SD (**2006**). Farnesol-induced apoptosis in *Aspergillus nidulans* reveals a possible mechanism for antagonistic interactions between fungi. **Mol Microbiol** 59: 753–764.

32. Semighini CP, Murray N, and Harris SD (**2008**). Inhibition of *Fusarium graminearum* growth and development by farnesol. **FEMS Microbiol Lett** 279: 259–264.

33. Shirtliff ME, Krom BP, Meijering RA, Peters BM, Zhu J, Scheper MA, Harris ML, and Jabra-Rizk MA (**2009**). Farnesol-induced apoptosis in *Candida albicans*. **Antimicrob Agents Chemother** 53: 2392–2401.

34. Kobayashi D, Kondo K, Uehara N, Otokozawa S, Tsuji N, Yagihashi A, and Watanabe N (**2002**). Endogenous reactive oxygen species is an important mediator of miconazole antifungal effect. **Antimicrob Agents Chemother** 46: 3113–3117.

35. Carmona-Gutierrez D, Eisenberg T, Buttner S, Meisinger C, Kroemer G, and Madeo F (**2010**). Apoptosis in yeast: triggers, pathways, subroutines. **Cell Death Differ** 17: 763–773.

36. Hamann A, Brust D, and Osiewacz HD (**2008**). Apoptosis pathways in fungal growth, development and ageing. **Trends Microbiol** 16: 276-283.

37. Hengartner MO (**2001**). Apoptosis. DNA destroyers. **Nature** 412: 27–29.

38. Clinical and Laboratory Standard Institute. (**2008**). Reference method for broth dilution Antifungal Susceptibility Testing of filamentous fungi. Approved standard. 2nd edition. M38-A2, CLSI, Wayne, PA, USA.

39. Madeo F, Frohlich E, and Frohlich KU (**1997**). A yeast mutant showing diagnostic markers of early and late apoptosis. **J Cell Biol** 139: 729–734.

40. Madeo F, Frohlich E, Ligr M, Grey M, Sigrist SJ, Wolf DH, and Frohlich KU (**1999**). Oxygen stress: a regulator of apoptosis in yeast. **J Cell Biol** 145: 757–767.

41. Madeo F, Herker E, Maldener C, Wissing S, Lachelt S, Herlan M, Fehr M, Lauber K, Sigrist SJ, Wesselberg S, and Frohlich KU (**2002**). A caspase-related protease regulates apoptosis in yeast. **Mol Cell** 9: 911–917.

42. Wu XZ, Cheng AX, Sun LM, Sun SJ, and Lou HX (**2009**). Plagiochin E, an antifungal bis(bibenzyl), exerts its antifungal activity through mitochondrial dysfunction-induced reactive oxygen species accumulation in *Candida albicans*. **Biochim Biophys Acta** 1790: 770–777.

6

Multiple metabolic requirements for size homeostasis and initiation of division in *Saccharomyces cerevisiae*

Shivatheja Soma, Kailu Yang, Maria I. Morales and Michael Polymenis*

Department of Biochemistry and Biophysics, Texas A&M University, College Station, TX 77843, USA.
* Corresponding Author: Michael Polymenis, 300 Olsen Boulevard; College Station, TX 77843-2128, USA;
E-mail: polymenis@tamu.edu

ABSTRACT Most cells must grow before they can divide, but it is not known how cells determine when they have grown enough so they can commit to a new round of cell division. Several parameters affect the timing of initiation of division: cell size at birth, the size cells have to reach when they commit to division, and how fast they reach that size. We report that *Saccharomyces cerevisiae* mutants in metabolic and biosynthetic pathways differ in these variables, controlling the timing of initiation of cell division in various ways. Some mutants affect the size at birth, size at initiation of division, the rate of increase in size, or any combination of the above. Furthermore, we show that adenylate kinase, encoded by *ADK1*, is a significant determinant of the efficiency of size control mechanisms. Finally, our data argue strongly that the cell size at division is not necessarily a function of the rate cells increase in size in the G1 phase of the cell cycle. Taken together, these findings reveal an unexpected diversity in the G1 cell cycle phenotypes of metabolic and biosynthetic mutants, suggesting that growth requirements for cell division are multiple, distinct and imposed throughout the G1 phase of the cell cycle.

Keywords: START, elutriation, protein synthesis, growth rate, TDA1.

INTRODUCTION

In proliferating cells, the G1 phase of any given cell cycle lasts from the end of the previous mitosis until the beginning of DNA synthesis. In unfavorable growth conditions, *Saccharomyces cerevisiae* cells stay longer in G1, delaying initiation of DNA replication [1-6]. Subsequent cell cycle transitions are less sensitive to growth limitations, and their timing does not vary greatly, even if growth conditions worsen. Thus, differences in the length of G1 account for most of the differences in total cell cycle, or generation times, between the same cells growing in different media [1-6]. However, it is not clear how cells determine what growth requirements have to be met and how they are monitored so that cells can commit to a new round of cell division, at a point in late G1 called START. How nutrient, metabolic or other "growth" inputs activate the cell division machinery remains obscure. Historically, mutations in essential metabolic genes that arrest cell division at or before START have not received much attention. Such mutants were thought to resemble nutritionally limited cells because their growth in size was inhibited [6, 7]. Overall, it is not known if growth and metabolic requirements for cell division reflect hierarchical pathways, perhaps converging on a few specific biosynthetic needs. Alternatively, metabolic requirements for division may be multiple, distinct and imposed at different times from cell birth until commitment to a new round of cell division at START.

Decades ago, a relationship between the size or mass of a cell and the timing of initiation of DNA replication was shown from bacterial [8] to mammalian cells [9]. A newborn budding yeast cell is smaller than its mother is, and it will not initiate cell division until it becomes bigger [1, 2, 6]. These observations are consistent with the existence of a critical size threshold for initiation of division in yeast [10, 11]. How this critical size is set in response to metabolic cues, however, is unclear. It has been reported that the amount of G1 cyclins, which activate START, depends on both cell size and growth rate [12]. Based on single-cell analyses, a recent report suggested that the rate of size increase in the G1 phase determines the critical size [13]. In that scenario, slow growing cells would have a smaller critical size. Variations of G1 length among different mutants, or growth in different nutrients, could arise from differences in the size at which different mutants may enter and exit G1 and differences in the rate at which cells traverse G1. Measuring these variables (birth size, rate of size increase, critical size) in metabolic and biosynthetic mutants, and the extent to which any of these variables

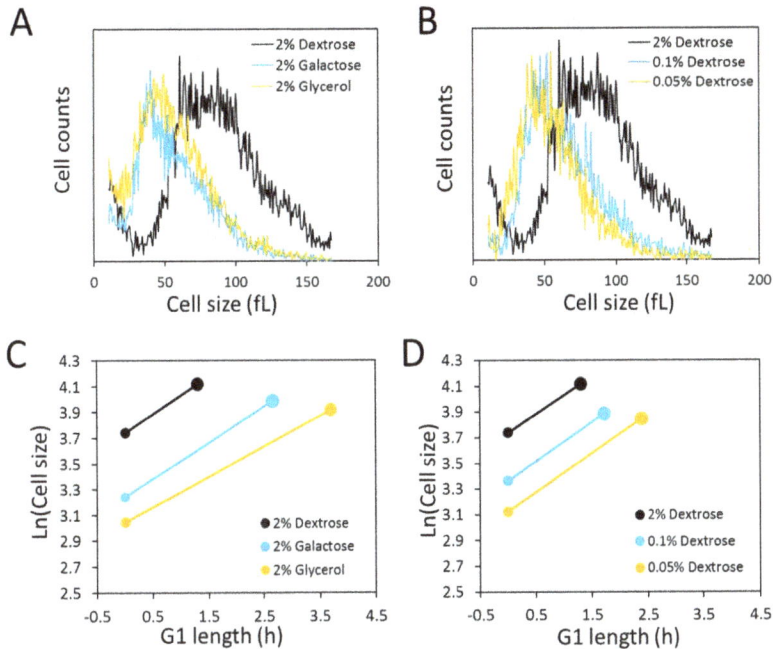

FIGURE 1. Nutrient control of size homeostasis and rate of size increase. (A) and (B) Cell size histograms of exponentially and asynchronously proliferating wild type diploid cells (strain BY4743), cultured in 1% $^w/_v$ yeast extract, 2% $^w/_v$ peptone and the indicated amount of the carbon source shown. The x-axis is cell size and on the y-axis is the number of cells. (C) and (D) Graphical representation of G1 variables in the growth conditions shown in A and B, using the values shown in Table 1. The x-axis is the calculated length of the G1 phase and the y-axis is the natural log of cell size at birth. The birth size at each condition is indicated with the smaller filled circle and the critical size with the larger filled circle. The length of the line connecting birth size with critical size is equal to the length of the G1 phase (T_{G1} in Table 1), and the slope of the line is equal to the specific rate of size increase (k in Table 1).

depends on one another is a necessary step towards deciphering the metabolic control of G1 progression and initiation of cell division.

Here, we identify nutritional requirements under which wild type cells adjust their critical size independently of the rate they increase in size in G1. We also show that cells lacking the kinase Tda1p specifically reduce their rate of size increase in response to different carbon sources, while their critical size remains unaffected, compared to wild type cells. Furthermore, from an analysis of mutants lacking enzymes of central metabolism or components of biosynthetic pathways, we identify several examples where birth size, rate of size increase, or critical size are affected independently of one another. Taken together, these results suggest that how cells set their critical size in not necessarily dependent on the rate cells increase in size in G1. Finally, the data we present are consistent with the notion that metabolic and biosynthetic requirements for division are multiple, distinct and imposed throughout G1, from cell birth until START.

RESULTS

Nutritional requirements and size homeostasis

As described in numerous reports in the past (e.g., see [14]), the poorer the carbon source in the medium used to culture *S. cerevisiae* cells, with galactose and glycerol being less favorable than glucose, the slower the population doubling time and the smaller the size of the cells. Recently, it was also reported that poorer carbon sources support a reduced rate of size increase in the G1 phase, causing a reduced size at the time of budding [13]. As expected, compared to the cells grown in glucose (2% $^w/_v$), cell size distributions of asynchronous cultures of diploid BY4743 cells shifted to the left in galactose (2% $^w/_v$) or glycerol (2% $^w/_v$) media (Fig. 1A). From these experiments, we also cal-

culated the daughter birth size ([15], and Materials and Methods). Using galactose or glycerol as a carbon source led to a significant reduction in daughter birth size (Table 1), compared to the birth size of cells cultured with glucose. We also found similar trends toward a smaller population mean and daughter birth size in cells cultured with glucose as a carbon source but with the concentration of glucose dropping from 2% $^w/_v$ to 0.1% $^w/_v$, or 0.05% $^w/_v$ (Figure 1B and Table 1).

To measure the rate of size increase and the critical size (defined here as the size at which half the cells in a synchronous population have budded) under all the above culture conditions, we then turned towards synchronous cultures obtained by centrifugal elutriation. An exponential mode of growth is thought to describe better the size increase of *S. cerevisiae* in G1 using single-cell photomicroscopy [10] or synchronous population monitoring by continuous volume measurements with a Coulter counter [16]. Therefore, to calculate the rate of size increase, we incorporated the obtained values of cell size measured with a channelyzer into an exponential function. Cells proliferating in media with galactose and glycerol as a carbon source had a reduced rate of size increase compared to cells proliferating in glucose-containing medium (Fig. 1C and Table 1). In accordance with Ferrezuelo et al [13], there was a concomitant decrease (≈15-20%) in the critical size of cells in galactose and glycerol media, compared to cells in glucose medium (Fig. 1C and Table 1). Note that there was also a substantial decrease in the daughter birth size (≈75-100%) in cells proliferating in galactose or glycerol media ((Figs. 1A, C and Table 1). Despite the reduced critical size, the smaller birth size and the reduced rate of size increase accounted for the much longer duration of the G1 phase in these carbon sources compared to growth in glucose (Fig. 1C and Table 1).

Similar experiments in media containing different concentrations of glucose revealed that limiting the concentration of glucose from 2% to 0.05% had no effect on the rate of size increase (Fig. 1D, and Table 1). Although the rate of size increase was unaffected, size homeostasis was altered significantly, to smaller daughter birth size (Fig. 1B and Table 1) and critical size (Fig. 1D and Table 1). The disproportionately greater reduction in birth size resulted in an increase in the length of the G1 phase in these cells (Fig. 1D and Table 1). Hence, at least within the range of glucose we used, these experiments provide an example where under physiological nutritional conditions critical size is set independently of the rate of size increase in G1.

The kinase Tda1p contributes to the control of the rate of size increase in response to carbon source

To examine further the relationship between the rate of size increase and critical size, we focused on Tda1p because we had previously shown that cells lacking Tda1p had a prolonged G1 phase without altered size homeostasis [17]. Here, we compared the birth size, rate of size increase and critical size of *TDA1/TDA1* vs. *tda1Δ/tda1Δ* cells in cultures with glucose, galactose or glycerol as a carbon source. We found that while in all carbon sources the daughter birth size and critical size of *tda1Δ* cells were similar to the corresponding values of *TDA1*[+] cells (Figs. 2A, C), their growth decreased disproportionately in galactose and glycerol media (Fig. 2B). These results suggest that Tda1p plays a role in the mechanisms that determine growth rate in response to carbon source, and that the putative control of critical size by the rate of size increase is not evident in cells lacking Tda1p.

Diverse G1 phenotypes of metabolic and biosynthetic mutants

Next, to test if growth rate can be modulated independently of critical size, we reasoned that mutations that alter growth rate ought to be examined for their effects on criti-

TABLE 1. G1 parameters in different nutrients of wild type and *tda1Δ/tda1Δ* cells[a].

Strain	Medium	Birth size (fL)	k (h^{-1})	Critical size (fL)	T_{G1} (h)
TDA1/TDA1	2% Dextrose	42.1±1.0[b]	0.28±0.01[c]	61.5±0.6[c]	1.35[d]
TDA1/TDA1	2% Galactose	25.5±0.4	0.27±0.01	53.9±4.7	2.77
TDA1/TDA1	2% Glycerol	21.0±1.3	0.23±0.01	50.3±2.1	3.80
TDA1/TDA1	0.1% Dextrose	28.9±0.8	0.29±0.01	48.8±3.1	1.81
TDA1/TDA1	0.05% Dextrose	22.7±1.3	0.29±0.02	46.7±4.2	2.49
tda1Δ/tda1Δ	2% Dextrose	41.8±1.3	0.28±0.02	60.4±1.4	1.31
tda1Δ/tda1Δ	2% Galactose	26.3±0.9	0.24±0.01	52.6±0.7	2.89
tda1Δ/tda1Δ	2% Glycerol	21.8±0.6	0.19±0.02	52.7±2.9	4.65

[a] The strains were in the homozygous diploid BY4743 background. They were examined in at least 3 independent experiments, and in each experiment a technical duplicate was evaluated. The cells were cultured in 1% $^w/_v$ yeast extract, 2% $^w/_v$ peptone and the indicated amount of the carbon source shown in each case.

[b] Birth size was calculated from the size distributions of exponentially proliferating asynchronous populations, as described previously [15]. The average of at least three independent measurements, with a technical duplicate for each measurement, and the associated standard deviation are shown in each case.

[c] The specific rate of size increase (k) and critical size were calculated from elutriated synchronous cultures as we described previously [17], assuming an exponential mode of growth. The average of at least three independent experiments and the associated standard deviation are shown in each case.

[d] These are G1 estimates from the formula: G1(hours)=Ln(Critical size/Birth size)/k. Note that these values reflect the G1 length of newborn daughter cells. For G1 length calculations, the errors (± sd) were not propagated.

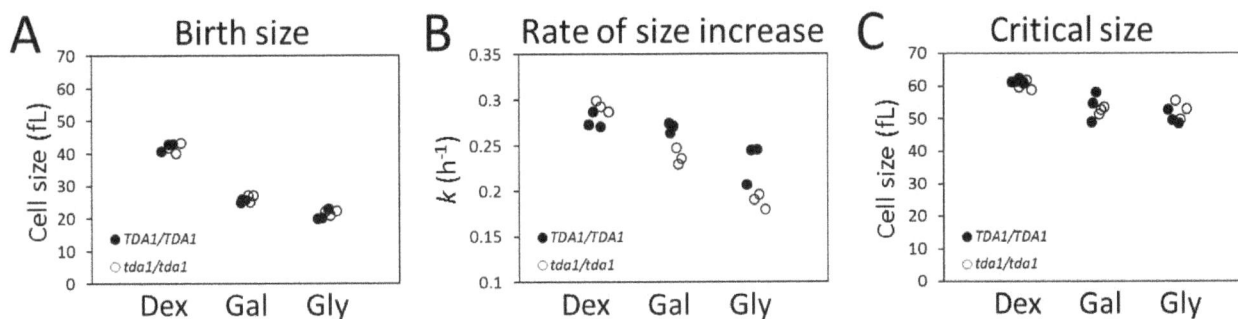

FIGURE 2: Loss of Tda1p reduces the rate of size increase but it does not affect the critical size or the birth size. (A) The birth size of *tda1Δ/tda1Δ* cells and their *TDA1/TDA1* counterparts (in the BY4743 background) was measured from three independent experiments in each case, cultured in 1% ^w/_v yeast extract, 2% ^w/_v peptone and 2% ^w/_v of either Dextrose (Dex), Galactose (Gal) or Glycerol (Gly). From synchronous, elutriated cultures (see Materials and Methods, and Table 1) of the same strains and media as in (A), we calculated the corresponding values for the specific rate of cell size increase constant k (in h^{-1}) shown in **(B)**, and the critical size values shown in **(C)**.

cal size. Consequently, to further test the deterministic role of the rate of size increase on setting the critical size, we analyzed a set of 18 mutants, each lacking a single gene product functioning in diverse metabolic and biosynthetic pathways. We chose 10 single gene deletions that impair different reactions of central metabolism (Fig. 3A). We also analyzed strains lacking the Rps0Bp and Rpl20Bp ribosomal proteins and three kinases, including Tda1p (Fig. 3A). Tor1p has a general, well-described pro-anabolic role [18], while Sch9p, the yeast S6 kinase ortholog, regulates ribosome biogenesis downstream of Tor1p [18]. Growth pathways must ultimately activate the cell division machinery. The components of the cell division machinery we included here are thought to function in the earliest steps of the switch that triggers cell division. Cln3p is an activating cyclin subunit of the major Cdk in yeast, Cdc28p (Cdk1p) [19]. Bck2p is a protein that functions in parallel with Cln3p to activate transcription of cell cycle genes [20]. Cells lacking both Cln3p and Bck2p are not viable [21]. Whi5p is a repressor of the late G1 transcriptional program. Commitment to division is marked molecularly by nuclear eviction of Whi5p [11]. We examined each gene deletion in the standard BY4741 background, using commercially available deletion strains [22]. To better interpret the obtained results and minimize artifacts due to suppressors accumulated during or after construction of these haploid strains in the BY4741 background, we independently constructed the same gene deletions in the Y7092 background ([23], see Materials and Methods). Together these two sets of mutants also enable construction of any desired double mutant combination in future experiments. The genotype of all the deletion strains in both backgrounds was verified, and we then measured their birth size, rate of size increase, and critical size in standard YPD medium with 2% ^w/_v dextrose (see Table 2).

Most mutants had an increased length of the G1 phase, consistent with the notion that growth and metabolism are required for initiation of cell division. Surprisingly, cells lacking ornithine decarboxylase, Spe1p, which catalyzes

the first step in polyamine biosynthesis [24], had a shortened G1 phase compared to wild type cells in both strain backgrounds (Table 2 and Fig. 3). The rate of size increase of *spe1Δ* cells was lower than the corresponding value of wild type cells (Table 2 and Fig. 3C). However, this effect was countered by the larger birth size (Table 2 and Fig. 3B) and slightly smaller critical size (Table 2 and Fig. 3D) of *spe1Δ* cells, shortening the overall length of the G1 phase. The only other case with a reduced length of the G1 phase was *whi5Δ* cells (see Table 2), which was expected given the well-established role of Whi5p as a repressor of START [11]. In agreement with previous results [25], the rate of size increase of *whi5Δ* cells was significantly reduced (Table 2 and Fig. 3C), but the shortened G1 of these cells is due to their greatly diminished critical size (Table 2 and Fig. 3D).

Our data suggest that although size homeostasis was affected in *bck2Δ* cells, displaying larger birth and critical sizes, the net effect in the duration of the G1 phase was minimal (Table 2). In the BY4741 background, the large birth size of *bck2Δ* cells was more pronounced, leading even to an apparent shortening of the G1 phase in that strain background (Table 2). Because cells lacking both Bck2p and Cln3p are not viable, Bck2p was thought to have a significant role at START, in parallel to Cln3p [21]. However, later work showed that Bck2p has a rather generic transcriptional role in early G1 [20], perhaps explaining the data we present here.

As mentioned earlier, the remaining mutants had a longer G1 phase. However, the mutants varied in their behavior not only quantitatively, but also qualitatively, due to different combinations of variables in each case accounting for the lengthening of the G1 phase (see Table 2 and Fig. 3). To determine if the single mutants represent distinct classes, we used principal component analysis. We found that three principal components accounted for >90% of the observed variance. We then used *k*-means clustering to assign the single mutants into the three clusters shown with different colors in Figs. 3B-D. Due to their very large birth and critical size, *adk1Δ* cells lacking adenylate kinase and *cln3Δ* cells were clustered together and separately from all other mutants (Fig. 3). The increase in the critical size we observed for *cln3Δ* cells was very significant, especially in the BY4741 background (Table 2). The value we obtained for *cln3Δ* cells (120 fL) was the average of two independent experiments (yielding 114 fL and 126 fL), and it was nearly three-fold higher than that of wild type cells (41.2 fL). In the Y7092 background the critical size of *cln3Δ* cells was still quite large (94 fL), two-fold greater than that of *CLN3+* Y7092 cells (46.9 fL; see Table 2). Hence, on average, loss of *CLN3* leads to a 2.5-fold larger critical size than wild type cells. Earlier elutriation experiments done in the W303 background (which is different from the S288c ancestry of the BY4741 and Y7092 strains) by the Nasmyth group ([26]; see Fig. 3 of that paper) also revealed that in YP Raffinose medium *cln3Δ* cells had to grow in size by more than 2.5-fold to reach the same budding index as wild type cells. It should be noted that the overall size in-

crease in *cln3Δ* cells is to a significant extent due to enlargement of the vacuolar compartment [27, 28]. The critical size enlargement of *adk1Δ* cells (60-70% in both the BY4741 and Y7092 strains; see Table 2) was substantial but not as dramatic as that of *cln3Δ* cells. Interestingly, however, despite their very large critical size, *adk1Δ* and *cln3Δ* cells displayed opposite trends in their rate of size increase. Compared to wild type cells, *adk1Δ* and *cln3Δ* cells had reduced vs. increased growth rate, respectively. Hence, this is one more example where the rate of size increase does not apparently determine the critical size.

Consistent with the known roles of the Tor1p and Sch9p kinases in ribosome biogenesis and protein synthesis [18], *tor1Δ* and *sch9Δ* cells clustered together with *rps0bΔ* and *rpl20bΔ* cells, characterized mostly by a significant reduction in both their birth size, and their rate of size increase (Fig. 3). Interestingly, cells lacking the major yeast hexokinase [29], Hxk2p, also clustered into the same group

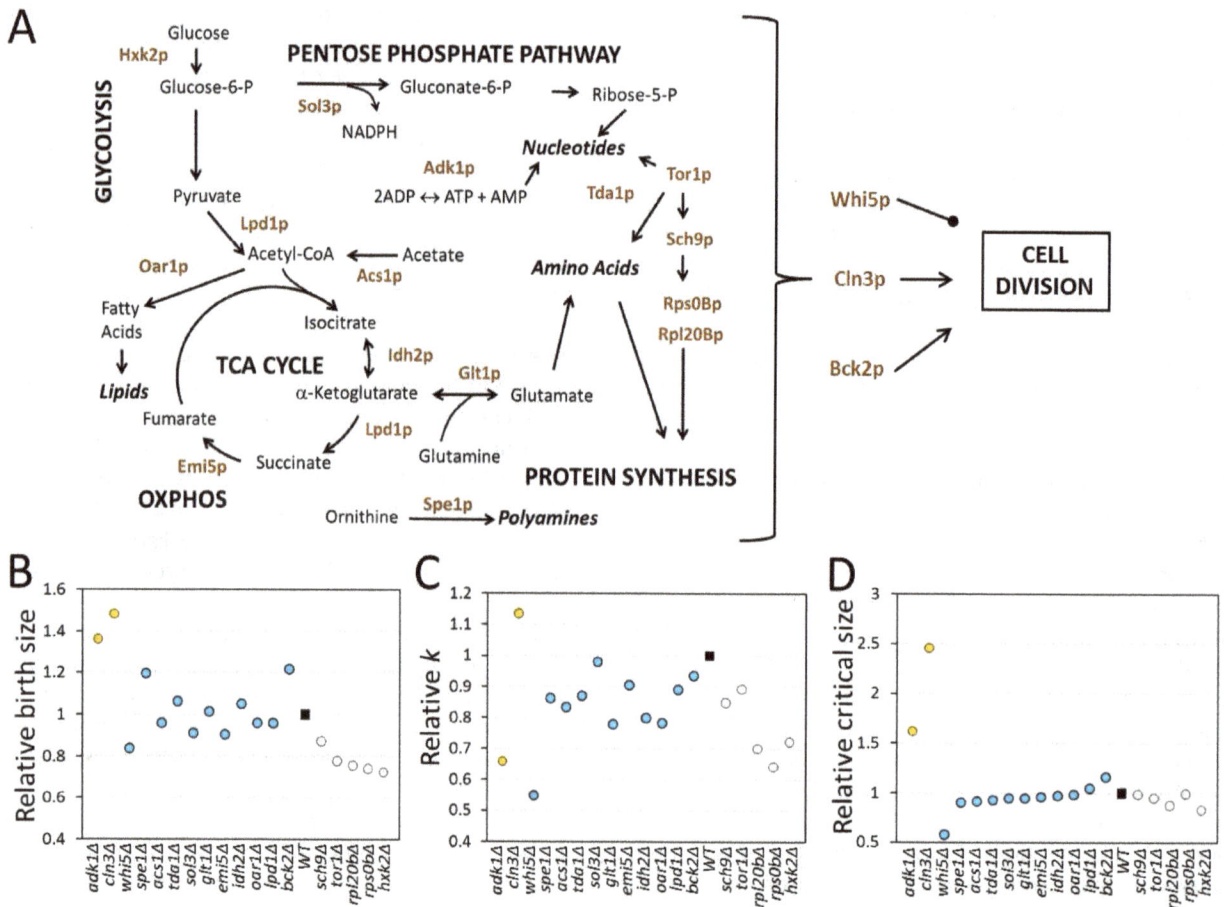

FIGURE 3: Diverse G1 phenotypes of metabolic and biosynthetic mutants. (A) Schematic overview of the reactions affected by the gene products we examined. This is a simplified view for clarity, missing numerous intervening reactions. (B) The birth size of each mutant shown on the x-axis was calculated for each deletion strain in the BY4741 and Y7092 background, shown in Table 2. For each gene deletion, the values from the two strain backgrounds were averaged, expressed relative to the corresponding value of the wild type, and shown on the y-axis. The gene deletions were group in three groups, based on principal component analysis and *k*-means clustering, using the R open source software, from the data shown in Table 2. The filled square is the wild type value. The relative specific rate of size increase (C) and critical size (D) are shown for each gene deletion, calculated and displayed as in (B), from the data shown in Table 2.

TABLE 2. G1 parameters of "growth" and cell cycle mutants[a].

ORF	Strain	Birth size (fL)	k (h^{-1})	Critical size (fL)	T_{G1} (h)
NA	BY4741	21.9	0.35	41.2	1.81
ACS1	acs1Δ::KanMX	20.2	0.3	39.1	2.17
ADK1	adk1Δ::KanMX	33	0.25	66.6	2.82
BCK2	bck2Δ::KanMX	28.3	0.35	49.9	1.64
CLN3	cln3Δ::KanMX	33.6	0.41	120	3.14
EMI5	emi5Δ::KanMX	18.8	0.29	40	2.62
GLT1	glt1Δ::KanMX	21.4	0.27	40.2	2.32
HXK2	hxk2Δ::KanMX	15.6	0.3	37.4	2.88
IDH2	idh2Δ::KanMX	24.1	0.27	40.1	1.92
LPD1	lpd1Δ::KanMX	21.1	0.33	46.1	2.34
OAR1	oar1Δ::KanMX	21.6	0.3	44.9	2.47
RPL20B	rpl20bΔ::KanMX	16.7	0.26	37.1	3.08
RPS0B	rps0bΔ::KanMX	15	0.25	44.5	4.37
SCH9	sch9Δ::KanMX	17	0.29	44.2	3.25
SOL3	sol3Δ::KanMX	22.4	0.3	40.8	1.98
SPE1	spe1Δ::KanMX	25.8	0.26	36.9	1.37
TDA1	tda1Δ::KanMX	24.4	0.33	42	1.64
TOR1	tor1Δ::KanMX	17.9	0.35	42.3	2.47
WHI5	whi5Δ::KanMX	17.3	0.19	25.5	2.04
NA	Y7092	19.5	0.35	46.9	2.53
ACS1	acs1Δ::NatMX	19.4	0.28	41.1	2.71
ADK1	adk1Δ::NatMX	23.7	0.21	76.3	5.54
BCK2	bck2Δ::NatMX	22.4	0.31	52.3	2.79
CLN3	cln3Δ::NatMX	27.9	0.39	94.1	3.14
EMI5	emi5Δ::NatMX	18.5	0.34	44.5	2.56
GLT1	glt1Δ::NatMX	20.4	0.27	43.3	2.77
HXK2	hxk2Δ::NatMX	14.3	0.2	35.2	4.48
IDH2	idh2Δ::NatMX	19.4	0.29	45.4	2.91
LPD1	lpd1Δ::NatMX	18.6	0.29	46.1	3.16
OAR1	oar1Δ::NatMX	18.2	0.25	41.5	3.31
RPL20B	rpl20bΔ::NatMX	14.6	0.23	39.8	4.35
RPS0B	rps0bΔ::NatMX	15.5	0.2	42.7	5.08
SCH9	sch9Δ::NatMX	18.9	0.3	42.5	2.73
SOL3	sol3Δ::NatMX	15.5	0.38	42.4	2.65
SPE1	spe1Δ::NatMX	23.7	0.34	42.8	1.73
TDA1	tda1Δ::NatMX	19.7	0.28	39.3	2.5
TOR1	tor1Δ::NatMX	14.3	0.28	40.8	3.81
WHI5	whi5Δ::NatMX	17.3	0.19	25.3	2

[a] The strains were examined in at least one experiment in each background, and in each case a technical duplicate was evaluated. The cells were cultured in 1% ʷ/ᵥ yeast extract, 2% ʷ/ᵥ peptone, 2% ʷ/ᵥ Dextrose. All the parameters were calculated as described in Table 1 and in the Materials and Methods section.

(shown in light gray in Figs. 3B-D). As in the previous examples we discussed, a drop in the rate of size increase was not necessarily correlated with a reduction in critical size (e.g., in sch9Δ cells and rps0bΔ cells, see Figs. 3C, D). Although chemically inhibiting the catalytic activity of Sch9p has been reported to decrease critical size [30], we found that a complete deletion of *SCH9* does not (see Table 2). Instead, sch9Δ cells are born small, explaining their small overall size phenotype reported previously [31], and sch9Δ cells also grow in size slower ([30], and Table 2). These phenotypes account for the long G1 of sch9Δ cells reported when *SCH9* was first identified ([32], and Table 2).

The remaining mutants all apparently clustered in the same group (shown in blue in Fig. 3). There was no clear and consistent relationship among the variables we examined. For example, spe1Δ cells were born large, had a reduced rate of size increase and yet their critical size was slightly smaller than normal (Table 2 and Fig. 3). Overall, at least within the set of mutants we examined, we did not observe a significant deterministic relationship between the rate of size increase and critical size. Furthermore, the behavior of metabolic mutants is not uniform at all, arguing for distinct and diverse growth requirements in the G1 phase of the cell cycle.

Adenylate kinase, Adk1p, has a major role in the efficiency of size control mechanisms

We next asked about the efficiency of size control mechanisms in the mutants we analyzed. Most of the mutants we queried had altered birth size, rate of size increase or critical size, and altered kinetics of G1 transit (see Fig. 3 and Table 2). However, despite altered size homeostasis and growth rate in many of these mutants, the question is

whether such mutants still maintain the mechanisms that enable them to grow enough in size in G1, at a level comparable to that of wild type cells, before initiating a new round of cell division. Plotting the logarithm of birth size against the relative growth in the G1 phase of the cell cycle displays the efficiency of cell size control mechanisms [10, 11]. In such plots, a slope of zero indicates no size control. Based on photomicroscopy of single cells, wild type budding yeast daughter cells display a slope of -0.7 in such graphs, indicative of an imperfect but still significant size control [10, 11]. We applied this methodology to all the synchronous daughter cell populations of the strains we examined (Fig. 4A). In some cases we noticed differences between the BY4741 and Y7092 backgrounds, also between the two parental strains. For this reason, for the data we show in Fig. 4A, the values we used were the average from the two strain backgrounds. The strains shown in red were clear outliers from the rest (Fig. 4A). This was not surprising for cln3Δ and whi5Δ cells, which represent known and prototypical examples of inefficient size control [10, 11]. Cells lacking Whi5p do not wait long enough, while cells lacking Cln3p wait too long, before initiating a new round of cell division, respectively. Interestingly, we found that cells lacking Adk1p also had very inefficient size control (Fig. 4A). Adenylate kinase is a key metabolic enzyme, catalyzing the rapid return of the adenine nucleotide pool to equilibrium if the level of ATP, ADP or AMP is altered. To our knowledge, this is the first time that a significant role for adenylate kinase in the efficiency of size control has been described, in any system. The remaining strains, however, displayed efficient size control, with an overall linear fit of a slope of -0.77 (Fig. 4A). This included bck2Δ cells, suggesting that loss of Bck2p does not appear

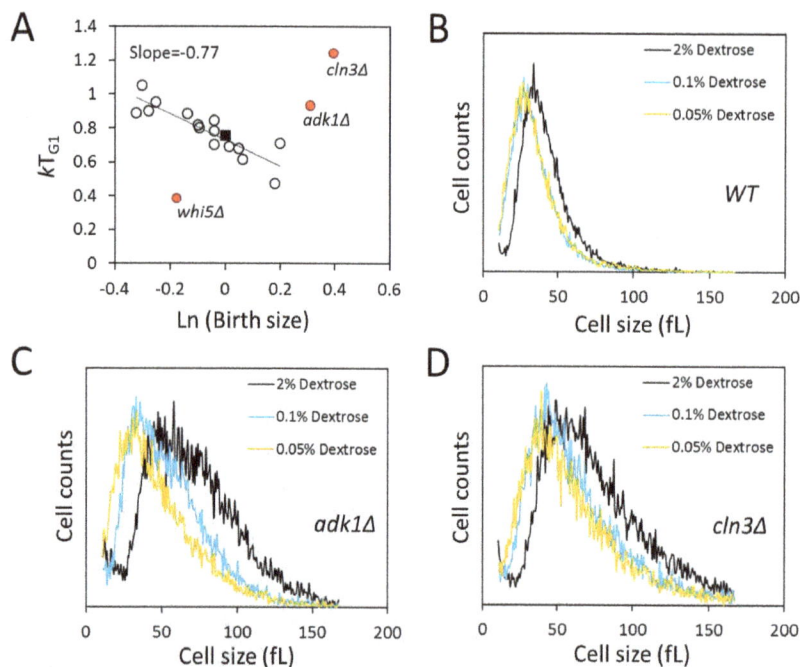

FIGURE 4: Adenylate kinase has a role in the efficiency of size control, but cells lacking Adk1p still adjust their size in response to nutrients. (A) In most mutants we examined (shown with open circles) size control operates efficiently. The filled square is the wild type value. On the x-axis is the natural logarithm of the normalized birth size values used in Fig. 3 (with the wild type values equal to one), which were obtained from the data in Table 2. These were plotted against their relative growth in size during the G1 phase (kT_{G1}, y-axis). The values we used were the average from the two strain backgrounds. The line is a linear fit obtained with the regression function of Microsoft Excel, from all the strains except those shown in red. **(B-D)** Cell size histograms of exponentially and asynchronously proliferating wild type haploid cells of the indicated genotype (all in the Y7092 background) cultured in 1% w/v yeast extract, 2% w/v peptone and the indicated amount of the carbon source shown. The x-axis is cell size and the y-axis is the number of cells.

to significantly compromise the efficiency of size control, at least not to the same extent that loss of Cln3p, Whi5p or Adk1p does (Fig. 4A). We conclude that although most metabolic and growth mutants we examined have altered G1 variables, they nonetheless displayed cell size control that appeared to be as efficient as that of wild type cells.

Next, we asked if cells lacking Adk1p still respond to the nutrient control of cell size homeostasis, with cells getting smaller as the concentration of glucose is reduced (see Fig. 1B). At all conditions tested $adk1\Delta$ and $cln3\Delta$ cells remained massively larger than their wild type counterparts (Fig. 4). Nonetheless, the size of these cells was also progressively reduced as glucose levels were reduced (Fig. 4C, D). Therefore, despite the inefficiency of size control in $adk1\Delta$ and $cln3\Delta$ cells, the nutrient control of size homeostasis is largely independent of Adk1p and Cln3p.

DISCUSSION

We discuss our results in the context of previous reports linking critical size with the rate of size increase, and we expand on the implications of our findings in regard with the cell cycle phenotypes of the mutants we examined.

Does the rate of size increase set the critical size for initiation of division? Although such a dependency may hold true in some cases, the following examples argue against that general rule proposed previously [13]. First, in wild type cells we identified nutritional interventions that completely dissociate these two parameters: Reducing the glucose content of the medium drastically reduces birth size and critical size, but not the rate of size increase (Fig. 1). Second, even when nutrients simultaneously reduce both the rate of size increase, and the critical size, we identified contexts that one parameter is disproportionately affected. Cells lacking Tda1p reduce their critical size to the same extent as wild type cells do when cultured in media with poorer carbon sources (Fig. 2C). However, at the same time there was a disproportionate reduction in the rate of size increase of $tda1\Delta$ cells (Fig. 2B), which did not lead to an even greater reduction in critical size (Fig. 2C). Third, the correlation between the rate of size increase and critical size was not evident at all from our analysis of many mutants (Fig. 3 and Table 2). For example, $adk1\Delta$, $whi5\Delta$ and $rps0b\Delta$ cells all had a similarly compromised rate of growth, yet their critical sizes diverged widely, from very large ($adk1\Delta$ cells), to very small ($whi5\Delta$ cells) or slightly larger than normal ($rps0b\Delta$). Note also that even in cases with similarly dispersed population size distributions, the rates of cell size increase could diverge in opposite directions. For example, the cell size distributions of $cln3\Delta$ and $adk1\Delta$ populations were remarkably similar, displaying large variance. However, the relationship between the rate of size increase and critical size trended in the opposite direction in the two mutants. The rate of size increase was moderately increased in $cln3\Delta$ cells (Figure 3 and Table 2). In contrast, the rate of size increase was significantly reduced in $adk1\Delta$ cells (Figure 3 and Table 2). Finally, it was recently reported that several aneuploid strains display a reduced rate of size increase and a larger than normal critical size

[33], providing yet another example of incongruence between the rate of size increase and critical size. Taken together, we think the physiological and genetic evidence we presented above argues against a general deterministic role of the rate of size increase in setting the critical size.

We think two major factors may account for the different conclusions we reached, compared to those of Ferrezuelo et al. [13], regarding the role of growth rate in setting the critical size. First, we examined a much broader array of nutritional and genetic interventions, including several gene products with distinct metabolic roles, under which the putative linkage between the rate at which cells increase in size and their critical size was clearly disrupted. Second, in accordance with previous reports [10, 16], we calculated the increase in size based on an exponential mode, which incorporates size differences.

Our data reveal a multitude of ways that biosynthetic and metabolic mutants affect G1 progression. Among the mutants we examined, their birth size, rate of size increase and critical size were affected in virtually any combination. The most straightforward interpretation of these findings is that growth requirements for cell division do not reflect a single hierarchical pathway. Instead, it is more likely that growth requirements are multiple and that they are imposed throughout the G1 phase. Metabolic mutants that affect cell division have not attracted much attention in the past. Historically, several screens for regulators of initiation of cell division interrogated cell size [30, 34-37]. Using only critical size mutants to identify mechanisms that determine the timing of initiation of cell division obviously does not allow the sampling of gene products that do not affect critical size. A prime such example is cells lacking Tor1p. The key growth signaling role of Tor1p is well established [18], yet the critical size of $tor1\Delta$ cells is normal (Fig. 3). The phenotype of other mutants is even more subtle. For example, loss of Tda1p affects neither the birth size nor the critical size, only the rate of size increase and that only on poor carbon sources (Fig. 2). Tda1p is a kinase of unknown function, originally identified as a modifier of topoisomerase I-induced DNA damage [37]. Interestingly, the human ortholog of Tda1p, NUAK1 (based on predictions by the P-POD program at http:// http://ppod.princeton.edu/), is an AMPK-related protein kinase, with roles in metabolic homeostasis, tumorigenesis and senescence [39]. Our findings regarding Tda1p's role in different carbon sources in yeast are perhaps consistent with a conserved growth-related function of these kinases.

Most of the loss-of-function metabolic and biosynthetic mutants we examined had a prolonged G1 phase. Despite the delay in G1 progression, in most cases size control was still operational (Fig. 4A). In other words, although in these strains G1 progression was delayed due to altered size homeostasis, rate of growth, or both, these mutants still "knew" how much they had to grow in size before initiating a new round of cell division. Interestingly, this was not the case for cells lacking Adk1p (in addition to mutants lacking the well-known START regulators Cln3p and Whi5p, see Fig. 4A). Adk1p has a central role in maintaining the equilibrium in the concentration of ATP, ADP and AMP in the cell.

The cellular energy charge, expressed as "half of the average number of anhydride-bound phosphate groups per adenine moiety" [40], is not altered by adenylate kinase. However, at any given value of energy charge, the actual proportions of ATP, ADP, and AMP, and the activity of any enzymes that respond to changes in those proportions are determined by adenylate kinase. Based on these considerations and the cell cycle phenotypes of adk1Δ cells we report, it is reasonable to speculate that Adk1p and possibly other proteins that respond to perturbations of nucleotide pools play a significant role in size control mechanisms.

In conclusion, with regard to when cells initiate division, our results suggest that "growth" mutants occupy a large and varied phenotypic space. Among the mutants we examined, there were numerous qualitative differences in G1 variables (see Table 2). The mechanistic basis of these differences is unclear at present and needs to await further experimentation. Nonetheless, a reasonable interpretation of these results is that metabolic and biosynthetic requirements for initiation of cell division are multiple and they are imposed throughout the G1 phase of the cell cycle. These growth requirements are likely the output of several metabolic pathways, acting perhaps in parallel. Defining the network arrangement of these metabolic outputs and how they impinge on the cell division machinery will illuminate the metabolic control of cell division.

MATERIALS AND METHODS
Strains and media
The strains we used are described in the corresponding Figures and Tables, and they were in the following backgrounds: BY4743 (MATa/α his3Δ1/his3Δ1 leu2Δ0/leu2Δ0 lys2Δ0/LYS2 MET15/met15Δ0 ura3Δ0/ura3Δ0); BY4741 (MATa his3Δ1 leu2Δ0 met15Δ0 ura3Δ0); Y7092 (MATα can1Δ::STE2pr-Sp_his5 lyp1Δ ura3Δ0 leu2Δ0 his3Δ1 met15Δ0) – a gift from Dr. C. Boone (Univ. of Toronto). Single gene deletion mutants in the BY4741 background were generated by the Yeast Deletion Project [22]. The corresponding deletions in the Y7092 background were constructed exactly as described previously [23]. The genotype of all strains was verified by PCR, for the presence of the replacement cassette and the absence of the corresponding ORF in each case.

Cell size measurements
All size measurements were performed with a Z2 Beckman Coulter Channelyzer. In experiments where the population mean was recorded, we measured the geometric mean of the cell size distribution of the population, using the AccuComp software package that accompanies the instrument. For each sample, we evaluated two cell dilutions, differing two-fold in the concentration of cells. The average of these two measurements was recorder for a single experiment. For birth size measurements of asynchronous, exponentially proliferating cells, we focused on the left of mode area of the distribution. From that area of the histogram we recorded the largest value of the 10% smallest cells, as we have described previously [15]. In all cases where we report either a birth size value or a population mean for any given strain and condition, we report the average of at least three independent experiments, each performed as we described above.

Elutriations
We used a J6-ME Beckman centrifugal elutriator to obtain highly synchronous, early G1 cells. We have described in detail elsewhere the methodology for elutriation experiments [17]. Briefly, for each experiment, we loaded a 250 ml culture in late exponential phase (for YPD cultures, the cell density was 2-5E+7 cells/ml) and collected the early G1 cell suspension at 2,400 rpm centrifugal speed and 40 ml/min, or 50 ml/min, pump speed for haploid, or diploid strains, respectively. The percentage of budded cells was typically 0-1%, and rarely exceeded 5%, with the exception of whi5Δ cells, which were typically 10-15% budded. The cell density was adjusted to about 1E+7 cells/ml. Every 20 min afterwards, aliquots were taken to measure the fraction of budded cells with a phase microscope and the cell size of the population as we described above. From these data, we plotted the natural logarithm of the cell size (y-axis) as a function of time (in hours, on the x-axis). From the slope of these graphs, we obtained the specific rate of size increase, k. We also plotted the fraction of budded cells (y-axis) as a function of cell size (in fL, on the x-axis). After the fraction of budded cells began to increase, we fitted the linear portion of these graphs to a straight line using the regression function of Microsoft Excel, and calculated the critical size for 50% of budded cells. To estimate the length of the G1 phase (T_{G1}), we used the exponential growth equation Ln(Critical size/Birth size)=kT_{G1} using the values of the corresponding variables calculated as we described above.

ACKNOWLEDGMENTS
This work was supported by a grant to M.P. from the National Science Foundation (MCB-0818248).

CONFLICT OF INTEREST
The authors declare no conflict of interest.

REFERENCES
1. Hartwell LH, Unger MW (**1977**). Unequal division in Saccharomyces cerevisiae and its implications for the control of cell division. **The Journal of cell biology** 75(2 Pt 1): 422-435.

2. Johnston GC, Pringle JR, Hartwell LH (**1977**). Coordination of growth with cell division in the yeast Saccharomyces cerevisiae. **Experimental cell research** 105(1): 79-98.

3. Carter BL, Jagadish MN (**1978**). Control of cell division in the yeast Saccharomyces cerevisiae cultured at different growth rates. **Experimental cell research** 112(2): 373-383.

4. Jagadish MN, Carter BL (**1977**). Genetic control of cell division in yeast cultured at different growth rates. **Nature** 269(5624): 145-147.

5. Blagosklonny MV, Pardee AB (**2002**). The restriction point of the cell cycle. **Cell cycle (Georgetown, Tex)** 1(2): 103-110.

6. Pringle JR, Hartwell, L.H. (**1981**). The Saccharomyces cerevisiae Cell Cycle. The Molecular and Cellular Biology of the Yeast Saccharomyces. Cold Spring Harbor Laboratory Press; pp 97-142.

7. Reed SI (**1980**). The Selection of S-Cerevisiae Mutants Defective in the Start Event of Cell-Division. **Genetics** 95(3): 561-577.

8. Donachie WD (**1968**). Relationship between cell size and time of initiation of DNA replication. **Nature** 219(5158): 1077-1079.

9. Killander D, Zetterberg A (**1965**). A quantitative cytochemical investigation of the relationship between cell mass and initiation of DNA synthesis in mouse fibroblasts in vitro. **Experimental cell research** 40(1): 12-20.

10. Di Talia S, Skotheim JM, Bean JM, Siggia ED, Cross FR (**2007**). The effects of molecular noise and size control on variability in the budding yeast cell cycle. **Nature** 448(7156): 947-951.

11. Turner JJ, Ewald JC, Skotheim JM (**2012**). Cell size control in yeast. **Current biology : CB** 22(9): R350-359.

12. Schneider BL, Zhang J, Markwardt J, Tokiwa G, Volpe T, Honey S, Futcher B (**2004**). Growth rate and cell size modulate the synthesis of, and requirement for, G1-phase cyclins at start. **Molecular and cellular biology** 24(24): 10802-10813.

13. Ferrezuelo F, Colomina N, Palmisano A, Gari E, Gallego C, Csikasz-Nagy A, Aldea M (**2012**). The critical size is set at a single-cell level by growth rate to attain homeostasis and adaptation. **Nature communications** 3:1012.

14. Tyson CB, Lord PG, Wheals AE (**1979**). Dependency of size of Saccharomyces cerevisiae cells on growth rate. **Journal of bacteriology** 138(1): 92-98.

15. Truong SK, McCormick RF, Polymenis M (**2013**). Genetic Determinants of Cell Size at Birth and Their Impact on Cell Cycle Progression in Saccharomyces cerevisiae. **G3** 3(9): 1525-1530.

16. Bryan AK, Engler A, Gulati A, Manalis SR (**2012**). Continuous and long-term volume measurements with a commercial Coulter counter. **PLoS one** 7(1): e29866.

17. Hoose SA, Rawlings JA, Kelly MM, Leitch MC, Ababneh QO, Robles JP, Taylor D, Hoover EM, Hailu B, McEnery KA, Downing SS, Kaushal D, Chen Y, Rife A, Brahmbhatt KA, Smith R, 3rd, Polymenis M (**2012**). A systematic analysis of cell cycle regulators in yeast reveals that most factors act independently of cell size to control initiation of division. **PLoS genetics** 8(3): e1002590.

18. Loewith R, Hall MN (**2011**). Target of rapamycin (TOR) in nutrient signaling and growth control. **Genetics** 189(4): 1177-1201.

19. Bloom J, Cross FR (**2007**). Multiple levels of cyclin specificity in cell-cycle control. **Nature reviews Molecular cell biology** 8(2): 149-160.

20. Bastajian N, Friesen H, Andrews BJ (**2013**). Bck2 acts through the MADS box protein Mcm1 to activate cell-cycle-regulated genes in budding yeast. **PLoS genetics** 9(5): e1003507.

21. Epstein CB, Cross FR (**1994**). Genes that can bypass the CLN requirement for Saccharomyces cerevisiae cell cycle START. **Molecular and cellular biology** 14(3): 2041-2047.

22. Giaever G, Chu AM, Ni L, Connelly C, Riles L, Veronneau S, Dow S, Lucau-Danila A, Anderson K, Andre B, Arkin AP, Astromoff A, El-Bakkoury M, Bangham R, Benito R, Brachat S, Campanaro S, Curtiss M, Davis K, Deutschbauer A, Entian KD, Flaherty P, Foury F, Garfinkel DJ, Gerstein M, Gotte D, Guldener U, Hegemann JH, Hempel S, Herman Z, et al. (**2002**). Functional profiling of the Saccharomyces cerevisiae genome. **Nature** 418(6896): 387-391.

23. Baryshnikova A, Costanzo M, Dixon S, Vizeacoumar FJ, Myers CL, Andrews B, Boone C (**2010**). Synthetic genetic array (SGA) analysis in Saccharomyces cerevisiae and Schizosaccharomyces pombe. **Methods in enzymology** 470:145-179.

24. Fonzi WA, Sypherd PS (**1987**). The gene and the primary structure of ornithine decarboxylase from Saccharomyces cerevisiae. **The Journal of biological chemistry** 262(21): 10127-10133.

25. Yang J, Dungrawala H, Hua H, Manukyan A, Abraham L, Lane W, Mead H, Wright J, Schneider BL (**2011**). Cell size and growth rate are major determinants of replicative lifespan. **Cell cycle (Georgetown, Tex)** 10(1): 144-155.

26. Dirick L, Böhm T, Nasmyth K (**1995**). Roles and regulation of Cln-Cdc28 kinases at the start of the cell cycle of Saccharomyces cerevisiae. **EMBO Journal** 14(19):4803-13.

27. Han BK, Aramayo R, Polymenis M (**2003**). The G1 cyclin Cln3p controls vacuolar biogenesis in Saccharomyces cerevisiae. **Genetics** 165(2):467-76.

28. Han BK, Bogomolnaya LM, Totten JM, Blank HM, Dangott LJ, Polymenis M (**2005**). Bem1p, a scaffold signaling protein, mediates cyclin-dependent control of vacuolar homeostasis in Saccharomyces cerevisiae. **Genes & development** 19(21):2606-18.

29. Walsh RB, Kawasaki G, Fraenkel DG (**1983**). Cloning of genes that complement yeast hexokinase and glucokinase mutants. **Journal of bacteriology** 154(2): 1002-1004.

30. Jorgensen P, Rupes I, Sharom JR, Schneper L, Broach JR, Tyers M (**2004**). A dynamic transcriptional network communicates growth potential to ribosome synthesis and critical cell size. **Genes & development** 18(20): 2491-2505.

31. Jorgensen P, Nishikawa JL, Breitkreutz BJ, Tyers M (**2002**). Systematic identification of pathways that couple cell growth and division in yeast. **Science** 297(5580): 395-400.

32. Toda T, Cameron S, Sass P, Wigler M (**1988**). SCH9, a gene of Saccharomyces cerevisiae that encodes a protein distinct from, but functionally and structurally related to, cAMP-dependent protein kinase catalytic subunits. **Genes & development** 2(5): 517-527.

33. Thorburn RR, Gonzalez C, Brar GA, Christen S, Carlile TM, Ingolia NT, Sauer U, Weissman JS, Amon A (**2013**). Aneuploid yeast strains exhibit defects in cell growth and passage through START. **Molecular biology of the cell** 24(9): 1274-1289.

34. Zhang J, Schneider C, Ottmers L, Rodriguez R, Day A, Markwardt J, Schneider BL (**2002**). Genomic scale mutant hunt identifies cell size homeostasis genes in S. cerevisiae. **Current biology : CB** 12(23): 1992-2001.

35. Carter BL, Sudbery PE (**1980**). Small-sized mutants of Saccharomyces cerevisiae. **Genetics** 96(3): 561-566.

36. Sudbery PE, Goodey AR, Carter BL (**1980**). Genes which control cell proliferation in the yeast Saccharomyces cerevisiae. **Nature** 288(5789): 401-404.

37. Prendergast JA, Murray LE, Rowley A, Carruthers DR, Singer RA, Johnston GC (**1990**). Size selection identifies new genes that regulate Saccharomyces cerevisiae cell proliferation. **Genetics** 124(1): 81-90.

38. Liu L, Ulbrich J, Muller J, Wustefeld T, Aeberhard L, Kress TR, Muthalagu N, Rycak L, Rudalska R, Moll R, Kempa S, Zender L, Eilers M, Murphy DJ (**2012**). Deregulated MYC expression induces dependence upon AMPK-related kinase 5. **Nature** 483(7391): 608-612.

39. Atkinson DE (**1968**). The energy charge of the adenylate pool as a regulatory parameter. Interaction with feedback modifiers. **Biochemistry** 7(11): 4030-4034.

7

Early manifestations of replicative aging in the yeast *Saccharomyces cerevisiae*

Maksim I. Sorokin[1,3], Dmitry A. Knorre[2,3], and Fedor F. Severin[2,3,]*

[1] Faculty of Bioengineering and Bioinformatics, Moscow State University, Vorobyevy Gory 1, Moscow, Russia
[2] Belozersky Institute of Physico-Chemical Biology, Moscow State University, Vorobyevy Gory 1, Moscow, Russia
[3] Institute of Mitoengineering, Moscow State University, Vorobyevy Gory 1, Moscow, Russia
* Corresponding Author: Fedor F. Severin, Belozersky Institute of Physico-Chemical Biology, Moscow State University, Vorobyevy Gory 1; Moscow 119992, Russia; E-mail: severin@belozersky.msu.ru

ABSTRACT The yeast *Saccharomyces cerevisiae* is successfully used as a model organism to find genes responsible for lifespan control of higher organisms. As functional decline of higher eukaryotes can start as early as one quarter of the average lifespan, we asked whether *S. cerevisiae* can be used to model this manifestation of aging. While the average replicative lifespan of *S. cerevisiae* mother cells ranges between 15 and 30 division cycles, we found that resistances to certain stresses start to decrease much earlier. Looking into the mechanism, we found that knockouts of genes responsible for mitochondria-to-nucleus (retrograde) signaling, *RTG1* or *RTG3*, significantly decrease the resistance of cells that generated more than four daughters, but not of the younger ones. We also found that even young mother cells frequently contain mitochondria with heterogeneous transmembrane potential and that the percentage of such cells correlates with replicative age. Together, these facts suggest that retrograde signaling starts to malfunction in relatively young cells, leading to accumulation of heterogeneous mitochondria within one cell. The latter may further contribute to a decline in stress resistances.

Keywords: yeast, aging, stress resistance, retrograde signaling.

INTRODUCTION

The budding yeast *Saccharomyces cerevisiae* is an example of a unicellular organism with markedly asymmetrical division: newly emerged daughter cells are normally significantly smaller than the mothers (see [1]). However, the asymmetry is not limited to the differences in volume: mother-to-daughter transport of a specific set of mRNAs prevents mating type switching by repression of the *HO* promoter in the daughter cells [2]. Further, it was shown that carbonylated or misfolded proteins and extrachromosomal rDNA circles [3] are retained in the mother cells. Recently, we showed that such transport is necessary to prevent high stress susceptibility of the daughter cells [4]. At the same time, multiple rounds of asymmetrical divisions result in subsequent decrease in viability in replicatively old cells. As a result, the chance of a spontaneous death of the yeast cell increases with each division cycle after production of approximately 15 buds [5,6]. One of the most pronounced consequences of asymmetrical redistri-

bution of cytoplasmic contents appears to be the mitochondrial morphology. In yeast, abnormal morphology of the mitochondrial network and increased reactive oxygen species production can be observed already after 6-8 divisions [7]. The latter is not surprising because the mother/daughter redistribution of mitochondria is not random: the mother cell tends to retain mitochondria with higher superoxide level [8], while mitochondria with intact aconitase (a Fe-S cluster-containing enzyme that is sensitive to ROS) are preferentially transported to the bud [9].

Here we report a number of changes in cellular physiology (i.e. a decrease in stress resistances) that occur even earlier than 6-8 cell divisions. We also found that the probability of harboring a mitochondrial fragment with significantly different transmembrane potential than in the rest of the mitochondrial network within same cell increases with each cell cycle. These early age-dependent changes might significantly affect the stress resistances of the cells.

RESULTS

Stress resistances change early during replicative aging

First, to test the age-dependence of stress resistances, exponentially growing yeast cells were subjected to a number of treatments (Table 1) and then stained with propidium iodide (PI) to visualize dead cells and with calcofluor white (CW) to visualize birth and bud scars. Then we counted the scars for each dead and live cell and calculated the percentage of PI positive cells for each aging class, where age = 1 corresponds to virgin daughter cells (with birth scars only), age = 2 corresponds to mother cells (with one birth scar and one bud scar), etc. As the proportion of cells with more than four scars did not exceed 6% of the total (Figure S1), we pooled them into one group (≥5) and analyzed it as a single aging class. We found that in case of heat shock the dead/alive cell ratios for different aging classes were statistically different (Figure 1A). The same results were obtained for hyperosmotic (Figure 1A) and acidic stresses (Figure S2). However, in the case of oxidative stress (menadione) and for stresses induced by high concentrations of ethanol or butanol, there were no statistically significant differences (Figure S2).

Virgin daughter cells have some unique properties. In particular, daughter cells have increased duration of G1 phase of cell cycle compared to mother cells [10] and demonstrate significantly lower resistances to acetic acid stress and heat shock [4]. Therefore, we excluded the daughter cells (age = 1) from the calculations and performed the statistical Kruskal–Wallis test for the mother cells only. It was found that for heat shock and hyperosmotic stress the p value is below 0.01 (Figure 1A), whereas for acetic acid stress the p value was 0.29 (Figure S2).

TABLE 1. Types of stresses used in this study.

Type of stress	Chemical agent (final concentration)	Duration
Osmotic stress	NaCl (2.5 M)	30 min
Alcohol stress	Ethanol (20%)	60 min
	Butanol (1%)	16 hours
Heat shock (47°C)	-	30 min
Oxidative stress	Menadione (1 mM)	2 hours
Acidic stress	Acetic acid (180 mM, media pH 3.5)	2 hours

Obviously the ability to extrude propidium iodide is not the only criterion of stress resistance. Another is the ability to start growing after the return to stress-free conditions. Thus, we decided to test whether this ability depends on replicative age. To do that, we synchronized yeast cells with nocodazole and stained the cell walls with FITC-ConA (Concanavalin A labeled with green fluorescent dye). FITC-ConA irreversibly binds to the cell thus allowing to distinguish between the old cells (stained) and the newly emerged (unstained ones). Then the cells were subjected to a mild stress: +42°C for 30 minutes. After removal of the stress, we incubated the cells at 30°C for 90 minutes, stained them with calcofluor white to visualize ages, and measured the sizes of the newly formed (FITC-ConA free) buds (Figure 1B). It appeared that cells with two scars have

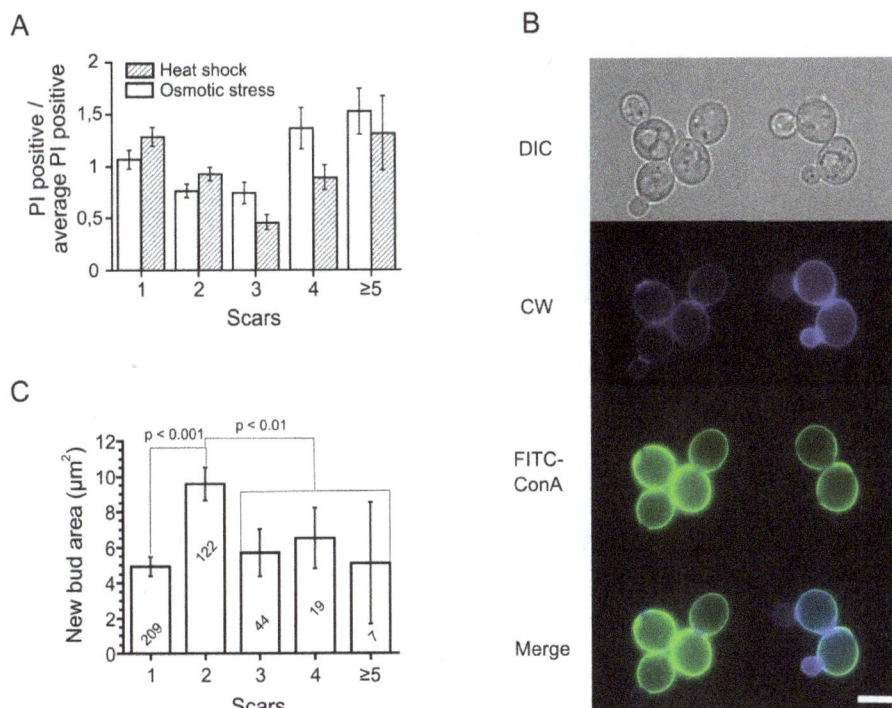

FIGURE 1: Stress resistances and regrowth of yeast cells depend on their replicative age. (A) The data are presented as percentages of dead cells in a particular aging class divided by the average percentages of dead cells in this experiment. Values higher than one show that the percentage of dead cell in this aging class is above average, and values below one represent the relatively resistant aging classes. p < 0.01 for all treatments, Kruskal-Wallis test. **(B, C)** Microscopy and quantification of regrowth. In C the exact numbers of cells analyzed are shown in each column. Bar = 5 μm.

larger buds compared to the other age classes (Figure 1C).

Together these data support the idea that the stress resistances of yeast cells which represent the major part of the population (ages 1-5) vary depending on their replicative age.

Mitochondrial function and stress resistance during replicative aging

What determines these early age-dependent changes in the stress resistances? It is known that mitochondrial morphology of the mother cells starts to change already after five or six cell cycles [7]. Thus, we decided to look in more detail at age-dependent changes in mitochondria. We used yeast cells expressing mitochondria-targeted GFP [11] and stained them with tetramethylrhodamine (TMR). TMR is a lipophilic cation that accumulates in mitochondria depending on their $\Delta\Psi$. We noticed that the mother cells have much higher levels of heterogeneity within a cell than the daughters and that the level of heterogeneity positively correlates with replicative age (Figures 2 A, B).

What are the possible links between the mitochondrial heterogeneity and the stress resistances? Malfunctioning mitochondria in yeast induce specific transcriptional changes in the nuclei that act to normalize the mitochondrial activities. Therefore, our finding that aging yeast cells contain heterogeneous mitochondria (meaning that some of them might be not in optimal conditions) may reflect the fact that this signaling pathway in such cells is compromised. The mitochondria-to-nucleus signaling is driven by

transcription factors of the retrograde pathway: *RTG1*, *RTG2*, and *RTG3* [12]. We found that the deletion of *rtg1* or *rtg3* affects old cells but not the young ones when subjected to severe heat stress (30 minutes at 47°C, Figure 2C). This indicates that cells with replicative age higher than four are more reliant on retrograde signaling.

We also compared the effects of preconditioning on the stress responses of cells of different ages. It appeared that the preconditioning increased survival of the young cells to such an extent, that proportion of PI-positive cells became similar between the cells with 1 to 4 scars (Figure 2D). The viability of the preconditioned cells with ≥5 scars also seemed to be higher than in control. However, the proportion of PI-positive cells of this aging class appeared to be different from the one for younger ages (Figure 2D). This supports the idea that in the old cells the ability to adapt to a changing environment is diminished.

Does this mean that the mitochondrial heterogeneity is a sign of a general decline of the cell? Recently we showed that differentiation of yeast cells requires two subpopulations of mitochondria [13]. This result implies that the ability of cells to differentiate in response to environmental signals increases with replicative age. To test this, we induced meiosis in diploid cells by plating them on solid medium containing 2% potassium acetate. As predicted, it appeared that the ability of cells to form spores positively correlates with age (Figure 3); at the same time, the diploid cells show similar stress resistance profile as haploids cell (Figure S3). It should be noted, however, that the latter

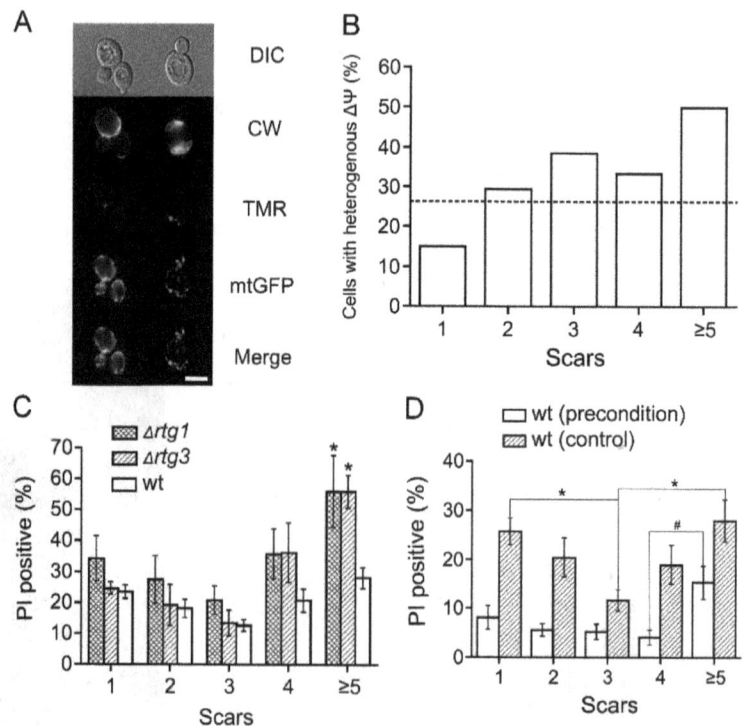

FIGURE 2: (A) Mitochondrial heterogeneity within a cell increases with age. Representative images of cells with uniform and heterogeneous subpopulations of mitochondria. Cells expressing mitochondrial GFP (mitoGFP) were stained with tetramethylrhodamine (TMR) and calcofluor white (CW). Bar = 5 μm. (B) Quantification of the results from (A). 114 cells were analyzed from 3 independent experiments. The difference between actual and expected (average proportion of cells with heterogeneous potential among the whole population) values is significant according to the chi-squared test (p < 0.05). Positive correlation is also significant according to Kendall's (p-value = 0.05, tau = 0.8) and Spearman's (p-value = 0.037, rho = 0.9) tests. (C) Knockouts of retrograde signaling genes decrease the survival of yeast cells with replicative ages above four. Cell were subjected to heat shock and then stained with propidium iodide (PI) and calcofluor white (see *Materials and methods*). The scars were calculated for each individual cell. The percentage of PI-positive (dead) cells was plotted as a function of replicative age (numbers of scars). * indicates p < 0.05 when compared to wt, Wilcoxon test. (D) Preconditioning decreases the difference in heat shock resistance between daughter cells and young mother cells. Yeast cells were incubated for 30 min at 37°C and then stressed with heat shock (47°C). Then the cells were stained with propidium iodide (PI) and calcofluor white (see *Materials and methods*). The percentages of PI-positive (dead) cells were plotted as a function of replicative age (numbers of scars). * indicates p < 0.01, # indicates p = 0.057.

effect could be due to correlation of sporulation efficiency and cell size [14], which also strongly correlates with cell age (Figure S4).

DISCUSSION

Our data suggest that the response to environmental challenges of yeast starts to change early during their replicative aging. In particular, these changes lead to a decreased resistance to certain stresses of the cells with 4-5 scars. This implies that under natural (as opposed to laboratory) conditions the cells are unlikely to reach replicative age of twenty or thirty. Does this mean that most of the works on yeast replicative aging (which were conducted under laboratory conditions) were studying laboratory artifacts? It was found that similar groups of genes regulate replicative lifespan in yeast and longevity in worms and mice (histone deacetylases, antioxidant enzymes, etc., see for review [15]). At the same time, similarly to yeast, functional decline of humans also starts at relatively early age. For example, a decrease in the number of immune T-cells in humans starts at the age of 15 years [16]. Therefore, our data suggest that yeast can serve as a model to study not only lifespan-regulating mechanisms, but also the mechanisms of early age-associated functional decline. Interestingly, fission yeast display similar aging phenotype. Recently, it was reported that while *S. pombe* does not seem to age replicatively under standard laboratory conditions, stresses force them to do so [17].

Our data also show that not all stress resistances show sharp age-dependence. Moreover, the ability for spore formation seems to increase during early aging. In other words, early aging in yeast displays features of both regular aging and differentiation. At the same time, it is well known that heterogeneity of individual cells in clonal microbial populations helps them to survive in randomly changing environment. For instance, a small fraction of bacterial cells (persisters) show slower growth rate and, at the same time, increased antibiotic resistance [18]. Early age-dependent differentiation of yeast cells has also been already reported. In the stationary phase of growth, *S. cerevisiae* forms two types of cells: quiescent and nonqui-

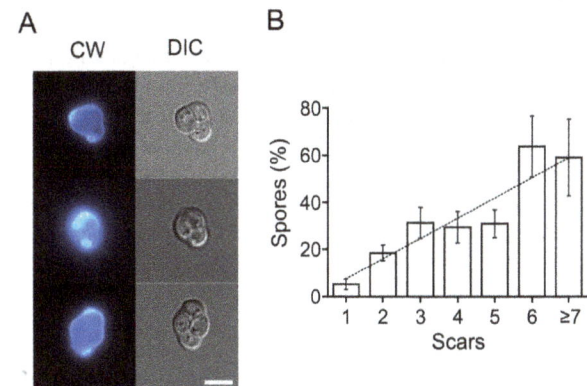

FIGURE 3: Sporulation ability is positively correlated with the number of completed cell cycles. (A) Representative images of yeast tetrads stained with calcofluor white (CW) and visualized under differential interference contrast (DIC). Bar = 5 μm. (B) Quantification of results. 1183 cells analyzed from 8 separate experiments. Kendall's tau = 0.54, Spearman's rho = 0.65, p < 0.001 both tests.

escent [19,20]. Quiescent cells are represented mainly by virgin daughters; they are arrested in the G1 phase of the cell cycle and have a higher chance to survive in the stationary phase. The nonquiescent cells are continuing to proliferate. Our data suggest that such differentiation is more complex than simply mother-daughter differences. Possibly, there is a distinct number (higher than two) of aging classes of cells, with each class showing a unique set of resistances to harsh conditions.

MATERIALS AND METHODS
Strains and growth conditions
In this study, we used strains of W303 genetic background strain (Table 2). If not indicated, exponentially growing yeast cells were prepared as following: prior to the treatments the cells were grown overnight at 30°C on solid YPD medium prepared according to [21] and transferred in liquid YPD for 3.5-4 hours until the optical density (OD^{550}) of the suspension reached 0.2 (4 x 10^6 cells per ml).

TABLE 2. Yeast strains used in this study.

Strain	Genotype	Source or reference
W303	*MATa ade2-101 his3-11 trp1-1 ura3-52 can1-100 leu2-3*	[22]
W303 (2n)	*ade2-101/ade2-101 his3-11/his3-11 trp1-1/trp1-1 ura3-52/ura3-52 can1-100/can1-100 leu2-3/leu2-3*	this study, mating W303 with isogenic mat alpha strain
Δrtg1	*MATa ade2-101 his3-11 trp1-1 ura3-52 can1-100 leu2-3 rtg1::KanMX4*	[13]
Δrtg3	*MATa ade2-101 his3-11 trp1-1 ura3-52 can1-100 leu2-3 rtg3::KanMX4*	[13]

Stress induction

Cell suspensions in liquid YPD were subjected to various types of stresses: 2.5 M NaCl for 30 minutes, 20% ethanol for 1 hour, 1% butanol for 16 hours, heat shock (47°C) for 30 minutes, 1 mM menadione for 2 hours, or 180 mM acetic acid (pH of the medium was set to 3.5) for 2 hours. The types of stresses used in this study are summarized in Table 1. Incubation of the cells for 30 minutes at 37°C was used as preconditioning for heat shock. Sporulation of diploid W303 cells was induced on solid potassium acetate-containing medium (2% potassium acetate, 2% agar) for 3 days.

New bud formation

The cells were synchronized with nocodazole (10 μg/ml), washed 4 times with YP, then moved to YPD containing FITC-conjugated Concanavalin A (FITC-ConA) and incubated at 30°C for 10 minutes and then at 42°C for 30 minutes. Then the cells were washed 2 times with YPD and incubated at 30°C in YPD for 90 minutes. After 90 minutes, the cells were transferred to YP media containing calcofluor white, as described in the *Microscopy* section.

Microscopy

After the stresses the cells were washed with YP containing Propidium Iodide (PI, 1 μg/ml) and calcofluor white (CW, 5 μM). PI-positive cells were counted as dead. Calcofluor white was used to visualize scars on the cell surface (birth scar and bud scars) for replicative age measurements. Cell viability, replicative age, and percent of spores were determined manually using an Olympus BX51 microscope. Photographs were taken with a DP30BW CCD camera. Cell size was measured manually using the ImageJ program.

Statistical analysis

All data are presented as average and standard error. Wilcoxon signed ranked unpaired test and Kruskal-Wallis test were used to compare datasets from different strains or conditions with the R software package. All viability experiments were performed independently 4 times, and each time at least 90 cells were analyzed unless indicated otherwise. When the proportions of the PI-positive cells were low, we counted the age distribution of PI-positive cells only and then extrapolated the result using the average survival in this experiment.

ACKNOWLEDGMENTS

We are very grateful to Svyatoslav Sokolov, Vitaly Kushnirov, Boris Feniouk, and Peter Kamenski for a valuable discussion of our work. This work was supported by the Russian Foundation for Basic Research grant 12-04-01412-a.

CONFLICT OF INTEREST

The authors declare no conflict of interest.

REFERENCES

1. Di Talia S, Wang H, Skotheim JM, Rosebrock AP, Futcher B, and Cross FR (**2009**). Daughter-specific transcription factors regulate cell size control in budding yeast. **PLoS Biol** 7(10): e1000221.

2. Bobola N, Jansen RP, Shin TH, and Nasmyth K (**1996**). Asymmetric accumulation of Ash1p in postanaphase nuclei depends on a myosin and restricts yeast mating-type switching to mother cells. **Cell** 84(5): 699–709.

3. Sinclair DA and Guarente L (**1997**). Extrachromosomal rDNA circles–a cause of aging in yeast. **Cell** 91(7): 1033–1042.

4. Knorre DA, Kulemzina IA, Sorokin MI, Kochmak SA, Bocharova NA, Sokolov SS, and Severin FF (**2010**). Sir2-dependent daughter-to-mother transport of the damaged proteins in yeast is required to prevent high stress sensitivity of the daughters. **Cell Cycle** 9(22): 4501–4505.

5. Ashrafi K, Sinclair D, Gordon JI, and Guarente L (**1999**). Passage through stationary phase advances replicative aging in Saccharomyces cerevisiae. **Proc Natl Acad Sci USA** 96(16): 9100–9105.

6. Kennedy BK, Austriaco NR Jr, and Guarente L (**1994**). Daughter cells of Saccharomyces cerevisiae from old mothers display a reduced life span. **J Cell Biol** 127(6 Pt 2): 1985–1993.

7. Lam YT, Aung-Htut MT, Lim YL, Yang H, and Dawes IW (**2011**). Changes in reactive oxygen species begin early during replicative aging of Saccharomyces cerevisiae cells. **Free Radic Biol Med** 50(8): 963–970.

8. McFaline-Figueroa JR, Vevea J, Swayne TC, Zhou C, Liu C, Leung G, Boldogh IR, and Pon LA (**2011**). Mitochondrial quality control during inheritance is associated with lifespan and mother-daughter age asymmetry in budding yeast. **Aging Cell** 10(5): 885–895.

9. Klinger H, Rinnerthaler M, Lam YT, Laun P, Heeren G, Klocker A, Simon-Nobbe B, Dickinson JR, Dawes IW, and Breitenbach M (**2010**). Quantitation of (a)symmetric inheritance of functional and of oxidatively damaged mitochondrial aconitase in the cell division of old yeast mother cells. **Exp Gerontol** 45(7-8): 533–542.

10. Lord PG and Wheals AE (**1981**). Variability in individual cell cycles of Saccharomyces cerevisiae. **J Cell Sci** 50: 361–376.

11. Westermann B and Neupert W (**2000**). Mitochondria-targeted green fluorescent proteins: convenient tools for the study of organelle biogenesis in Saccharomyces cerevisiae. **Yeast** 16(15): 1421–1427.

12. Liu Z and Butow RA (**2006**). Mitochondrial retrograde signaling. **Annu Rev Genet** 40: 159–185.

13. Starovoytova AN, Sorokin MI, Sokolov SS, Severin FF, and Knorre DA (**2013**). Mitochondrial signaling in Saccharomyces cerevisiae pseudohyphae formation induced by butanol. **FEMS Yeast Res** 13(4): 367–374.

14. Calvert GR and Dawes IW (**1984**). Cell size control of development in Saccharomyces cerevisiae. **Nature** 312(5989): 61–63.

15. Wasko BM and Kaeberlein M (**2013**). Yeast replicative aging: a paradigm for defining conserved longevity interventions. **FEMS Yeast Res**: n/a–n/a.

16. Pawelec G and Larbi A (**2008**). Immunity and ageing in man: Annual Review 2006/2007. **Exp Gerontol** 43(1): 34–38.

17. Coelho M, Dereli A, Haese A, Kühn S, Malinovska L, Desantis ME, Shorter J, Alberti S, Gross T, and Tolić-Nørrelykke IM (**2013**). Fission Yeast Does Not Age under Favorable Conditions, but Does So after Stress. **Curr Biol** 23(19): 1844–1852.

18. Balaban NQ, Merrin J, Chait R, Kowalik L, and Leibler S (**2004**). Bacterial persistence as a phenotypic switch. **Science** 305(5690): 1622–1625.

19. Allen C, Büttner S, Aragon AD, Thomas JA, Meirelles O, Jaetao JE, Benn D, Ruby SW, Veenhuis M, Madeo F, and Werner-Washburne M (**2006**). Isolation of quiescent and nonquiescent cells from yeast stationary-phase cultures. **J Cell Biol** 174(1): 89–100.

20. Aragon AD, Rodriguez AL, Meirelles O, Roy S, Davidson GS, Tapia PH, Allen C, Joe R, Benn D, and Werner-Washburne M (**2008**). Characterization of differentiated quiescent and nonquiescent cells in yeast stationary-phase cultures. **Mol Biol Cell** 19(3): 1271–1280.

8

Polyamines directly promote antizyme-mediated degradation of ornithine decarboxylase by the proteasome

R. Roshini Beenukumar[1,#], Daniela Gödderz[1,2,#], R. Palanimurugan[1,3], and R. Jürgen Dohmen[1,*]

[1] Institute for Genetics, University of Cologne, Biocenter, Zülpicher Str. 47a, D-50674 Cologne, Germany.

[2] Present address: Karolinska Institute, Department for Cell- and Molecular Biology, Von Eulers väg 3, 171 77 Stockholm. E-mail: daniela.godderz@ki.se

[3] Present address: Center for Cellular and Molecular Biology (CCMB), Uppal Road, Hyderabad 500007, India. E-mail: murugan@ccmb.res.in

[#] These authors contributed equally to this study.

* Corresponding Author: R. Jürgen Dohmen, University of Cologne, Biocenter, Institute for Genetics; Zülpicher Str. 47a, D-50674 Cologne; E-mail: j.dohmen@uni-koeln.de

ABSTRACT Ornithine decarboxylase (ODC), a ubiquitin-independent substrate of the proteasome, is a homodimeric protein with a rate-limiting function in polyamine biosynthesis. Polyamines regulate ODC levels by a feedback mechanism mediated by ODC antizyme (OAZ). Higher cellular polyamine levels trigger the synthesis of OAZ and also inhibit its ubiquitin-dependent proteasomal degradation. OAZ binds ODC monomers and targets them to the proteasome. Here, we report that polyamines, aside from their role in the control of OAZ synthesis and stability, directly enhance OAZ-mediated ODC degradation by the proteasome. Using a stable mutant of OAZ, we show that polyamines promote ODC degradation in *Saccharomyces cerevisiae* cells even when OAZ levels are not changed. Furthermore, polyamines stimulated the *in vitro* degradation of ODC by the proteasome in a reconstituted system using purified components. In these assays, spermine shows a greater effect than spermidine. By contrast, polyamines do not have any stimulatory effect on the degradation of ubiquitin-dependent substrates.

Keywords: antizyme, ODC, polyamines, proteasome, ubiquitin.

Abbreviations:
ODC - ornithine decarboxylase,
OAZ - ODC antizyme,
CP - core particle,
RP - regulatory particle,
ODS - ODC degradation signal,
CryoEM - cryo electron microscopy,
GST - glutathione S-transferase,
ATP - Adenosine triphosphate,
DTT - Dithiothreitol,
Spd - spermidine,
Spm - spermine,
SD - synthetic dextrose.

INTRODUCTION

The ubiquitin-proteasome system, with the 26S proteasome as its central player, provides a key regulatory protein degradation mechanism in eukaryotes [1]. The 26S proteasome is a ~2.6 MDa multi-subunit protease that utilizes ATP for substrate degradation. It consists of two sub-complexes, the barrel-shaped catalytic core particle (CP) and the 19S regulatory particle (RP) [2]. The CP is composed of four heptameric rings, two outer α rings and two inner β rings. The latter harbor the proteolytic active sites residing in the β1, β2 and β5 subunits [3]. Using biochemical experiments, the RP was characterized to consist of two sub-complexes, the base with its hexameric ATPase ring and ubiquitin receptor subunits (Rpn10 and Rpn13), and the lid including the Rpn11 deubiquitylating activitiy [4]. Targeting of substrates to the proteasome is mainly

achieved by polyubiquitin-tagging, which leads to recognition by intrinsic ubiquitin receptors of the proteasome or by proteasome-associated shuttling factors [1]. Recent CryoEM studies have shed light on the details of ubiquitin-dependent substrate recognition and engagement by the proteasome [5-8]. Upon substrate recognition through polyubiquitin binding by the receptor, an unstructured region of the substrate gets threaded through the ATPase pore thereby engaging the proteasome. Translocation of the substrate leads to a conformational change that promotes removal of the polyubiquitin chain from the substrate by Rpn11. Subsequently, the rest of the substrate is translocated into the CP [9].

Interestingly, certain substrates of the proteasome do not require ubiquitylation for their recognition and degradation. The mechanism of ubiquitin-independent proteasomal targeting still remains unclear. It has been speculated that presence of an unstructured domain in these proteins is sufficient for proteasome association [10]. The mammalian thymidylate synthase, yeast Rpn4, and Ornithine decarboxylase (ODC) are some of the best-studied ubiquitin-independent proteasome substrates [11]. The ubiquitin-independent nature of ODC degradation is conserved from yeast to humans [12].

ODC is a rate-limiting enzyme in the biosynthesis of essential cellular polycations called polyamines. Spermidine and spermine are the most important polyamines, which, in fungi and animals, derive from ornithine [13]. Polyamines are ubiquitous molecules involved in a variety of cellular functions ranging from DNA stabilization, regulation of gene expression and protein synthesis, to ion channel function and cell cycle progression [14]. Increased ODC activity and elevated polyamine content have been implicated in several cancers [15-17]. Recent phase II clinical trials have shown that the ODC inhibitor difluoromethylornithine (DFMO) reduced prostate polyamine levels in patients at risk for invasive prostate cancer [18]. Another phase III clinical trial of a combinatorial DFMO and sulindac (a nonsteroidal anti-inflammatory drug) therapy for colorectal

cancer showed a significant interaction between dietary polyamines and the treatment [19]. The biosynthesis of polyamines in yeast begins with the decarboxylation of ornithine to putrescine by ODC (Spe1). Putrescine is then converted to spermidine by spermidine synthase (Spe3), and spermidine to spermine by spermine synthase (Spe4).

Cellular polyamine levels are tightly controlled by a homeostatic control of the enzymes involved in their synthesis or catabolism [13, 20, 21]. ODC levels are mainly regulated at the post-translational level in a mechanism that involves ODC antizyme (OAZ). Antizyme forms a heterodimer with ODC and targets it for ubiquitin-independent proteasomal degradation [22]. In yeast, the binding of antizyme to ODC is necessary for the exposure of its N-terminal degron called ODS (ODC Degradation Signal) which is essential for its degradation [23]. OAZ levels in the cells are in turn regulated by polyamines. OAZ mRNA is unusual as it has a stop codon after about one third of its coding sequence. For synthesis of full-length OAZ, the ribosome undergoes a +1 frameshift at this internal stop codon, a process that is conserved from yeast to humans [20, 21]. We have recently shown that polyamine binding to the nascent yeast Oaz1 polypeptide prevents a pile up of ribo-

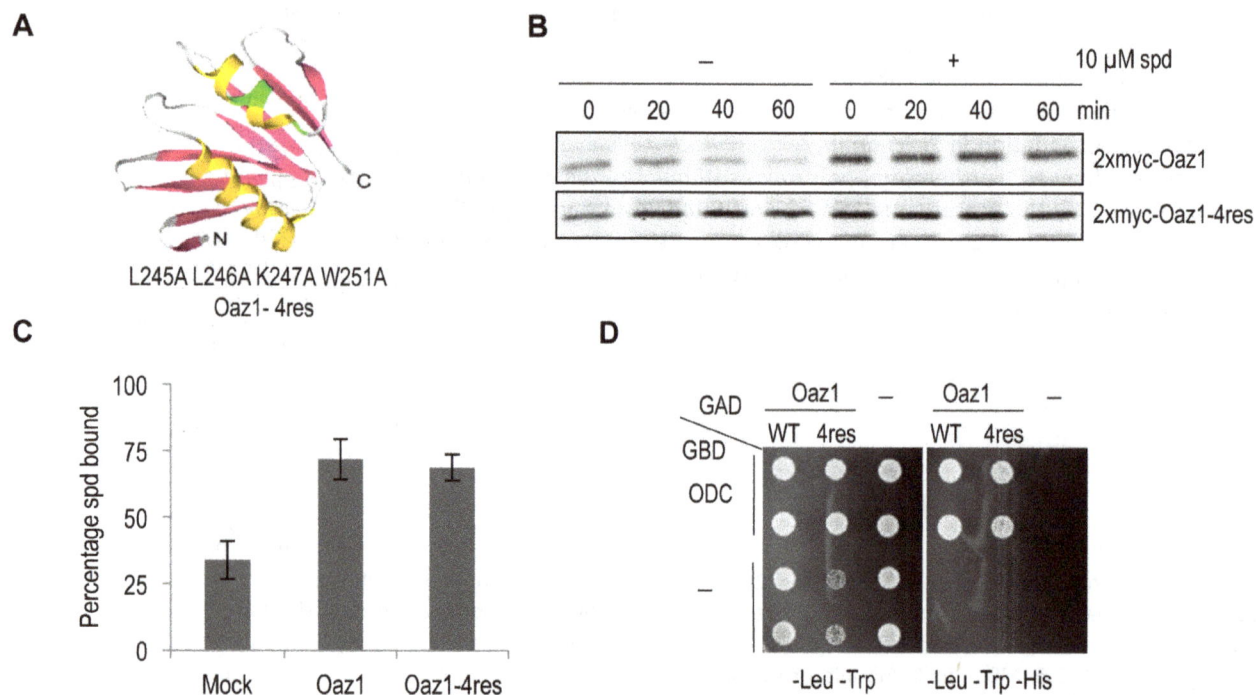

FIGURE 1: The antizyme mutant Oaz1-4res is metabolically stable in yeast cells. (A) Ribbon diagram of the NMR structure of rat antizyme [PDBcode 1ZO0] with β-sheets shown in pink and α-helices in yellow. Four amino acid changes (L245A L246A K247A W251A) were introduced into the yeast antizyme homologue at corresponding positions (indicated in green). The diagram was prepared using 3D molecular viewer. (B) Pulse-chase analysis of ^{35}S-radiolabelled Oaz1 (upper panel) or Oaz1-4res (lower panel) in cells grown in the presence or absence of 10 µM spermidine (spd) showing that the stability of Oaz1-4res is independent of polyamine addition. (C) Spermidine binding assay showing the retention of [^{3}H]-spermidine by 6His-Oaz1 or 6His-Oaz1-4res. Error bars, s.d.; n = 3. (D) Yeast two-hybrid analysis showing the interaction of Gal4 DNA binding domain (GBD) fused to ODC (GBD-ODC) with Gal4 transcription activation domain (GAD) fused to either Oaz1 or Oaz1-4res (GAD-Oaz1 or GAD-Oaz1-4res). Interaction of the two separated Gal4 domains via the polypeptides fused to them leads to reconstitution of the Gal4 transcriptional activator which controls expression of the HIS3 gene in the reporter strain used. Interaction-mediated functional reconstitution of Gal4 can be monitored as growth on medium lacking histidine in this strain.

somes on the *OAZ* mRNA thereby promoting the completion of Oaz1 synthesis [24]. Aside from this translational regulation, polyamines also control Oaz1 degradation. Oaz1 in yeast undergoes ubiquitin-dependent degradation by the proteasome, which is inhibited by high cellular polyamine concentrations [21]. Together, these two mechanisms lead to an increase of OAZ levels when intracellular polyamine concentrations are high, which in turn promotes ubiquitin-independent degradation of ODC.

Here we report that polyamines, in addition to the indirect enhancement of ODC degradation in *S. cerevisiae* by up-regulation of antizyme levels described above, promote ODC degradation by a third mechanism. Both *in vivo* and *in vitro* experiments show that polyamines directly stimulate antizyme-dependent ODC degradation by the proteasome. These findings identify a novel mechanism, by which a small organic metabolite controls ubiquitin-independent degradation of an enzyme involved in its synthesis.

RESULTS

A stable mutant of yeast antizyme

A mutant of antizyme with four amino acid replacements (Oaz1-4res) was generated in a targeted mutagenesis approach initially aimed to identify polyamine binding sites in antizyme. Note that in frame versions of the *OAZ1* gene [21] were used in all experiments to avoid any effects of polyamines on translational decoding of the mRNA. Selected residues in an α helix near the C-terminal end of antizyme were mutated to alanines (Fig. 1A). Pulse chase analyses showed that Oaz1-4res was a stable protein irrespective of the presence or absence of polyamines in the growth media, whereas degradation of wild-type Oaz1 was inhibited by polyamines as reported previously (Fig. 1B) [21]. We next asked whether the stable mutant Oaz1 protein binds polyamines. Oaz1-4res bound [^{3}H]-spermidine with a similar efficiency as its wild-type counterpart (Fig. 1C). Yeast two-hybrid analysis showed that Oaz1-4res also retained its ability to bind ODC although a slight reduction in binding was observed (Fig. 1D). It should be noted, however, that the reporter readout in this assay (histidine prototrophic growth) does not necessarily reflect strength of interaction.

In vitro characterization of ubiquitin-independent degradation of ODC

Using an *in vitro* assay with mouse ODC and purified 26S proteasome, Hoyt *et al.* have reproduced key *in vivo* features of ubiquitin-independent ODC degradation [25]. Moreover, another *in vitro* study using yeast ODC and Oaz1 has shown that antizyme promotes ODC degradation in a ubiquitin-independent and ATP-dependent manner [26]. Subsequent *in vivo* experiments revealed that binding of ODC monomers to antizyme is required to expose an N-terminal degron of yeast ODC called ODS (ODC Degradation Signal) [23]. In the present study, we reconstituted ODS-dependent degradation of ODC *in vitro*. 26S proteasomes were affinity-purified using anti-Flag beads from a yeast strain with a Flag-His$_6$-tagged β4/Pre1 subunit.

Native-PAGE analysis showed that these preparations mainly yielded active forms of the proteasome in its singly (SC) or doubly capped (DC) form, i.e. CP with one or two RPs [27] (Fig. 2A). ODC-2xHa or ΔODS-ODC-2xHa were co-purified from *Escherichia coli* cells as heterodimers together with 6His-Oaz1, which was pulled down with Ni-NTA beads binding to its 6xHis tag. The purified proteins were first characterized by SDS-PAGE and Coomassie staining. ODC-2xHa co-purified with 6His-Oaz1 as a double band (Fig. 2B). The different ODC-Oaz1 heterodimers were mixed with 26S proteasomes in a buffer supplemented with ATP, and incubated at 30°C for specific time periods followed by SDS-PAGE analysis. As expected, ODC was degraded over time whereas antizyme remained stable (Fig. 2C; lanes 4-6). In the control without 26S proteasomes, in contrast, ODC was not degraded (Fig. 2C; lanes 1-3). Around 75% inhibition of degradation was observed upon addition of epoxomicin, a selective proteasome inhibitor [28] (Fig. 2C; lanes 7-9). In a similar experiment using the ΔODS variant of ODC, only around 30% degradation was observed compared to the 80% degradation observed for the full-length ODC (Fig. 2D). These results show that ODS is critical for efficient degradation of ODC in line with the *in vivo* data reported earlier [23].

Polyamines directly enhance ODC degradation by the proteasome in yeast cells

Polyamines regulate cellular ODC levels by two known mechanisms, namely by inducing decoding of antizyme mRNA, and by inhibiting ubiquitin-dependent antizyme degradation [21, 24]. Here, we explored a third possible mechanism of ODC regulation. We investigated whether polyamines would directly influence antizyme-dependent ODC degradation. To address this *in vivo*, we employed yeast cells expressing the stable antizyme mutant (Oaz1-4res) described above. Important for this approach was that Oaz1-4res levels were not altered by polyamine addition (Fig. 1). Steady-state levels of ODC were determined in a strain expressing either wild-type Oaz1 or Oaz1-4res and grown in the presence of difluromethylornithine (DFMO) with or without additional spermidine. DFMO, an inhibitor of ODC and anti-cancer drug, was used to minimize the internal cellular polyamine levels [29]. As observed previously, ODC levels dropped upon spermidine addition in cells with wild-type Oaz1, which went along with elevated Oaz1 levels (Fig. 3A, lanes 1-2) [21]. Interestingly, however, ODC levels were also lowered further upon spermidine addition in cells with Oaz1-4res in spite of unchanged levels of the latter protein (Fig. 3A, lanes 3-4). Both versions of Oaz1 were expressed from in frame (if) variants of the *OAZ1* gene that are not subject to polyamine regulation of decoding involving a ribosomal frameshift event [24]. These data indicated that polyamines directly promote antizyme-mediated ODC degradation.

Next, we wanted to verify the conclusion derived from the experiments performed with the Oaz1-4res mutant with structurally wild-type Oaz1. This was important to exclude any influence of the four mutations present in Oaz1-4res on proteasomal targeting of ODC. To this end,

ODC degradation was studied in a strain with a copper-inducible, P$_{CUP1}$ promoter-driven in frame (if), but otherwise wild-type version of the antizyme gene (*OAZ1-wt*). With increasing copper concentrations, Oaz1 levels increased and ODC levels declined confirming that ODC levels are inversely correlated with those of Oaz1 (Fig. 3B). Addition of spermidine caused markedly increased Oaz1

levels due to the inhibition of its ubiquitin-dependent degradation [21]. Importantly for this analysis, the steady state levels of Oaz1 after full copper induction (500 µM) in the absence of added spermidine were similar to the levels in the absence of copper induction after addition of 50 µM spermidine. In other words, now that similar Oaz1 levels were established in the absence or presence of spermidine

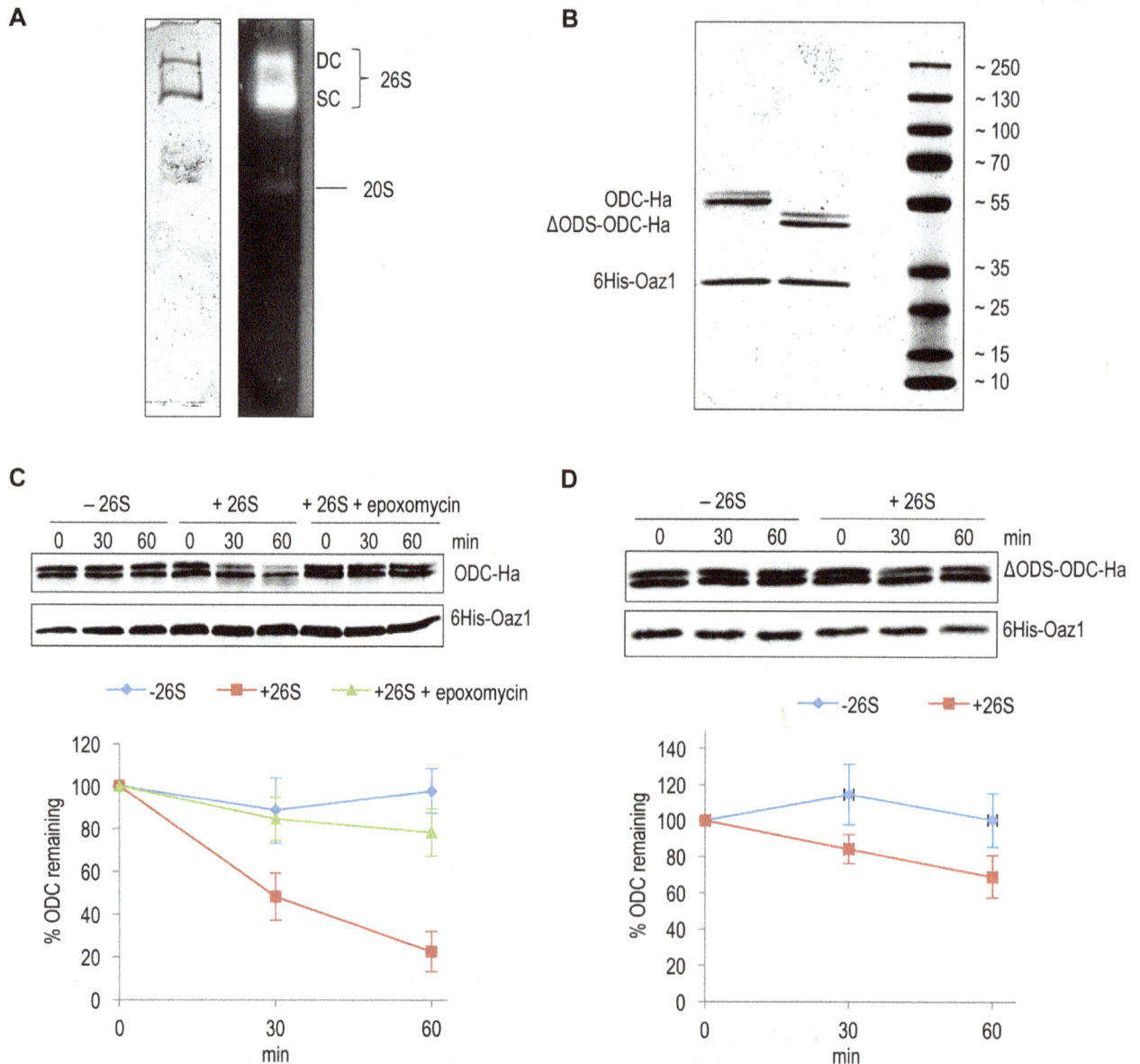

FIGURE 2: *In vitro* characterization of antizyme-mediated and ODS-dependent degradation of ODC. (A) Flag-tagged proteasome affinity-purified from yeast cells was analyzed by native-PAGE and Coomassie staining (left) or activity staining by overlay with the fluorogenic peptide Suc-LLVY-AMC (right). Proteasomes in this preparation were either doubly-capped (DC) with RPs on both sides of the core particle (CP), or singly-capped (SC) with only one RP attached to the CP. 20S CPs without any RPs attached to them were also present in the preparation. (B) SDS-PAGE analysis and Coomassie staining of 6His-Oaz1 co-purified from *E. coli* cells as heterodimer either with full length ODC (ODC-2xHa) or with N-terminally truncated ODC lacking the first 47 residues (ΔODS-ODC-2xHa). (C) *In vitro* degradation assays with purified 26S proteasomes and purified ODC-Oaz1 heterodimer as a substrate showing the degradation of ODC over time. 50 ng of ODC-Oaz1 heterodimer and 3 µg of 26S were mixed in 15 µL resulting in starting concentrations, respectively, of 40 nM and 80 nM. As controls, otherwise identical samples were assayed without 26S proteasome (-26S), or with the proteasome inhibitor epoxomycin. Two bands (full-length ODC-2xHa and a truncated derivative) were detected with the anti-Ha antibody, and both forms showed similar turnover rates. Therefore both bands were quantified together. In the resulting graph, values for the 0 time points were set to 100%. Error bars, s.d.; *n* = 3. (D) Experiments were performed as described for Fig. 2B, except that ΔODS-ODC-2xHa was used instead of full length ODC.

by altering the transcriptional induction of *OAZ1* gene expression with copper, direct effects of the polyamine on ODC degradation by the proteasome could be investigated. Indeed, the presence of spermidine caused a remarkable decrease in ODC levels in spite of similar Oaz1 levels established under these conditions (Fig. 3B; compare lanes 7

and 8). These data thus confirmed, now with a structurally wild-type Oaz1 (Fig. 3B), that spermidine exerts a direct effect on ODC targeting to the proteasome as observed before with the Oaz1-4res mutant protein (Fig. 3A).

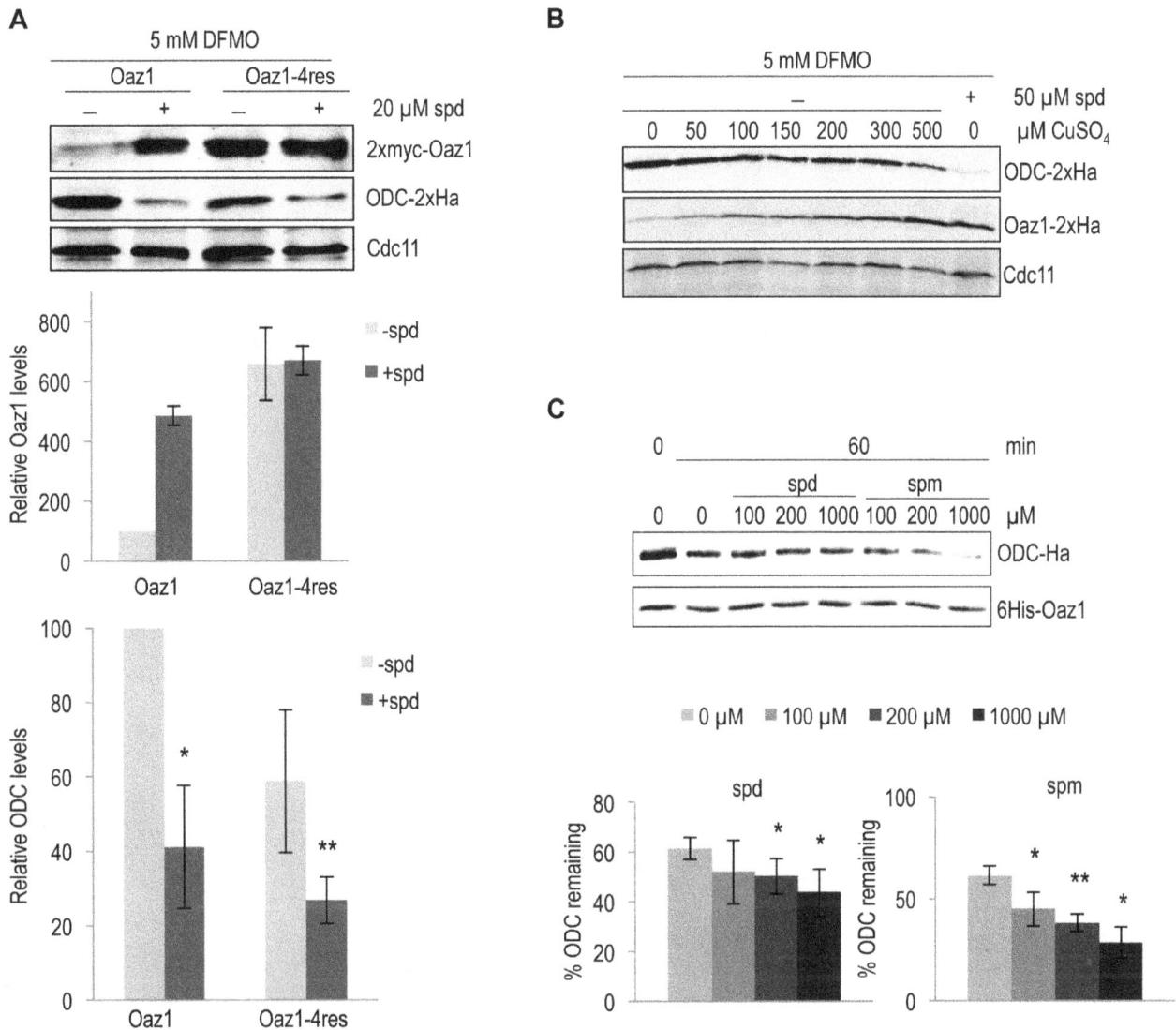

FIGURE 3: Degradation of ODC by the proteasome is directly enhanced upon polyamine addition. (A) *S. cerevisiae spe1-Δ oaz1-Δ* cells were transformed with plasmids encoding ODC-2xHa and either wild-type 2xMyc-Oaz1(wt) or its 4res mutant variant. The latter proteins were encoded by in frame versions of the gene (*OAZ1-if*) that did not require ribosomal frameshifting during decoding. Western blot analysis of the steady state levels of ODC (middle panel) in the presence of either Oaz1 or Oaz1-4res (top panel) from yeast cells grown in the presence of 5 mM DFMO. 20 μM spermidine (spd) was added as indicated. Also shown is the quantification of the 2xmyc-Oaz1 (wild-type and 4res) and ODC-2xHa signals normalized to Cdc11 levels. Levels are given relative to, respectively, Oaz1 and ODC levels obtained with cells expressing wild-type Oaz1 in the absence of spd, which were set to 100%. Error bars, s.d.; *n* = 2. Significance values were calculated by paired T test comparing ODC levels to those with wild-type Oaz1 and without spd; P ≤ 0.05 (*) and P ≤ 0.01 (**). **(B)** Western blot analysis of the steady state levels of ODC-2xHa and Oaz1-2xHa in a *spe1-Δ oaz1-Δ* strain in the presence of 5 mM DFMO. The plasmid-encoded ODC gene (*SPE1*) was expressed from its own promoter, whereas *OAZ1-if*(wt) was expressed from the copper-inducible P$_{CUP1}$ promoter. CuSO$_4$ was added to the medium as indicated. The last lane shows ODC and Oaz1 levels in cells grown without CuSO$_4$ but with 50 μM spd showing decreased ODC levels compared to other lanes in spite of similar Oaz1 levels. **(C)** *In vitro* degradation of ODC was assayed with 0.06 μg of 26S proteasome (1.6 nM) and 100 ng of ODC-Oaz1 heterodimer (80 nM) in a volume of 15 μL and varying concentrations of either spd or spermine (spm) as indicated. The graph shows the quantification of ODC-2xHa signals. Error bars, s.d.; *n* = 3. Signficance values were calculated by paired T test comparing ODC levels to the samples incubated for 60 min without spd; P ≤ 0.05 (*) and P ≤ 0.01 (**).

Polyamines directly enhance ODC degradation by the proteasome *in vitro*

To further validate our *in vivo* data, we used the *in vitro* ODC degradation system described above to study the direct effects of polyamines on ODC degradation. Consistent with the *in vivo* results, increased degradation of ODC was observed with increasing concentrations of either spermidine or spermine (Fig. 3C). Spermine showed a greater effect on ODC degradation than spermidine. This finding is compatible with the higher binding affinity of Oaz1 observed for spermine compared to spermidine [24].

To investigate the specificity of the observed effect of polyamines on ODC degradation, we asked if polyamines have any general effect on proteolytic degradation. To address this possibility *in vivo*, we used two well characterized ubiquitin-dependent substrates, an N-end rule substrate (Ub-R-e^k-Ha-Ura3) and a Ubiquitin Fusion Degradation (UFD) pathway substrate (Ub-V76-e^k-Ha-Ura3) [30, 31]. No significant effect on degradation of these two substrates was observed upon polyamine addition to polyamine-depleted cells (Fig. 4A). Additionally, the chymotrypsin-like activity of purified 26S proteasome was measured in the presence of increasing spermine concentration. A small reduction in proteasome activity was observed with polyamine addition (Fig. 4B). Taken together, the results presented above demonstrate that polyamines directly and specifically enhance ODC degradation by the proteasome.

Polyamines do not alter the affinity of antizyme for ODC

To understand the mechanism behind the effect of polyamines on ODC degradation, we tested if polyamines changed the affinity of the ODC-antizyme interaction. To address this question, we performed co-pull down assays using epitope-tagged variants of ODC and Oaz1 expressed in *E. coli*. GST-Oaz1 bound beads were exposed to *E. coli* cell extracts overexpressing ODC-Flag in the presence or absence of spermine. Western blot analysis after GST pull down showed no significant difference in ODC-Flag binding between the samples with and without spermine (Fig. 5). This data suggested that polyamines promote ODC degradation without altering ODC-antizyme heterodimer interactions.

Both spermidine and spermine promote antizyme and ODC degradation *in vivo*

As spermine binds antizyme better than spermidine [24] and also shows a greater effect on the enhancement of ODC degradation *in vitro*, we asked whether spermine is the major mediator of ODC regulation in yeast cells. We, therefore, compared the effect of spermidine and spermine on antizyme stabilization and ODC degradation in wild-type and *spe4-Δ* strains. *SPE4* encodes spermine synthase, an enzyme that mediates the conversion of spermidine to spermine. Hence, *spe4-Δ* cells are devoid of spermine [32]. Antizyme degradation was similarly inhibited in both WT and *spe4-Δ* cells upon addition of spermidine or spermine (Fig. 6A, top panel). When compared to spermine, addition of spermidine had a stronger effect on the inhibition of

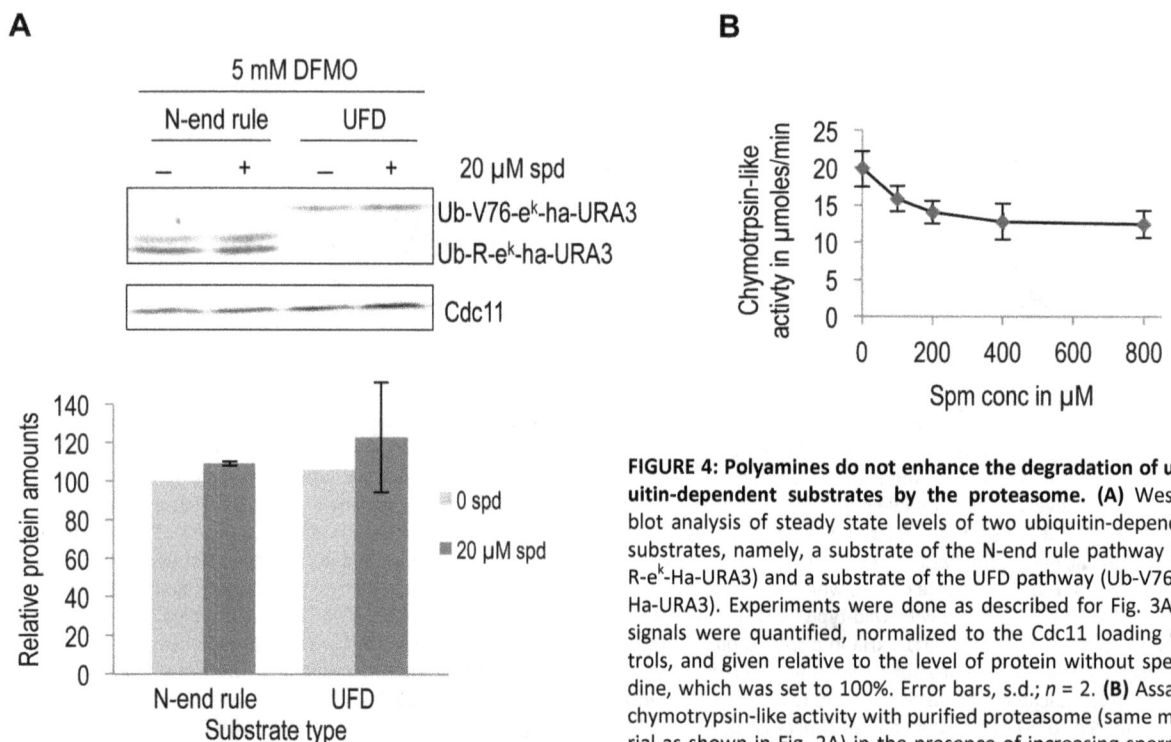

FIGURE 4: Polyamines do not enhance the degradation of ubiquitin-dependent substrates by the proteasome. (A) Western blot analysis of steady state levels of two ubiquitin-dependent substrates, namely, a substrate of the N-end rule pathway (Ub-R-e^k-Ha-URA3) and a substrate of the UFD pathway (Ub-V76- e^k-Ha-URA3). Experiments were done as described for Fig. 3A. Ha signals were quantified, normalized to the Cdc11 loading controls, and given relative to the level of protein without spermidine, which was set to 100%. Error bars, s.d.; n = 2. (B) Assay of chymotrypsin-like activity with purified proteasome (same material as shown in Fig. 2A) in the presence of increasing spermine (spm) concentrations. Error bars, s.d.; n = 4.

antizyme degradation in both strains. Similarly, spermidine had a stronger (stimulatory) effect on ODC degradation than spermine (Fig. 6A, middle panel). Since we used cells expressing *OAZ1-if* constructs, these assays only monitored the effects of polyamines on Oaz1 protein stability and its capacity to mediate ODC degradation by the proteasome. The results suggest that both spermidine and spermine are capable of mediating ODC regulation in yeast cells. The relatively weaker effect of spermine on ODC targeting *in vivo* contrasts with its relatively stronger effect *in vitro* and is likely due to a lower uptake efficiency of spermine by yeast cells [33, 34].

In mammals, when cellular polyamine levels are high, they are acetylated leading to their breakdown or export from the cells [35]. High cellular polyamine levels, in addition, lead to antizyme synthesis and hence the down-regulation of ODC. We therefore asked whether acetylated polyamines might be responsible for down-regulating ODC by binding to antizyme. We performed a competition assay with [³H]-spermidine and various acetylated spermidine derivatives for Oaz1 binding. In this assay, mono-acetylated spermidine derivatives showed a clearly reduced binding to antizyme when compared to unmodified spermidine, and di-acetylspermidine showed no competition at all (Fig. 6B). These findings suggest that acetylated polyamines do not participate in the feedback regulation of ODC.

DISCUSSION

Polyamines enhance degradation of ODC

We report the identification of an additional mechanism by which polyamines influence the post-translational control of ODC, which adds a new dimension to the polyamine-mediated feedback regulation of ODC. The canonical regulation of ODC levels by polyamines takes place via up-regulation of antizyme by promoting translational decoding of its mRNA, which involves a ribosomal frameshifting event [12, 20]. We have previously shown that antizyme binds directly to polyamines and that co-translational binding of the nascent yeast antizyme (Oaz1) to polyamines promotes completion of its synthesis [24]. Polyamines, in addition, inhibit the ubiquitin-dependent degradation of Oaz1 [21]. Together, these findings established that poly-amine binding leads to an increase in the levels of Oaz1 by promoting its synthesis and stability [12]. Higher Oaz1 levels, in turn, mediate a more rapid turnover of ODC. In the present work, we show that polyamines, both *in vivo* and *in vitro*, moreover have a direct effect on ODC by stimulating its antizyme-mediated and ubiquitin-independent degradation by the proteasome.

The difficulty in demonstrating this novel direct effect of polyamines on ODC turnover *in vivo* was to separate this mechanism from the role of polyamines in regulating the cellular levels of Oaz1. The effect of polyamines on decoding of the *OAZ1* mRNA could be eliminated by employing constructs lacking the ribosomal frameshifting site (Figure 3A and B). The identification of the metabolically stable Oaz1-4res mutant provided us with an additional new tool to investigate a direct effect of polyamines on ODC degra-

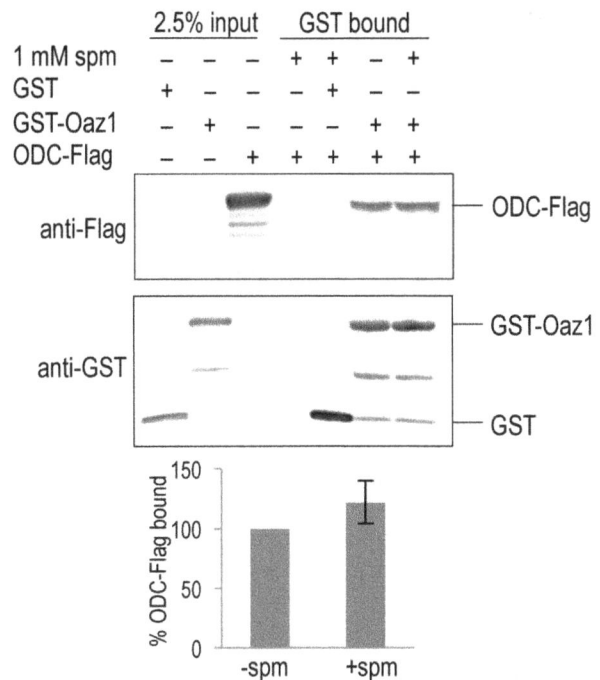

FIGURE 5. Spermine does not affect the affinity of antizyme to ODC. Co-pull down of Oaz1 and ODC in the presence or absence of 1 mM spermine (spm). Extracts from *E. coli* cells expressing the indicated tagged proteins were subjected to GST-pull down and subsequent quantitative anti-Flag western blotting for ODC-Flag detection and anti-GST for GST-Oaz1 detection. ODC-Flag signals after elution were normalized to GST-Oaz1 signals providing ODC-Flag values obtained in the absence of spm, the mean of which was set to 100%. Values obtained in the presence of spm are given in % of those obtained in its absence. Error bars, s.d.; *n* = 3.

dation. The Oaz1-4res mutant retains the ability to bind polyamines as well as to interact with ODC and target it for ubiquitin-independent degradation, but the stability of Oaz1-4res itself is not influenced by polyamines (Fig. 1). In yeast cells, wherein the internal polyamine levels were reduced by the ODC inhibitor DFMO, Oaz1-4res promoted ODC degradation more efficiently upon polyamine addition even though the levels of Oaz1-4res remained unaffected indicating that polyamine directly influence Oaz1-mediated degradation of ODC by the proteasome (Fig. 3A). Importantly, we could confirm this conclusion with an additional experimental approach that employed a structurally wild-type Oaz1, stability of which is influenced by polyamines. This could be achieved by appropriately adjusting Oaz1 synthesis using a copper-regulated promoter (Fig. 3B). Both experimental *in vivo* assays indicated that direct stimulation of Oaz1-mediated ODC degradation plays a critical role in efficiently lowering ODC levels when polyamine concentrations go up. This conclusion could be derived from the observation that even high levels of Oaz1 protein were not reducing ODC levels to the same extent when polyamine concentrations were low as when they were high (Figs. 3A and 3B). Together, these findings therefore suggest that the direct stimulation of ODC degradation by polyamines is a physiologically significant part in the negative feedback regulation of ODC in addition to the positive effect of polyamines on Oaz1 synthesis and stability. *In vitro* assays with purified 26S proteasome and ODC-antizyme heterodimer provided independent evidence supporting the direct effect of polyamines on ODC degradation observed *in vivo*. The *in vitro* effect, however, seemed smaller than expected from the *in vivo* results. A possible explanation is that the experimental *in vitro* conditions might not precisely reflect the biochemical milieu and environment in which this process occurs *in vivo*.

In the *in vitro* experiments, spermine showed a greater stimulatory effect on ODC degradation than spermidine (Fig. 3). This effect can be related to a higher binding affinity of spermine to antizyme in comparison to spermidine [24]. The stronger effect of spermine on ODC targeting compared to spermidine *in vitro*, contrasts with its relatively weaker effect *in vivo*. The weaker effect of spermine added to the yeast culture medium is probably due to a lower uptake efficiency for this polyamine compared to spermidine [33, 34]. Aside from the difference in the uptake efficiencies for the different polyamines, another difficulty in comparing the magnitude of the effects observed *in vitro* and *in vivo* is the absence of solid data on the intracellular concentration of free forms of these polyamines. Nonetheless, a clear and specific stimulatory effect of polyamines on ODC degradation by the proteasome was observed both *in vivo* and *in vitro*.

At this point, the exact mechanism by which polyamines directly enhance ODC degradation remains unclear. We tested whether polyamines enhance proteasome activity or the degradation of ubiquitin-dependent substrates. Polyamines slightly inhibited chymotrypsin-like activity of the proteasome *in vitro*, and had no effect on the degradation of ubiquitin-dependent substrates in DFMO-treated yeast cells (Fig. 4). Because they bind antizyme, polyamines might enhance ODC-antizyme heterodimer formation. In pull down experiments, however, polyamines did not have any detectable effect on the binding of ODC to Oaz1 (Fig. 5).

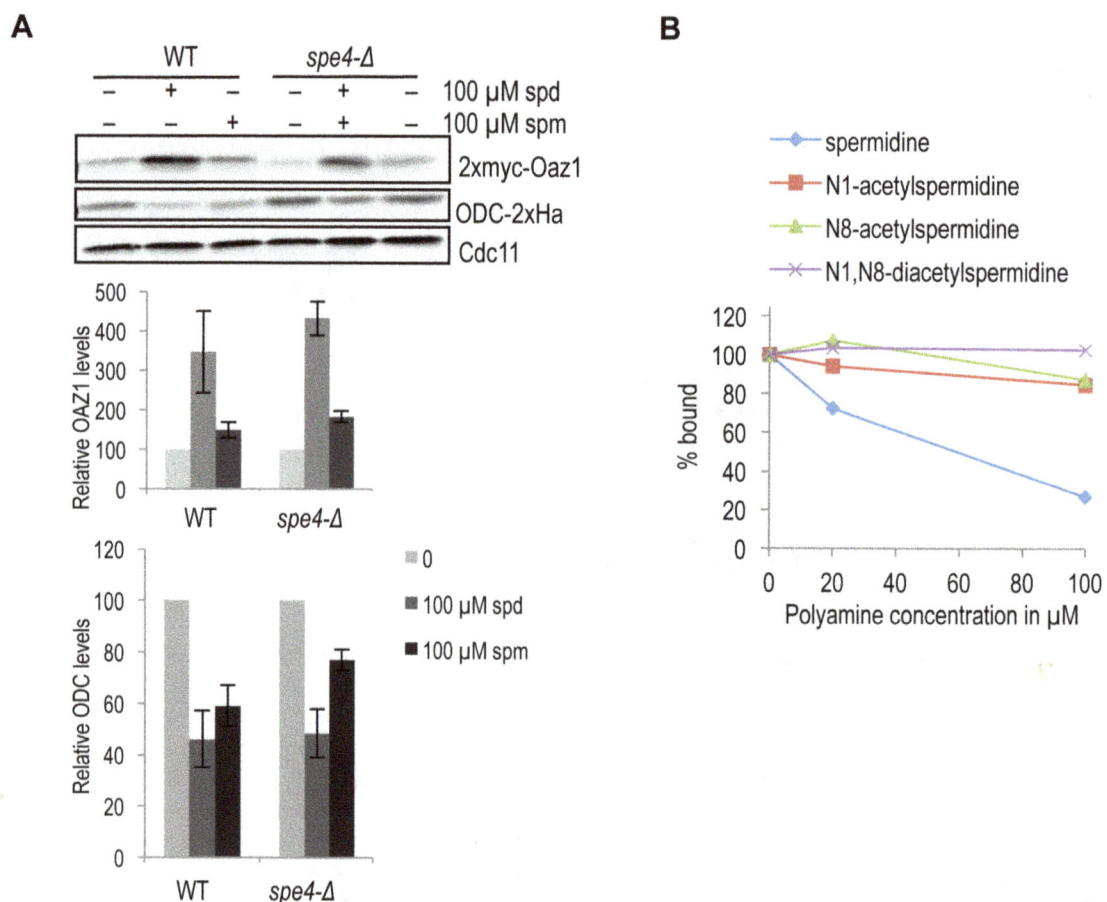

FIGURE 6: Role of polyamine subtypes and their modification in the targeting of ODC. (A) Western blot analysis comparing steady state levels of Oaz1 and ODC in either the wild type or a strain lacking spermine synthase (*spe4Δ*), grown with or without polyamine supplementation as indicated. Cells were transformed with plasmids expressing 2xmyc-Oaz1 from an in frame version of the *OAZ1* gene, and with a plasmid encoding ODC-2xHa. The graph shows the results of a quantification of myc (upper part) and Ha signals (lower part) normalized to the Cdc11 loading control. Levels are given relative to the respective levels of the same proteins in cells grown without polyamine addition, which was set to 100%. Error bars, s.d.; *n* = 3. **(B)** Acetylation of spermidine inhibits its binding to antizyme. *In vitro* binding assay showing the competition between [^{3}H]-spermidine and different species of acetylated spermidine for binding to 6xHis-tagged antizyme purified from *E. coli*.

It remains a possibility that this binding assay is not sensitive enough to capture physiologically relevant small differences in binding affinity. Another possibility is that polyamine binding to the ODC-Oaz1 complex promotes ODC degradation by the proteasome without altering binding affinity between ODC and Oaz1. Two scenarios could be envisioned. Binding of polyamines to the complex might either cause a conformational change in ODC resulting in a better exposure and binding to the proteasome of its N-terminal unstructured domain [23]. Another possibility is that polyamine binding to Oaz1 increases its affinity to an additional binding site in the proteasome [23, 36]. Further studies including structural analyses will be required to resolve this issue.

Role of polyamine subtypes and their modification in ODC targeting

Since spermine showed a greater effect on ODC degradation in vitro, we asked whether spermine might be the main factor driving ODC degradation in vivo. To test this, we used a strain (spe4-Δ) lacking the enzyme spermine synthase, which is unable to convert spermidine into spermine. Since there were no significant differences in ODC or Oaz1 levels detectable between wild-type and spe4-Δ cells (Fig. 6B), we conclude that formation of spermine from spermidine is not critical for ODC targeting in vivo.

Catabolism and export of polyamines is known to be initiated by their acetylation [35]. Since these mechanisms are relevant at high polyamine concentrations that are also known to trigger ODC down-regulation, we tested whether acetylation of polyamines had an effect on their binding to antizyme. We observed that acetylation of spermidine clearly inhibited its binding to antizyme. This is likely due to the neutralization of the positive charges on polyamines by the acetyl groups. These findings indicate that acetylation of polyamines is not enhancing their roles in promoting ODC degradation.

The findings of the present work, together with previous observations, establish that polyamines act at three levels in a negative feedback loop that controls ODC. Polyamine binding to Oaz1 stimulates its synthesis, inhibits its ubiquitin-dependent degradation, and directly promotes Oaz1-mediated degradation of ODC by the proteasome. The latter is, to the best of our knowledge, the first example wherein a small natural compound directly promotes the degradation of a protein by the proteasome.

MATERIALS AND METHODS
Yeast media, strains and plasmids
Yeast rich (YPD) and synthetic (S) minimal media with 2% dextrose (SD) were prepared as described [37]. Spermidine (Sigma-Aldrich), spermine (Sigma-Aldrich), DFMO (kindly provided by Dr. Patrick Woster), or $CuSO_4$ were added to the media at various concentrations as indicated. Strains and plasmids used in this study are listed in tables S1 and S2, respectively, in the supplementary information.

Analysis of protein levels, stability and interactions
Analysis of steady state proteins levels in S. cerevisiae cells by SDS-PAGE and immunoblotting was performed as described [21]. Proteins were detected using either anti-mouse or anti-rabbit IgG secondary antibodies coupled to near-infrared fluorophores (Rockland). The blots were scanned and the signal intensities quantified using the Odyssey Infrared Imaging System (Li-COR Biosciences). For detection of epitope tags, we used the following monoclonal antibodies: The Ha epitope was detected with 16B12 monoclonal mouse antibody (Covance), Cdc11 was detected with polyclonal rabbit antibody (Santa Cruz), and the myc epitope with 9B11 mouse monoclonal antibody. For co-immunoprecipitation, proteins were extracted from E. coli BL21 codon$^+$ cells harboring either pDG241 (GST-Oaz1), pGEX-2TX (GST), pDG273 (ODC-Flag) or pUC19 (mock) by glass bead lysis in ice-cold lysis buffer (50 mM Na-HEPES (pH 7.5), 5 mM EDTA, 1% Triton X-100) containing 'Complete' Protease-Inhibitor cocktail (Roche). Total protein amounts were equilibrated between GST-Oaz1 and GST lysate using the mock lysate. 800 µg of total proteins were incubated with 100 µl of glutathione beads (GE Healthcare) at 4°C for 2 h. From this step onwards, 1 mM spermine was added to certain tubes as indicated in Fig. 5. The beads were washed two times with lysis buffer, and further incubated after the addition of ODC-Flag lysate at 4°C for 2 h. Bound proteins were eluted by incubation with 125 µl elution buffer [25 mM Glutathione (Sigma-Aldrich), 20 mM NaOH in lysis buffer] at 4°C for 90 min. The samples were then analyzed by SDS-PAGE and western blotting as described above. The GST and Flag epitopes were detected using rabbit polyclonal antibody (Santa Cruz) and mouse monoclonal M2 antibody (Sigma-Aldrich), respectively. Pulse chase analysis [38] and yeast two-hybrid assays [39] were carried out as described earlier.

Protein purification
26S proteasomes were purified as described previously [40] from yeast strain MO24, in which the PRE1 gene, encoding the 20S core particle subunit Pre1, has been stably modified to express a C-terminally Flag-6His tagged version. The purity and activity of the proteasomes were analyzed by Native-PAGE [27] followed by either Coomassie staining or in-gel chymotrypsin-like activity degradation assay [27]. 6His-OAZ1 and 6His-OAZ1-4res proteins were affinity-purified from E. coli strain Rosetta (Merck) transformed with pDG240 (6His-Oaz1) or pDG246 (6His-Oaz1-4res) respectively, as described earlier [24]. 6His-Oaz1 and ODC-2xHa or 6His-Oaz1-ΔODS-ODC-2xHa were co-expressed in E. coli strain Rosetta and Ni-affinity purified as described above with a few variations. The lysis buffer used was buffer B (25 mM Na-HEPES, pH 7.8, 5 mM $MgCl_2$, 25 mM KCl, 10% glycerol). After elution of the protein, imidazole was removed using NAP™-5 (GE Healthcare) columns. The purity of the eluted proteins was evaluated by SDS-PAGE followed by Coomassie staining as described above.

Proteasomal peptidase activity assay
The chymotrypsin-like activity was measured using Suc-LLVY-AMC as a substrate as described earlier [41] with the following modifications. The reactions were carried out using 0.06 µg of purified 26S proteasome and varying amounts of spermine (as shown in Fig. 4B) in 90 µL of buffer B supplemented with 1 mM ATP and 1 mM DTT. 10 µL of a 1 mg/mL Suc-LLVY-AMC solution was added to this mixture.

In vitro degradation assay

Degradation assays were performed as described earlier [40] with the following variations. The amounts of 26S and substrate used are as indicated in Figs. 2 and 3. The reaction buffer (buffer B) was supplemented with 1 mM ATP and 1 mM DTT. To inhibit proteasomal activity; proteasomes were pre-treated with 100 µM epoxomycin (Enzo life sciences) at 30°C for 45 min before adding to the degradation assays. Wherever indicated (refers to Fig. 3C), the reactions were supplemented with either spermidine or spermine. The degradation reactions were carried out at 30°C for various time points as indicated in Figs. 2 and 3 followed by SDS-PAGE and Western blot analysis. ODC-2xHa was detected with 16B12 mouse monoclonal antibody and 6His-Oaz1 with anti-Oaz1 polyclonal antibody.

Polyamine binding assay

Polyamine binding was measured as described earlier [24, 42]. The [^{3}H]-spermidine competition assay was performed using 10 µM of [^{3}H]-spermidine and the indicated concentrations (Fig. 6B) of the following acetyl polyamines; N1-acetylspermidine (Wako), N8-acetylspermidine (Sigma-Aldrich) and N1, N8-diacetylspermidine (Wako).

ACKNOWLEDGMENTS

We thank Patrick Woster for providing DFMO. R.R.B. and D.G. were supported by the International Graduate School in Development Health and Disease, Cologne, Germany. This work was supported by a grant (Do 649/4-3) to R.J.D. from the Deutsche Forschungsgemeinschaft (DFG).

CONFLICT OF INTEREST

The authors declare no conflict of interest.

REFERENCES

1. Hershko A, Ciechanover A (**1998**). The ubiquitin system. **Annu Rev Biochem** 67: 425-479.

2. Tomko RJ, Jr., Hochstrasser M (**2013**). Molecular architecture and assembly of the eukaryotic proteasome. **Annu Rev Biochem** 82: 415-445.

3. Baumeister W, Walz J, Zuhl F, Seemuller E (**1998**). The proteasome: paradigm of a self-compartmentalizing protease. **Cell** 92: 367-380.

4. Glickman MH, Rubin DM, Coux O, Wefes I, Pfeifer G, Cjeka Z, Baumeister W, Fried VA, Finley D (**1998**) A subcomplex of the proteasome regulatory particle required for ubiquitin-conjugate degrada- tion and related to the COP9-signalosome and eIF3. **Cell** 94, 615– 623.

5. Sledz P, Unverdorben P, Beck F, Pfeifer G, Schweitzer A, Forster F, Baumeister W (**2013**). Structure of the 26S proteasome with ATP-gammaS bound provides insights into the mechanism of nucleotide-dependent substrate translocation. **Proc Nat Acad Sci U S A** 110: 7264-7269.

6. Matyskiela ME, Lander GC, Martin A (**2013**). Conformational switching of the 26S proteasome enables substrate degradation. **Nature Struct Mol Biol** 20: 781-788.

7. Lander GC, Estrin E, Matyskiela ME, Bashore C, Nogales E, Martin A (**2012**). Complete subunit architecture of the proteasome regulatory particle. **Nature** 482: 186-191.

8. Beck F, Unverdorben P, Bohn S, Schweitzer A, Pfeifer G, Sakata E, Nickell S, Plitzko JM, Villa E, Baumeister W, Förster F (**2012**). Near-atomic resolution structural model of the yeast 26S proteasome. **Proc Natl Acad Sci U S A** 109: 14870-14875.

9. Lander GC, Martin A, Nogales E (**2013**). The proteasome under the microscope: the regulatory particle in focus. **Curr Opin Struct Biol** 23: 243-251.

10. Inobe T, Matouschek A (**2014**). Paradigms of protein degradation by the proteasome. **Curr Opin Struct Biol** 24: 156-164.

11. Erales J, Coffino P (**2014**). Ubiquitin-independent proteasomal degradation. **Biochimn Biophys Acta** 1843: 216-221.

12. Palanimurugan R, Kurian L, Hegde V, Hofmann K, Dohmen RJ (**2014**). Co-translational Polyamine Sensing Co-translational polyamine sensing by Nascent ODC Antizyme ODC antizyme. In: Ito K, editor Regulatory Nascent Polypeptides. Springer Japan; pp 203-222.

13. Wallace HM (**2009**). The polyamines: past, present and future. **Essays Biochem** 46: 1-9.

14. Ramani D, De Bandt JP, Cynober L (**2014**). Aliphatic polyamines in physiology and diseases. **Clin Nutr** 331: 14-22.

15. Rounbehler RJ, Li WM, Hall MA, Yang CY, Fallahi M, Cleveland JL (**2009**). Targeting Ornithine Decarboxylase Impairs Development of MYCN-Amplified Neuroblastoma. **Cancer Res** 69: 547-553.

16. Evageliou NF, Hogarty MD (**2009**). Disrupting Polyamine Homeostasis as a Therapeutic Strategy for Neuroblastoma. **Clin Cancer Res** 15: 5956-5961.

17. Nowotarski SL, Woster PM, Casero RA (**2013**). Polyamines and cancer: implications for chemotherapy and chemoprevention. **Expert Rev Mol Med** 15: e3.

18. Meyskens FL, Jr., Simoneau AR, Gerner EW (**2014**). Chemoprevention of prostate cancer with the polyamine synthesis inhibitor difluoromethylornithine. **Recent Results Cancer Res** 202: 115-120.

19. Raj KP, Zell JA, Rock CL, McLaren CE, Zoumas-Morse C, Gerner EW, Meyskens FL (**2013**). Role of dietary polyamines in a phase III clinical trial of difluoromethylornithine (DFMO) and sulindac for prevention of sporadic colorectal adenomas. **Br J cancer** 108: 512-518.

20. Matsufuji S, Matsufuji T, Miyazaki Y, Murakami Y, Atkins JF, Gesteland RF, Hayashi S (**1995**). Autoregulatory frameshifting in decoding mammalian ornithine decarboxylase antizyme. **Cell** 80: 51-60.

21. Palanimurugan R, Scheel H, Hofmann K, Dohmen RJ (**2004**). Polyamines regulate their synthesis by inducing expression and blocking degradation of ODC antizyme. **EMBO J** 23: 4857-4867.

22. Coffino P (**2001**). Regulation of cellular polyamines by antizyme. **Nature Rev Mol Cell Biol** 2: 188-194.

23. Gödderz D, Schäfer E, Palanimurugan R, Dohmen RJ (**2011**). The N-terminal unstructured domain of yeast ODC functions as a transplantable and replaceable ubiquitin-independent degron. **J Mol Biol** 407: 354-367.

24. Kurian L, Palanimurugan R, Gödderz D, Dohmen RJ (**2011**). Polyamine sensing by nascent ornithine decarboxylase antizyme stimulates decoding of its mRNA. **Nature** 477: 490-494.

25. Hoyt MA, Zhang M, Coffino P (**2003**). Ubiquitin-independent mechanisms of mouse ornithine decarboxylase degradation are conserved between mammalian and fungal cells. **J Biol Chem** 278: 12135-12143.

26.	Porat, Z., Landau, G., Bercovich, Z., Krutauz, D., Glickman, M. & Kahana, C. (2008). Yeast antizyme mediates degradation of yeast ornithine decarboxylase by yeast but not by mammalian proteasome: new insights on yeast antizyme. **J Biol Chem** 283: 4528-4534.

27. Elsasser S, Schmidt M, Finley D (**2005**). Characterization of the Proteasome Using Native Gel Electrophoresis. **Methods Enzymol** 398: 353-363.

28. Meng L, Mohan R, Kwok BH, Elofsson M, Sin N, Crews CM (**1999**). Epoxomicin, a potent and selective proteasome inhibitor, exhibits in vivo antiinflammatory activity. **Proc Natl Acad Sci U S A** 96: 10403-10408.

29. Fozard JR, Part ML, Prakash NJ, Grove J (**1980**). Inhibition of murine embryonic development by alpha-difluoromethylornithine, an irreversible inhibitor of ornithine decarboxylase. **Eur J Pharmacol** 65: 379-391.

30. Varshavsky A (**1996**). The N-end rule: functions, mysteries, uses. **Proc Natl Acad Sci U S A** 93: 12142-12149.

31. Ghislain M, Dohmen RJ, Levy F, Varshavsky A (**1996**). Cdc48p interacts with Ufd3p, a WD repeat protein required for ubiquitin-mediated proteolysis in Saccharomyces cerevisiae. **EMBO J** 15: 4884-4899.

32. Hamasaki-Katagiri N, Katagiri Y, Tabor CW, Tabor H (**1998**). Spermine is not essential for growth of Saccharomyces cerevisiae: identification of the SPE4 gene (spermine synthase) and characterization of a spe4 deletion mutant. **Gene** 210: 195-201.

33. Erez O, Kahana C (**2001**). Screening for modulators of spermine tolerance identifies Sky1, the SR protein kinase of Saccharomyces cerevisiae, as a regulator of polyamine transport and ion homeostasis. **Mol Cell Biol** 21: 175-184.

34. Kim SK, Jin YS, Choi IG, Park YC, Seo JH (2015) Enhanced tolerance of Saccharomyces cerevisiae to multiple lignocellulose-derived inhibitors through modulation of spermidien content. Metab Eng 25: 46-55.

35. Casero RA, Jr., Pegg AE (**1993**). Spermidine/spermine N1-acetyltransferase--the turning point in polyamine metabolism. **FASEB J** 7: 653-661.

36. Li X, Stebbins B, Hoffman L, Pratt G, Rechsteiner M, Coffino P (**1996**). The N terminus of antizyme promotes degradation of heterologous proteins. **J Biol Chem** 271: 4441-4446.

37. Dohmen RJ, Stappen R, McGrath JP, Forrova H, Kolarov J, Goffeau A, Varshavsky A (**1995**). An essential yeast gene encoding a homolog of ubiquitin-activating enzyme. **J Biol Chem** 270: 18099-18109.

38. Ramos PC, Höckendorff J, Johnson ES, Varshavsky A, Dohmen RJ (**1998**). Ump1p Is Required for Proper Maturation of the 20S Proteasome and Becomes Its Substrate upon Completion of the Assembly. **Cell** 92: 489-499.

39. James P, Halladay J, Craig EA (**1996**). Genomic libraries and a host strain designed for highly efficient two-hybrid selection in yeast. **Genetics** 144: 1425-1436.

40. Ha SW, Ju D, Xie Y (**2012**). The N-terminal domain of Rpn4 serves as a portable ubiquitin-independent degron and is recognized by specific 19S RP subunits. **Biochem Biophys Res Commun** 419: 226-231.

41. Dohmen RJ, London M, Glanemann C, Ramos P (**2005**). Assays for Proteasome Assembly and Maturation. **In: Patterson C, Cyr D, editors. Ubiquitin-Proteasome Protocols. Humana Press**; pp 243-254.

42. R Palanimurugan, Dohmen RJ (**2012**). Ultrafiltration-based in vitro assay for determining polyamine binding to proteins. **Protocol Exchange.**

9

INO1 transcriptional memory leads to DNA zip code-dependent interchromosomal clustering

Donna Garvey Brickner, Robert Coukos and Jason H. Brickner*

Department of Molecular Biosciences, Northwestern University, Evanston, IL USA 60201.

* Corresponding Author: Jason H. Brickner, E-mail: j-brickner@northwestern.edu

ABSTRACT   Many genes localize at the nuclear periphery through physical interaction with the nuclear pore complex (NPC). We have found that the yeast *INO1* gene is targeted to the NPC both upon activation and for several generations after repression, a phenomenon called epigenetic transcriptional memory. Targeting of *INO1* to the NPC requires distinct *cis*-acting promoter DNA zip codes under activating conditions and under memory conditions. When at the nuclear periphery, active *INO1* clusters with itself and with other genes that share the GRS I zip code. Here, we show that during memory, the two alleles of *INO1* cluster in diploids and endogenous *INO1* clusters with an ectopic *INO1* in haploids. After repression, *INO1* does not cluster with GRS I - containing genes. Furthermore, clustering during memory requires Nup100 and two sets of DNA zip codes, those that target *INO1* to the periphery when active and those that target it to the periphery after repression. Therefore, the interchromosomal clustering of *INO1* that occurs during transcriptional memory is dependent upon, but mechanistically distinct from, the clustering of active *INO1*. Finally, while localization to the nuclear periphery is not regulated through the cell cycle during memory, clustering of *INO1* during memory is regulated through the cell cycle.

Keywords: epigenetic inheritance, transcriptional memory, nuclear pore, DNA zip code, interchromosomal clustering.

Abbreviations:
GRS - gene recruitment sequence,
LacO - Lac operator,
MRS - memory recruitment sequence,
NPC - nuclear pore complex,
Nups - nuclear pore proteins,
TetO - Tet operator.

INTRODUCTION

Eukaryotic genomes are spatially organized. Chromosomes compact, form intrachromosomal loops and interact with each other and with subnuclear structures [1]. Such interactions lead to stereotypical arrangements of chromosomes with respect to each other and with respect to nuclear landmarks.

Individual genes often change their position when induced or repressed. A well-studied phenomenon that illustrates this point and that serves as an excellent model is the movement of yeast genes from the nucleoplasm to the nuclear periphery upon activation [2, 3]. Inducible genes such as *INO1*, *GAL1-10*, *GAL2*, *TSA2*, *HSP104* and *HXT1* move to the nuclear periphery and physically interact with the NPC upon activation [2-7]. Mutations in nuclear pore proteins (Nups) block targeting to the periphery [6, 8-10] and genome-wide ChIP experiments in yeast, flies, and mammalian cells indicates that hundreds to thousands of genes interact with NPCs or nuclear pore proteins [3, 11-14]. Thus, interaction with the NPC leads to changes in gene positioning.

Interaction of yeast genes with the NPC and positioning to the nuclear periphery requires small *cis*-acting DNA elements in their promoters [6, 7, 10]. For example, two elements called GRS I and GRS II in the *INO1* promoter are necessary for targeting to the NPC (Figure 1A; ref. [6]). These elements function as *DNA zip codes*: they are both necessary for *INO1* targeting and, when inserted at an ectopic site in the genome, they are sufficient to induce repositioning to the nuclear periphery and interaction with the NPC. The GRS I element binds to the Put3 transcription factor, which is required for GRS I-mediated positioning [7]. Thus, genomes *encode* subnuclear positioning through transcription factor binding sites that function as DNA zip codes.

In addition to promoting interaction with the NPC, DNA zip codes like GRS I promote interchromosomal clustering of genes [7]. The two alleles of *INO1* in diploid cells cluster together upon activation. In haploids, *INO1* clusters with another GRS I-targeted gene, *TSA2* and with GRS I inserted at an ectopic locus [7]. Mutations in GRS I, loss of Nups or loss of Put3 disrupt interchromosomal clustering. Therefore, DNA zip codes such as the GRS I, are necessary and

A

D

B

C

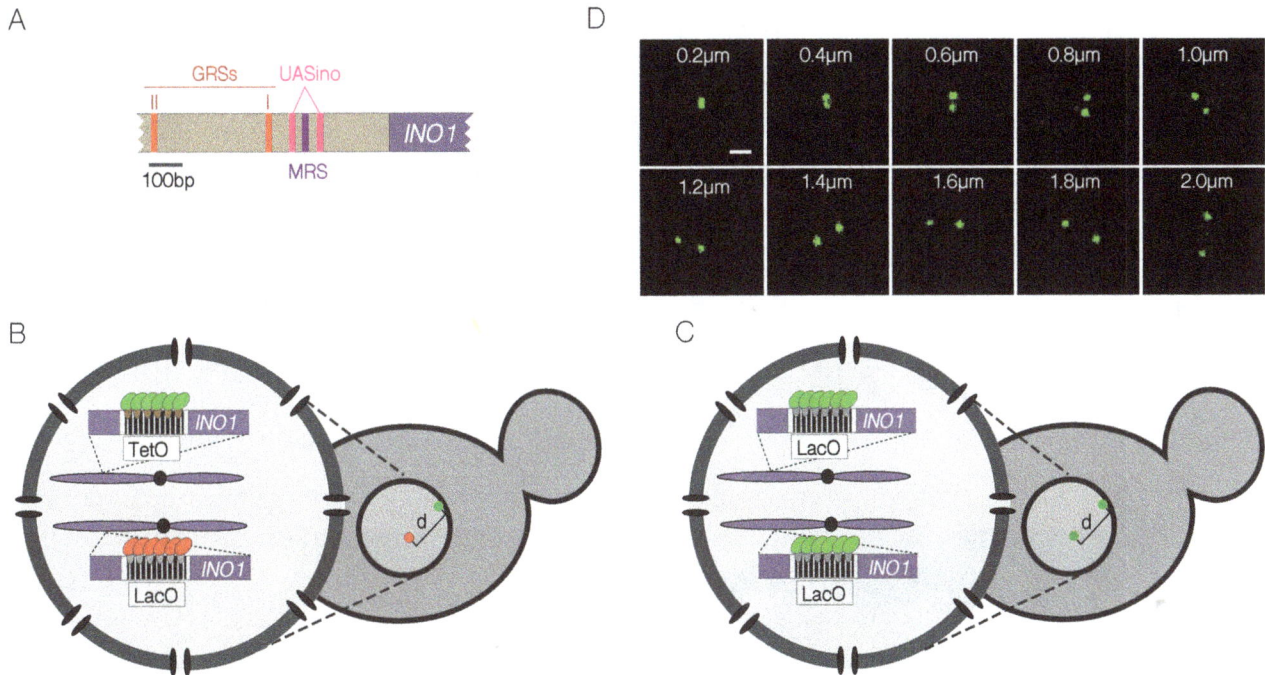

FIGURE 1: Experimental system. (A) Schematic of the *INO1* promoter, with the relevant regulatory elements and DNA zip codes highlighted. GRS: Gene Recruitment Sequence [6]; MRS: Memory Recruitment Sequence [10]; UAS$_{INO}$: Upstream Activating Sequence regulated by inositol. (B and C) Experimental setups for studying interchromosomal clustering using two different repressor arrays (B) or two identical arrays (C). (D) Representative confocal micrographs of cells having two GFP-marked arrays. Scale bar = 1μm.

sufficient to induce interchromosomal clustering of active loci through interaction with transcription factors and the NPC.

Upon repression, the *INO1* gene remains associated with the NPC for several generations, a phenomenon called epigenetic transcriptional memory [8]. This interaction involves a different *cis*-acting DNA zip code (the Memory Recruitment Sequence, MRS; Figure 1A) and different nuclear pore proteins (e.g. Nup100; refs [10], [15]). Mutations in the GRS I and II elements do not affect targeting to the NPC during memory and mutations in the MRS do not affect targeting to the NPC during activation, suggesting that these two mechanisms are independent [8]. Here, we show that transcriptional memory also leads to interchromosomal clustering. Clustering during memory requires both previous clustering of active *INO1* and the MRS zip code. However, unlike under activating conditions, *INO1* does not cluster with the GRS I at an ectopic site under memory conditions. Therefore, *INO1* clusters with different partners under activating and memory conditions, suggesting that interchromosomal clusters are remodeled upon repression. Finally, unlike targeting to the periphery [16], *INO1* clustering is regulated through the cell cycle under both activating and memory conditions. These results show that interchromosomal clustering of *INO1* during transcriptional memory is zip code-dependent but represents a molecular event that is distinct from targeting to the NPC.

RESULTS

INO1 clustering during transcriptional memory

To monitor clustering of *INO1*, we utilized several experimental systems: diploid strains in which one allele of *INO1* is marked with an array of 128 Lac Operator (LacO) repeats and the other allele is marked with an array of 112 Tet Operator (TetO) repeats (Figure 1B), diploid strains in which both alleles of *INO1* are marked with LacO (Figure 1C) or haploid strains in which *INO1* is marked with TetO and other sites (i.e. *URA3* or *GAL1*) are marked with LacO (similar to the system shown in Figure 1B). The distance between the two genes can be measured in each cell in a population (Figure 1D) and clustering can be assessed by comparing the distribution of distances between the two loci in the population or by measuring the fraction of cells in which the two loci are ≤ 0.55 μm apart [7].

We previously showed that *INO1* clusters with an ectopic copy of *INO1* integrated near the *URA3* locus in a haploid cell upon recruitment to the nuclear periphery [7]. To test if this clustering is maintained during memory, we compared the distances between *INO1-TetO* and *URA3:INO1-LacO* under long-term repressing conditions (overnight, +inositol) and under memory conditions (-inositol → +inositol, 3h). Under long-term repressing conditions, *INO1* and *URA3:INO1* do not obviously cluster together, showing a broad distribution with a mean distance of 0.85 ± 0.38 μm and 20% ≤ 0.55 μm (Figure 2A). This is similar to the distribution and clustering observed for two unrelated loci in haploid nuclei [7]. However, under

A

INO1 vs URA3:INO1 — mean distance: repressed $0.85 \pm 0.38\mu m$; memory $0.64 \pm 0.33\mu m$

Percent of cells vs Distance between loci (μm); $P = 0.0001$

Percent clustered ($<.55\mu m$): rep / mem; $P = 0.001$

B Intra-replicate subsampling

	n = 100	r50.1	r50.2	r50.3	r40.1	r40.2	r40.3
n = 100		0.59	0.76	0.57	0.75	0.46	0.85
r50.1			0.48	0.94	0.88	0.27	0.56
r50.2				0.50	0.60	0.74	0.91
r50.3					0.84	0.26	0.55
r40.1						0.37	0.65
r40.2							0.68
r40.3							

Legend: $P \leq 10^{-4}$; $P \leq 0.005$; $P \leq 0.05$; $P \geq 0.05$

C Inter-replicate subsampling — replicate 1

replicate 2 (rows)

	n = 100	r50.1	r50.2	r50.3	r40.1	r40.2	r40.3
n = 100	0.38	0.95	0.18	0.71	0.97	0.10	0.37
r50.1	0.44	0.89	0.26	0.79	0.87	0.15	0.41
r50.2	0.46	0.93	0.23	0.80	0.95	0.15	0.41
r50.3	0.18	0.56	0.09	0.89	0.56	0.05	0.19
r40.1	0.63	0.95	0.38	0.66	0.97	0.23	0.55
r40.2	0.43	0.85	0.22	0.81	0.89	0.13	0.41
r40.3	0.89	0.55	0.73	0.33	0.56	0.51	0.95

Legend: $P \leq 10^{-4}$; $P \leq 0.005$; $P \leq 0.05$; $P \geq 0.05$

D Inter-condition subsampling — memory

repressing (rows)

	n = 100	r50.1	r50.2	r50.3	r40.1	r40.2	r40.3
n = 100	4x10-8	3x10-5	5x10-6	5x10-7	2x10-4	2x10-5	0.001
r50.1	1x10-5	5x10-4	1x10-4	2x10-5	0.001	3x10-4	0.006
r50.2	2x10-6	1x10-4	3x10-5	4x10-6	4x10-4	6x10-5	0.003
r50.3	4x10-5	9x10-4	3x10-4	5x10-5	0.003	5x10-4	0.015
r40.1	0.001	0.008	0.003	0.001	0.014	0.005	0.042
r40.2	5x10-5	4x10-5	8x10-6	8x10-7	1x10-4	2x10-5	8x10-4
r40.3	0.001	0.009	0.004	1x10-3	0.019	0.005	0.058

Legend: $P \leq 10^{-4}$; $P \leq 0.005$; $P \leq 0.05$; $P \geq 0.05$

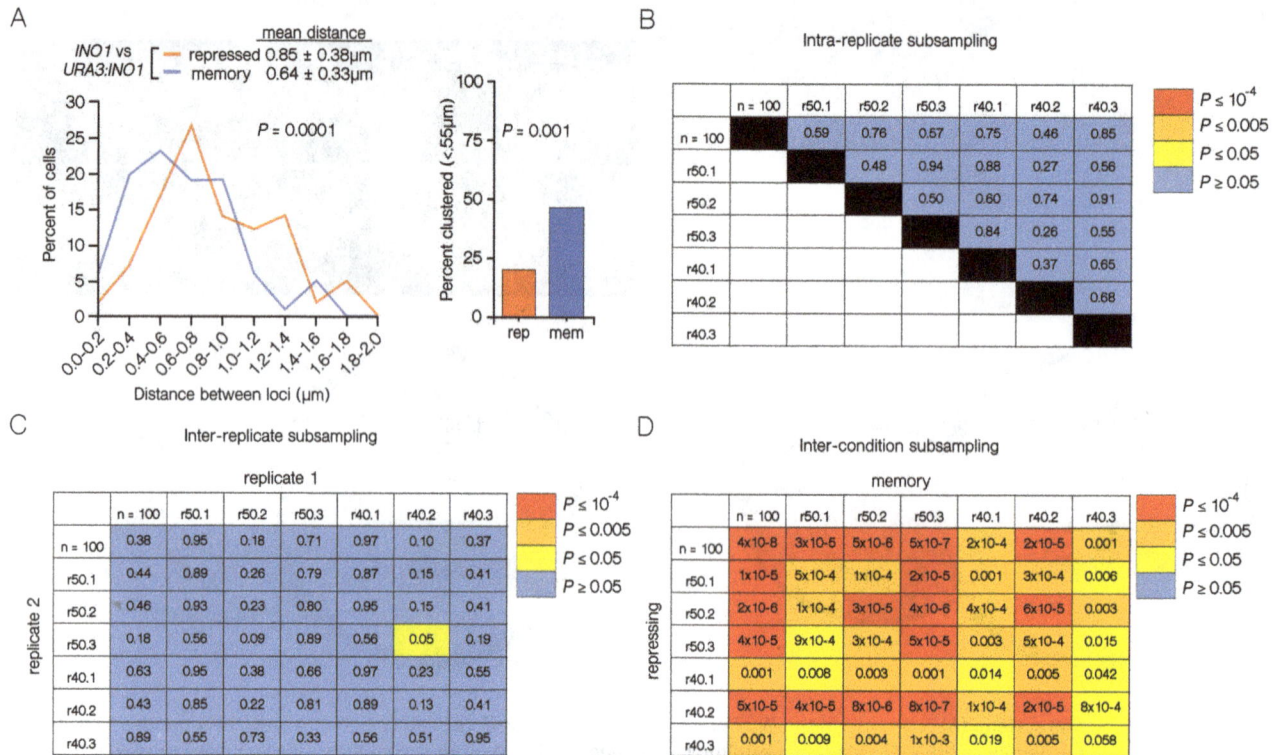

FIGURE 2: *INO1* transcriptional memory leads to interchromosomal clustering. (A) Haploid cells having the endogenous *INO1* gene marked with the TetO and *URA3:INO1* marked with the LacO, expressing GFP-TetR and mRFP-LacI [8] were grown under *INO1* repressing (+ inositol) or memory (- inositol → + inositol, 3h) conditions, fixed and processed for immunofluorescence against GFP and mRFP. **Left:** The distribution of distances between the two loci in ~100 cells, binned into 0.2 μm bins. *P* values were calculated using a Wilcoxon Rank Sum Test. **Right:** the fraction of cells in which the two loci were ≤ 0.55 μm. *P* values were calculated using a Fisher Exact Test. Note: the distribution of the repressed condition has been previously published [7] and is shown only for comparison to the distribution under the experimental (memory) condition. **(B-D)** Subsampling analysis. Full datasets (n = 100) or randomly generated subsamples of 50 or 40 measurements (r50 or r40, respectively) were compared pairwise using a Wilcoxon Rank Sum test. The numbers in each cell are the *P* values, color-coded as described in the legend. **(B)** A biological replicate compared with itself. **(C)** Two biological replicates compared with each other. **(D)** Distributions from repressing and memory conditions compared with each other.

memory conditions, there is a significant shift in the distribution to shorter distances (mean = 0.64 ± 0.33 μm; *P* = 0.0001, Wilcoxon Rank Sum test) and an increase in the fraction of cells in which the two loci are ≤ 0.55 μm (47%; *P* = 0.001, Fisher Exact test; Figure 2A). Thus, endogenous *INO1* remains clustered with an ectopic copy of *INO1* under memory conditions. For the experiments that follow (Figure 3, 4A), this distribution served as a control for comparison.

To assess the variance and sample size, we subjected the data to additional analysis. First, we collected three random subsamples (of 50 or 40 cells each; labeled r50 and r40 in Figure 2) from the data that were used to generate the distribution of distances under memory conditions (n = 100 cells) and compared them to the total dataset and to each other using a Wilcoxon Rank Sum test. This analysis showed that there is no significant difference between the total data and subsets of the data of ≥ 40 cells (Figure 2B).

To assess the variance between biological replicates, we performed the analysis above using total datasets (n ≥ 100 cells) or random subsamples from two independent biological replicates (Figure 2C). Of the 49 comparisons, only one (a subset of 40 compared with a subset of 50) was significantly different (*P* = 0.05; Figure 2C). In every comparison in which one of the two datasets was complete (n ≥ 100) or in which both datasets contained ≥ 50 cells, no significant differences were observed (Figure 2C). This suggests that data from ≥ 100 cells are oversampled and sufficient to avoid Type I errors (i.e. incorrect rejection of the null hypothesis that two datasets are the same).

Finally, to assess statistical power and sensitivity, we used random subsampling to compare two datasets from different conditions (repressing vs. memory conditions; Figure 3D). This analysis revealed that all sets of measurements containing ≥ 40 measurements were sufficient to reveal statistically significant difference and that the significance of the difference was greater when more cells were analyzed (Figure 3D). Therefore, comparing two datasets of 40 cells is sufficient to avoid Type II errors (failing to reject the null hypothesis that the two datasets are same), so we have measured ~ 100 cells per experiment in the work described here.

INO1 clustering during transcriptional memory is specific and MRS-dependent

To confirm that clustering under memory conditions is specific, we examined the positioning of *INO1* (marked with TetO) with respect to *GAL1* (marked with LacO) after simultaneously repressing both genes (Figure 3A). Both *INO1* and *GAL1* exhibit transcriptional memory, localizing at the nuclear periphery for several generations after repression [8]. However, unlike *INO1* and *URA3:INO1*, *INO1* and *GAL1* did not cluster under memory conditions (Figure 3A; 15% ≤ 0.55 μm). Therefore, the clustering of *INO1* with itself under memory conditions is specific.

The MRS zip code is specifically required for localization of *INO1* at the nuclear periphery during memory and has no role in targeting of active *INO1* to the nuclear periphery [10]. Mutation of the MRS disrupts peripheral localization under memory conditions. Therefore, we asked if the MRS is necessary for clustering by measuring the distances between *INO1* and *URA3:INO1* having a mutation in the MRS element (*mrs INO1*) under repressing and memory conditions (Figure 3B). Mutation of the MRS disrupted *INO1* clustering during memory (13% ≤ 0.55 μm), indicating that the clustering of *INO1* with *URA3:INO1* after repression requires the MRS zip code.

The MRS is both necessary and sufficient to promote targeting to the nuclear periphery under memory conditions [10]. To test if the MRS is sufficient to induce cluster-ing with *INO1* during memory, we compared the positions of *INO1* and *URA3:MRS50* (a 50 bp element including the MRS; ref. [10]) during memory. The MRS50 was not suffi-cient to induce clustering with the endogenous *INO1* gene (Figure 3C; 6% ≤ 0.55 μm). This suggests that additional information besides the MRS is provided by the *INO1* gene to promote clustering during transcriptional memory.

A hierarchy of DNA zip codes controls *INO1* clustering during transcriptional memory

To explore what other signals might be important for clus-tering of *INO1* during transcriptional memory, we tested other DNA zip codes. Targeting of active *INO1* to the nucle-ar periphery requires the GRS I and GRS II zip codes and mutation of both elements blocks targeting of active *INO1* to the nuclear periphery and disrupts interchromosomal clustering (Figure 1A and ref. [7]). However, these muta-tions do not affect targeting of *INO1* to the nuclear periph-ery during memory, indicating that MRS-mediated target-ing to the nuclear periphery is independent of GRS-mediated targeting [10]. Because the MRS was necessary but not sufficient to induce clustering during memory, we hypothesized that the interchromosomal interactions of active genes might be required for the persistent clustering after repression.

To test this proposal, we examined the interaction be-tween wild type *INO1* and *URA3:INO1* having mutations in

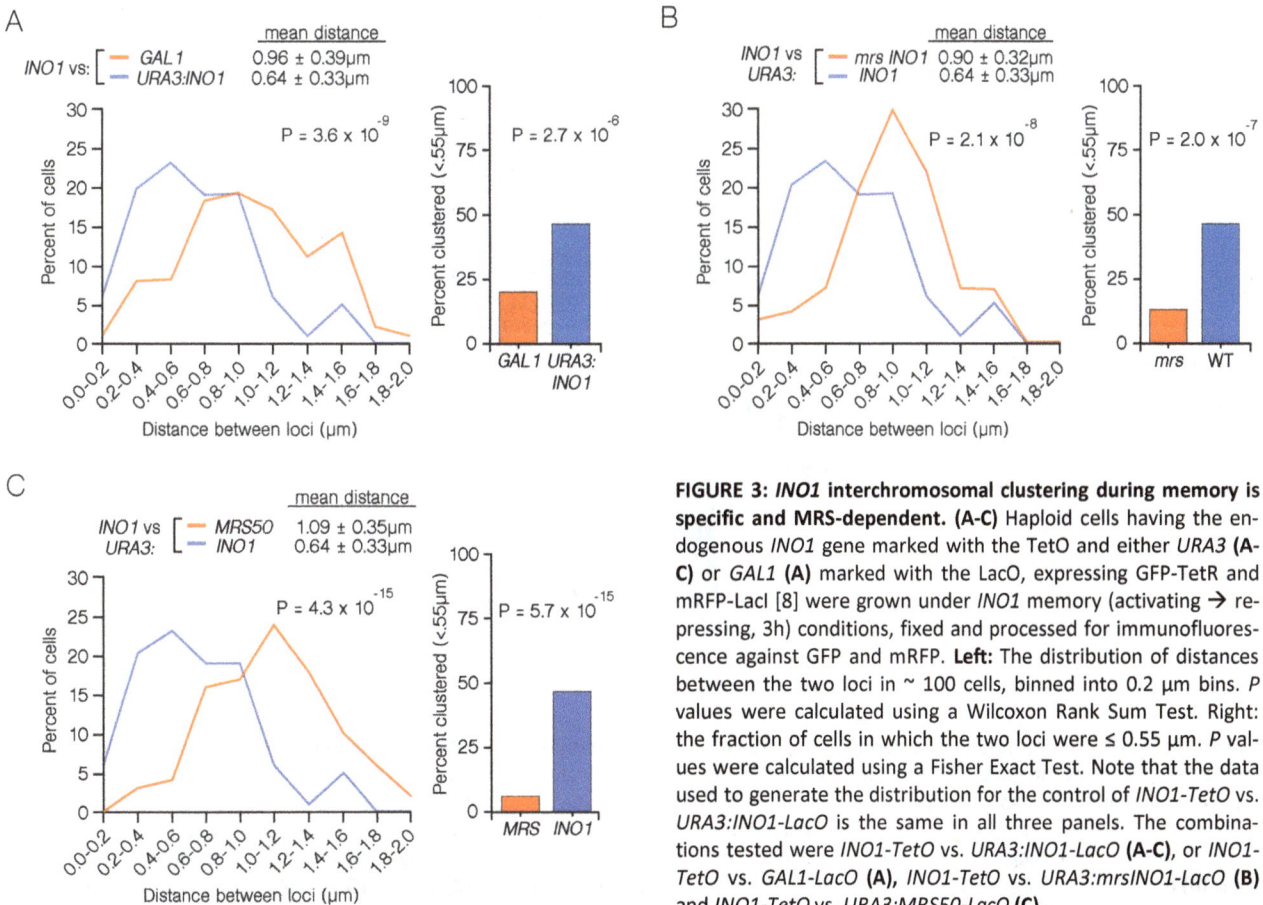

FIGURE 3: *INO1* interchromosomal clustering during memory is specific and MRS-dependent. (A-C) Haploid cells having the en-dogenous *INO1* gene marked with the TetO and either *URA3* (A-C) or *GAL1* (A) marked with the LacO, expressing GFP-TetR and mRFP-LacI [8] were grown under *INO1* memory (activating → re-pressing, 3h) conditions, fixed and processed for immunofluores-cence against GFP and mRFP. **Left:** The distribution of distances between the two loci in ~ 100 cells, binned into 0.2 μm bins. *P* values were calculated using a Wilcoxon Rank Sum Test. Right: the fraction of cells in which the two loci were ≤ 0.55 μm. *P* val-ues were calculated using a Fisher Exact Test. Note that the data used to generate the distribution for the control of *INO1-TetO* vs. *URA3:INO1-LacO* is the same in all three panels. The combina-tions tested were *INO1-TetO* vs. *URA3:INO1-LacO* (A-C), or *INO1-TetO* vs. *GAL1-LacO* (A), *INO1-TetO* vs. *URA3:mrsINO1-LacO* (B) and *INO1-TetO* vs. *URA3:MRS50-LacO* (C).

the GRS I and GRS II (*grs1,2 INO1*). In this strain, clustering is lost under both activating [8] and memory conditions (Figure 4A; 17% ≤ 0.55 μm). Therefore, even though targeting to the nuclear periphery during transcriptional memory is GRS I- and GRS II-independent, clustering of *INO1* during memory is GRS I- and GRS II-dependent. This suggests that targeting to the nuclear periphery and interchromosomal clustering during memory must represent distinct molecular events.

Although both GRS I and GRS II are capable of mediating targeting to the nuclear periphery, GRS I directs interchromosomal clustering of *INO1* with itself and GRS II does not [8]. This suggests that targeting to the periphery is necessary, but not sufficient, to promote interchromosomal clustering. Because the loss of GRS I and GRS II disrupted clustering of *INO1* during transcriptional memory, we next asked if this is due to loss of peripheral targeting under activating conditions or due to loss of interchromosomal clustering. Mutants that lack the GRS I - binding protein Put3 still target *INO1* to the nuclear periphery normally (due to GRS II function), but fail to cluster [8]. We tested the effect of loss of Put3 on *INO1* clustering during memory (Figure 4B). As expected, clustering of active *INO1* with *URA3:INO1* was disrupted in *put3Δ* mutants (Figure 4B). Consistent with the idea that clustering of active *INO1* being required for clustering of recently repressed *INO1*, loss of Put3 also disrupted clustering of *INO1* with *URA3:INO1* during memory (Figure 4B).

This result led us to ask if the GRS I alone, which is sufficient to induce clustering with active *INO1*, or GRS II, which is not [8], could also induce clustering with *INO1* during memory conditions. Consistent with our previously published work, active *INO1* clusters strongly with *URA3:GRS I* (ref. [7]; Figure 4C; 58% ≤ 0.55 μm) but not with *URA3:GRS II* (Figure 4D; 13% ≤ 0.55 μm). *INO1* did not cluster with either *URA3:GRS I* (Figure 4C; 11% ≤ 0.55 μm) or *URA3:GRS II* (Figure 4D; 15% ≤ 0.55 μm) under memory conditions. Therefore, neither the GRS I nor the GRS II are

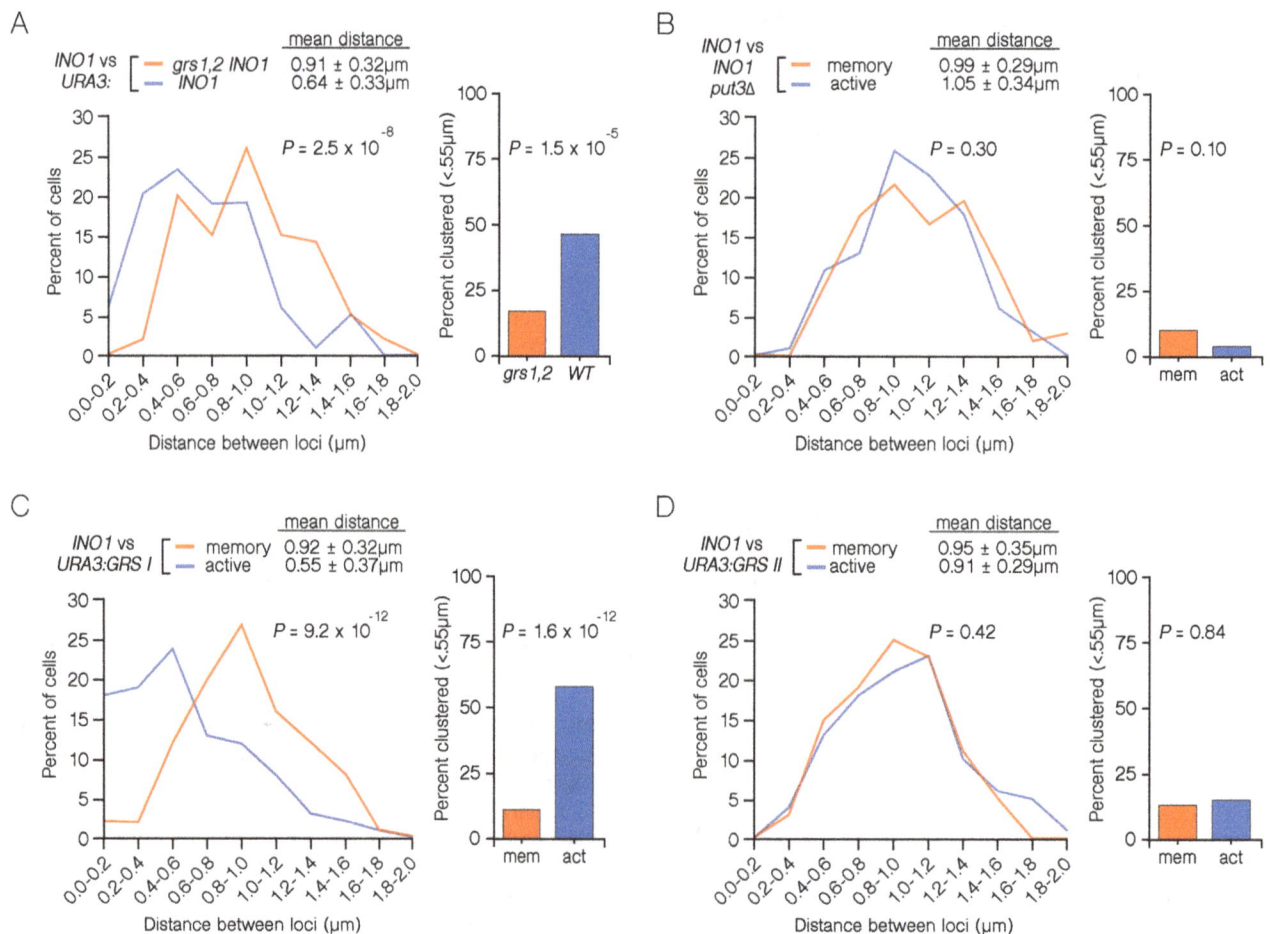

FIGURE 4: *INO1* **interchromosomal clustering during transcriptional memory requires clustering of active** *INO1*. **(A and B)** Haploid cells having the LacO array integrated at *URA3* and the TetO array integrated at *INO1,* and expressing GFP-TetR and mRFP-LacI were fixed and processed for immunofluorescence against GFP and mRFP. **Left:** The distribution of distances between the two loci in ~ 100 cells, binned into 0.2 μm bins. *P* values were calculated using a Wilcoxon Rank Sum Test. **Right:** the fraction of cells in which the two loci were ≤ 0.55 μm. *P* values were calculated using a Fisher Exact Test. **(A)** *INO1-TetO* vs. *URA3:INO1-LacO* or *URA3:grs1,2 INO1-LacO* grown under memory conditions. **(B)** *INO1-TetO* vs. *URA3:GRS I-LacO* under either activating or memory conditions. **(C)** *INO1-TetO* vs *URA3:GRS II-LacO* under activating or memory conditions. **(D)** *INO1-TetO* vs. *URA3:INO1-LacO put3Δ* cells under activating or memory conditions.

sufficient to induce clustering with *INO1* during memory and the clustering of *INO1* with *URA3:GRS I* is not stably maintained after *INO1* repression.

Interchromosomal clustering of *INO1* during transcriptional memory requires Nup100

The interaction of *INO1* with the NPC is different under activating and memory conditions. The interaction of the NPC with these two states requires different *cis*-acting DNA zip codes and different Nups [10]. One such protein is Nup100; mutants lacking Nup100 target *INO1* to the periphery under activating conditions but not under memory conditions [10].

To explore the role of the nuclear pore in clustering of *INO1* during transcriptional memory, we measured the distances between *INO1* alleles (marked with LacO and TetO; Figure 1B) in wild type or *nup100Δ/nup100Δ* diploid cells grown under activating, repressing and memory conditions. Under both activating or memory conditions in the wild type (*NUP100/NUP100*) diploid cells, *INO1* clustered to very similar extent (Figures 5A and 5B). Nup100 is dispensable for clustering of active *INO1* alleles; in diploid

cells lacking Nup100, the two active alleles of *INO1* clustered together (Figure 5C; mean distance = 0.74 ± 0.43 µm; 51% ≤ 0.55 µm). However, Nup100 is specifically required for *INO1* clustering during memory; under memory conditions, *INO1* clustering was lost in *nup100Δ/nup100Δ* diploid cells (Figure 5D; mean distance = 1.11 ± 0.52 µm; 18% ≤ 0.55 µm).

Cell cycle regulation of interchromosomal clustering of *INO1* during memory

Positioning of active *INO1*, *GAL1* and *HSP104* at the nuclear periphery is regulated through the cell cycle. Immediately after the initiation of DNA replication, all three of these genes reposition to the nucleoplasm for ~ 30 minutes before returning to the periphery during mitosis [16, 17]. The clustering of active *INO1* with itself is maintained during S-phase in the nucleoplasm [7]. Thus, the cell cycle regulation of peripheral localization is uncoupled from interchromosomal clustering.

In contrast to the active genes, during transcriptional memory, *INO1* and *GAL1* remain at the nuclear periphery throughout the cell cycle [16]. Thus, it was unclear if clus-

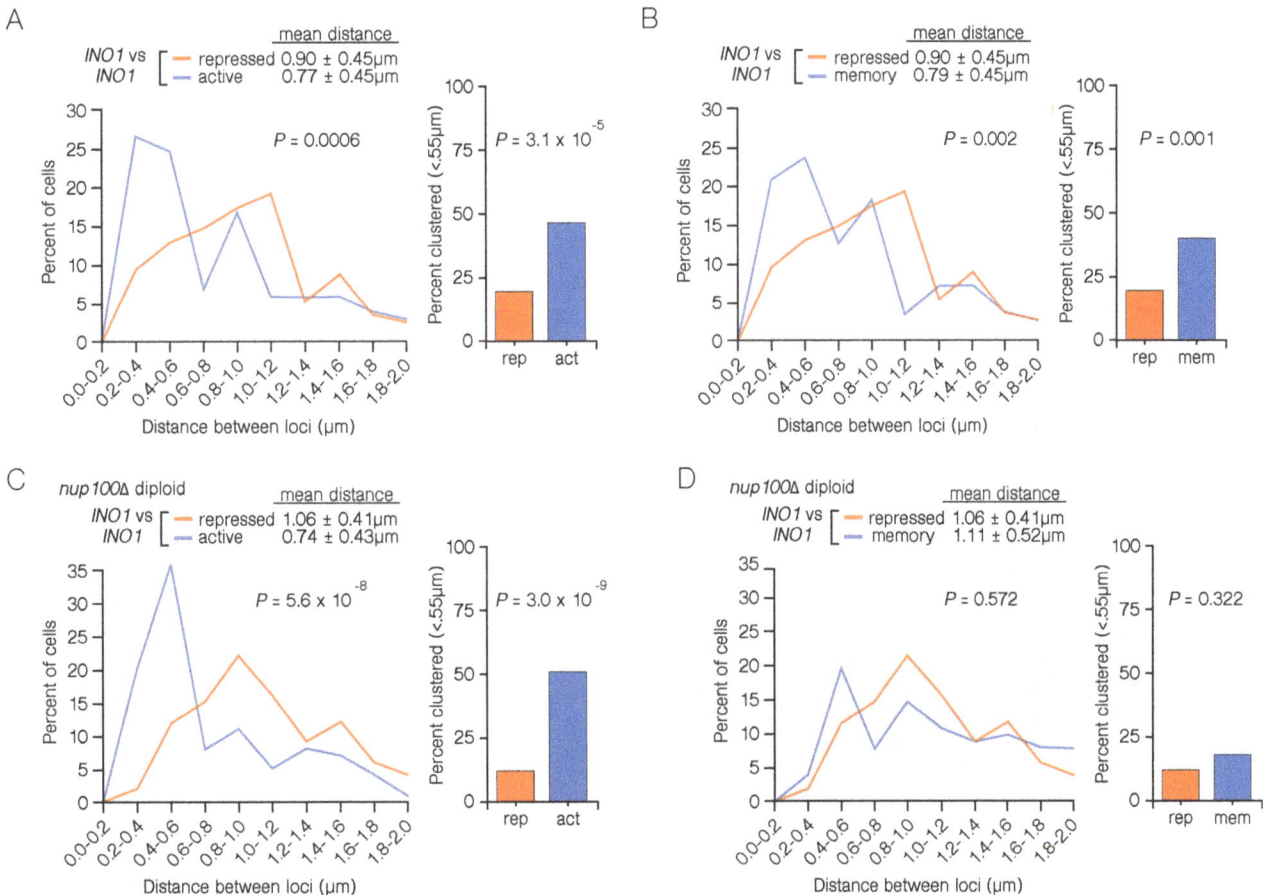

FIGURE 5: Nup100 is specifically required for *INO1* clustering during memory. Diploid cells having both alleles of *INO1* marked with the LacO array and expressing GFP-LacI were grown under repressing, activating or memory conditions. **Left:** The distribution of distances between the two loci in ~ 100 cells, binned into 0.2 µm bins. *P* values were calculated using a Wilcoxon Rank Sum Test. **Right:** the fraction of cells in which the two loci were ≤ 0.55 µm. *P* values were calculated using a Fisher Exact Test. **(A and B)** *NUP100/NUP100* diploids. **(C and D)** *nup100Δ/nup100Δ* diploids.

tering would also be maintained through the cell cycle. We tested this idea by comparing the % of unbudded (G1), small budded (S) and large budded (G2) cells in an asynchronous population in which the two alleles were ≤ 0.55 μm (Figure 6A). We utilized a diploid strain in which both copies of INO1 were marked with LacO arrays bound to GFP-LacI (Figure 1C and 1D). This experimental setup allows rapid 3D mapping of the two alleles in live cells. However, because both spots are the same color, it is not possible to score cells in which the spots are unresolvable (i.e. only a single spot is observed), which includes some cells in which the two loci are ≤ 0.2 μm apart. Despite this limitation, we could observe clustering of INO1 alleles under both activating (Figure 6A) and memory conditions (Figure 6B) using this system.

Distances between the two alleles of INO1 were measured in unbudded G1, small budded S-phase and large budded G2 cells under memory conditions (Figure 6C). In unbudded cells and in small budded, we observed high level clustering (57% ≤ 0.55 μm; Figure 6D). However, clus-

tering of INO1 was lost in large budded cells (9% ≤ 0.55 μm; Figure 6D). In these cells, which have not yet undergone nuclear division, each allele of recently repressed INO1 remains at the nuclear periphery [16], paired with its sister chromatid. However, the two alleles of INO1 do not cluster together, suggesting that clustering during memory is uncoupled from gene positioning to the periphery.

DISCUSSION

Interchromosomal clustering is a common phenomenon in eukaryotic cells. In budding yeast, as in many organisms, the 32 telomeres cluster into several foci at the nuclear periphery [18]. Likewise, the ~ 250 tRNA genes, scattered throughout the genome, form two clusters, one in the nucleolus and the other near the spindle pole body [19]. In flies, Polycomb repressed loci cluster together into Polycomb bodies [20, 21]. And in mammalian cells, co-regulated genes frequently co-localize within the nucleus [22-24]. Thus, cells frequently utilize interchromosomal clustering as a mechanism of spatial, and perhaps tran-

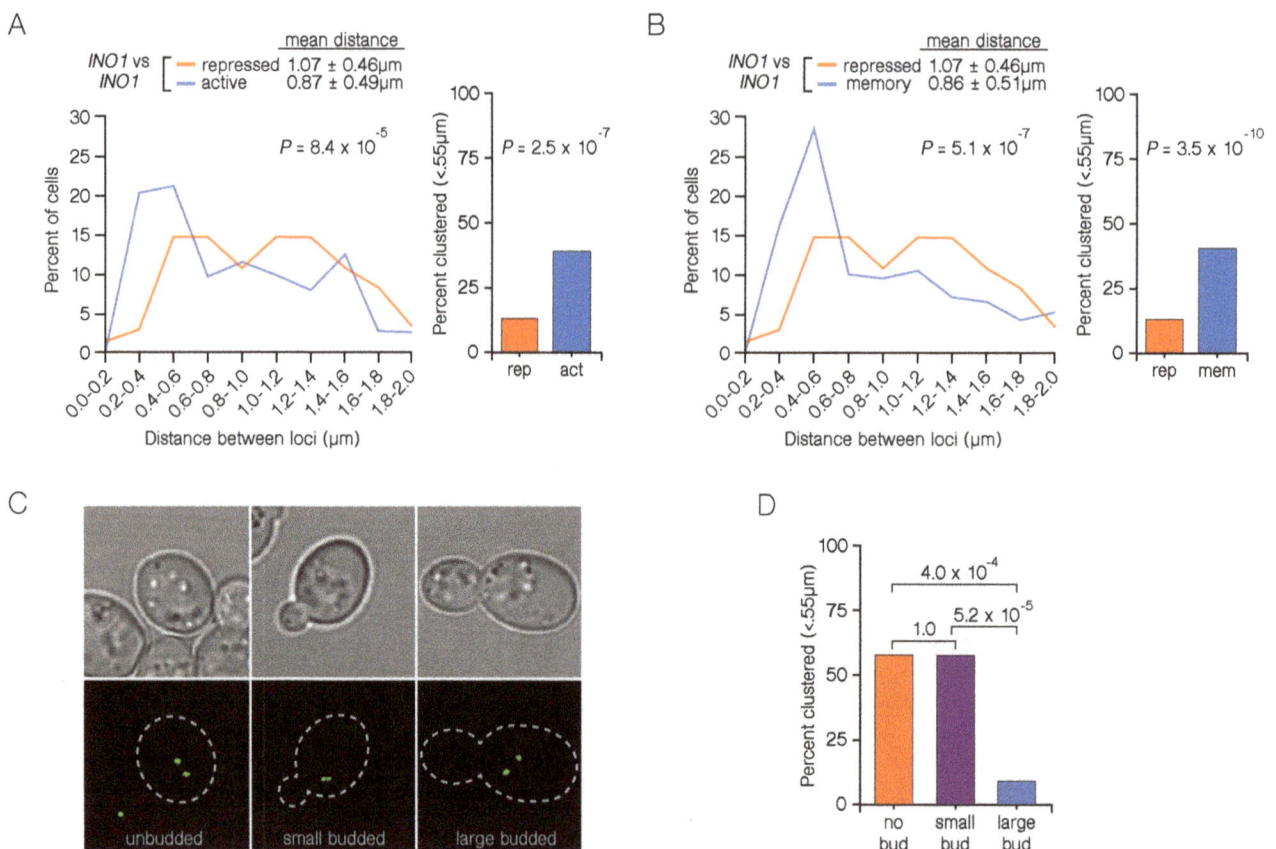

FIGURE 6: Interchromosomal clustering of recently repressed INO1 is regulated through the cell cycle. (A and B) An asynchronous population of cells having INO1:TetO and INO1:LacO, expressing GFP-TetR, GFP-LacI and Pho88-mCherry (ER/nuclear envelope membrane protein) were grown under repressing, activating or memory conditions. **Left:** The distribution of distances between the two loci in ~ 100 cells, binned into 0.2 μm bins. P values were calculated using a Wilcoxon Rank Sum Test. **Right:** the fraction of cells in which the two loci were ≤ 0.55 μm. P values were calculated using a Fisher Exact Test. **(C)** Bright field (top) and green fluorescence (bottom) channels of typical cells used to measure distances in cells with different bud morphologies. The outline of the cell above is overlaid on the green channel (hatched line). **(D)** Cells were scored for both their bud morphology and the distance between the two loci and the fraction of each class of cells that was ≤ 0.55 μm was determined. P values were calculated using the Fisher Exact Test.

scriptional, control.

Our previous work has established another mechanism by which interchromosomal clustering is facilitated. Interaction of active yeast genes with the NPC leads to both repositioning to the nuclear periphery and clustering with other loci that share the same zip code [7]. DNA zip codes are both necessary and sufficient to induce this type of interchromosomal clustering. It is unclear if clustering reflects targeting of genes to the same portion of the nuclear envelope or homotypic interactions between genes after they are targeted to an imprecise location on the envelope. Nups are essential for clustering, suggesting that targeting to the NPC is a prerequisite for clustering. However, once formed, the clusters can persist in the nucleoplasm [7]. This argues either that clustering is maintained by Nups that dissociate from the NPC, as has been shown to occur in many cell types, or that the NPC serves as a site of assembly for clusters but is not required after they are assembled.

Here we have explored a new model for interchromosomal clustering between genes that are poised for future activation. Of the genes that interact with the NPC when they are active, a small subset shows transcriptional memory, maintaining the interaction after repression (Figure 7A; ref. [8]). The Nups and the DNA zip codes required for memory are distinct from those required for peripheral localization and clustering of active genes (Figure 7A; ref. [10]). Here, we find that the *INO1* gene remains clustered with itself after repression. However, the molecular mechanism of clustering during memory is very different from the mechanism of clustering of active *INO1*.

Active *INO1* clusters both with itself and with other GRS I-containing genes [7]. Although we cannot rule out that *INO1* clusters with other loci during memory, this mechanism is more selective than GRS I-mediated clustering (Figure 7B). The MRS zip code is necessary, but not sufficient (when inserted at an ectopic site), to induce clustering with *INO1* during memory. Additional information besides the MRS is required for clustering.

In addition to the MRS, the GRS zip codes are required for *INO1* clustering during memory, suggesting that clustering during memory requires previous clustering during the activating condition (Figure 7B). In other words, clustering during transcriptional memory reflects the previous clustering of active *INO1*. This is surprising because the mechanisms of targeting to the nuclear periphery when *INO1* is active or during memory are completely independent [10]. Therefore, although interchromosomal clustering requires targeting to the nuclear periphery and is lost in cells in which either the zip code (MRS) or Nups (i.e. Nup100) are mutated, it has additional requirements and represents a different molecular mechanism.

Consistent with this idea, the regulation of interchromosomal clustering through the cell cycle does not perfectly mirror peripheral localization. Whereas targeting of active *INO1* and other genes is lost briefly during S-phase, peripheral localization of *INO1* during transcriptional memory is not regulated by the cell cycle [16]. In contrast, the interchromosomal clustering of *INO1* alleles during

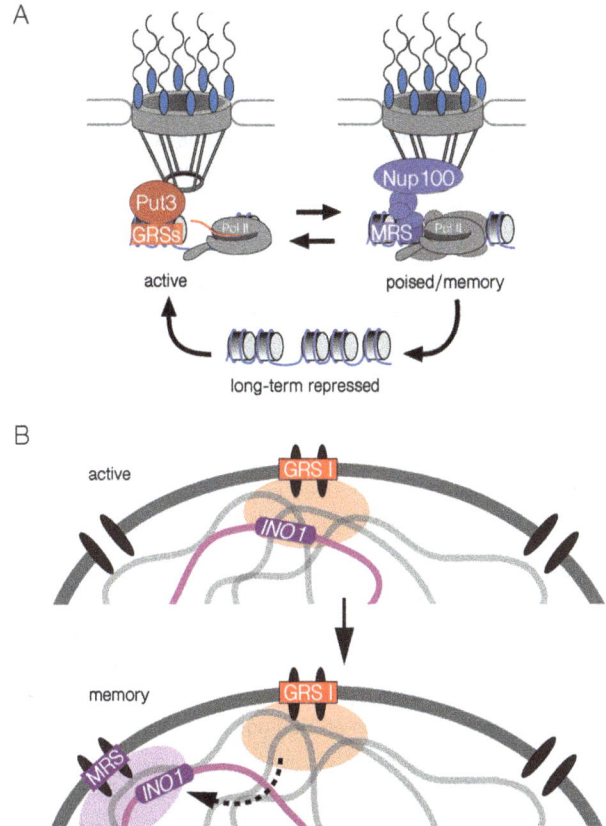

FIGURE 7: Model for zip code-dependent *INO1* targeting to the NPC and interchromosomal clustering. (A) The long-term repressed gene is positioned in the nucleoplasm and both the active and recently repressed memory state of the gene are positioned at the nuclear periphery through interaction with the NPC. The GRS elements control targeting to the NPC under activating conditions. The Put3 transcription factor binds the GRS I zip code and is required for GRS I-mediated peripheral targeting and interchromosomal clustering [7]. The MRS element controls targeting to the NPC under memory conditions and requires Nup100 [10]. **(B)** The *INO1* gene clusters with other GRS I-containing loci under activating conditions (top) and this is a prerequisite for clustering with itself (and potentially other loci) in an MRS-dependent cluster for several generations after repression, during transcriptional memory.

memory is lost during G2. This cell cycle regulation may reflect the coordination of interchromosomal clustering with chromosome condensation and chromosome segregation.

Most of the interchromosomal clustering events described previously are associated with either silenced sites like subtelomeric regions and Polycomb repressed sites or expressed sites like tRNA genes or induced genes. Here we show that a third class of genes, those that are repressed but poised for future induction, can also cluster in the nucleus. This interaction is specific to a moment in the life history of the *INO1* gene: *INO1* clusters with itself for 3-4 generations after repression, after which this is lost. Thus, *INO1* epigenetic transcriptional memory, which leads to heritable changes in chromatin structure and binding of poised RNA Polymerase II to the promoter, also leads to heritable interchromosomal clustering.

MATERIALS AND METHODS
Chemicals, yeast strains and growth conditions
Chemicals were from Sigma Aldrich, molecular biology enzymes were from New England Biolabs and yeast growth media components were from Sunrise Science Products. For experiments involving scoring peripheral localization in live cells, *PHO88* was C-terminally tagged with mCherry using PCR-based integration [25]. Briefly, the *PHO88* termination codon was replaced by homologous recombination with a PCR product encoding a C-terminal translational fusion to mCherry, along with the *His5+* gene from *S. pombe*. PCR primers were designed to incorporate 45 base pairs of sequences upstream and downstream of the termination codon. All yeast strains used in this study are described in Table 1. Media, transformations and growth conditions are described in [26, 27].

Microscopy
For all experiments, cells were maintained at $OD_{600} < 0.5$. For inositol starvation experiments, strains were grown overnight in SDC-inositol in the presence or absence of 100 µM *myo*-inositol. To induce memory, inositol was added to 100 µM and the cells were grown for an additional 3h. Immunofluores-cence microscopy was carried out as described [7, 27]. For live cell experiments, cells were concentrated by brief centrifugation and imaged immediately. The images were captured as 0.34 µm thick z-stacks with a Leica SP5 II Line Scanning Confocal Microscope with 100 × 1.44NA (oil immersion) objective using an Argon 488 nm and Diode Pumped Solid State 561 nm lasers in the Northwestern Biological Imaging Facility as described [28]. Distances between the centers of each dot for cells in which the dots were in the same z slice were measured using LAS LiTE software.

ACKNOWLEDGMENTS
This work was supported by NIH grant GM080484 (JHB). JHB is the Sorretta and Henry Shapiro Research Professor in Molecular Biology. RC was supported by Undergraduate Research Grants from the Program in Biological Sciences and Northwestern University as well as a Krieghbaum Award. The authors thank members of the Brickner lab for helpful discussions and feedback on the manuscript.

CONFLICT OF INTEREST
The authors declare no conflict of interest.

TABLE 1. Yeast strains used in this study.

Strain	Genotype	Figures
DBY326	MATa ade2-1 can1-100 his3-11,15 leu2-3,112 trp1-1 ura3-1 INO1:TetO-Nat LEU2:TetR-GFP TRP1:LacI-RFP GAL1:p6LacO128	Figure 3A
DBY336	MATa ade2-1 can1-100 his3-11,15 leu2-3,112 trp1-1 ura3-1 INO1:TetO-Nat LEU2:TetR-GFP TRP1:LacI-RFP URA3:p6LacO128-INO1	Figs 2A, 3A-C, 4A
DBY339	MATa ade2-1 can1-100 his3-11,15 leu2-3,112 trp1-1 ura3-1 INO1:TetO-Nat LEU2:TetR-GFP TRP1:LacI-RFP URA3:p6LacOGRS$_{241-270}$	Figure 4C
DBY348	MATa/MATα ade2-1/ade2-1 can1-100/can1-100 his3-11,15/his3-11,15 leu2-3,112/ LEU2:TetR-GFP trp1-1/ TRP1:LacI-RFP ura3-1/URA3:p6LacO128-INO1 INO1/INO1:TetO-Nat	Figure 3C
DBY354	MATa ade2-1 can1-100 his3-11,15 leu2-3,112 trp1-1 ura3-1 HIS3:LacI-GFP Sec63-myc:TRP1 LEU2:LacI-GFP URA3:p6LacO128 grs1,2mtINO1:p6LacO128	Figure 4A
DBY355	MATa ade2-1 can1-100 his3-11,15 leu2-3,112 trp1-1 ura3-1 INO1:TetO-Nat LEU2:TetR-GFP TRP1:LacI-RFP URA3:p6LacO-mrsmtINO1	Figure 3B
DBY362	MATa/MATα ade2-1/ade2-1 can1-100/can1-100 his3-11,15/ HIS3:LacI-GFP leu2-3,112/ LEU2:TetR-GFP trp1-1/trp1-1 ura3-1/ URA3:p6LacO128-MRS50 INO1/INO1:TetO-Nat SEC63/SEC63-13myc::Kan^R	Figure 3C
DBY515	MATa/MATα ade2-1/ade2-1 can1-100/can1-100 his3-11,15/ HIS3:LacI-GFP leu2-3,112/ LEU2:LacI-GFP trp1-1/trp1-1 ura3-1/ ura3-1 INO1:p6LacO128/INO1:p6LacO128	Figure 5A and B
DBY766	MATa/MATα ade2-1/ade2-1 can1-100/can1-100 his3-11,15/his3-11,15 LEU2:TetR-GFP/LEU2:LacI-GFP trp1-1/trp1-1 ura3-1/ura3-1 PHO88/PHO88-mCherry:SpHis5$^+$ INO1:TetO-Nat/INO1:p6LacO128	Figure 6
DBY768	MATa/MATα ade2-1/ADE2 can1-100/can1-100 his3-11,15/HIS3:LacI-GFP leu2-3,112/leu2-3,112 trp1-1/TRP1:Sec63-13myc ura3-1/ura3-1 nup100Δ::Kan^R/nup100Δ::Kan^R INO1:p6LacO128/INO1:p6LacO128	Figure 5C and D
DBY796	MATa ade2-1 can1-100 his3-11,15 leu2-3,112 trp1-1 ura3-1 INO1:TetO-Nat LEU2:TetR-GFP HIS3:LacI-GFP URA3:GRS II-p6LacO128	Figure 4D
DBY798	MATa ade2-1 can1-100 his3-11,15 leu2-3,112 trp1-1 ura3-1 INO1:TetO-Nat LEU2:TetR-GFP HIS3:LacI-GFP URA3:p6LacO128-INO1 put3Δ	Figure 4B

REFERENCES

1. Meldi L & Brickner JH (**2011**). Compartmentalization of the nucleus. **Trends Cell Biol** 21: 701-708.

2. Brickner JH and Walter P (**2004**). Gene recruitment of the activated INO1 locus to the nuclear membrane. **PLoS Biol.** 2:e342.

3. Casolari JM, Brown CR, Komili S, West J, Hieronymus H and Silver PA (**2004**). Genome-wide localization of the nuclear transport machinery couples transcriptional status and nuclear organization. **Cell** 117, 427-439.

4. Dieppois G, Iglesias N and Stutz F (**2006**). Cotranscriptional recruitment to the mRNA export receptor Mex67p contributes to nuclear pore anchoring of activated genes. **Mol Cell Biol.** 26:7858-7870.

5. Taddei A, Van Houwe G, Hediger F, Kalck V, Cubizolles F, Schober H and Gasser SM (**2006**). Nuclear pore association confers optimal expression levels for an inducible yeast gene. **Nature** 441:774-778.

6. Ahmed S, Brickner DG, Light WH, Cajigas I, McDonough M, Froyshteter AB, Volpe T and Brickner JH (**2010**). DNA zip codes control an ancient mechanism for gene targeting to the nuclear periphery. **Nat Cell Biol** 12:111-118.

7. Brickner DG, Ahmed S, Meldi L, Thompson A, Light WH, Young M, Hickman TL, Chu F, Fabre E and Brickner JH (**2012**). Transcription factor binding to a DNA zip code controls interchromosomal clustering at the nuclear periphery. **Dev Cell** 22:1234-1246.

8. Brickner DG, Cajigas I, Fondufe-Mittendorf Y, Ahmed S, Lee PC, Widom J and Brickner JH (**2007**). H2A.Z-mediated localization of genes at the nuclear periphery confers epigenetic memory of previous transcriptional state. **PLoS Biol** 5:e81.

9. Cabal GG, Genovesio A, Rodriguez-Navarro S, Zimmer C, Gadal O, Lesne A, et al. (**2006**). SAGA interacting factors confine sub-diffusion of transcribed genes to the nuclear envelope. **Nature** 441:770-773.

10. Light WH, Brickner DG, Brand VR and Brickner JH (**2010**). Interaction of a DNA zip code with the nuclear pore complex promotes H2A.Z incorporation and INO1 transcriptional memory. **Mol Cell** 40:112-125.

11. Casolari JM, Brown CR, Drubin DA, Rando OJ. and Silver PA (**2005**). Developmentally induced changes in transcriptional program alter spatial organization across chromosomes. **Genes Dev** 19:1188-1198.

12. Brown CR, Kennedy CJ, Delmar VA, Forbes DJ and Silver PA (**2008**). Global histone acetylation induces functional genomic reorganization at mammalian nuclear pore complexes. **Genes Dev** 22:627-639.

13. Capelson M, Liang Y, Schulte R, Mair W, Wagner U and Hetzer MW (**2010**). Chromatin-bound nuclear pore components regulate gene expression in higher eukaryotes. **Cell** 140:372-383.

14. Kalverda B, Pickersgill H, Shloma VV and Fornerod M (**2010**). Nucleoporins directly stimulate expression of developmental and cell-cycle genes inside the nucleoplasm. **Cell** 140:360-371.

15. Light WH, Freaney J, Sood V, Thompson A, D'Urso A, Horvath C and Brickner JH (**2013**). A conserved role for human Nup98 in altering chromatin structure and promoting epigenetic transcriptional memory. **PLoS Biology** 11:e1001524.

16. Brickner DG and Brickner JH (**2010**). Cdk phosphorylation of a nucleoporin controls gene localization through the cell cycle. **Mol Biol Cell** 21:3421-3432.

17. Brickner DG and Brickner JH (**2011**). Gene positioning is regulated by phosphorylation of the nuclear pore complex by Cdk1. **Cell Cycle** 10: 392-395.

18. Gotta M, Laroche T, Formenton A, Maillet L, Scherthan H and Gasser SM (**1996**). The clustering of telomeres and colocalization with Rap1, Sir3, and Sir4 proteins in wild-type Sac*charomyces cerevisiae*. **J Cell Biol** *134*, 1349-1363.

19. Thompson M, Haeusler RA, Good, PD and Engelke DR (**2003**). Nucleolar clustering of tRNA genes. **Science** 302:1399-1401.

20. Lanzuolo C, Roure V, Dekker J, Bantignies F and Orlando V (**2007**) Polycomb response elements mediate the formation of chromosome higher order structures in the bithorax complex. **Nat Cell Biol** 10:1167-1174.

21. Tolhuis B, Blom M, Kerkhoven RM, Pagie L, Tenissen H, Nieuwland M, Simonis M, de Laat W, van Lohuizen M and van Steensel B (**2011**). Interactions among Polycomb domains are guided by chromosome architecture. **PLoS Genet** 3:e1001343.

22. Brown JM, Leach J, Reittie JE, Atzberger A, Lee-Prudhoe J, Wood WG, Higgs DR, Iborra FJ and Buckle VJ (**2006**). Coregulated human globin genes are frequently in spatial proximity when active. **J Cell Biol** *172*, 177-187.

23. Schoenfelder S, Sexton T, Chakalova L, Cope NF, Horton A, Andrews S, Kurukuti S, Mitchell JA, Umlauf D, Dimitrova DS, Eskiw CH, Luo Y, Wei CL, Ruan Y, Beiker JJ and Fraser P (**2010**). Preferential associations between co-regulated genes reveal a transcriptional interactome in erythroid cells. **Nat Genet** *42*, 53-61.

24. Xu M and Cook PR (**2008**). Similar active genes cluster in specialized transcription factories. **J Cell Biol** 181, 615-623.

25. Longtine MS, McKenzie A, 3rd, Demarini DJ, Shah NG, Wach A, Brachat A, Phillipsen P and Pringle J (**1998**). Additional modules for versatile and economical PCR-based gene deletion and modification in Saccharomyces cerevisiae. **Yeast** 14: 953–961.

26. Amberg DC, Burke DJ and Strathern J (**2005**). Methods in Yeast Genetics: A Cold Spring Harbor Laboratory Course Manual. **Cold Spring Harbor Press, Cold Spring Harbor, NY, USA.**

27. Brickner DG, Light W and Brickner JH (**2010**). Quantitative localization of chromosomal loci by immunofluorescence. **Meth Enzymol** 470: 571-582.

28. Egecioglu DE, D'Urso A, Brickner DG, Light WH and Brickner JH (2014) Approaches to studying subnuclear organization and gene-nuclear pore interactions. **Methods Cell Biol** 122:463-485.

10

The lysosomotropic drug LeuLeu-OMe induces lysosome disruption and autophagy-independent cell death in *Trypanosoma brucei*

Hazel Xinyu Koh[1,2], Htay Mon Aye[1], Kevin S. W. Tan[2,]* and Cynthia Y. He[1,]*

[1] Department of Biological Sciences, National University of Singapore.
[2] Department of Microbiology, National University of Singapore.
* Corresponding Authors: Kevin S. W. Tan, E-mail: kevin_tan@nuhs.edu.sg; Cynthia Y. He, E-mail: dbshyc@nus.edu.sg

ABSTRACT Background: *Trypanosoma brucei* is a blood-borne, protozoan parasite that causes African sleeping sickness in humans and nagana in animals. The current chemotherapy relies on only a handful of drugs that display undesirable toxicity, poor efficacy and drug-resistance. In this study, we explored the use of lysosomotropic drugs to induce bloodstream form *T. brucei* cell death via lysosome destabilization. Methods: We measured drug concentrations that inhibit cell proliferation by 50% (IC_{50}) for several compounds, chosen based on their lysosomotropic effects previously reported in *Plasmodium falciparum*. The lysosomal effects and cell death induced by L-leucyl-L-leucyl methyl ester (LeuLeu-OMe) were further analyzed by flow cytometry and immunofluorescence analyses of different lysosomal markers. The effect of autophagy in LeuLeu-OMe-induced lysosome destabilization and cytotoxicity was also investigated in control and autophagy-deficient cells. Results: LeuLeu-OMe was selected for detailed analyses due to its strong inhibitory profile against *T. brucei* with minimal toxicity to human cell lines *in vitro*. Time-dependent immunofluorescence studies confirmed an effect of LeuLeu-OMe on the lysosome. LeuLeu-OMe-induced cytotoxicity was also found to be dependent on the acidic pH of the lysosome. Although an increase in autophagosomes was observed upon LeuLeu-OMe treatment, autophagy was not required for the cell death induced by LeuLeu-OMe. Necrosis appeared to be the main cause of cell death upon LeuLeu-OMe treatment. Conclusions: LeuLeu-OMe is a lysosomotropic agent capable of destabilizing lysosomes and causing necrotic cell death in bloodstream form of *T. brucei*.

Keywords: Trypanosoma brucei, LeuLeu-OMe, lysosome, lysosomotropic, necrosis, autophagy.

Abbreviations:
4-HT - 4-hydroxytamoxifen,
CPZ - chlorpromazine,
CQ - chloroquine,
DSP - desipramine,
HAT - African sleeping sickness or Human African Trypanomiasis,
LeuLeu-OMe - L-leucyl-L-leucyl methyl ester,
PMZ - promethazine.

INTRODUCTION

African sleeping sickness (also known as Human African Trypanomiasis, or HAT) is endemic in 36 countries in sub-Saharan Africa, endangering life and economy of ~ 60 million people living in the area [1]. The disease has two characteristic stages: the early stage is due to a hemolymphatic infection, and the late stage to an infection of the central nervous system. Clinical manifestation in the early stage includes fever, headache and pains in joints. In the late stage, the manifestations are more severe and symptoms may include personality changes, motor and sensory abnormalities, insomnia, cerebral edema, and coma [2, 3]. Infection is lethal if left untreated [4].

The causative agents of the disease are parasitic protozoa of the species *Trypanosoma brucei*. They live and multiply in blood and tissue fluids of their mammalian hosts (the bloodstream form) and are transmitted to humans through the bite of an infected tsetse fly of the *Glossina* genus. Though the number of infection cases has dropped significantly during the past years due to a series of control activities, new tools for vector control, diagnosis, and case treatment are needed towards HAT elimination declared by the World Health Organization in 2014 [5-7]. Presently, there are only a handful of drugs available for treating HAT. Some of these drugs are plagued by various problems such as side effects, high cost, acute toxicities, poor oral bioavailability, long treatment, low efficacies, and drug resistance [6]. Identification of new targets for the development of anti-trypanosomal drugs is of importance.

Lysosomes are membrane-bounded, acidic organelles containing proteases and other hydrolytic enzymes that are responsible for degradation of macromolecule derived from endocytosis and/or autophagy pathways. They are essential for normal cellular functions and malfunction of lysosomes has been implicated in many human diseases [8-10]. A large gamut of stimuli such as oxidative stress, ultraviolet exposure and lysosomotropic detergents have been found to induce lysosomal membrane permeabilization in mammalian cells, which is distinctly characterized by the rupture of lysosomal membranes and the translocation of lysosomal components, including enzymes, from the lysosomal lumen to the cytosol. Perhaps the best described is the mechanism of lysosomotropic detergents [11-13]. These agents are described as basic amphiphilic amines that can attain concentrations several hundred-fold higher within the lysosomes than in the cytosol [14]. These amines can freely diffuse across membranes in their uncharged form but becomes trapped in their protonated form when they are localized in acidic vesicles. Accumulation of the protonated form above a certain threshold concentration results in osmotic swelling and acquisition of detergent-like properties which induces lethal lysosomal destabilization and cell death [12]. Lysosomotropic agents were shown to trigger a variety of death-associated morphologies ranging from classical apoptosis to necrosis in mammalian cells. The type of cell death depends on the type of lethal stimulus, the extent of lysosomal membrane permeabilization, the amount and the type of enzymes released into the cytoplasm [11].

The lysosomes of parasitic unicellular protozoa, such as *Plasmodium* and Trypanosomatids, share similar properties and functions to those of mammalian cells, and lysosomal membrane permeabilization-induced cell death have been observed and characterized in *Plasmodium* and *Leishmania* parasites [15-18]. Ch'Ng *et al.* [18] demonstrated that a group of lysosomotropic compounds including chloroquine (CQ), L-leucyl-L-leucyl methyl ester (LeuLeu-OMe), chlorpromazine (CPZ), promethazine (PMZ), desipramine (DSP) and 4-hydroxytamoxifen (4-HT), all disrupted the digestive vacuole of the malaria parasite, *Plasmodium falciparum*, causing lysosomal membrane permeabilization and triggered downstream programmed cell death pathways such as mitochondria dysregulation and DNA degradation. Among them, LeuLeu-OMe was also shown to target the lysosomal system of intracellular and isolated amastigotes of *Leishmania amazonensis* [15]. The targets of the esters have been identified as acid-phosphatase-positive megasomes, lysosome-like organelles that are electron-dense. This was supported by the observation that incubation with L-leucine methyl ester (Leu-OMe) resulted in swelling and fusion of megasomes, decreased electron density of the internal contents and the release of acid phosphatase into the medium [16]. Adade *et al.* [17] also demonstrated that Leu-OMe was toxic to all three developmental forms of *Trypanosoma cruzi*, targeting in particu-

TABLE 1. IC_{50} of selected lysosomotropic drugs in *T. brucei*.

Drugs	IC_{50} for *T. brucei* (μm)	IC_{50} for mammalian cells* (μm)	Selectivity Index (IC_{50} mammalian* / IC_{50} *T. brucei*)
LeuLeu-OMe	16.8 (14.6-19.4)	125-5000	7.44-298
CQ	29.9 (20.5 - 43.8)	148.3	5
CPZ	9.21 (7.80-10.9)	20.9	2.27
PMZ	13.4 (9.7-18.4)	38.7	2.89
DSP	8.8 (6.37-12.2)	10.7	1.22
4HT	6.23 (3.02-12.8)	6.6	1.06

The inhibitory concentration (IC_{50}) values of 6 lysosomotropic drugs on bloodstream form *T. brucei* were measured in this study and compared with those reported in mammalian cells (13, 42, 43). *Note that mammalian IC_{50} values for CQ, CPZ, PMZ, DSP and 4HT were measured on H9c2 rat-cardiomyocyte-derived cells, whereas mammalian IC_{50} for LeuLeu-OMe was estimated as the concentration observed to induce necrosis in 8 human cell lines including human neuroblastoma cell line SH-SY5Y, human immortalized keratinocytes HaCaT, human adenocarcinoma cell line HeLa, human hepatoma cell line HepG2, human colon carcinoma cell line CaCo-2, and human embryonic kidney fibroblasts HEK293, human breast carcinoma cell line MCF-7 and normal human dermal fibroblasts, NHDF, cytotoxic T lymphocytes and NK cells. LeuLeu-OMe = L-leucyl-L-leucyl methyl ester; CQ = Chloroquine; CPZ = Chlorpromazine; PMZ = Promethazine; DSP = Desipramine; 4HT = 4-hydroxytamoxifen. The IC_{50} values for *T. brucei* were shown as mean (95% confidence interval) from >=4 experiments for each drug.

lar acidic compartments in the parasite. They showed a decrease in overall acridine orange fluorescence in the organelles by flow cytometry and a dispersed fluorescence throughout the cell cytoplasm by confocal microscopy. The effect of lysosomtropic drugs is yet to be analyzed in *T. brucei,* the trypanosomatid most accessible to molecular genetic manipulations.

In this study, LeuLeu-OMe-induced lysosomal morphological change was characterized in *T. brucei.* Results indicated that LeuLeu-OMe caused destabilization of the lysosome and necrotic cell death in the parasite. Although autophagosome formation was observed upon LeuLeu-OMe treatment, autophagy did not contribute to LeuLeu-OMe-indued cytotoxicity. These results support lysosomal destabilization as a novel therapeutic approach for future anti-trypanosome drug design.

RESULTS
Cytotoxicity of LeuLeu-OMe
Several lysosomotropic compounds including LeuLeu-OMe, CQ, CPZ, PMZ, DSP and 4-HT, have been found to exhibit parasiticidal effects by targeting the vacuolar compartment in *Plasmodium falciparum* [18]. To evaluate if similar lysosomotropic effects can be induced in bloodstream form *T. brucei,* cells were incubated with the above-mentioned compounds at various concentrations for 24 h and the cell viability was monitored using flow cytometry system. The IC_{50} values were then calculated for each compound (Table 1) and compared with their IC_{50} against mammalian cells based on published results. Notably, LeuLeu-OMe and CQ both recorded a higher selectivity index (> 5.0) while the other drug compounds were less selective. LeuLeu-OMe, which has the highest selectivity index, was therefore chosen for further studies.

The cytotoxicity of LeuLeu-OMe was further evaluated in cells treated with 30 μM LeuLeu-OMe, which corresponds to the IC_{90} concentration. In this experiment, cell density were monitored every 2 h and the results normalized to cell concentration at t = 0 to obtain growth index (Figure 1). The cells were viable and able to proliferate

moderately until 2 h post LeuLeu-OMe treatment, significant cell death was observed between 4 h and 8 h of treatment. As a control, cells with 1% DMSO (solvent for LeuLeu-OMe) added to the cultivation medium, proliferated normally with a doubling time of ~ 8 h (Figure 1, inset). Due to the rapid cell death observed in the first 8 h of LeuLeu-OMe treatment, most subsequent analyses described in this study were restricted to this time period.

Lysosome destabilizing effects of LeuLeu-OMe
To determine if LeuLeu-OMe has lysosome destabilizing effects in *T. brucei,* p67, an integral lysosome membrane protein, were used to monitor lysosome morphology by immunofluorescence assays during the course of 30 μM LeuLeu-OMe treatment. Most control cells contained one or two major puncta representing lysosomes in different cell cycle stages (Figure 2A), similar to a previous report [19]. Upon LeuLeu-OMe treatment, the lysosome staining by p67 became more diffused and fragmented as shown by two-dimensional projection of serial optical sections (sMovie 1 and sMovie 2 in Supplemental Materials). The images taken at t = 0 h, 2 h and 4 h post-LeuLeuOMe treatment were then quantitated, and the percentage of cells containing fragmented lysosomes (i.e. > 2 puncta in each cell) was found to increase from 12% at t = 0, to > 40% at 2 h and 4 h (Figure 2B). Interestingly, the increased lysosmal fragmentation was also accompanied with an increase in p67 expression (Figure 2C). However, the protein level of trypanopain, a lysosome luminal protein, remained unchanged during the course of LeuLeu-OMe treatment. Other membrane bound organelles, including ER, Golgi apparatus, glycosomes and acidocalcisomes, all appeared normal when labeled with anti-BiP, anti-GRASP, anti-SKL and anti-TbVP1 respectively (data not shown).

To further evaluate lysosome structural integrity, LeuLeu-OMe-treated cells were also labeled with anti-trypanopain. We reasoned that if lysosome membrane integrity was compromised during drug treatment, the staining of the soluble lysosome luminal trypanopain would no longer be confined and may be found in cytosol.

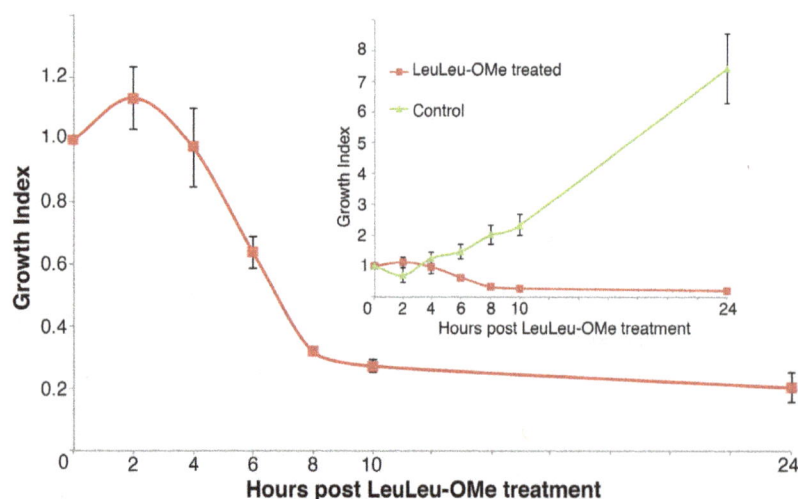

FIGURE 1: Cytotoxicity of LeuLeu-OMe to bloodstream form *T. brucei.* Cells were incubated with or without 30 μM LeuLeu-OMe for up to 24 h. Samples were taken at the indicated time points and monitored for cell growth. Decreased cell number was observed after 2 h of LeuLeu-OMe treatment, while control cells continued to proliferate (see inset for comparison). Growth index was calculated as described in Methods. The results were presented as mean ± SD from 3 independent experiments.

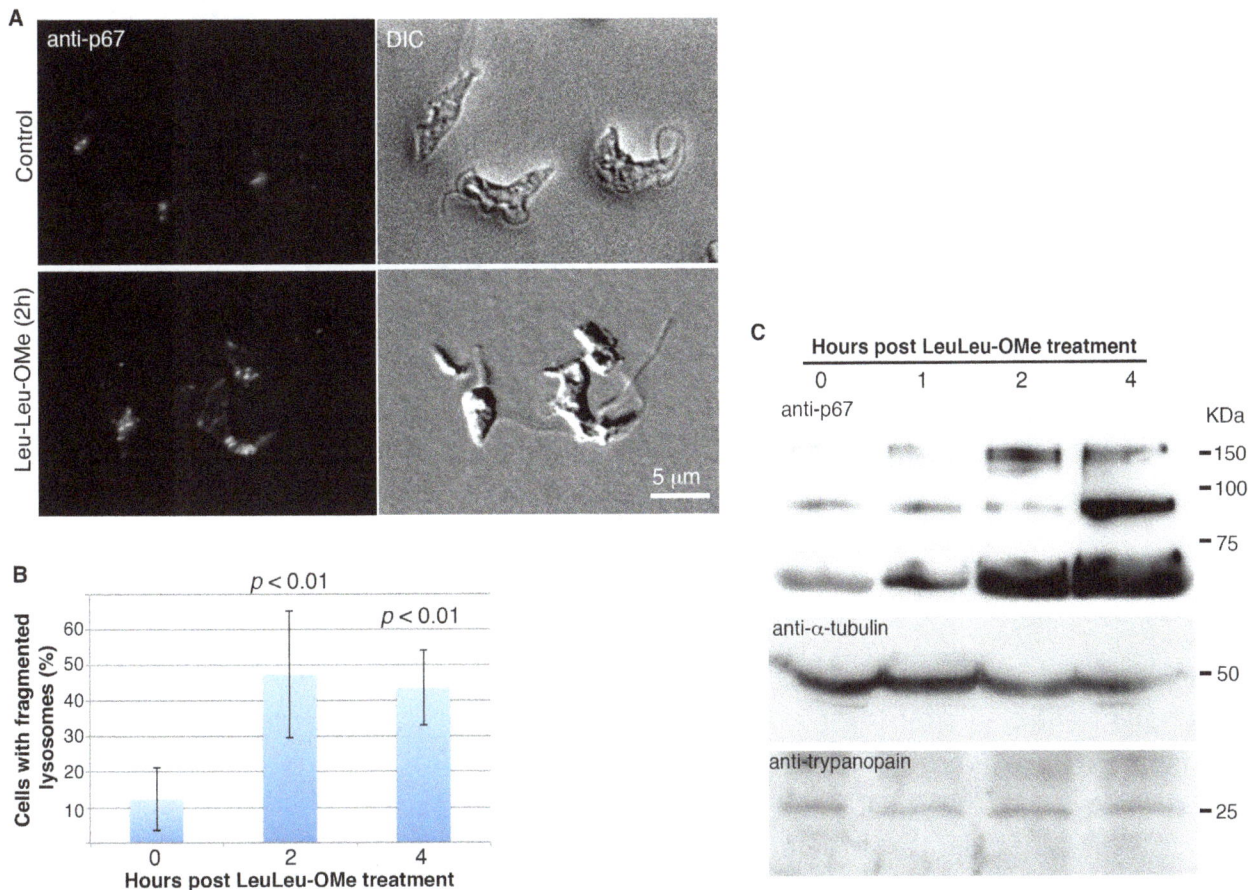

FIGURE 2: Lysosome destabilizing effects of LeuLeu-OMe. (A) *T. brucei* cells were incubated with or without 30 µM LeuLeu-OMe, and samples were taken at t = 0 h, 1 h, 2 h and 4 h for immunofluorescence assays with anti-p67. Representative images of control and LeuLeu-OMe-treated cells are maximum intensity projection of serial optical sections through the entire cells. **(B)** Quantitation of images as shown in (A) revealed an increase in lysosome fragmentation at 2 h and 4 h post LeuLeu-OMe treatment. The quantitation results are presented as mean ± SD from 3 independent experiments. **(C)** Samples treated as described in (A) were also processed for immunoblotting analyses with anti-p67 and anti-trypanopain. Anti-α-tubulin was used as a loading control.

Indeed, upon LeuLeu-OMe treatment, more cells were found to contain weak and diffused anti-trypanopain staining distributed throughout the cell (Figure 3A). As the expression level of trypanopain did not change during the course of LeuLeu-OMe treatment (see Figure 2C), the change from localized to diffused staining suggested release of lysosomal trypanopain into the cytosol, providing a convenient marker to evaluate lysosome structure integrity during LeuLeu-OMe treatment and other conditions (Figure 3B).

Due to disruption of lysosomal integrity, LeuLeu-induced cell death is likely caused by the release of lysosomal cathepsins into the cytosol. To test this possibility, LeuLeu-OMe-induced cell death was further evaluated in the presence of cathepsin inhibitor Z-Phe-Ala [20]. Treatment of cells with Z-Phe-Ala alone, up to 40 µM, had little effect on parasite growth by 9h. Prolonged treatment (24h) showed a dose-dependent inhibition of cell proliferation (Figure S1). Treatment of cells with LeuLeu-OMe together with 10 µM Z-Phe-Ala complete rescued the rapid cell death induced by LeuLeu-OMe at IC_{90} (Figure 4A), supporting a role of cathepsins in LeuLeu-OMe-induced cell death.

Lysosome acidity is required for LeuLeu-OMe cytotoxicity
Lysosomotropic detergents are known for their ability to accumulate in vesicles that possess an acidic interior (low pH). Raising the pH of the lysosome has been shown to decrease the uptake of these detergents, indicating the importance of a low pH for their entry into lysosomes [12]. To verify if the acidic pH of lysosomes was required for LeuLeu-OMe cytotoxicity in *T. brucei*, cells were pre-incubated with 100 nM monensin (an ionphore and a Na+/H+ antiporter) for 30 min, which was sufficient to neutralize the acid pH and inhibit lysotracker accumulation in lysosomes (data not shown). At this concentration, monensin alone had no observable effects on cell viability, but significantly reduced cytotoxicity and lysosome destabilizing effects in LeuLeu-OMe treated cells (Figure 3B and Figure 4B). These results are consistent with the effects of LeuLeu-OMe on lysosomes, and supported a role of acidic pH in LeuLeu-OMe action.

A

B

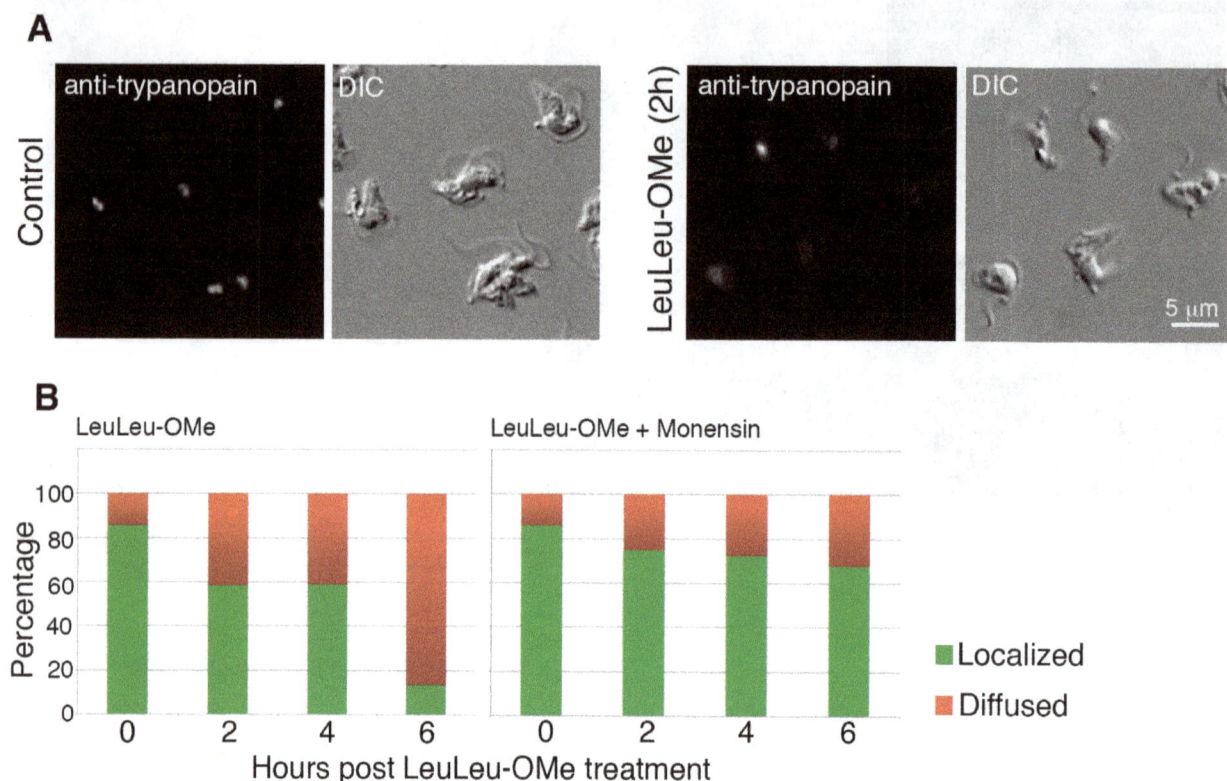

FIGURE 3: LeuLeu-OMe-induced lysosome destabilization revealed by anti-trypanopain staining. (A) Control and LeuLeu-OMe-treated cells were fixed and stained with anti-trypanopain. The staining was localized to one or two puncta in control cells, but became diffused in LeuLeu-OMe-treated cells. Control and drug-treated cells were processed for immunostaining using same conditions. Note that LeuLeu-OMe-treated cells were exposed longer to capture the weaker anti-trypanopain staining in the cytosol. **(B)** Cells with localized (green) or diffused (red) anti-trypanopain staining were quantitated over the course of LeuLeu-OMe treatment (left), and in cells pre-treated with monensin (right). At least 200 cells were counted for each time point.

LeuLeu-OMe-induced cell death in *T. brucei*

To further investigate how LeuLeu-OMe induced *T. brucei* cell death, control and cells treated with 30 µM LeuLeu-OMe were evaluated for necrotic and apoptotic cell death, using propidium iodide (PI) and Annexin-FITC stains. Cells positively stained for both PI and Annexin-FITC were found to increase from 14.4% in the control population to 17.5% at 2 h, 36.4% at 4 h and 43.3% at 6 h of LeuLeu-OMe treatment (Figure 5A), indicating an increase of necrotic cells in a time-dependent fashion. Increase in apoptotic cells (positively stained with Annexin-FITC but negative for PI) was not observed during LeuLeu-OMe treatment. To rule out the possibility that the lack of apoptosis may be due to rapid cell killing by high concentration of LeuLeu-OMe (30 µM, which corresponded to IC_{90}), the above same experiments were repeated in cells treated with 16 µM LeuLeu-OMe, which corresponded to IC_{50} (Figure S2A). While cell death was slowed at the lower drug concentration, we still did not observe an increase in apoptotic cells.

Autophagy-dependant cell death has been previously reported in many protozoan pathogens, including *T. brucei* and *Plasmodium*, upon drug treatment and starvation stresses [21-24]. To evaluate if autophagy may be involved in LeuLeu-OMe-induced cell death in *T. brucei*, cells stably expressing YFP-TbATG8.2 were incubated with 30 µM (IC_{90}) LeuLeu-OMe, and autophagosome formation was monitored by fluorescence microscopy. In control cells, YFP::TbATG8.2 exhibited a cytoplasmic distribution, similar to previously observed in procyclic *T. brucei* [21]. Upon starvation stress, YFP::TbATG8.2 relocated to autophagosomes that appear as punctate structures throughout the cell (Figure 5B). A steady increase in autophagosome numbers was observed in LeuLeu-OMe-treated cells in a time-dependant fashion (Figure 5C).

To determine the effect of autophagy on LeuLeu-OMe-induced cell death, a stable RNAi cell line targeting both Atg8 homologs in *T. brucei*, TbAtg8.1 and TbAtg 8.2 [21], was constructed. Efficient RNAi was shown by immunoblotting with anti-ATG8.2 antibody (Figure 6A), and TbAtg8.1/8.2 double RNAi had little effect on cell proliferation under normal growth conditions (Figure 6B). Upon treatment with 30 µM (IC_{90}) or 16 µM (IC_{50}) LeuLeu-OMe, the uninduced control and cells induced for TbAtg8.1/8.2-RNAi did not exhibit significant differences in cell growth (Figure 6C and supplementary Figure S2B), suggesting that autophagy did not play a major role in LeuLeu-OMe-induced cell death.

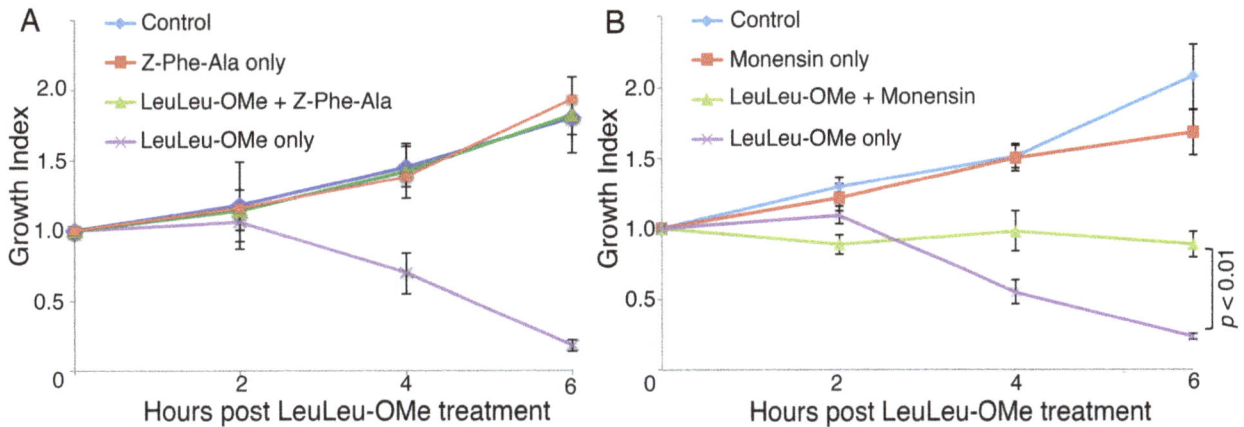

FIGURE 4: Treatment with a cathepsin inhibitor or monensin reduces LeuLeu-OMe-induced cell death. (A) Cells were treated with 10 μM Z-Phe-Ala or 30 μM LeuLeu-OMe, alone or in combination. **(B)** Cells pre-treated with 100 nM monensin for 30 min was then cultivated in the presence or absence of 30 μM LeuLeu-OMe. While monensin treatment alone had no observable effects on parasite growth (compared to control), monensin treatment significantly reduced LeuLeu-OMe-induced cytotoxicity. Results were presented as mean ± SD from 3 independent experiments. p value for the samples treated with LeuLeu-OMe only or LeuLeu-OMe + monensin at 6 h time point is indicated next to the bracket.

DISCUSSION

Lysosomes have long been deemed as a waste bin where macromolecules derived from outside of the cell through endocytosis or from within the cell through autophagy are degraded and recycled. Full of proteases and other hydrolytic enzymes, the lysosome is also considered a 'dangerous' place. While the presence of lysosome associated membrane proteins such as LAMP1 and LAMP2 can protect lysosome itself from the hydrolytic enzymes [25], damage to lysosome integrity may allow these enzymes to leak into the cytosol and cause unintended damage to other cellular components. This damaging potential of lysosomes have long been noted and been explored as a mechanism to induce cell death [11]. In particular, the action of lysosomotropic detergents has been well documented in mammalian cells ever since the 1980s, when lysosomotropic compounds were chemically synthesized to mimic the lysosome accumulation and lysosomotropic properties of amines [26]. Cell death induced by lysosomotropic activities is of particular interest and offers a potential new drug target in protozoan parasites, which are single-celled eukaryotic pathogens living inside mammalian hosts.

We have previously reported evidence that several lysosomotropic drugs can cause digestive vacuole (similar to lysosomes) destabilization in the malaria parasite, *P. falciparum*, and that digestive vacuolar membrane permeabilization triggers the program cell death pathway [18]. In this study, we have extended this work to another important human and animal parasite *T. brucei*.

We have focused on a well-characterized lysosomotropic detergent, LeuLeu-OMe, which shows potent inhibition of *T. brucei* with IC$_{50}$ in the micromolar range and is the most selective of 6 lysosomotropic agents tested in this study. Similar to that reported in mammalian cells as well as other organisms, LeuLeu-OMe treatment led to lysosome fragmentation and release of lysosomal lumenal pro-

tein trypanopain. The use of two different lysosomal markers, p67, which is a LAMP-like lysosome membrane glycoprotein [27], and trypanopain, which is a major cathepsin-L type protease [28], provided a highly reproducible and reliable method to examine lysosome integrity. It is interesting to note that the levels of different p67 glycosylation forms increased significantly during the course of LeuLeu-OMe treatment. This up-regulation may be due to the cell's compensatory effect in response to the lysosome destabilizing activity of LeuLeu-OMe, considering the role of p67 in lysosome structure maintenance [27]. In contrast, trypanopain level remained unchanged during the treatment, though its cellular distribution changed from a localized pattern in control cells, to a diffused distribution in LeuLeu-OMe treated cells, consistent with a role of LeuLeu-OMe in compromising lysosome integrity.

Both necrotic and apoptotic cell death have been described following treatment with lysosomotropic drugs, and the exact type of cell death is likely dependent on the extent of lysosome destabilization and thus influencing the amount and type of enzymes that are released into the cytoplasm [11]. An example is the sphingolipid sphingosine, which is a lipophilic weak base. Upon protonation, sphingosine accumulates within acidic compartments where it may act as a detergent. Limited doses of sphingosine induce a cascade of lethal events including lysosomal membrane permeabilization, caspase activation, as well as the dissipation of the mitochondrial membrane potential. On the other hand, high doses rapidly cause lysosomal rupture culminating in rapid necrosis [29]. Other examples include the lysosomotropic detergents N-dodecylimidazole and O-methyl-serine-dodecylamide hydrochloride, both induce lysosome destabilization followed by caspase-3-dependent apoptosis at low doses and necrosis at high doses [30, 31]. It is unlikely that the lack of apoptosis was due to rapid killing by LeuLeu-OMe, as lower dose of LeuLeu-OMe that

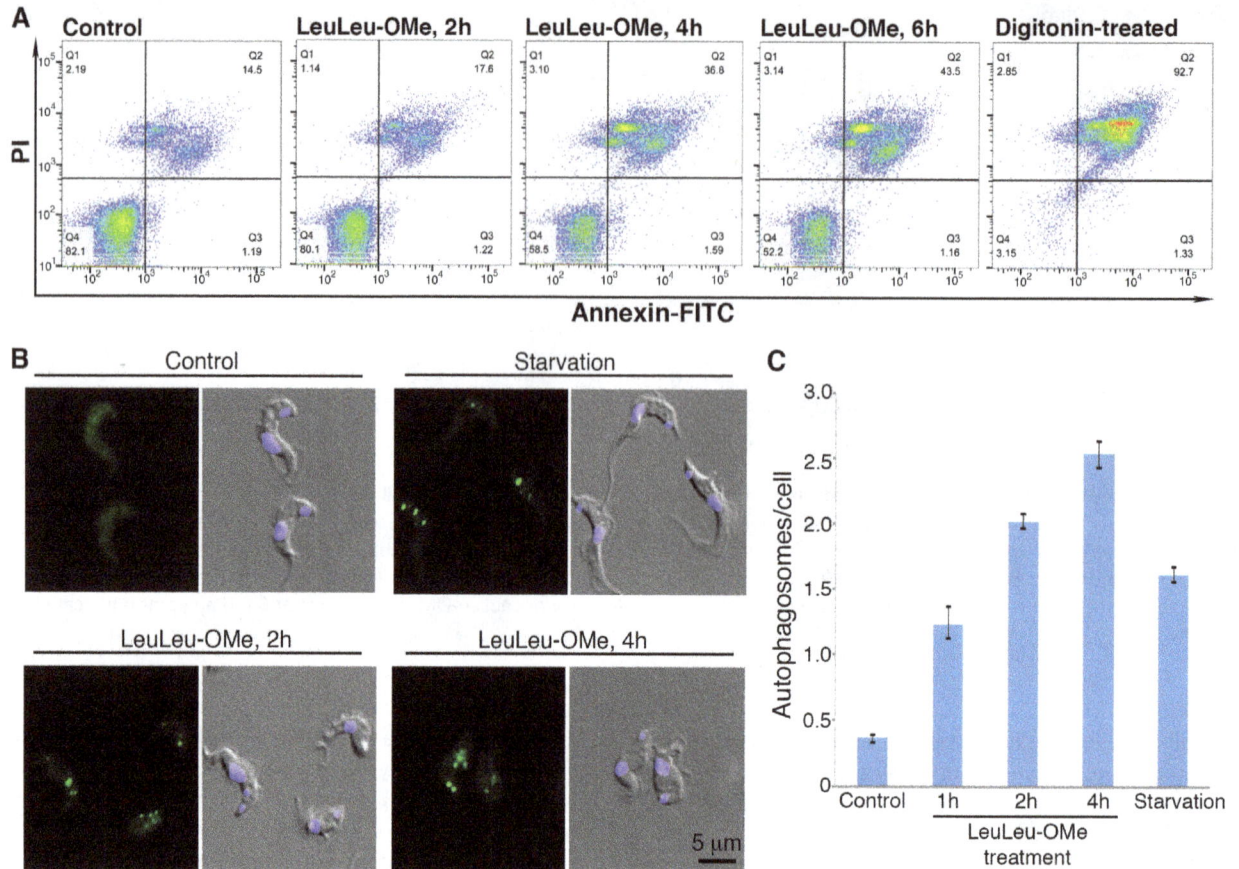

FIGURE 5: LeuLeu-OMe induces necrotic cell death and autophagy. (A) Cells treated with 30 μM LeuLeu-OMe were fixed and stained with PI and annexin-FITC, and analyzed by flow cytometry. Digitonin-permeabilized cells were used as a positive control for the stains. **(B, C)** Cell stably expressing YFP-TbATG8.2 were cultivated in medium with or without 30 μM LeuLeu-OMe. Autophagosome formation was monitored by relocalization of YFP-TbATG8.2 from a cytoplasmic distribution in control cells to a punctate structures using fluorescence microscopy. Autophagosome formation was quantitated and the results are shown as mean ± SD from 3 independent experiments. Cells starved in cytomix for 2 h were used as a positive control.

killed the cells at slower kinetics did not lead to apoptosis either. Notably, the presence of apoptosis in *T. brucei* and other protozoan parasites has long been controversial. The lack of key apoptotic molecules such as caspases in the parasite genomes and inconsistency in detection of apoptotic phenotypes suggest that these organisms do not possess the regulated apoptosis pathway as those described in higher eukaryotes [32].

In the recent years, the presence of a conserved autophagy pathway and its molecular machineries has been demonstrated in *T. brucei* and many other protozoan parasites. Although the physiological trigger and function of autophagy in these single-celled organisms remains to be experimentally tested, a possible role of autophagy in cell death has been proposed based on molecular evidence in *T. brucei* and *Toxoplasma gondii* [21, 24]. It is therefore interesting to note in our study that a significant increase in autophagosome numbers was found upon LeuLeu-OMe treatment, though depletion of autophagy had little effects on the progress of cell death. The increased autophagosome formation observed during LeuLeu-OMe may repre-

sent either true autophagy induction, or inhibited autophagosome fusion to the damaged lysosomes. In either case, as lysosome functions, which are the final steps of the autophagy pathway, are inhibited by LeuLeu-OMes, any downstream function of autophagy would also be compromised. This may provide an explanation for the lack of a role for autophagy in LeuLeu-OMe induced cell death.

While neither apoptosis nor autophagy had a role in LeuLeu-OMe-induced cell death, nonspecific action of released lysosomal cathepsins in cytosol following lysosomal disintegration is likely the main cause of cell death. Consistent with this, co-treatment of cells with a cathepsin inhibitor completely reversed the rapid cell death induced by LeuLeu-OMe.

LeuLeu-OMe is a 2-amino acid compound known to have immunosuppressive activity. The drug has been commonly used to decrease the incidence of graft versus host disease (GVHD) via cytotoxic cell depletion *ex vivo* following blood transplant in animal models such as mice and dogs [33, 34]. LeuLeu-OMe has also been effective in preventing GVHD in humans [35, 36]. However, LeuLeu-

OMe has never been directly administered into animals or humans, and thus its immunosuppressive activity has not been validated in animals. Furthermore, necrotic cell death is not the preferred mode of killing because it might cause inflammation. However, LeuLeu-OMe was the most selective of the 6 lysosomotropic drugs tested in this study. With its anti-trypanosome mechanisms more extensively studied, structure activity relationship can be better explored to optimize leads with better efficacy and tolerability profiles as a lysosome-destabilizing agent.

MATERIALS AND METHODS

Cell lines and culture

Bloodstream form *T. b. brucei* single marker cell line [37], a derivative of the 427 strain, was maintained at 37°C, 5% CO_2, and > 90% humidity in HMI-9 medium supplemented with 10% (v/v) fetal bovine serum (Hyclone). Cells were sub-cultured by diluting into fresh medium every 24 h to maintain exponential growth (with cell density between 1×10^5 and 1×10^6 cells/ml).

Plasmid construction and transfection

A pZJM construct previously described [21] was used for inducible RNAi of both TbAtg8.1 and TbAtg8.2. Stable RNAi cell lines were generated by electroporating the pZJM-Atg8.1/8.2 vector into the bloodstream form cells using an Amaxa Nucleofector with programme X-001 (Lonza), and selection with 4 μg/ml phleomycin.

Inducible RNAi

To monitor the proliferation of TbAtg8.1/8.2-RNAi cells, 1 μg/ml tetracycline (Sigma) was added to induce RNAi. Cell density was measured every 24 h using a haemocytometer. The uninduced cells were used as a negative control. During the course of measurements, the cells were maintained in exponential growth phase at a density of 1×10^5 to 1×10^6/ml by dilution with fresh medium with or without tetracycline. For western blot analyses, protein samples (from 1×10^7 cells) were fractionated by 12% sodium dodecyl sulphate polyacrylamide gel electrophoresis (SDS-PAGE). Following transfer, the immunoblots were probed with anti-Atg8.2 antibody (1/1000 dilution) [21]. The same blots were also probed with anti-PFR2 [38] as loading controls.

Drug preparations

Stock solution of 4-hydroxytamoxifen (4-HT) (Sigma) was dissolved in ethanol and stored at -20°C. Stock solutions of chloroquine diphosphate (CQ), desipramine (DSP), chlorpromazine (CPZ) and promethazine (PMZ) (all from Sigma), were prepared fresh before each experiment, by dissolving in phosphate buffered saline (PBS) and filter sterilization. Stock solutions of L-leucyl-L-leucyl methyl ester (Sigma) was prepared freshly before each experiment, by dissolving in DMSO. Stock solution of Z-Phe-Ala fluoromethyl ketone (Z-Phe-Ala; Sigma) was prepared in DMSO. Vehicle controls consisted of equivalent amounts, all < 1%, of DMSO (LeuLeu-OMe), PBS (CQ, CPZ, PMZ) and ethanol (4-HT). Monensin (Sigma) was prepared at a final concentration of 100 nM in ethanol. The concentration of monensin was optimized to be the lowest concentration that inhibited lysotracker accumulation in lysosomes, yet did not obviously affect cell growth after 6 hours (data not shown).

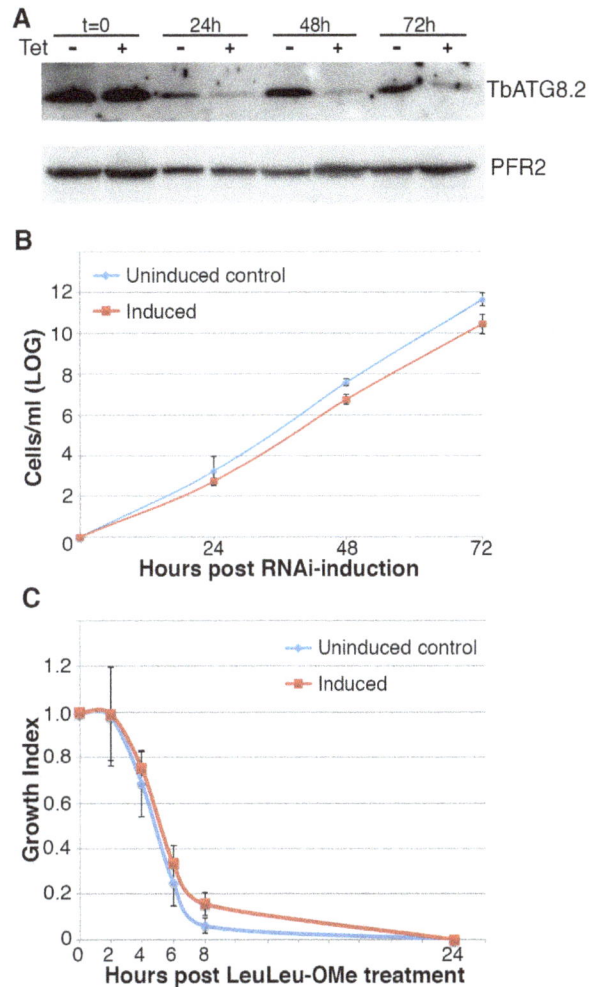

FIGURE 6. Autophagy is not required for LeuLeu-OMe-induced cell death. (A, B) Cells were stably transfected with pZJM-TbAtg8.1/8.2 for inheritable and inducible RNAi of TbATG8.1 and TbATG8.2 expression. The cells were induced with tetracycline, and samples were taken at t = 0 h, 24 h, 48 h and 72 h post-induction for immunoblots (A) and cell counting (B). Anti-TbATG8.2 antibody was used to monitor TbATG8.1/8.2-RNAi efficiency, and anti-PFR2 (a component of paraflagellar rod) was used as a loading control. **(C)** Cells induced for TbAtg8.1/8.2-RNAi or not were also treated with 30 μM LeuLeu-OMe, and cell growth monitored as described in Figure 1. Results shown in (B) and (C) are presented as mean ± SD from 3 independent experiments.

IC₅₀ measurements

Parasite cultures diluted with fresh medium to 10^5 cells/ml were incubated with various concentrations of test substances for 24 hours in a 96 well microtiter plate in a final volume of 200 μl. Serial dilutions were made on each drug to create working solutions. The drug concentration range used for LeuLeu-OMe is 1 nM - 1 mM, CQ at 1 nM - 3 mM, CPZ at 1 nM - 100 μM, PMZ at 1 nM - 200 μM, DSP at 1 nM - 250 μM and 4-HT at 1 nM - 100 μM. To determine cell viability at the 24 h time point, cell density was measured using a Guava® flow cytometry system (Millipore, USA). Solvent controls contained DMSO, PBS or ethanol that were diluted in cell cultures to give final concentrations not exceeding 1%. No less than four independent experiments were performed for each data set. Plotting of the dose-response curve as well as the computation of the IC_{50} and IC_{90} (drug concentration that inhibits cell growth by 90%) were performed with GraphPad Prism 5 demo version using a four-parameter logistic curve (variable slope). The selectivity index (SI) values were calculated using the ratio:

$$SI = IC_{50} \text{ for mammalian cell}/IC_{50} \text{ for } T. \text{ brucei}$$

Growth index

To monitor cytotoxicity by LeuLeu-OMe and other treatments, cell samples taken at various time points of the treatments were counted using a haemocytometer. Growth index was calculated using the formula below:

Growth index = Cell density at a time point / cell density at t=0

PI/Annexin assay

To monitor LeuLeu-OMe-incuded cell death, LeuLeu-OMe-treated (30 μM) and untreated parasites were washed and resuspended in 0.5 ml binding buffer (10 mM HEPES, 140 mM NaCl, 2.5 mM $CaCl_2$, 10 mM glucose, pH 7.4) and then incubated for 15 min at room temperature with propidium iodide (PI, 5 μg/ml) and annexin V-FITC (0.75 μg/ml). Cells were then analyzed by a Becton-Dickinson LSR Fortessa flow cytometry system (BD Biosciences). Cells permeabilized with 6 μM digitonin for 5 min were used as positive control for PI and annexin V dyes. Compensation of spectral overlap was performed using single-stained control cells with the FlowJo software.

Fluorescence microscopy

Parasite cultures diluted with fresh medium to approximately 10^5 cells/ml were incubated with 30 μM of LeuLeu-Ome. Samples of control and drug treated cells were taken at t = 0 h, 2 h, 4 h and 6 h for immunofluorescence analyses. The samples were fixed in situ with 4% paraformaldehyde for 15 min, washed twice with PBS, attached to poly-L-lysine coverslips, permeabilized in 0.125% Triton X-100 (w/v in PBS) for 15 min, blocked with 3% BSA (w/v in PBS) for 1 hour and then processed for antibody labelling. Antibodies directed against trypanopain, a soluble cysteine cathepsin in the lysosome lumen [28] or p67, a lysosomal membrane protein [39] were used as lysosome markers. Anti-GRASP, anti-SKL, anti-VP1 and

anti-BiP were used to label the Golgi [40], the glycosomes [41], the acidocalcisomes [42], and the endoplasmic reticulum (ER) [43], respectively. To image lysosomes, Z-stack images (a total of nine images/stack, with step size of 0.3 μm) were acquired over the entire depth of the cells containing lysosome staining, using Observer Z1 (Zeiss, Germany) equipped with a 63 X NA1.4 objective and a CoolSNAP HQ^2 CCD camera (Photometrics). Images were processed with ImageJ and Adobe Photoshop.

Autophagosome formation in YFP:TbAtg8.2 cells

YFP:TbATG8.2 fusion as described previously [21] was subcloned in pHD1034 vector, which allows protein overexpression in the bloodstream form parasites. Stable clones were obtained through selection with 1 μg/ml puromycin. To monitor autophagosome formation, YFP:TbATG8.2 cells were diluted with fresh medium to approximately 10^5 cells/ml in the presence or absence of 30 μM LeuLeu-OMe. Cell samples were taken at 1 h, 2 h and 4 h post drug treatments, fixed in situ with 4% paraformaldehyde for 15 min, washed twice with PBS and attached to poly-L-lysine coverslips. Starvation control contained cells incubated with cytomix (2 mM EGTA, 120 mM KCl, 0.15 mM $CaCl_2$, 10 mM K_2HPO_4/KH_2PO_4, 25 mM HEPES, 5 mM $MgCl_2.6H_2O$, 0.5% Glucose, 100 μg/ml BSA, 1 mM Hypoxanthine, pH 7.6) at 37°C for 2 hours. At least three independent experiments were performed for each condition.

Statistics

The statistical significance of results from experimental groups in comparison with control groups was determined by the Student's t test. 2 sample equal variance and $p < 0.05$ was considered to be statistically significant.

ACKNOWLEDGMENTS

We wish to thank Drs. Ch'ng Jun Hong, Li Feng-Jun and Anais Brasseur for their advices and technical assistance. This work was funded by generous grants from the National Medical Research Council (NMRC/1310/2011) and Singapore Ministry of Education (MOE2013-T2-1-092).

CONFLICT OF INTEREST

The authors do not have any commercial or other associations that might constitute a conflict of interest.

REFERENCES

1. Simarro PP, et al. (2012) Estimating and mapping the population at risk of sleeping sickness. **PLoS Neglected Tropical Diseases** 6(10):e1859.

2. Kato CD, et al. (2015) Clinical Profiles, Disease Outcome and Co-Morbidities among T. b. rhodesiense Sleeping Sickness Patients in Uganda. **PloS One** 10(2):e0118370.

3. Morrison LJ (2011) Parasite-driven pathogenesis in Trypanosoma brucei infections. **Parasite Immunology** 33(8):448-455.

4. Welburn SC & Maudlin I (2012) Priorities for the elimination of sleeping sickness. **Advances in Parasitology** 79:299-337.

5. Simarro PP, Diarra A, Ruiz Postigo JA, Franco JR, & Jannin JG (2011) The human African trypanosomiasis control and surveillance programme of the World Health Organization 2000-2009: the way forward. **PLoS Neglected Tropical Diseases** 5(2):e1007.

6. Steinmann P, Stone CM, Sutherland CS, Tanner M, & Tediosi F (2015) Contemporary and emerging strategies for eliminating human African trypanosomiasis due to Trypanosoma brucei gambiense: review. **Tropical Medicine & International Health** 20(6):707-18.

7. Holmes P (2014) First WHO meeting of stakeholders on elimination of gambiense. Human African Trypanosomiasis. **PLoS Neglected Tropical Diseases** 8(10):e3244.

8. Coutinho MF, Matos L, & Alves S (2015) From bedside to cell biology: a century of history on lysosomal dysfunction. **Gene** 555(1):50-58.

9. Maxfield FR (2014) Role of endosomes and lysosomes in human disease. **Cold Spring Harbor Perspectives in Biology** 6(5):a016931.

10. Appelqvist H, Waster P, Kagedal K, & Ollinger K (2013) The lysosome: from waste bag to potential therapeutic target. **Journal of Molecular Cell Biology** 5(4):214-226.

11. Boya P & Kroemer G (**2008**) Lysosomal membrane permeabilization in cell death. **Oncogene** 27(50):6434-6451.

12. Miller DK, Griffiths E, Lenard J, & Firestone RA (**1983**) Cell killing by lysosomotropic detergents. **The Journal of Cell Biology** 97(6):1841-1851.

13. Nadanaciva S, et al. (**2011**) A high content screening assay for identifying lysosomotropic compounds. **Toxicology in vitro : an international journal published in association with BIBRA** 25(3):715-723.

14. de Duve C, et al. (**1974**) Commentary. Lysosomotropic agents. **Biochemical Pharmacology** 23(18):2495-2531.

15. Rabinovitch M, Zilberfarb V, & Ramazeilles C (**1986**) Destruction of Leishmania mexicana amazonensis amastigotes within macrophages by lysosomotropic amino acid esters. **The Journal of Experimental Medicine** 163(3):520-535.

16. Antoine JC, Jouanne C, & Ryter A (**1989**) Megasomes as the targets of leucine methyl ester in Leishmania amazonensis amastigotes. **Parasitology** 99 Pt 1:1-9.

17. Adade CM, Figueiredo RC, De Castro SL, & Soares MJ (**2007**) Effect of L-leucine methyl ester on growth and ultrastructure of Trypanosoma cruzi. **Acta Tropica** 101(1):69-79.

18. Ch'ng JH, Liew K, Goh AS, Sidhartha E, & Tan KS (**2011**) Drug-induced permeabilization of parasite's digestive vacuole is a key trigger of programmed cell death in Plasmodium falciparum. **Cell Death & Disease** 2:e216.

19. Alexander DL, Schwartz KJ, Balber AE, & Bangs JD (**2002**) Developmentally regulated trafficking of the lysosomal membrane protein p67 in Trypanosoma brucei. **Journal of Cell Science** 115(Pt 16):3253-3263.

20. Steverding D, et al. (**2012**) Trypanosoma brucei: chemical evidence that cathepsin L is essential for survival and a relevant drug target. **International journal for parasitology** 42(5):481-488.

21. Li FJ, et al. (**2012**) A role of autophagy in Trypanosoma brucei cell death. **Cellular microbiology** 14(8):1242-1256.

22. Yin J, Ye AJ, & Tan KS (**2010**) Autophagy is involved in starvation response and cell death in Blastocystis. **Microbiology** 156(Pt 3):665-677.

23. Hain AU & Bosch J (**2013**) Autophagy in Plasmodium, a multifunctional pathway? **Computational and Structural Biotechnology Journal** 8:e201308002.

24. Ghosh D, Walton JL, Roepe PD, & Sinai AP (**2012**) Autophagy is a cell death mechanism in Toxoplasma gondii. **Cellular Microbiology** 14(4):589-607.

25. Eskelinen EL (**2006**) Roles of LAMP-1 and LAMP-2 in lysosome biogenesis and autophagy. **Molecular Aspects of Medicine** 27(5-6):495-502.

26. Firestone RA, Pisano JM, & Bonney RJ (**1979**) Lysosomotropic agents. 1. Synthesis and cytotoxic action of lysosomotropic detergents. **Journal of Medicinal Chemistry** 22(9):1130-1133.

27. Peck RF, et al. (**2008**) The LAMP-like protein p67 plays an essential role in the lysosome of African trypanosomes. **Molecular Microbiology** 68(4):933-946.

28. Mbawa ZR, Webster P, & Lonsdale-Eccles JD (**1991**) Immunolocalization of a cysteine protease within the lysosomal system of Trypanosoma congolense. **European Journal of Cell Biology** 56(2):243-250.

29. Kagedal K, Zhao M, Svensson I, & Brunk UT (**2001**) Sphingosine-induced apoptosis is dependent on lysosomal proteases. **The Biochemical Journal** 359(Pt 2):335-343.

30. Li W, et al. (**2000**) Induction of cell death by the lysosomotropic detergent MSDH. **FEBS Letters** 470(1):35-39.

31. Zhao M, Antunes F, Eaton JW, & Brunk UT (**2003**) Lysosomal enzymes promote mitochondrial oxidant production, cytochrome c release and apoptosis. **European Journal of Biochemistry** 270(18):3778-3786.

32. Proto WR, Coombs GH, & Mottram JC (**2013**) Cell death in parasitic protozoa: regulated or incidental? **Nature reviews. Microbiology** 11(1):58-66.

33. Charley M, Thiele DL, Bennett M, & Lipsky PE (**1986**) Prevention of lethal murine graft versus host disease by treatment of donor cells with L-leucyl-L-leucine methyl ester. **The Journal of Clinical Investigation** 78(5):1415-1420.

34. Raff RF, et al. (**1988**) L-leucyl-L-leucine methyl ester treatment of canine marrow and peripheral blood cells. Inhibition of proliferative responses with maintenance of the capacity for autologous marrow engraftment. *Transplantation* 46(5):655-660.

35. Pecora AL, et al. (**1991**) Characterization of the in vitro sensitivity of human lymphoid and hematopoietic progenitors to L-leucyl-L-leucine methyl ester. **Transplantation** 51(2):524-531.

36. Thiele DL & Lipsky PE (**1986**) The immunosuppressive activity of L-leucyl-L-leucine methyl ester: selective ablation of cytotoxic lymphocytes and monocytes. **Journal of Immunology** 136(3):1038-1048.

37. Wirtz E, Leal S, Ochatt C, & Cross GA (**1999**) A tightly regulated inducible expression system for conditional gene knock-outs and dominant-negative genetics in Trypanosoma brucei. **Molecular and Biochemical Parasitology** 99(1):89-101.

38. Brasseur A, et al. (**2014**) The bi-lobe-associated LRRP1 regulates Ran activity in Trypanosoma brucei. **Journal of Cell Science** 127(22):4846-4856.

39. Kelley RJ, Alexander DL, Cowan C, Balber AE, & Bangs JD (**1999**) Molecular cloning of p67, a lysosomal membrane glycoprotein from Trypanosoma brucei. **Molecular and Biochemical Parasitology** 98(1):17-28.

40. He CY, et al. (**2004**) Golgi duplication in Trypanosoma brucei. **The Journal of Cell Biology** 165(3):313-321.

41. Keller GA, et al. (**1991**) Evolutionary conservation of a microbody targeting signal that targets proteins to peroxisomes, glyoxysomes, and glycosomes. **The Journal of Cell Biology** 114(5):893-904.

42. Lemercier G, et al. (**2002**) A vacuolar-type H+-pyrophosphatase governs maintenance of functional acidocalcisomes and growth of the insect and mammalian forms of Trypanosoma brucei. **The Journal of Biological Chemistry** 277(40):37369-37376.

43. Bangs JD, Uyetake L, Brickman MJ, Balber AE, & Boothroyd JC (**1993**) Molecular cloning and cellular localization of a BiP homologue in Trypanosoma brucei. Divergent ER retention signals in a lower eukaryote. **Journal of Cell Science** 105 (Pt 4):1101-1113.

11

DNA damage checkpoint adaptation genes are required for division of cells harbouring eroded telomeres

Sofiane Y. Mersaoui, Serge Gravel, Victor Karpov, and Raymund J. Wellinger*

Dept of Microbiology and Infectious Diseases, Faculty of Medicine and Health Sciences, Université de Sherbrooke, 3201, Rue Jean Mignault, Sherbrooke, J1E 4K8, Canada.
* Corresponding Author: Raymund J. Wellinger, E-mail: Raymund.Wellinger@Usherbrooke.ca

ABSTRACT In budding yeast, telomerase and the Cdc13p protein are two key players acting to ensure telomere stability. In the absence of telomerase, cells eventually enter a growth arrest which only few can overcome via a conserved process; such cells are called survivors. Survivors rely on homologous recombination-dependent mechanisms for telomeric repeat addition. Previously, we showed that such survivor cells also manage to bypass the loss of the essential Cdc13p protein to give rise to Cdc13-independent (or cap-independent) strains. Here we show that Cdc13-independent cells grow with persistently recognized DNA damage, which does not however result in a checkpoint activation; thus no defect in cell cycle progression is detectable. The absence of checkpoint signalling rather is due to the accumulation of mutations in checkpoint genes such as *RAD24* or *MEC1*. Importantly, our results also show that cells that have lost the ability to adapt to persistent DNA damage, also are very much impaired in generating cap-independent cells. Altogether, these results show that while the capping process can be flexible, it takes a very specific genetic setup to allow a change from canonical capping to alternative capping. We hypothesize that in the alternative capping mode, genome integrity mechanisms are abrogated, which could cause increased mutation frequencies. These results from yeast have clear parallels in transformed human cancer cells and offer deeper insights into processes operating in pre-cancerous human cells that harbour eroded telomeres.

Keywords: telomeres, DNA damage checkpoints, chromosome capping.

Abbreviations:
ALT - alternative lengthening of telomeres,
DSB - double strand break,
HR - homologous recombination,
HU - hydroxyurea,
TRF - terminal restriction fragment,
ts - temperature sensitivity,
wt – wild type.

INTRODUCTION

Telomeres are essential for genome stability in all organisms with linear chromosomes; they play multiple roles in chromosome end protection, chromosome end replication and distinguishing chromosome ends from double strand breaks (DSBs) [1, 2]. Indeed, chromosome ends and DSBs superficially share a great deal of similarity as both are physical ends of DNA molecules. However, functional telomeres do not activate checkpoints and they are not subjected to DNA repair activities such as homologous recombination (HR) or end-to-end fusions [3, 4]. These features are provided by the unique structures and organization of the nucleoprotein complexes located at the ends of chromosomes [5, 6]. Telomere structure and the molecular functions required for the above activities are also highly conserved, suggesting a common evolutionary origin. Telomeric DNA consists of short tandem DNA repeats, which generally create a G-rich strand that makes up the 3' end

of the chromosome. This strand also protrudes beyond the 5' end, forming a single stranded "G-tail" [7-9].

Chromosomes of *Saccharomyces cerevisiae* cells share these features and serve as an excellent model for studying telomere biology. There is an approximately 300 bps double-stranded portion and a short single-stranded DNA portion of characteristic repeats at each chromosome end. These repeats are associated with specialized proteins that are essential for telomeric functions [1]. Specifically, telomeric capping in yeast is assured by a heterotrimeric complex composed of Cdc13p, Stn1p and Ten1p (the CST-complex; [1, 10-13]. Of these, Cdc13p specifically recognizes a short telomeric G-rich DNA substrate that can be the terminal G-tail and all three genes are essential. Hypomorphic alleles of *CDC13* exist and the *cdc13-1* allele, for example, confers temperature sensitivity (ts) to cells [14]. In cells with the *cdc13-1* allele that are incubated at restrictive temperatures (>26°C), the C-rich strand of telomeric

DNA is degraded, yielding extensive single-stranded DNA that can reach into subtelomeric DNA [15]. These ends become recognized as sites of DNA damage which triggers a robust Rad9-dependent cell-cycle checkpoint response and, eventually, cell death [13]. Cdc13p is also involved in allowing the complete replication of telomeres by recruitment of telomerase to telomeres [16].

Telomerase is the ribonucleoprotein that elongates telomeres during S-phase to counteract the shortening of telomeric DNA that occurs due to the 'end replication problem' [17-19]. A loss of telomerase leads to progressive telomere shortening, the so-called "ever shorter telomeres", or "EST" phenotype and a concomitant loss of the telomeric capping function [20-22]. Upon outgrowth of such cultures, the majority of cells ceases to divide after approximately 60-80 generations.

However, a small proportion of cells can regain the ability to divide and to maintain chromosome ends by telomerase-independent mechanisms; these cells are called survivors [1, 23]. There are two major types of survivors which have one of two different arrangements of telomeric and subtelomeric DNA [24]. In type I survivors, a complex subtelomeric repeat element called Y' spreads to all telomeres and only a very short telomeric repeat tract remains. Type II survivors on the other hand are defined by the presence of very long and heterogeneous telomeric repeat tracts [24]. These two types of cells can also be distinguished by the genetic requirements for the pathways involved [25]. Survivor cells still require Cdc13p-mediated chromosome capping, but Cdc13-independent survivors can be generated [26-28]. These latter cells do grow without the canonical organization of chromosomal ends; their telomeres yield a new pattern of terminal restriction fragments (TRFs) with a complete absence of discrete bands. Moreover, there is a high amount of telomeric and subtelomeric single-stranded DNA at chromosomal ends in Cdc13-independent survivors [28]. Previous evidence also suggests that telomeres in these cells essentially are elongated and maintained by HR mechanisms [28, 29]. Finally, cell growth without the capping protein Cdc13p is possible if genes involved in DSB processing (EXO1, SGS1) and certain checkpoint genes (RAD9, RAD24) are deleted simultaneously [29].

It has been speculated that particularly type II survivor yeast cells resemble human cancerous cells that replenish their telomeric repeat DNA via alternative telomerase-independent mechanisms (alternative lengthening of telomeres, or ALT), [30, 31]. A commonality between yeast survivor and human ALT cells is that both are thought to amplify their telomeric DNA through recombination-dependent DNA replication [32]. This probably occurs through a mechanism involving extrachromosomal circular DNA containing telomeric repeat sequences [28, 33]. However, in virtually all human transformed cells, genomic integrity mechanisms are severely hampered and cells divide with ongoing genomic instability [34].

Here we show that Cdc13-independent survivor cells (cap-independent cells) do grow even though DNA damage foci can be observed to persist on telomeres. However, this apparent damage does not result in checkpoint signalling or cell cycle arrest. Our results show that this is because the Mec1-branch of the damage signalling pathway was abrogated via mutations in central checkpoint genes, mainly RAD24 or MEC1. We also report that CDC5, PTC2 and TID1 are required for the initial generation of cap-independent cells. Mutations in either of these genes significantly reduce the ability of survivor cells to overcome the loss of Cdc13p and resume growth. These results therefore reveal intriguing similarities between yeast cells dividing in the absence of Cdc13p and human cancerous ALT-cells. Both display absence of cell division controls and continued cell divisions, despite ongoing telomere instability. We therefore hypothesize that this yeast system represents a useful tool for investigating the early phases of human cancerous cell growth.

RESULTS

Permanent detection of telomeric DNA damage but no checkpoint activation in Cdc13-independent survivors

Previous analyses of telomeres in Cdc13-independent survivors showed that their TRFs are extremely heterogeneous in length (see Fig. S1, lanes 8-10; and [28]). In order to obtain a more precise assessment of the terminal sequences on their chromosomes, we cloned and sequenced 17 independent terminal DNA fragments. 10 of those 17 harboured potentially functional telomeric repeat tracts (> 50 bps of repeat DNA), one had a critically short tract (35 bps) and six had tracts that were too short for even a single binding site for Rap1p, the major yeast protein binding double-stranded telomeric repeat tracts [1]. In fact, two of the 17 clones had no detectable telomeric G-rich sequences and ended with a subtelomeric Y' element (Fig. 1A).

Cells with such short or absent telomeric repeat tracts and potentially exposed single-stranded Y'-DNA at chromosome ends display a strong DNA damage checkpoint [35]. We therefore verified whether the DNA damage checkpoint was activated in Cdc13-independent survivor cells, by assessing the level of phosphorylation of Rad53p, a major DNA damage checkpoint effector kinase in budding yeast ([36]; Fig. 1B). Rad53p migrates as a single discreet band in unperturbed growing cells, while DSBs caused by the addition of the radiomimetic phleomycin (a derivative of bleomycin) causes DNA damage checkpoint activation and robust phosphorylation of Rad53p as detected in the form of a retarded Rad53p band (Fig. 1B, lane 2). In Cdc13-independent survivors, no phosphorylated Rad53p is detected in unperturbed cells and phleomycin addition only causes a very partial retardation of Rad53p, as if only a partial phosphorylation was possible (Fig. 2B, lanes 5, 6).

Similar results were obtained when phosphorylation of the checkpoint effector kinase Chk1p was assessed (Fig. S2). DNA damage, once detected, causes the activation of two sensor kinases, namely Mec1p and Tel1p (ATR and ATM in mammals [37, 38]), and phleomycin induced damage is sensed mainly via Mec1p. We thus reasoned that the very partial Rad53p phosphorylation detected in Cdc13-independent survivors could be due to a loss of the Mec1

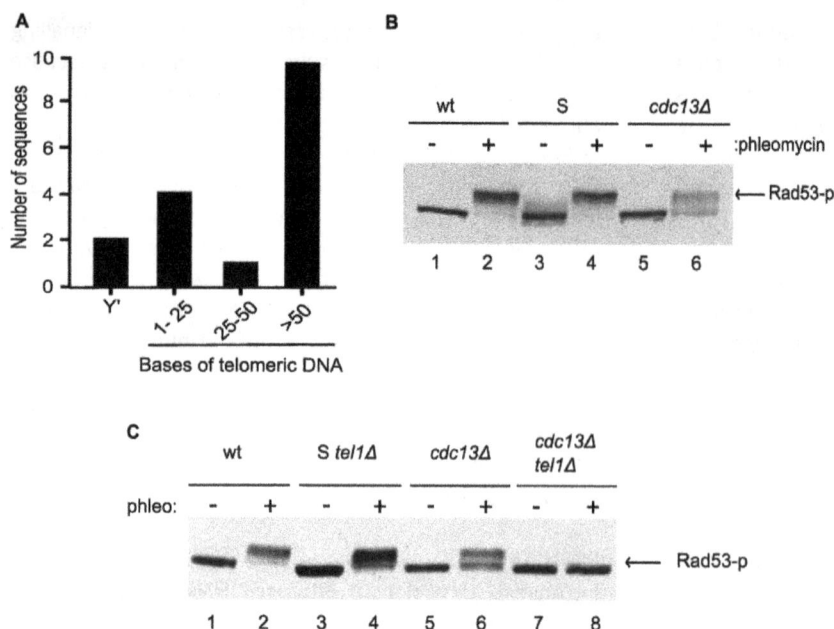

FIGURE 1: Mec1-dependent Rad53 phosphorylation is defective in cdc13Δ cells. (A) Terminal chromosomal regions were amplified by PCR and sequenced. Telomeric DNA (in bases) present on the amplified fragments is indicated. A total of 17 independent sequences were obtained, with two of them ending in the subtelomeric Y' element. **(B)** Exponentially growing wild-type (wt), Survivor (S) or cdc13Δ (cdc13Δ) cells were left untreated or were treated with phleomycin for 2 h. Protein extracts were prepared and analyzed by western blotting with an anti-Rad53 antibody. **(C)** As in (B) except that cells also lacked the *TEL1* gene as indicated in the Survivors (S) and Cdc13-independent (cdc13Δ) cells. Cells were left untreated or were treated with phleomycin for 2 h.

branch of the sensor kinase pathway. We therefore constructed Cdc13-independent survivor strains that lacked Tel1p and assessed the degree of Rad53p phosphorylation induced by phleomycin. While survivor strains with a *tel1Δ* allele are able to strongly phosphorylate Rad53p (Fig. 1C, lane 4), Cdc13-independent survivors harbouring the *tel1Δ* allele were unable to do so (Fig. 1C, lane 8).

As DSBs are usually recognized rather efficiently, we wondered whether telomeres in Cdc13-independent cells had developed an alternative, telomeric DNA-independent capping mechanism or whether they were simply ignored by the DNA damage/repair machinery. To address this, we used fluorescence microscopy to visualize proteins involved in the recognition and repair of DNA DSBs. DNA end resection at DSB generates single-stranded DNA (ssDNA) that is rapidly bound by the RPA complex, and the focal localization of RPA on DNA damage sites can be visualized in living cells via YFP-tagged Rfa1p [39]. As a positive control, we treated cells with the radiomimetic drug zeocin and, as expected, zeocin induced the formation of Rfa1-YFP foci in wild-type cells (Fig. 2A, B). Strikingly, in Cdc13-independent survivor cells, Rfa1-YFP foci were readily detectable, even in the absence of zeocin treatment (Fig. 2).

Quantification indicates that ssDNA is present in more than 80% of unperturbed S/G2 cells and more than 50% of cells in G1. As an additional read-out for the presence of DNA damage recognition, we carried-out the same analysis with Rad52-YFP. In untreated wt (wild type) cells, Rad52-YFP foci are rare and zeocin treatment increased the fraction of cells displaying Rad52-YFP foci to about 80-90 % (Fig. 2). Again in stark contrast with these results, the majority of untreated Cdc13-independent cells already had Rad52-YFP foci and there was no detectable increase upon zeocin treatment.

If uncapped telomeres were the source of Rfa1-YFP foci in Cdc13-independent survivor cells, they should rapidly disappear following reintroduction of *CDC13*. We thus transformed Cdc13-independent survivor cells with a plasmid that contained the wt *CDC13* gene and assessed telomeric restriction fragments and RPA-foci. TRF patterns in Cdc13-independent survivor cells with Cdc13p re-expressed reverted to a typical survivor pattern (compare Fig. 2C with Fig. S1) and the appearance of Rfa1-YFP foci reverted back to a level observed in *CDC13* wt cells (Fig. 2D). These results indicate that re-establishing a Cdc13-dependent capping system in cdc13Δ cells eliminates the presence of the detected DNA damage and suggest that telomere uncapping is indeed the source of the RFA and Rad52p foci in cdc13Δ mutants. Thus, in Cdc13-independent survivors, uncapped and resected telomeres persist throughout the cell cycle and are bound by proteins that would allow ongoing DNA repair. However, Mec1-mediated DNA damage signalling is by and large abrogated, allowing for cell divisions to continue and hence, culture growth.

A functional checkpoint is incompatible with growth in the absence of CDC13

The data described above indicate that uncapped telomeres and apparent DNA damage foci are constitutively present in a large fraction of the Cdc13-independent cells. Given the virtual absence of DNA damage signalling, we considered that inactivation of Mec1-mediated checkpoint signalling could have been caused by reduced or dysregulated DNA resection. To test this, an HO endonuclease-induced DSB was created in cells in such a way that no homologous sequences for recombinational repair were present next to the break. The rate of DNA resection next to the HO site was monitored by denaturing slot blot analysis using a probe complementary to the processed DNA strand [40].

In this assay, signal loss for the resected strand was very similar in *CDC13* cells and cells that harboured a cdc13Δ allele (Fig. S3A, S3B). As an additional assay for

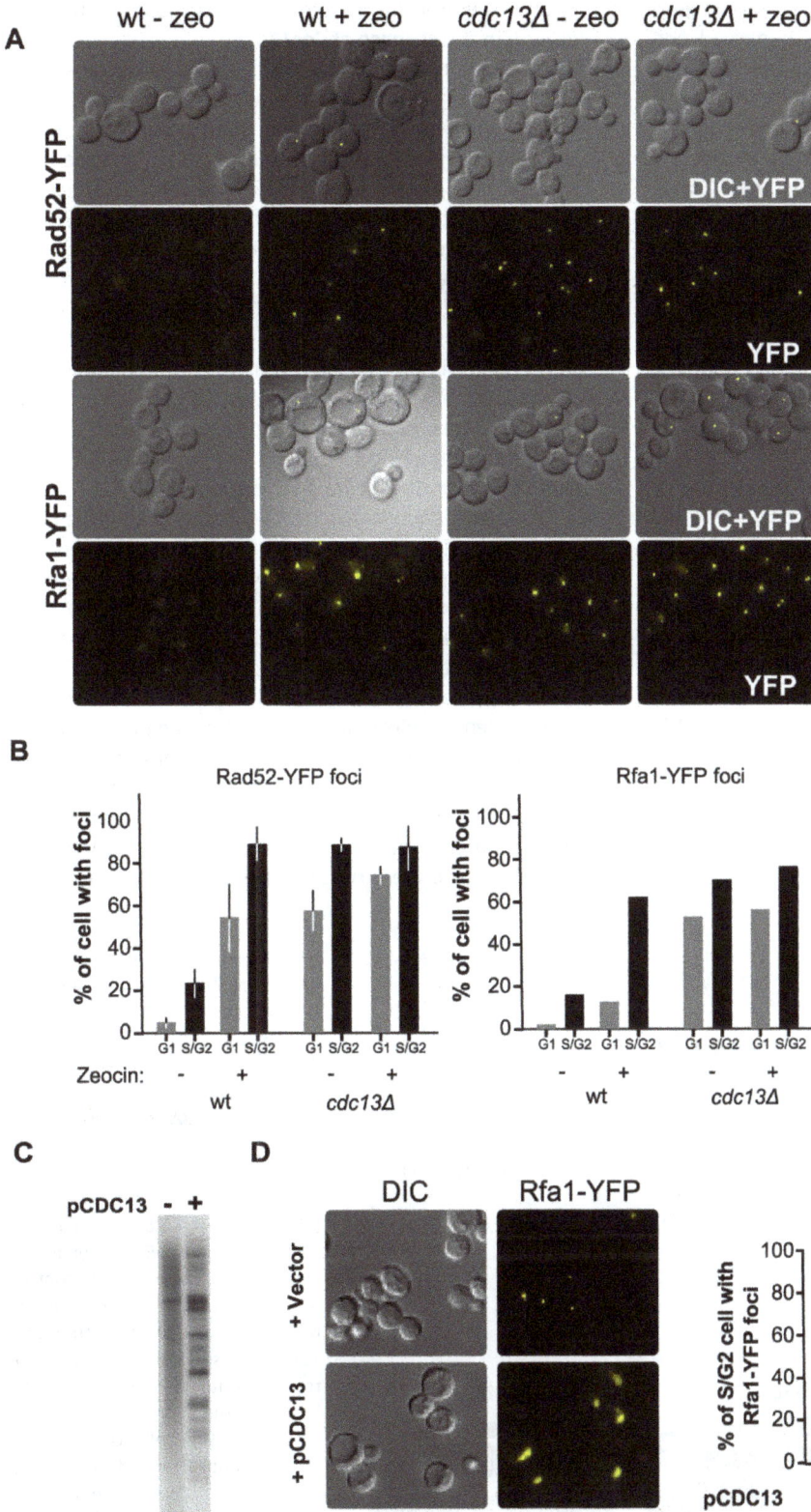

FIGURE 2: Rad52p and Rfa1p proteins form DNA damage foci in untreated *cdc13Δ* cells. (A) Wild-type (wt) or *cdc13Δ* cells expressing the Rfa1-YFP fusion protein (RWY801) or the Rad52-YFP protein (CPY821) were analysed by fluorescence microscopy untreated or treated with zeocin for 3 h. Representative examples of the images obtained are shown. **(B)** Quantification of cells with foci. Cells were morphologically divided into G1 cells (unbudded cells) and G2M cells (large budded cells). The Rad52-YFP experiment was performed in 3 independent biological replicates, and the Rfa1-YFP experiment was performed with at least 150 cells for each condition. **(C)** Southern blot with DNA derived from MLY122 using a telomeric DNA probe. Cells were growing in the absence of Cdc13p (lane marked -) or the wt *CDC13* gene was introduced via plasmid p*CDC13* (lane marked +). Note that re-introduction of the protein CDC13 restores survivor type II TRF phenotype. **(D)** The re-expression of *CDC13* eliminates the formation of Rfa1-YFP foci in *cdc13Δ* cells. Strain CPY821 was grown to become Cdc13-independent and then an empty vector or pCDC13 was introduced. Representative examples of images obtained as in (A) are shown in the left panel. Quantification of the data is shown in the right panel. >100 cells were counted for each sample.

DNA resection and ensuing repair, we assessed completion of mating-type switching (a process based on HR) by Southern blot analyses. As observed in the resection assay, no significant difference between wt and cells that lack Cdc13p could be detected (Fig. S3C). These physical assays show that the dynamics and efficiency of DNA end-processing and HR repair remained virtually unaffected in cells growing without Cdc13p.

An alternative hypothesis to explain the absence of Mec1p-mediated DNA damage signalling is that the checkpoint signalling proteins are themselves impaired, for instance by having acquired a mutation. In this case a forced re-expression of the functional wt allele of a mutated gene would cause a growth arrest in cells with perceived permanent DNA damage. We therefore cloned the wt alleles of genes encoding essential components of the Mec1-checkpoint pathway into a vector that allowed a carbon source dependent expression of the protein (pGal; Fig. 3A). Twelve individual and independently generated Cdc13-independent survivor strains were then transformed with those plasmids and the resulting growth behaviour was assessed on plates that induced expression of the particular gene (Gal-plates).

Remarkably, each of the twelve strains showed a growth arrest with one of the plasmids used. An example is shown in Fig. 3B, where strain MSY053 grew well on Gal-plates when *MEC1, RAD17* or *MEC3* were overexpressed, yet they failed to grow when *RAD24* was overexpressed (Fig. 3B, right). These cells did grow on glucose plates (Fig. 3B, left) and Rad24p overexpression was readily tolerated in wt cells or regular survivor cells that contained a wt copy of *CDC13* (Fig. S4, top plate). The genomic *RAD24* locus of strain MSY053 was sequenced and found to contain a frameshift mutation that caused a loss of function. In summary, in the 12 independently obtained Cdc13-independent survivor strains, sensitivity to Rad24p expression was uncovered seven times, Mec1p six times, Rad17p two times and to Mec3p one time. These results thus further confirm that persistent telomere uncapping is sensed as DNA damage in cells without the Cdc13 protein and that in the absence of Cdc13p, cells can only grow with an inactivated Mec1-branch of checkpoint signalling.

Factors required for adaptation to DSBs are required for growth in the absence of Cdc13p

It has been observed that wt cells are only very rarely able to overcome a complete loss of Cdc13p but actual survival rates are unknown. We used fluctuation analyses to measure survival rates of wt cells and of cells that have overcome a telomerase deficiency (survivor cells). Only about 1 in 10^9 - 10^{10} cells survived an abrupt loss of Cdc13p and created a growing culture in telomerase-positive cells, whereas in cultures of survivor cells the rate is at least 1000 fold higher (Fig. 4A, [28]). This difference may be explained by the fact that telomerase-mediated telomere maintenance is not required in survivors and that upon loss of Cdc13p, cells only have to adapt to chromosome capping loss. Loss of telomeric capping is thought to have comparable effects as induction of a number of DSBs at the same time, but direct data on this is lacking.

As a consequence of telomere loss, mammalian and budding yeast cells arrest the cell cycle for prolonged periods of time before resuming growth, even in the presence of persistent damage. The latter process has been dubbed 'adaptation' and *CDC5, TID1* and *PTC2* genes are key elements required for adaptation to occur in yeast [37]. We therefore examined whether the generation of Cdc13-independent survivors is dependent on adaptation genes. Fluctuation analyses show that survivor cells that also harbour a deletion of *TID1* or *PTC2* (*cdc13Δ tcl1Δ tid1Δ* cells or *cdc13Δ tcl1Δ ptc2Δ* cells), as well as survivors that harbour an adaptation negative *CDC5* allele (*cdc13Δ tcl1Δ cdc5-ad* cells) only generate survivors extremely rarely, at a rate of

A

Cdc13-independent cells   +   pGal-XXX

Yc- URA +
Glu (2%)

YEP +
Gal (2%)
(Induction)

XXX= MEC3, MEC1, RAD17, RAD24   Growth test

B

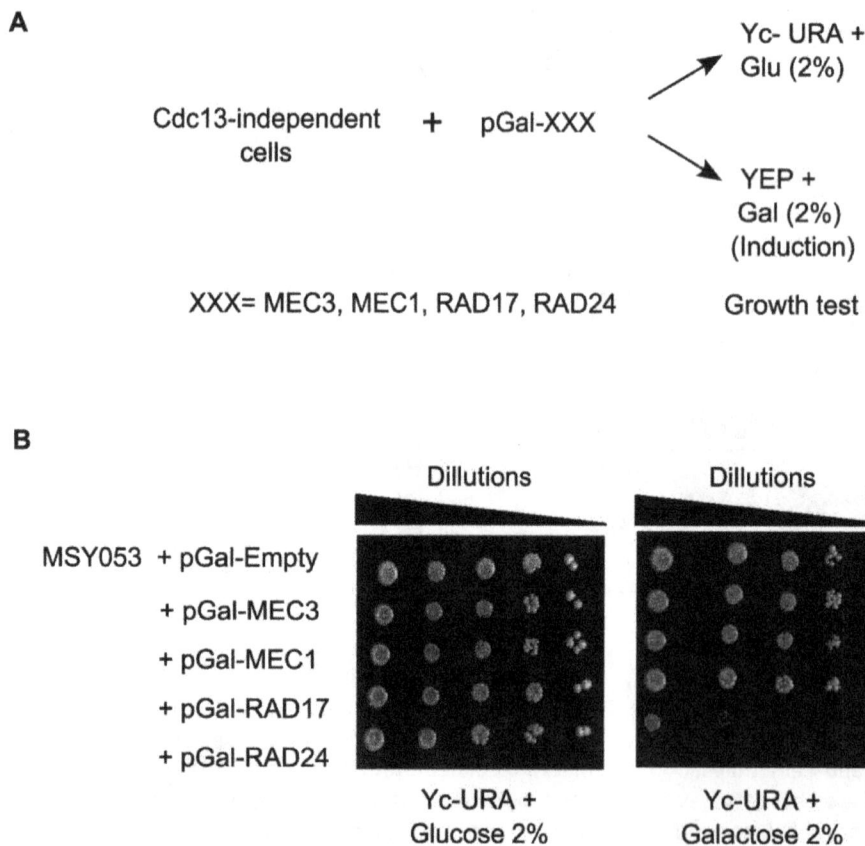

Dillutions

Dillutions

MSY053 + pGal-Empty
+ pGal-MEC3
+ pGal-MEC1
+ pGal-RAD17
+ pGal-RAD24

Yc-URA +
Glucose 2%

Yc-URA +
Galactose 2%

FIGURE 3: An abrogated Mec1-branch of the checkpoint pathway is required for Cdc13-independent cell growth. (A) Schematic representation of the functional complementation test: 12 independent strains of *cdc13Δ* cells were used in this experiment. Each cap-independent *cdc13Δ* strain was transformed with five constructs (plasmids): pGal-*MEC1*, pGal-*MEC3*, pGal-*RAD17*, pGal-*RAD24* or empty vector pGal-Empty as a control. Serial dilutions of cultures of the resulting strains were spotted onto YEP+ Glucose (2%) for growth control and onto YEP+ Galactose (2%) to induce the expression of indicated genes. **(B)** Cells of a Cdc13-independent strain stop growing when the mutated checkpoint gene is complemented by the corresponding wild-type construct. The results shown in this figure were derived with strain MSY053 which holds a mutation in *RAD24* in the genome. ▪

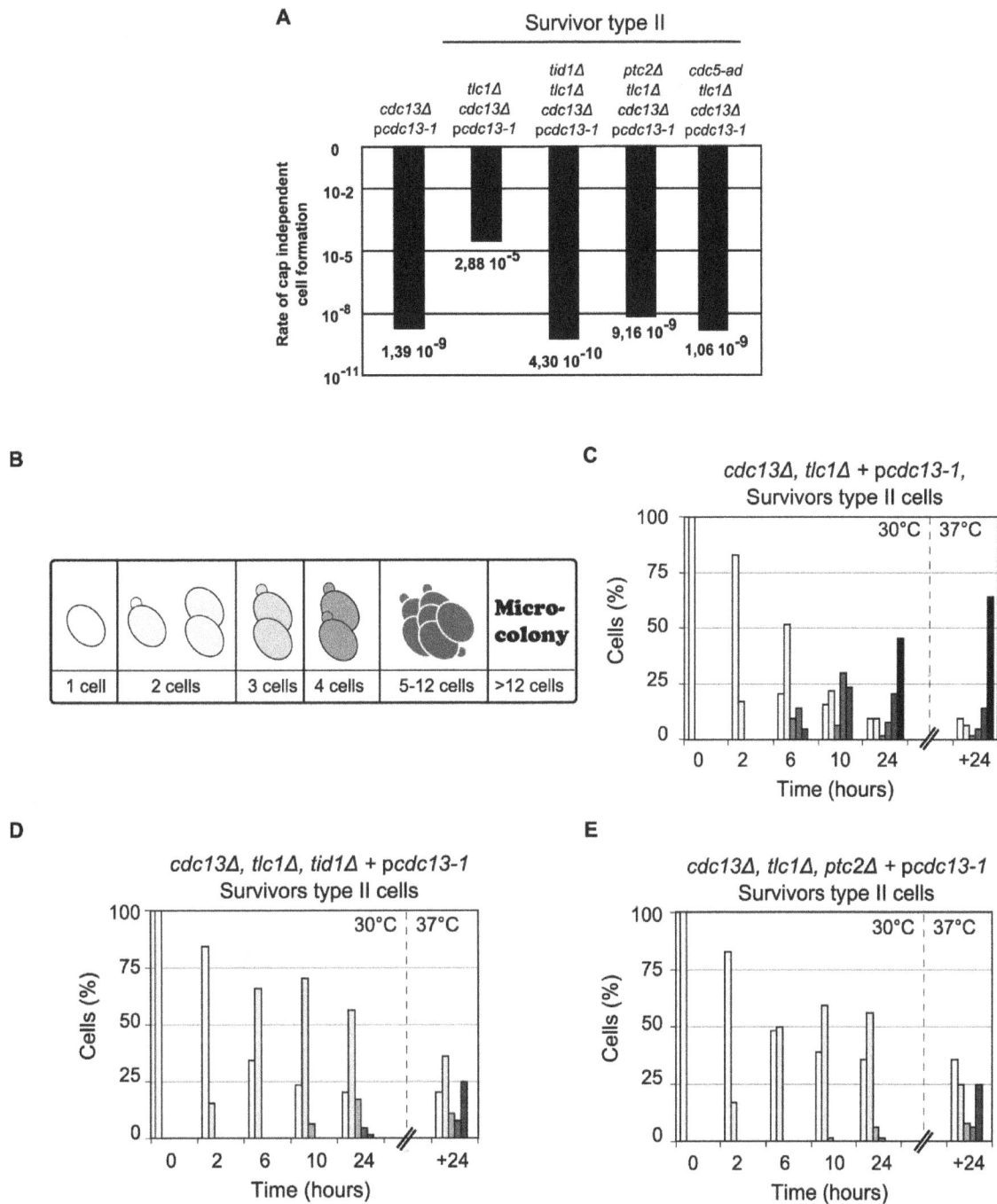

FIGURE 4: The formation of *cdc13*-independent survivors (*cdc13Δ*) requires DSB adaptation genes. (A) The rate of successful colony formation was calculated by fluctuation test foAr five strains: (i) telomerase positive controls, *cdc13Δ* + pcdc13-1 (VKY20), (ii) type II survivors: *tlc1Δ*, *cdc13Δ* + pcdc13-1 (MLY112), (iii) type II survivors harbouring a deletion of *TID1: tid1Δ*, *tlc1Δ*, *cdc13Δ* + pcdc13-1 (VKY19), (iv) type II survivors harbouring a deletion of *PTC2: ptc2Δ, tlc1Δ, cdc13Δ* + pcdc13-1 (VKY12), or (v) type II survivors harbouring the *cdc5-ad* allele: *cdc5-ad, tlc1Δ, cdc13Δ* + pcdc13-1 (MSY421). (B) Schematic representation of standardization for morphological classification used to analyse cell cycle progression of the strains of interest. (C) (D) (E) Single round unbudded cells (64 cells) from three strains described above (ii) (iii) (iv) were identified and arrayed on YEPD plates and incubated at 30°C (restrictive temperature for the *cdc13-1* allele). Morphology and growth of cells were inspected microscopically at 2, 6, 10 and 24 h after incubation. Plates were then incubated at 37°C for another 24 h. Colony morphology was recorded as outlined in (B) and results plotted as the percentage of the initial group. (C) Cell cycle progression for the survivor type II *tlc1Δ, cdc13Δ* + pcdc13-1 (MLY112); (D) and (E) cell cycle progression for both adaptation deficient mutants *tid1Δ, tlc1Δ, cdc13Δ* + pcdc13-1 (VKY19), *ptc2Δ, tlc1Δ, cdc13Δ* + pcdc13-1 (VKY12).

about 5 x 10^{-9} to 5 x 10^{-10} (Fig. 4A). Similarly, when the morphology of cells and the generation of micro-colonies was examined in a single cell assay after inactivation of Cdc13p via a temperature upshift to restrictive temperatures, survivor cells (cdc13Δ tcl1Δ + pcdc13-1) showed clear signs of adaptation; after 24 h at 30°C, more than 50% of the cells had grown into micro-colonies of 5 cells or more and at least 60% of them became micro-colonies with more than 12 cells after another 24 h at 37°C (Fig. 4B, C). In contrast, more than 50% of adaptation-negative survivors (cdc13Δ tcl1Δ tid1Δ + pcdc13-1 cells or cdc13Δ tcl1Δ ptc2Δ + pcdc13-1 cells) were still single cells or small budded cells after 24 h at 30°C and only a small minority formed 5-12

cell colonies (Fig. 4D, E). Finally, none of the 64 adaptation-negative (either ptc2Δ or tid1Δ) survivor cells generated a viable colony. This was expected given the very low rate determined by fluctuation tests. We conclude that the TID1, PTC2 and CDC5 genes are required in yeast for both, adaptation to persistent DNA damage after DSB induction as well as for the ability of survivor cells to generate growing colonies after the loss of telomere capping.

A reversible component to adaptation to telomere uncapping

The results thus suggest that adaptation to telomere uncapping requires known adaptation genes, that telomere

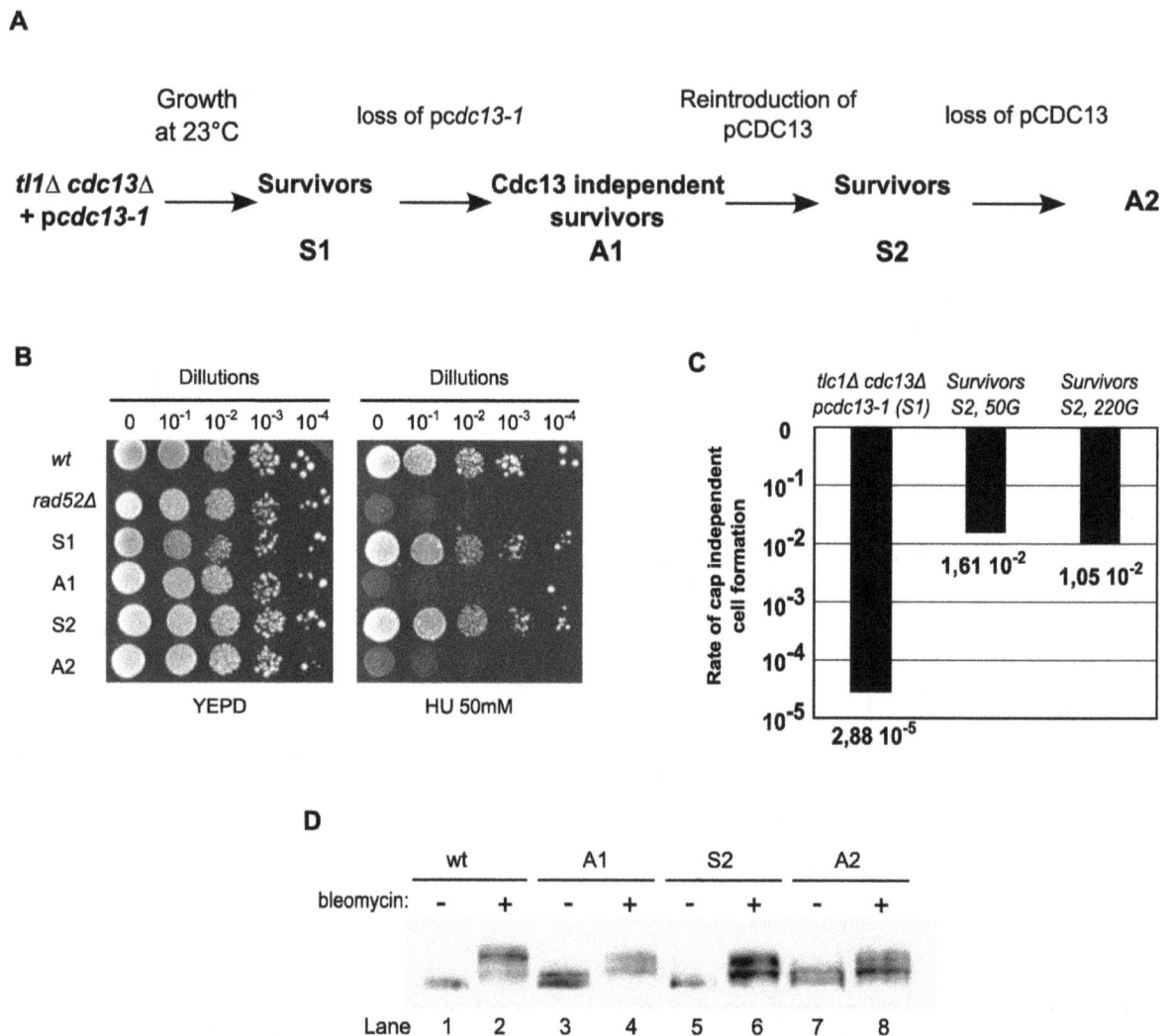

A

B

C

D

FIGURE 5: The process of generating Cdc13-independent survivors (cdc13Δ-cells) comprises a reversible component. (A) Schematic representation of the procedure used to generate S2 survivors and second generation cdc13Δ-A2 of Cdc13-independent survivors. S1 survivors and cdc13Δ-A1 were generated as described in Fig. 1. S2 survivor cells were derived from cdc13-A1. These cells were grown for up to 220 generations (220G) after transformation with the plasmid carrying the cdc13-1 allele. For cdc13Δ-A2 cells (MLY123), the loss of the pCDC13 plasmid was verified by growing cells on FOA synthetic medium. (B) Cells identified with symbols as in (A) were spotted onto YEPD plates with or without hydroxyurea (HU) and incubated for 72 h at 23°C. A haploid rad52Δ strain was used as control for HU sensitivity. (C) Rate of successful cdc13-independent cell formation (calculated by fluctuation test) for survivors S1 (MLY112) and survivors S2 (MLY122 + pcdc13-1). S2 survivors were grown for 50 or 220 generations after pcdc13-1 introduction and before the fluctuation test for plasmid loss was performed. (D) Western blot of whole cell protein extracts prepared from strains indicated with the same symbols as in (A)(top). Cell treatment with bleomycin is indicated with − and +, and the blot was probed with an anti-Rad52 antibody as in Fig. 1.

maintenance is carried out by recombination, and that at least one gene of the DNA checkpoint signalling machinery has to be inactivated by mutation. However, it was unclear whether this adapted state, once achieved as a metabolic state, is stable and remains active in cells or whether it is reversible. In order to assess this question, we generated cells that passed through the survivor state and had adapted to Cdc13p-loss. We then re-introduced the Cdc13p capping protein into these cells to let them grow with the capping protein (see schematic in Fig. 5A). The TRF patterns of such cells changed and again became typical for survivor cells [28].

Adapted cdc13Δ cells were sensitive to agents causing replication stress, such as hydroxyurea (HU). However after Cdc13p re-introduction they reverted to an insensitive phenotype similar to wt and pre-adaptation survivors (Fig. 5B, [28]). We eventually challenged these latter cells to a Cdc13p-loss a second time. In this experiment, we call survivors before the first Cdc13-loss 'S1-cells' and those generated by re-introduction of Cdc13p into Cdc13-independent cells 'S2-cells' (see Fig. 5A). We then compared the rates of adaptation of S1 or S2 cells by fluctuation analyses (Fig. 5C). Remarkably, even though S2 cells had been derived from adapted cells, only about 1 % of them formed viable colonies after the second loss of Cdc13p. While this rate was about 1000 fold higher than that observed for the original transition of S1 cells to adapted cells (Fig. 4A, 5C), it still far from complete and indicated that the reintroduction of Cdc13p into adapted cells is associated with a reversion of at least part of the phenotype.

This low adaptation rate did not change with longer outgrowth of S2 cells; after 220 generations of growth in the S2 state, the rate of adaptation remained about 1 % (Fig. 5C). Similar to the transition from S1 to the first adaptation state, the transition from S2 to adapted state also resulted in an increased sensitivity to hydroxyurea (HU) (Fig. 5B). However, the capacity to fully phosphorylate Rad53p did not recover after reintroduction of Cdc13p into adapted A1 cells (S2 cells; Fig. 5D, lane 6). The signalling also remained partial after the second loss of Cdc13p (Fig. 5D, lane 8), as would be expected if this part of the phenotype was genetically determined. We conclude that the very low rate of colony formation after a first loss of Cdc13p from survivor cells (about 3×10^{-5}) can be explained by the fact that a mutation in one of the Mec1-branch checkpoint genes is required. A second Cdc13p loss after a temporary reintroduction of it still is poorly tolerated and only about one percent of cells survive, suggesting that a metabolic, non-genetic, component of the adaptation state does reverse upon introduction of Cdc13p.

DISCUSSION

Checkpoint adaptation was originally described as the ability of S. cerevisiae cells to overcome a sustained checkpoint arrest due the presence of irreparable DNA damage [41-43]. Subsequently, mechanisms to abrogate a prolonged checkpoint arrest were also reported to operate in

Xenopus laevis and human cells and the well conserved genetic requirements for the process suggested a common evolutionary origin ([44, 45]; see below). Although cells undergoing checkpoint adaptation almost invariantly die in subsequent cell cycles, owing to rampant genome instability, some cells do divide a limited number of times. However, it remained unclear how these cells managed to pass through the cell divisions and whether checkpoint abrogation is permanent or temporary. A conceptually similar situation arises in the etiology of malignant human cells. Current evidence strongly suggests that precancerous cells, very early on, undergo a phase of high level genome instability that is due to dysfunctional telomeres [46]. Once this serious bottleneck is overcome, cancerous cells have invariably activated a mechanism to maintain telomeric repeats, which is almost always achieved by a reactivation of telomerase, and they have inactivated genome surveillance mechanisms, in most cases at least including TP53 [34]. Experimental setups that allow a systematic study of the chain of events happening in human cells when passing from normal to pre-cancerous therefore promise to yield invaluable insights into the very early etiology of cellular transformation.

Budding yeast cells maintain telomeres via a constitutively active telomerase, but cells can be engineered to lose telomerase and thus, in this respect, phenotypically become more like human somatic cells [1]. For example, yeast cells without telomerase endure telomere shortening eventually leading to crisis and growth arrest, when at least some telomeres are dysfunctional [1]. Yeast survivor cells are defined as the fraction of telomerase negative yeast cells that overcome this short telomere crisis by replenishing telomeric repeats by HR. Previously, we showed that again only a fraction of such survivor cells are able to survive a loss of functional telomere capping as well, generating so called cap-independent survivors [28]. These latter cells are able to divide but are very sensitive to genotoxic compounds and they have a significantly reduced ability for DNA damage signalling.

The characteristics of the cap-independent survivor cells reported here are similar in several ways to early transformed human cells. Most significantly, cap-independent survivors have acquired mutations in at least one gene in the canonical DSB signalling pathway, which is governed by Mec1p in yeast (ATR in humans; Fig. 3, S4). As a consequence, the downstream effectors for cell cycle arrest are not activated and cell divisions continue even in the continued presence of DNA damage. The fact that cells stopped growing when we re-established the pathway by transforming cells with wt copies of the mutated genes demonstrates that these continued cell divisions are indeed dependent on an abrogated damage signalling pathway (Fig. 3). A conceptually similar effect has been reported in human and mouse cancer cells in which the reversion of a mutated TP53 allele leads to a re-establishment of genome surveillance mechanisms and tumour cell death [47-49]. As in human cancer cells, the physical recognition of a DSB and its actual repair appear virtually unaffected in the cap-independent yeast cells (Figs. S3). Remarkably, a very high level of DNA damage foci are observed in cap-independent survivor cells in any phase of the cell cycle (Fig. 2). In normal cells, this level of DNA damage would invariably lead to a prolonged cell cycle arrest in G2/M [39]. However, this is not the case in Cdc13p-independent cells due to the loss of damage signalling.

The results strongly suggest that the DNA damage foci detected in these experiments are on chromosome ends and that DNA repair activity, i.e. HR, is ongoing on those sites. This idea is supported by the observations that a significant fraction of telomeric restriction fragments cloned from these cells does not contain any telomeric sequences anymore (Fig. 1). Those sites are therefore indistinguishable from any other DSB in the genome and are presumably recognized as such. Furthermore, when we introduced one additional specific DSB via an induced endonucleolytic cleavage, that DSB gets resected and eventually repaired via HR with similar kinetics to normal cells (Fig. S3). Thus, DNA damage in these cells is recognized by the DNA repair machinery and HR particularly, can be active.

The reason why Cdc13p-independent cells remain viable and the cultures grow with abundant DNA damage foci present is that in these cells, telomeric DNA (either subtelomeric complex repeats or terminal short repeats) becomes amplified and chromosome ends eventually are all composed of extensive head to tail arrays of repeated elements [1]. Therefore, the genome remains, by and large, intact and functional, even though significant amounts of terminal DNA may be lost due to end-degradation from these repeated elements.

A similar dynamic behaviour of DNA adjacent to telomeres has also been described in *Drosophila melanogaster* [50]. In this organism, chromosomes end with telomere specific retroelements, i.e. arrays of repeated and complex DNA elements, and there are neither short direct repeats nor evidence for a telomerase enzyme. As expected therefore, the terminal elements suffer gradual sequence losses with each cell division. However, the relatively limited overall losses are confined to the most distal parts of the chromosomes and these are balanced by the occasional acquisition of large repeat units [50]. Despite this very dynamic telomere maintenance mechanism, the genome in *D. melanogaster* cells remains intact, as it does in the cap-independent yeast cells reported here.

During these studies of cap-independent yeast cells, we further discovered that genes required for checkpoint adaptation, namely *CDC5wt*, *PTC2* and *TID1/RDH54*, are also required for generating dividing survivor cells that harbour a non-functional *cdc13* allele (Fig. 4). Ptc2p is a type 2C phosphatase and has been described as being directly involved in deactivating the DNA damage signalling pathway by dephosphorylating proteins, in particular Rad53p [51]. The cellular regulation of this phosphatase and how its checkpoint-abrogating activity is induced remain unclear however. The Tid1/Rdh54p protein is involved in DSB repair, by mitotic and meiotic HR, and it is itself phosphorylated by the Mec1p branch of the checkpoint signalling pathway after DNA damage [52]. The precise roles for Tid1/Rdh54p in checkpoint adaptation are poorly defined though.

Consistent with the above, the adaptation characteristic *cdc5-ad* allele [42] caused an indistinguishable phenotype as the losses of Tid1/Rdh54p or Ptc2p. Although the precise roles of these checkpoint adaptation genes in allowing survivor cells to recover from inactivating Cdc13p are not known, they are not simply suppressors of the ts phenotype conferred by the *cdc13-1* allele. We could not detect any differences in the temperature dependent growth characteristics between *cdc13-1* cells as compared to *cdc13-1 tid1Δ* or *cdc13-1 ptc2Δ* cells (Fig. S5).

Given the importance of these above genes in checkpoint adaptation, collectively our data underscore that cap-independent growth of yeast cells relies on the ability of cells to shut off checkpoint signalling. One might have suspected that once checkpoint signalling is abrogated by a mutation in one of the essential genes of the pathway (*MEC1*, *MEC3*, *RAD24*, or *RAD17*, see Fig. 3), checkpoint adaptation may not be necessary anymore and cells would readily divide and grow in Cdc13p-inactivating conditions (37°C with a *cdc13-1* allele). However, when we tested this idea by challenging *rad24* cells with a Cdc13p-loss a second time, only a small percentage of cells succeed to become dividing cultures (Fig. 5).

An efficient generation of cap-independent cells was reported to occur in cells that were engineered to lack the *RAD9* gene as well as deletions of at least two additional genes involved in pathways that generate single-stranded DNA, namely *SGS1* and *EXO1* [29]. It is therefore tempting to speculate that checkpoint adaptation genes are required to deal with the excess single-stranded DNA that is generated after a loss of telomeric capping. This function could thus be in addition to causing the loss of phosphorylation on key checkpoint signalling proteins, but it remains unclear what this could entail. Our data therefore support the hypothesis that checkpoint adaptation might be a cancer promoting mechanism in humans and that polo-like kinases, which are required for checkpoint adaptation in yeast and human cells, could be potent anti-cancer targets [53]. Further investigation of the mechanistic details of adaptation to telomeric cap loss in yeast thus has the potential to reveal new anti-cancer targets.

Finally, Cdc13p-independent cells are sensitive to the replication-interfering drug hydroxyurea (HU), yet this sensitivity is reversed upon re-introduction of wt Cdc13p (Fig. 5) and therefore is not due to mutations in checkpoint signalling genes. In parallel, upon Cdc13p-introduction, DNA damage foci are also lost in these cells (Fig. 2). In order to explain the reversible sensitivity to HU, we speculate that in cap-independent survivors, a limiting component of the DNA repair machinery is tied up at uncapped telomeres and a large amount of additional DNA damage very rapidly becomes toxic, because it is left unrepaired. Upon reintroduction of Cdc13p, capping is re-established, single stranded DNA at chromosome ends is lost and the cells regain full potential to deal with drug-induced DNA damage. This hypothesis predicts that sub-lethal interference with telomere capping in human cancerous cells would sensitize these cells to DNA damaging agents. Approaches for therapeutic uncapping are being pursued [54, 55] and our results suggest that a key to their success will be to combine telomere uncapping reagents with standard DNA damage treatments.

MATERIAL AND METHODS

Yeast Strains and plasmids

All strains and plasmids are presented in Supplementary Tables 1 and 2, respectively. *Cdc13Δ* strains were derived from the diploid strain UCC3535 [56], in which one allele of the CDC13 gene was disrupted by the natR gene, removing coding sequences + 57 to + 2361 with respect to the initiation codon of *CDC13*.

Survivor and cap-independent cell generation

Haploid MLY100 cells containing plasmid p*cdc13-1* were grown at 23°C for approximately 150 generations and type II survivor cells were identified by Southern-blot. Individual colonies from these type II survivors were incubated on non-selective YEPD plates at 23°C for 48 h. The colonies obtained were re-plated on YEPD every 48-72 h at gradually higher temperatures until reaching 37°C (28°C for 48 h, 30°C for 48-72 h, then 34°C for 24-72 h and finally 24 h at 37°C). The colonies obtained are passaged two times on FOA-containing plates to ensure complete loss of the plasmid p*cdc13-1*. These *cdc13Δ* - cells were then able to grow at all temperatures.

Colony growth tests (spot dilution tests)

Strains of interest were grown in liquid media at various temperatures (as indicated in figures) until reaching exponential growth phase (OD_{660nm}= 0.7 − 1.0). Cultures were then serially diluted in order to provide from 10 (minimum) to 100 000 (maximum) cells per volume plated (10 µl). After mixing and separating cells by vortexing, 10 µl aliquots of a complete dilution series (1/10 dilution factor between spots) were spotted on petri dishes containing various selective media. The resulting plates were incubated for 72 to 150 h at appropriate constant temperature until single colonies were clearly visible for positive control strains.

Colony growth analysis

5 ml liquid cultures of strains of interest were grown at 23°C with continuous agitation in appropriate media (e.g. YEPD plate is used for the *cdc13Δ* strains). After the cultures reached OD_{660nm} = 0.5, 100 µl of the culture were spread on a marked area of a pre-heated (30°C) YEPD plates. Individual round unbudded cells were arrayed on YEPD plates as quickly as possible (max. time was 15 min) using a micromanipulator stage on a microscope. 16 cells were arrayed on one plate in a typical experiment, and a total of 64 cells of the same strain were arrayed. The plates with arrayed cells were then incubated at 30°C. Morphology and growth of cells were inspected microscopically every 2 h and colony growth progression was evaluated by counting cells as follows:
- Round unbudded cells were scored as single cells, still in G1.
- Mother cells with a bud or cells showing a dumbbell morphology were scored as 2 cells.
- Round unbudded cells attached to a mother cell with a bud were scored as 3 cells.

The plates were further incubated at 30°C up to 24 h and cells were counted again at this point. This incubation at 30°C was followed by incubation at 37°C for another 24 h. The number of cells was again registered at this point. The plates were left at 37°C for another 72 h in order to evaluate whether complete colonies could form and photos were taken at this point.

Quantification of the rate of cdc13-independent cell generation (fluctuation test)

Yeast strains were pre-grown in 5 ml synthetic medium lacking uracil at 23°C until saturation. 2×10^6 cells were inoculated in 5 ml YEPD liquid medium and incubated at 30°C in a rotary drum with constant rotation for exactly 24 h. 10^4 cells were plated on YEPD plates and one plate was incubated at 23°C (for viability control) and another at 37°C (number of cap-independent cells control) for at least 72 h. The number of well-developed colonies was counted on each plate. This procedure was performed 20 times for each strain using independent colonies for their initial culture inoculation. To calculate the adaptation rate to telomere deprotection for different strains, we used the equation proposed by Luria and Delbrück [57]:

$$A= [-\ln(N_0/N)]/M$$

A: Adaptation rate (number of adaptation events per cell division)
N$_0$: Number of experiments that resulted in 0 adapted cells (number of plates with 0 colonies after incubation at 37°C)
N: Number of experiments (number of plates)
M: Number of cells entering each experiment (number of cells plated per plate)

Colonies from these type II survivors were incubated in liquid YEPD media at 23°C and 30°C until dense cultures were achieved (approximately 10 generations of growth) and then assayed for frequencies of generating Cdc13-independent cells by plating onto YEPD at 37°C and FOA plates. A total viable cell count was derived from the same cultures plated on YEPD and grown at 23°C.

Drug sensitivity assays

For growth sensitivity assays on plates, exponentially growing cultures were 10-fold serially diluted and spotted onto YEPD or synthetic complete plates containing 0.01% MMS or 50 mM hydroxyurea. For acute exposure to MMS, exponentially growing cells were mock treated or treated with MMS by adding 0.01% MMS to the corresponding cultures for 90 min. Strains used for western blotting of Rad53p phosphorylation were treated or not with 5 µg/mL of phleomycin (phleo) of bleomycin.

SDS PAGE and Western blot analysis

Protein extracts were prepared using a modified trichloroacetic acid method and proteins were separated by 8% SDS–PAGE as described in [28]. Western blotting was performed using an in house polyclonal anti-Rad53p antibody (kindly provided by D. Durocher, Samuel Lunenfeld Research Institute, Toronto, Canada) or with a commercial polyclonal anti-Rad53p antibody (Abcam, N° ab 104232, with using a 1:500 dilution) and signals were revealed using horseradish peroxidase-conjugated anti-rabbit antibodies with the enhanced chemiluminescence (ECL) detection kit (GE-Healthcare).

Terminal transferase and PCR telomere amplification

Purified genomic DNA was treated with terminal transferase to add a poly-C tail to chromosome ends as described [58]. 100 ng of genomic DNA were heat denatured and tailed in 10 µl of 20 mM Tris–HCl pH 7.8, 50 mM KAc, 10 mM MgAc2, 1 mM dCTP with 1 U of terminal deoxynucleotidyl transferase for 30 min at 37°C. Following tailing, the enzyme was inactivated by incubating the samples for 10 min at 65°C and 5 min at 94°C. The tailing reaction was performed in 0.2 ml PCR-tubes using a thermal cycler. By using a poly-G primer and a primer hybridizing in the subtelomeric Y' region terminal PCR products were obtained and sequenced.

ACKNOWLEDGMENTS

We thank D. Durocher (Lunenfeld Tanenbaum Res. Institute, Toronto Canada) for generously providing antibodies and J. Haber (Brandeis University, Waltham USA) for yeast strains. We are particularly indebted to M. Lisby (Copenhagen University, Copenhagen Denmark) who generously accepted RJW in his lab and in whose lab all the *in vivo* localization studies were performed. Members of the Wellinger lab, in particular C. Prud'homme, are thanked for valuable input and some strain constructions. This work was supported by a grant from the Canadian Institutes for Health Research (CIHR, grant # 110982) and RJW holds the Canada Reserch Chair in Telomere Biology.

CONFLICT OF INTEREST

The authors declare no conflict of interest.

REFERENCES

1. Wellinger RJ, Zakian VA (**2012**). Everything you ever wanted to know about Saccharomyces cerevisiae telomeres: beginning to end. **Genetics** 191(4): 1073-1105.

2. de Lange T (**2009**). How telomeres solve the end-protection problem. **Science** 326(5955): 948-952.

3. Wang RC, Smogorzewska A, de Lange T (**2004**). Homologous Recombination Generates T-Loop-Sized Deletions at Human Telomeres. **Cell** 119(3): 355-368.

4. van Steensel B, Smogorzewska A, de Lange T (**1998**). TRF2 protects human telomeres from end-to-end fusions. **Cell** 92(3): 401-413.

5. Gilson E, Geli V (**2007**). How telomeres are replicated. **Nat Rev Mol Cell Biol** 8(10): 825-838.

6. Hug N, Lingner J (**2006**). Telomere length homeostasis. **Chromosoma** 115(6): 413-425.

7. LeBel C, Wellinger RJ (**2005**). Telomeres: what's new at your end? **J Cell Sci** 118(Pt 13): 2785-2788.

8. McElligott R, Wellinger RJ (**1997**). The terminal DNA structure of mammalian chromosomes. **Embo J** 16(12): 3705-3714.

9. Wellinger RJ, Wolf AJ, Zakian VA (**1993**). Saccharomyces telomeres acquire single-strand TG1-3 tails late in S phase. **Cell** 72(1): 51-60.

10. Gao H, Cervantes RB, Mandell EK, Otero JH, Lundblad V (**2007**). RPA-like proteins mediate yeast telomere function. **Nature Structural & Molecular Biology** 14(3): 208-214.

11. Grandin N, Damon C, Charbonneau M (**2001**). Ten1 functions in telomere end protection and length regulation in association with Stn1 and Cdc13. **Embo J** 20(5): 1173-1183.

12. Grandin N, Reed SI, Charbonneau M (**1997**). Stn1, a new Saccharomyces cerevisiae protein, is implicated in telomere size regulation in association with Cdc13. **Genes Dev** 11(4): 512-527.

13. Garvik B, Carson M, Hartwell L (**1995**). Single-stranded DNA arising at telomeres in cdc13 mutants may constitute a specific signal for the RAD9 checkpoint. **Mol Cell Biol** 15(11): 6128-6138.

14. Hartwell LH, Mortimer RK, Culotti J, Culotti M (**1973**). Genetic Control of the Cell Division Cycle in Yeast: V. Genetic Analysis of cdc Mutants. **Genetics** 74(2): 267-286.

15. Booth C, Griffith E, Brady G, Lydall D (**2001**). Quantitative amplification of single-stranded DNA (QAOS) demonstrates that cdc13-1 mutants generate ssDNA in a telomere to centromere direction. **Nucleic Acids Res** 29(21): 4414-4422.

16. Nugent CI, Hughes TR, Lue NF, Lundblad V (**1996**). Cdc13p: a single-strand telomeric DNA-binding protein with a dual role in yeast telomere maintenance. **Science** 274(5285): 249-252.

17. Greider CW, Blackburn EH (**1987**). The telomere terminal transferase of Tetrahymena is a ribonucleoprotein enzyme with two kinds of primer specificity. **Cell** 51(6): 887-898.

18. Greider CW, Blackburn EH (**1985**). Identification of a specific telomere terminal transferase activity in Tetrahymena extracts. **Cell** 43(2 Pt 1): 405-413.

19. Watson JD (**1972**). Origin of concatemeric T7 DNA. **Nat New Biol** 239(94): 197-201.

20. Lendvay TS, Morris DK, Sah J, Balasubramanian B, Lundblad V (**1996**). Senescence mutants of Saccharomyces cerevisiae with a defect in telomere replication identify three additional EST genes. **Genetics** 144(4): 1399-1412.

21. Lundblad V, Szostak JW (**1989**). A mutant with a defect in telomere elongation leads to senescence in yeast. **Cell** 57(4): 633-643.

22. Singer MS, Gottschling DE (**1994**). TLC1: template RNA component of Saccharomyces cerevisiae telomerase. **Science** 266(5184): 404-409.

23. Lundblad V, Blackburn EH (**1993**). An alternative pathway for yeast telomere maintenance rescues est1- senescence. **Cell** 73(2): 347-360.

24. Teng SC, Zakian VA (**1999**). Telomere-telomere recombination is an efficient bypass pathway for telomere maintenance in Saccharomyces cerevisiae. **Mol Cell Biol** 19(12): 8083-8093.

25. Le S, Moore JK, Haber JE, Greider CW (**1999**). RAD50 and RAD51 define two pathways that collaborate to maintain telomeres in the absence of telomerase. **Genetics** 152(1): 143-152.

26. Zubko MK, Lydall D (**2006**). Linear chromosome maintenance in the absence of essential telomere-capping proteins. **Nature Cell Biology** 8(7): 734-740.

27. Petreaca RC, Chiu H-C, Eckelhoefer HA, Chuang C, Xu L, Nugent CI (**2006**). Chromosome end protection plasticity revealed by Stn1p and Ten1p bypass of Cdc13p. **Nature Cell Biology** 8(7): 748-755.

28. Larrivee M, Wellinger RJ (**2006**). Telomerase- and capping-independent yeast survivors with alternate telomere states. **Nat Cell Biol** 8(7): 741-747.

29. Ngo H-P, Lydall D (**2010**). Survival and Growth of Yeast without Telomere Capping by Cdc13 in the Absence of Sgs1, Exo1, and Rad9. **PLoS genetics** 6(8): e1001072.

30. Bryan TM, Englezou A, Dalla-Pozza L, Dunham MA, Reddel RR (**1997**). Evidence for an alternative mechanism for maintaining telomere length in human tumors and tumor-derived cell lines. **Nat Med** 3(11): 1271-1274.

31. Bryan TM, Englezou A, Gupta J, Bacchetti S, Reddel RR (**1995**). Telomere elongation in immortal human cells without detectable telomerase activity. **Embo J** 14(17): 4240-4248.

32. McEachern MJ, Haber JE (**2006**). Break-induced replication and recombinational telomere elongation in yeast. **Annu Rev Biochem** 75(111-135.

33. Henson JD, Cao Y, Huschtscha LI, Chang AC, Au AY, Pickett HA, Reddel RR (**2009**). DNA C-circles are specific and quantifiable markers of alternative-lengthening-of-telomeres activity. **Nat Biotechnol** 27(12): 1181-1185.

34. Hanahan D, Weinberg RA (**2011**). Hallmarks of cancer: the next generation. **Cell** 144(5): 646-674.

35. Lydall D, Weinert T (**1995**). Yeast checkpoint genes in DNA damage processing: implications for repair and arrest. **Science** 270(5241): 1488-1491.

36. Pellicioli A, Lucca C, Liberi G, Marini F, Lopes M, Plevani P, Romano A, Di Fiore PP, Foiani M (**1999**). Activation of Rad53 kinase in response to DNA damage and its effect in modulating phosphorylation of the lagging strand DNA polymerase. **EMBO J** 18(22): 6561-6572.

37. Harrison JC, Haber JE (**2006**). Surviving the breakup: the DNA damage checkpoint. **Annu Rev Genet** 40(209-235.

38. Melo JA, Cohen J, Toczyski DP (**2001**). Two checkpoint complexes are independently recruited to sites of DNA damage in vivo. **Genes Dev** 15(21): 2809-2821.

39. Lisby M, Barlow JH, Burgess RC, Rothstein R (**2004**). Choreography of the DNA Damage ResponseSpatiotemporal Relationships among Checkpoint and Repair Proteins. **Cell** 118(6): 699-713.

40. Sugawara N, Haber JE (**1992**). Characterization of double-strand break-induced recombination: homology requirements and single-stranded DNA formation. **Mol Cell Biol** 12(2): 563-575.

41. Lee SE, Moore JK, Holmes A, Umezu K, Kolodner RD, Haber JE (**1998**). Saccharomyces Ku70, mre11/rad50 and RPA proteins regulate adaptation to G2/M arrest after DNA damage. **Cell** 94(3): 399-409.

42. Toczyski DP, Galgoczy DJ, Hartwell LH (**1997**). CDC5 and CKII control adaptation to the yeast DNA damage checkpoint. **Cell** 90(6): 1097-1106.

43. Sandell LL, Zakian VA (**1993**). Loss of a yeast telomere: arrest, recovery, and chromosome loss. **Cell** 75(4): 729-739.

44. Syljuasen RG, Jensen S, Bartek J, Lukas J (**2006**). Adaptation to the Ionizing Radiation-Induced G2 Checkpoint Occurs in Human Cells and Depends on Checkpoint Kinase 1 and Polo-like Kinase 1 Kinases. **Cancer Research** 66(21): 10253-10257.

45. Yoo HY, Kumagai A, Shevchenko A, Shevchenko A, Dunphy WG (**2004**). Adaptation of a DNA replication checkpoint response depends upon inactivation of Claspin by the Polo-like kinase. **Cell** 117(5): 575-588.

46. De Lange T (**2005**). Telomere-related genome instability in cancer. **Cold Spring Harb Symp Quant Biol** 70:197-204.

47. Yu X, Vazquez A, Levine AJ, Carpizo DR (**2012**). Allele-specific p53 mutant reactivation. **Cancer cell** 21(5): 614-625.

48. Xue W, Zender L, Miething C, Dickins RA, Hernando E, Krizhanovsky V, Cordon-Cardo C, Lowe SW (**2007**). Senescence and tumour clearance is triggered by p53 restoration in murine liver carcinomas. **Nature** 445(7128): 656-660.

49. Ventura A, Kirsch DG, McLaughlin ME, Tuveson DA, Grimm J, Lintault L, Newman J, Reczek EE, Weissleder R, Jacks T (**2007**). Restoration of p53 function leads to tumour regression in vivo. **Nature** 445(7128): 661-665.

50. Pardue ML, DeBaryshe PG (**2003**). Retrotransposons provide an evolutionarily robust non-telomerase mechanism to maintain telomeres. **Annu Rev Genet** 37:485-511.

51. Leroy C, Lee S, Vaze M, Ochsenbien F, Guerois R, Haber J, Marsolierkergoat M (**2003**). PP2C Phosphatases Ptc2 and Ptc3 Are Required for DNA Checkpoint Inactivation after a Double-Strand Break. **Molecular Cell** 11(3): 827-835.

52. Ferrari M, Nachimuthu BT, Donnianni RA, Klein H, Pellicioli A (**2013**). Tid1/Rdh54 translocase is phosphorylated through a Mec1- and Rad53-dependent manner in the presence of DSB lesions in budding yeast. **DNA Repair (Amst)** 12(5): 347-355.

53. Strebhardt K (**2010**). Multifaceted polo-like kinases: drug targets and antitargets for cancer therapy. **Nature reviews Drug discovery** 9(8): 643-660.

54. Mender I, Gryaznov S, Dikmen Z, Wright WE, Shay JW (**2014**). Induction of Telomere Dysfunction Mediated By the Telomerase Substrate Precursor 6-Thio-2'-Deoxyguanosine. **Cancer Discovery** 5(1): 82-95.

55. Patry C, Bouchard L, Labrecque P, Gendron D, Lemieux B, Toutant J, Lapointe E, Wellinger R, Chabot B (**2003**). Small interfering RNA-mediated reduction in heterogeneous nuclear ribonucleoparticule A1/A2 proteins induces apoptosis in human cancer cells but not in normal mortal cell lines. **Cancer Res** 63(22): 7679-7688.

56. Wellinger RJ, Ethier K, Labrecque P, Zakian VA (**1996**). Evidence for a new step in telomere maintenance. **Cell** 85(3): 423-433.

57. Luria SE, Delbruck M (**1943**). Mutations of Bacteria from Virus Sensitivity to Virus Resistance. **Genetics** 28(6): 491-511.

58. Forstemann K, Hoss M, Lingner J (**2000**). Telomerase-dependent repeat divergence at the 3' ends of yeast telomeres. **Nucleic Acids Res** 28(14): 2690-2694.

12

Micafungin induced apoptosis in *Candida parapsilosis* independent of its susceptibility to micafungin

Fazal Shirazi[1], Russel E. Lewis[1,2], Dimitrios P. Kontoyiannis[1,*]

[1] Department of Infectious Diseases, Infection Control and Employee Health, The University of Texas M D Anderson Cancer Center, Houston, TX, USA.

[2] Current Address: Department of Medical and Surgical Sciences, University of Bologna, Bologna, Italy.

* Corresponding Author: Dimitrios P. Kontoyiannis, MD, ScD, Department of Infectious Diseases, Infection Control and Employee Health, The University of Texas MD Anderson Cancer Center, 1515 Holcombe Boulevard; Houston, TX 77030, USA; E-mail: dkontoyi@mdanderson.org

ABSTRACT We hypothesized that the cell wall inhibitor micafungin (MICA) induces apoptosis in both MICA-susceptible (MICA-S) and MICA–non-susceptible (MICA-NS) *Candida parapsilosis*. Antifungal activity and apoptosis were analyzed in MICA-S and MICA-NS *C. parapsilosis* strains following exposure to micafungin for 3 h at 37°C in RPMI 1640 medium. Apoptosis was characterized by detecting phosphatidylserine externalization (PS), plasma membrane integrity, reactive oxygen species (ROS) generation, mitochondrial membrane potential changes, adenosine triphosphate (ATP) release, and caspase-like activity. Apoptosis was detected in MICA exposed (0.25 to 1 mg/L) susceptible *C. parapsilosis* strains and was associated with apoptosis of 20-52% of analyzed cells versus only 5-30% of apoptosis in MICA-NS cells exposed to micafungin (0.5 to 2 mg/L; P = 0.001). The MICA antifungal activity was correlated with apoptotic cells showing increased dihydrorhodamine-123 staining (indicating ROS production), Rh-123 staining (decreased mitochondrial membrane potential), elevated ATP, and increased metacaspase activity. In conclusion, MICA is pro-apoptotic in MICA-S cells, but still exerts apoptotic effects in MICA –NS *C. parapsilosis*.

Keywords: apoptosis, micafungin, metacaspase, reactive oxygen species.

Abbreviations:
MIC - minimum inhibitory concentration,
MICA - micafungin,
MICA-NS - MICA–non-susceptible,
MICA-S - MICA-susceptible,
PS - phosphatidylserine,
ROS - reactive oxygen species.

INTRODUCTION

Candida parapsilosis is a common cause of invasive candidiasis, especially in the setting of catheter associated blood stream infection [1-3]. The ability of *Candida* spp. to form biofilms on catheters has made candidiasis difficult to treat due to increased resistance to antifungal agents [4]. Echinocandins play a significant role in the treatment of invasive candidiasis [5]. These agents are non-toxic and exert potent fungicidal activity against *Candida* spp. through the inhibition of (1,3)-β-D glucan synthesis, a major constituent of the fungal cell wall [6]. In addition, they are the most promising antifungal agents for lock therapy strategies [4]. Although the effects of echinocandin activity have been extensively investigated by standard microbiological endpoints, an improved understanding of the mechanistic basis of micafungin (MICA)-induced cell death in *Candida* may provide new insights into effective antifungal strategies, especially in the era of increasing echinocandin resistance in *Candida* species.

A series of morphologic and biochemical features like exposure of phosphatidylserine (PS), DNA fragmentation, reactive oxygen species (ROS) accumulation and mitochondrial depolarization set apoptosis apart from necrosis [1]. Apoptotic responses have been studied in many higher eukaryotes but also have been observed in lower eukaryotes, including yeasts and filamentous fungi [7-10]. As other lower eukaryotes, *Candida* spp. exhibit apoptotic markers that are similar to mammalian cells, including externalization of phosphatidylserine, reactive oxygen species (ROS) accumulation, decreased mitochondrial membrane potential, and DNA condensation and fragmentation [3, 11]. Furthermore, apoptosis can be induced in *C. albicans* by oxidative stress, intracellular acidification and antifungal agents such as amphotericin B and caspofungin [1, 12-15]. However, it is unclear whether apoptotic effects in *Candida* are related to the degree of echinocandin susceptibility. Therefore, we hypothesized that the echinocandin MICA, although more pro-apoptotic in micafungin-susceptible (MI-

CA-S) *C. parapsilosis* cells, would also induce apoptosis in micafungin –non-susceptible (MICA-NS) *C. parapsilosis* cells.

RESULTS

Micafungin induces more pronounced apoptosis in MICA-susceptible *C. parapsilosis* planktonic cells versus a non-susceptible strain

The MICA concentrations used for apoptotic studies were based on the susceptibility of the two *C. parapsilosis* isolates to MICA, minimum inhibitory concentration (MIC) for MICA-S strain being 1 mg/L and for MICA-NS 2 mg/L. To assess whether MICA activity occurs through induction of apoptosis, we exposed *C. parapsilosis* planktonic cells to MICA concentration at MIC or sub-MIC level, as it was reported that concentration at MIC or above MIC leads to necrosis, hence annexin-V and propidium iodide (PI) staining were performed following exposure to different concentrations of MICA. For the MICA-S *C. parapsilosis* strain, up to 52% of planktonic cells exposed to micafungin (Table 1) showed apoptosis, whereas for the MICA-NS *C. parapsilosis* strain, only 30% of MICA - treated planktonic cells were apoptotic (P = 0.001) (Table 1).

ROS play an important role as an early initiator of apoptosis in yeasts and other filamentous fungi. MICA-S *C. parapsilosis* planktonic cells exhibited a 2.4 - fold increase in ROS levels compared to untreated cells following MICA exposure. In contrast, a 1.5 fold increase in fluorescence in the MICA-NS *C. parapsilosis* planktonic cell exposed to MICA versus untreated cells (P = 0.006) (Table 1, Fig. 1A). Similarly, MICA exposure was associated with decreases in mitochondrial membrane potential in both MICA-S and MICA-NS strains of *C. parapsilosis* (3.8 fold vs 2.0 fold increase in fluorescence, P = 0.0025, Table 1).

Micafungin damages the cell membrane of *C. parapsilosis* planktonic cells, especially in MICA-S strains

A rapid efflux of ATP due to membrane damage by MICA was detected after the incubation of *C. parapsilosis* planktonic cells with MICA for 3 h at 37°C (Fig. 1 B). Specifically, we observed higher degree of ATP release among MICA-S *C. parapsilosis* strain following MICA exposure (0.5 - 1 mg/L) as compared to MICA-NS strain (Fig. 1B). This increased ATP release was consistent with increased ROS generation and decreased mitochondrial membrane potential of cells incubated with MICA.

Micafungin activates metacaspases (caspase-like activity) in *C. parapsilosis* planktonic cells, especially in MICA-S strains

Researchers have identified orthologs of mammalian caspases called metacaspases in yeast and filamentous fungi, which are activated in the early stages of apoptosis [8, 15]. In fungi and plants, metacaspases (caspase-like) activity can be assessed using the *in situ* detection marker CaspACE FITC-VAD-FMK [9, 10]. To confirm the presence of metacaspase activation, MICA - pretreated (0.5 - 1 mg/L) *C. parapsilosis* planktonic cells were incubated with CaspACE FITC-VAD-FMK, which fluoresces green when it binds to active metacaspases. Indeed, the planktonic cells of MICA-S and MICA-NS *C. parapsilosis* treated with MICA showed green fluorescence, suggesting that apoptosis in these cells occurs via metacaspase activation irrespective of MICA susceptibility (Figure 1 C, D).

To further support the concept that caspase-like activity are involved in apoptosis, we treated *C. parapsilosis* planktonic cells with 0.5 to 1.0 mg/L MICA for 3 h in the presence or absence of the caspase 1 inhibitor Z-VAD-FMK at a concentration of 40 μM. MICA treated cells were assessed for the metacaspase activity in Z-VAD-FMK - treated and untreated samples of the *C. parapsilosis* planktonic cells. *C. parapsilosis* susceptible planktonic cells treated with MICA in combination with Z-VAD-FMK exhibited less fluorescence in cells (3.38 fold), than did cells treated with only MICA (13.74 fold). Similarly, *C. parapsilosis* non-susceptible cells showed decrease in fluorescence (1.41 fold) in Z-VAD-FMK - treated cells than in untreated cells (11.8 fold) (Figure 1D).

TABLE 1. Apoptotic cells and fold changes in fluorescence intensity of MICA-S and MICA-NS *C. parapsilosis* strains treated with micafungin.

Strains	Apoptotic Cells					
	Protoplast %		Fold change			
C. parapsilosis	Annexin V	PI	ROS		ΔΨ$_m$	
			-NAC	+NAC	-NAC	+NAC
Isolate-1 (MICA-S)						
Control	5.0 ± 0.01	-	-	-	-	-
0.25 mg/L	24.0 ± 2.0	-	1.8 ± 0.01	1.5 ± 1.00	2.8 ± 0.1	1.9 ± 0.01
0.5 mg/L	52.0 ± 3.0	2.0 ± 0.01	2.4 ± 0.2	1.68 ± 0.20	3.8 ± 0.4	2.48 ± 0.01
1.0 mg/L	19.0 ± 1.0	9.0 ± 1.0	1.9 ± 0.01	1.61 ± 0.21	2.5 ± 0.2	1.8 ± 0.01
Isolate-6 (MICA-NS)						
Control	2.0 ± 0.01	1.0 ± 0.01	-	-	-	-
0.5 mg/L	7.0 ± 1.0	-	1.5 ± 0.01	1.0 ± 0.02	1.9 ± 0.01	1.4 ± 0.10
1.0 mg/L	30.0 ± 2.0	-	1.4 ± 0.01	1.3 ± 0.01	1.98 ± 0.01	0.89 ± 0.01
2.0 mg/L	11.0 ± 1.0	6.0 ± 0.01	1.6 ± 0.01	1.1 ± 0.20	2.0 ± 0.01	1.4 ± 0.16

'-' Not detected (0% of cells showed particular apoptotic marker).

PI, propidium iodide; ROS, reactive oxygen species; ΔΨm, mitochondrial membrane potential; NAC, *N*-acetyl cysteine.

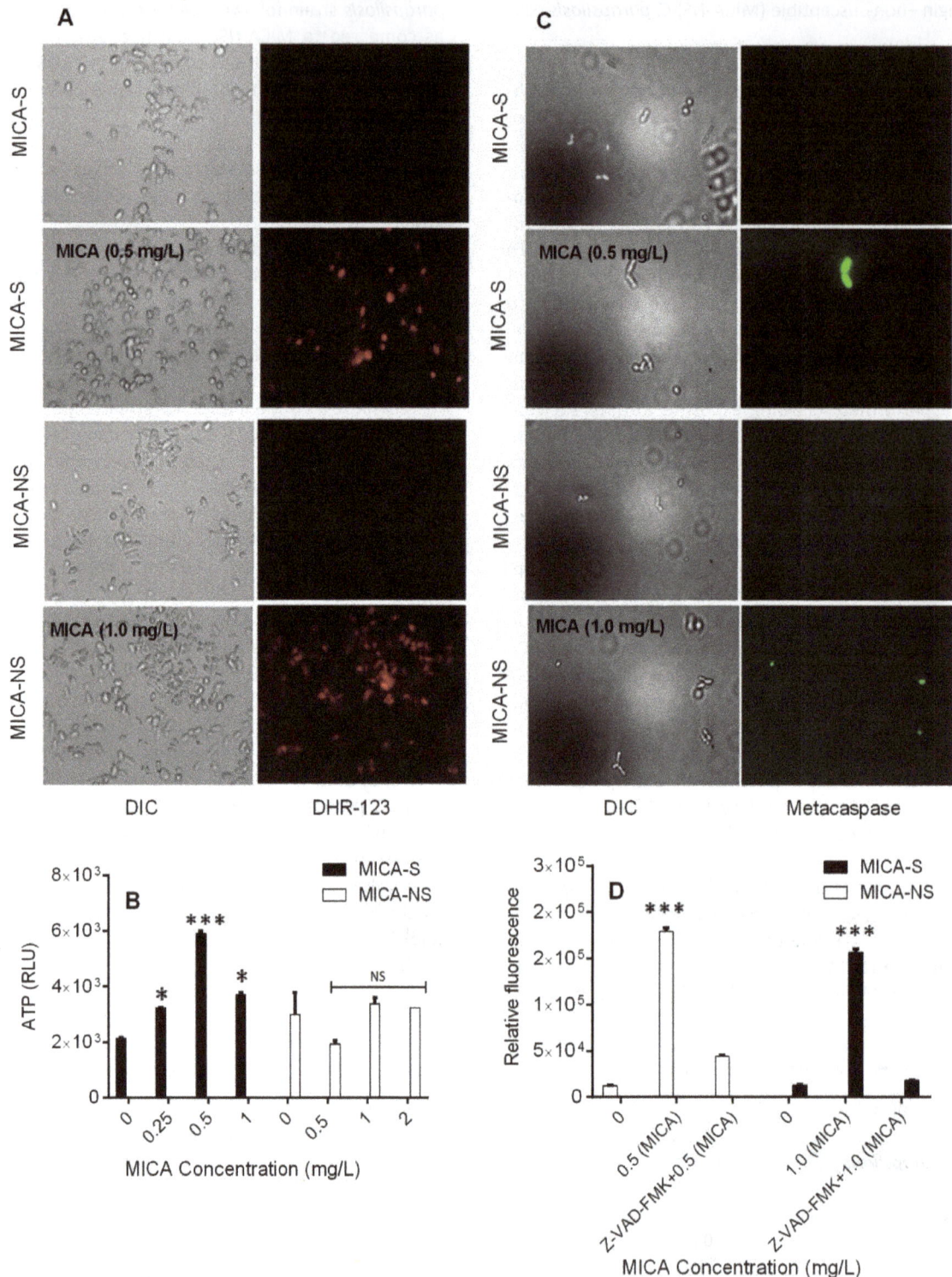

FIGURE 1: Intracellular ROS accumulation, ATP release and activation of caspase like activity in *C. parapsilosis* (MICA-S and MICA-NS isolates 1 and 6, respectively) cells treated with micafungin for 3 h at 37°C. (A) Fluorescence images of MICA-S and MICA-NS strains of *C. parapsilosis* treated with micafungin and stained for intracellular ROS with DHR-123. **(B)** ATP release assay indicating *C. parapsilosis* cell membrane disruption and plasma membrane leakage after micafungin treatment. **(C)** Fluorescent images of MICA-S and MICA-NS *C. parapsilosis* strains treated with micafungin and stained with caspase activity detection marker CaspACE FITC-VAD-FMK. **(D)** Relative fluorescence of *C. parapsilosis* MICA-S and MICA-NS cells treated with micafungin with or without caspase-1 inhibitor Z-VAD-FMK, and stained with caspase activity detection marker CaspACE FITC-VAD-FMK. DIC, Differential Interference Contrast; *Cp*, *C. parapsilosis*; RLU, relative light units; *P < 0.05; ***P < 0.0001; NS (non-significant), P > 0.05 (compared with untreated controls).

Finally, to further elucidate the ROS role in apoptosis, we examined whether the ROS scavenger NAC reverses apoptosis in *C. parapsilosis* S and NS planktonic cells. At 37°C, the levels of intracellular ROS were markedly decreased from 2.4 fold to 1.6 fold in susceptible cells and 1.6 to 1.0 fold in non-susceptible cells, after the addition of ROS scavenger *N*-acetyl-cysteine (NAC, 40 mM), compared to NAC unexposed cells (Table 1). Similar observation was evidenced in mitochondrial potential of *C. parapsilosis* S and NS planktonic cells treated with MICA in the presence of NAC (Table 1).

DISCUSSION

We found that MICA induced cellular apoptosis in both MICA-S and MICA-NS strains. After exposure to sub-inhibitory concentrations of MICA, PS was exposed on the outer surface of the plasma membrane, accompanied by an increase in intra-cellular ROS levels and depolarization of the mitochondrial membrane, which was marked by ATP and cytochrome *c* release. Hao *et al.* [13] reported that caspofungin exerts its antifungal activity against *C. albicans* by apoptosis. In concordance with these reports, our results have described that dying *C. parapsilosis* cells exhibit markers of apoptosis following MICA exposure (0.25 - 2 mg/L). Our results show that apoptosis markers in *C. parapsilosis* were detectable following even sub-MIC MICA concentrations in both MICA-S and MICA-NS strains. In addition, a higher percentage of MICA-S *C. parapsilosis* cells underwent apoptosis compared to MICA-NS strains following MICA exposure. This phenomenon could be attributed to a more complex glucan matrix in the cell walls of the MICA-NS strains, which might account for MICA-NS strains resistance to antifungal drugs, as described previously [16].

Caspases, cysteine-aspartic acid proteases, when cleaved, induce apoptosis. In *C. albicans*, a putative caspase encoded by metacaspase (Ca*MCA1*) has been shown to be involved in apoptosis [12, 13]. In our study, we used CaspACE FITC-VAD-FMK, which detects activated caspases, to show that *C. parapsilosis* caspases are activated in response to MICA exposure. These data are consistent with the hypothesis that apoptosis of *C. parapsilosis* proceeds through metacaspase-dependent pathways, although further studies are needed to demonstrate how metacaspase activity contributes to this process.

In conclusion, MICA initiate apoptosis of *C. parapsilosis* cells at concentrations below the minimum inhibitory concentration in both MICA-S and MICA-NS strains, although more susceptible and non-susceptible *C. parapsilosis* strains will need to be tested to confirm generalization of our findings. The mechanism of the apoptotic pathways induced by MICA needs further clarification, as do the roles of apoptosis in determining the therapeutic efficacy of these echinocandins *in vivo*. Given the differences in the architecture of the death-regulating machineries of fungi and higher mammals [7], apoptotic pathways may represent important targets for novel antifungal drug development and should be further investigated as a new antifungal approach.

MATERIALS AND METHODS

Echinocandins

Pure MICA powder (5 mg/mL stock solution) was obtained from Astellas Pharma US, Inc. The stock dilutions were prepared in distilled water and stored at -80°C until use.

Isolates and growth conditions

The *C. parapsilosis* isolates used in this study were from blood cultures of cancer patients with candidemia at The University of Texas MD Anderson Cancer Center, Houston, TX. Specifically, we used one MICA-S and one MICA-NS *C. parapsilosis* isolate. The isolates were grown on yeast peptone dextrose (YPD) agar plates overnight at 37°C. The cells were then collected and washed twice in phosphate-buffered saline and counted using a hemocytometer (Hausser Scientific). Next, 10^6 cells/mL of each isolate were resuspended in liquid YPD and incubated at 37°C for 12 h until reaching mid-log phase, and recounted with a hemocytometer to achieve final testing inocula.

Susceptibility testing

We performed broth microdilution susceptibility testing according to the Clinical and Laboratory Standards Institute-approved document M27-A3, in 96 well microtitration plates in RPMI 1640 medium (Corning Inc., New York, NY) containing serial two fold dilutions of micafungin and a final inoculum of 5×10^3 cells/mL of each isolate. The minimum inhibitory concentration (MIC) of MICA for each isolate was determined visually 24 h after incubation at 37°C as the lowest concentration that resulted in a prominent decrease in turbidity (reduction of > 50% of growth) and the results were analyzed according to guidelines set for *Candida* susceptibility to echinocandins [17, 18].

Detection of apoptosis and necrosis with annexin V and propidium iodide (PI) double staining of *C. parapsilosis* cells

The apoptosis marker phosphatidylserine (PS) is located on the inner leaflet of the lipid bilayer of the cytoplasmic membrane, and is translocated to the outer leaflet at the onset of apoptosis, where it can be identified by annexin V staining [9, 10, 19]. PI staining identifies necrotic cells, as it does not permeate cells in intact membranes. Therefore, staining patterns differentiate between live cells (Annexin V-/PI-), cells undergoing early apoptosis (Annexin V+/PI-) and necrotic cells (Annexin V+/PI+) [19, 20]. Cells treated with micafungin (0.25-2 mg/L), were digested with a lysing enzyme mixture (0.25 mg/mL of chitinase, 15 U of lyticase, and 20 mg/mL lysing enzyme; Sigma-Aldrich) for 3 h at 30°C. After digestion, cells were stained with the annexin V - fluorescein isothiocyanate (FITC) (50 mg/L, BD Pharmingen, USA) and PI (50 mg/L) at room temperature (RT) for 15 min and then observed under a Nikon Microphot SA fluorescence microscope to assess the externalization of the apoptosis marker PS as previously described [9, 10].

Detection of intracellular ROS in *C. parapsilosis* cells

Intracellular ROS levels in *C. parapsilosis* cells were measured as previously described [9, 10]. *C. parapsilosis* cells were treated with MICA (0.25-2 mg/L) for 3 h at 37°C. Cells were then spiked with dihydrorhodamine (DHR)-123 (5 mg/L). After incubation for 2 h at RT, cells were harvested after centrifugation at 13,000 x *g* for 5 min and observed with a Nikon Microphot SA fluorescence microscope (excitation, 488 nm; emission, 520 nm). For quantitative assays, fluorescence intensity values were recorded using a POLARstar Galaxy microplate reader (excitation, 490 nm; emission, 590 nm; BMG LABTECH, Offenburg, Germany). The same experiment was conducted in the presence of NAC at a concentration of 40 mM.

Mitochondrial membrane potential ($\Delta\Psi_m$) measurements in *C. parapsilosis* cells

Mitochondrial membrane depolarization was assessed by staining with rhodamine (Rh)-123, a fluorescent dye that is distributed in the mitochondrial matrix as previously described [9, 10]. Briefly, the cells were exposed to micafungin (0.25-2 mg/L) for 3 h at 37°C and harvested via centrifugation, washed twice, then resuspended in phosphate-buffered saline. Rh-123 was added to the final concentration of 10 µM and was incubated for 30 min in the dark at RT. Fluorescence intensity was recorded as described above. The same experiment was conducted in the presence of NAC at a concentration of 40 mM.

Adenosine triphosphate (ATP) release assay in *C. parapsilosis* cells

We assessed the severity of MICA-induced mitochondrial and plasma membrane damage by the amount of cellular ATP released into the medium as described by Ben-Ami *et al.* [21]. The *C. parapsilosis* cells were then counted with a hemocytometer (Hausser Scientific), and suspended in RPMI 1640 medium at 10^6 cells/mL. After 12 h at 37°C, the medium was removed by centrifugation at 13,000 x g for five min and cells were re-suspended in MICA-containing (0.25 - 2 mg/L) or drug-free RPMI 1640 medium. After 3 h of incubation, the cells were removed by centrifugation as described above and the ATP released in the supernatants was assayed by using the CellTiter-Glo luminescent kit (Promega). Data were recorded with a microplate luminometer (SpectraMax M5; Molecular Devices).

Detection of metacaspase activity in *C. parapsilosis* cells

Activation of metacaspases was detected with the CaspACE FITC-VAD-FMK *in situ* Marker (Promega) [9, 10]. The cells were

pretreated with MIC (0.5 - 1 mg/L) for 3 h at 37°C. Cells were harvested, washed in phosphate-buffered saline, and then re-suspended in 10 µM CaspACE FITC-VAD-FMK solution. After 2 h of incubation at RT, cells were washed twice and re-suspended in phosphate-buffered saline. Samples were mounted and viewed in a Nikon Microphot SA fluorescence microscope (excitation, 488 nm; emission, 520 nm).

Inhibition of apoptosis was performed by incubating *C. parapsilosis* planktonic cells with MICA, in the presence or absence of the caspase-1 inhibitor Z-VAD-FMK (Sigma) and the antioxidant *N*-acetyl-cysteine (NAC) at final concentrations of 40 µM and 40 mM, respectively. After incubation for 3 h, cells were analysed for ROS accumulation, mitochondrial potential and metacaspase activity.

Statistical analysis

For all assays, three independent experiments were performed on different days in triplicate. Multiple treatment groups were compared using Kruskall-Wallis test and post-hoc paired comparisons were compared using Dunnett's tests. Calculations were made with InStat (GraphPad Software). All results are expressed as means ± standard deviations. Two-tailed P values of less than 0.05 were considered statistically significant.

ACKNOWLEDGMENTS

D.P.K. acknowledges the Frances King Black Endowed Professorship for Cancer Research. This research was supported in part by a Research Grant from Astellas Pharma Inc.

CONFLICT OF INTEREST

D.P.K. has received research support and honoraria from Pfizer; Astellas Pharma US, Inc.; Merck & Co., Inc.; T2 Biosystems. F.S. and R.E.L. have no conflicts of interest.

REFERENCES

1. Phillips AJ, Sudbery I, and Ramsdale M (**2003**). Apoptosis induced by environmental stresses and amphotericin B in *Candida albicans*. **Proc Natl Acad Sci U. S. A.** 100:14327-14332.

2. Ostrosky-Zeichner L, Casadevall A, Galgiani JN, Odds FC and Rex JH (**2010**). An insight into the antifungal pipeline: selected new molecules and beyond. **Nat Rev Drug Discov** 9:719-727.

3. Nguyen LN, Cesar GV, Le GTT, Silver DL, Nmrichter L, and Nosanchuk JD (**2012**). Inhibition of *Candida parapsilosis* fatty acid synthase (Fas2) induces mitochondrial cell death in serum. **PLoS Pathog** 8:e1002879.

4. Walraven CJ, Lee SA (**2013**). Antifungal lock therapy. Antimicrob Agents Chemother 57: 1-8.

5. Andes DR, Safdar N, Baddley JW, Playford G, Reboli AC, Rex JH, Sobel JD, Pappas PG, and Kullberg BJ (**2012**). Impact of treatment strategy on outcomes in patients with candidemia and other forms of invasive candidiasis: a patient-level quantitative review of randomized trials. **Clin Infect Dis** 54:1110-1122.

6. Georgopapadakou NH and Tkacz JS (**1995**). The fungal cell wall as a drug target. **Trends Microbiol** 3:98-104.

7. Almeida B, Silva A, Mesquita A, Sampaio-Marques B, Rodrigues F,

and Ludovico P (**2008**). Drug-induced apoptosis in yeast. **Biochimica et Biophysica Acta -Mol Cell Res** 1783:1436-1448.

8. Ramsdale M (**2008**). Programmed cell death in pathogenic fungi. **Biochim Biophys Acta** 1783:1369-1380.

9. Shirazi F and Kontoyiannis DP (**2013**). Mitochondrial respiratory pathways inhibition in *Rhizopus oryzae* potentiates activity of posaconazole and itraconazole via apoptosis. **PLoS One** 8:e63393.

10. Shirazi F, Pontikos MA, Walsh TJ, Albert N, Lewis RE, and Kontoyiannis DP (**2013**). Hyperthermia sensitizes *Rhizopus oryzae* to posaconazole and itraconazole action through apoptosis. **Antimicrob Agents Chemother** 57:4360-4368.

11. Munoz AJ, Wanichthanarak K, Meza E and Petranovic D (**2012**).Systems biology of yeast cell death. **FEMS Yeast Res** 12:249-265.

12. Chin C, Donaghey F, Helming K, McCarthy M, Rogers S, and Austriaco N (**2014**). Deletion of AIF1 but not of YCA1/MCA1 protects *Saccharomyces cerevisiae* and *Candida albicans* cells from caspofungin-induced programmed cell death. **Microbial Cell** 1:58-63.

13. Hao B, Cheng S, Clancy CJ, and Nguyen MH (**2013**). Caspofungin kills *Candida albicans* by causing both cellular apoptosis and necrosis. **Antimicrob Agents Chemother** 57:326-332.

14. Al-Dhaheri RS and Douglas LJ (**2010**). Apoptosis in *Candida* biofilms exposed to amphotericin B. **J Med Microbiol** 59:149-157.

15. Hamann A, Brust D, and Osiewacz HD (**2008**). Apoptosis pathways in fungal growth, development and ageing. **Trends Microbiol** 16:276-283.

16. Taff HT, Nett JE, Zarnowski R, Ross KM, Sanchez H, Cain MT, Hamaker J, Mitchell AP, and Andes DR (**2012**). A *Candida* biofilm-induced pathway for matrix glucan delivery: implications for drug resistance. **PLOS Pathog** 8:e1002848.

17. Clinical and Laboratory Standards Institute (**2008**). Reference method for broth dilution antifungal susceptibility testing of yeast. 3rd ed. M27-A3. Clinical and Laboratory Standards Institute document M27-A2. Clinical and Laboratory Standards Institute, Wayne, PA.

18. Espinel-Ingroff A1, Arendrup MC, Pfaller MA, Bonfietti LX, Bustamante B, Canton E, Chryssanthou E, Cuenca-Estrella M, Dannaoui E, Fothergill A, Fuller J, Gaustad P, Gonzalez GM, Guarro J, Lass-Flörl C, Lockhart SR, Meis JF, Moore CB, Ostrosky-Zeichner L, Pelaez T, Pukinskas SR, St-Germain G, Szeszs MW, and Turnidge J (**2013**). Interlaboratory variability of Caspofungin MICs for Candida spp. Using CLSI and EUCAST methods: should the clinical laboratory be testing this agent? **Antimicrob Agents Chemother** 57:5836-5842.

19. Ben-Ami R, Lewis RE, Tarrand J, Leventakos K, and Kontoyiannis DP (**2010**). Antifungal activity of colistin against Mucorales species *in vitro* and in a murine model of *Rhizopus oryzae* pulmonary infection. **Antimicrob Agents Chemother** 54:484-490.

20. Madeo F, Fröhlich E, Ligr M, Grey M, Sigrist SJ, Wolf DH, and Fröhlich KU (**1999**). Oxygen stress: a regulator of apoptosis in yeast. **J Cell Biol** 145:757-767.

21. Cho J and Lee DG (**2011**). Oxidative stress by antimicrobial peptide pleurocidin triggers apoptosis in *Candida albicans*. **Biochimie** 93:1873-1879.

13

Identification of SUMO conjugation sites in the budding yeast proteome

Miguel Esteras[1], I-Chun Liu[1], Ambrosius P. Snijders[2], Adam Jarmuz[1] and Luis Aragon[1,*]
[1] Cell Cycle Group, MRC London Institute of Medical Sciences, Du Cane Road, London W12 0NN, UK.
[2] Protein Analysis and Proteomics Platform, The Francis Crick Institute, 1 Midland Road, London NW1 1AT, UK.
* Corresponding Author:
Luis Aragon, Cell Cycle Group, MRC London Institute of Medical Sciences, Du Cane Road, London W12 0NN, UK;
E-mail: luis.aragon@csc.mrc.ac.uk

ABSTRACT Post-translational modification by the small ubiquitin-like modifier (SUMO) is an important mechanism regulating protein function. Identification of SUMO conjugation sites on substrates is a challenging task. Here we employed a proteomic method to map SUMO acceptor lysines in budding yeast proteins. We report the identification of 257 lysine residues where SUMO is potentially attached. Amongst the hits, we identified already known SUMO substrates and sites, confirming the success of the approach. In addition, we tested several of the novel substrates using SUMO immunoprecipitation analysis and confirmed that the SUMO acceptor lysines identified in these proteins are indeed bona fide SUMOylation sites. We believe that the collection of SUMO sites presented here is an important resource for future functional studies of SUMOylation in yeast.

Keywords:
SUMO, proteome, budding yeast, mass spectrometry, site-specific SUMOylation.

Abbreviations:
aa – amino acid,
SUMO – small ubiquitin-related modifier,
MS – mass spectrometry,
NTA - nitrilotriacetic acid.

INTRODUCTION

SUMO (small ubiquitin-related modifier) is a 10-kDa highly conserved protein modifier that reversibly conjugates to specific lysine residues on many target proteins [1]. The functional consequences of SUMO modification include changes in protein stability, localization, DNA-binding, or protein interactions. These SUMO effects can be mediated by providing a new binding interface, masking existing binding sites, or inducing conformational changes on the substrates [2]. SUMO is covalently conjugated to its substrates sequentially by the action of an E1 activating enzyme, an E2 conjugating enzyme, and an E3 ligase [1]. Similar to ubiquitin, SUMO itself can be further SUMOylated via addition of SUMO moieties to lysine residues on SUMO [3]. SUMO can also be released from substrates by SUMO proteases (Ulp1 and Ulp2 in yeast).

A single SUMO gene is found in *S. cerevisiae* (*SMT3*), *C. elegans* (*SMO-1*) and *D. melanogaster* (*smt3*). In contrast, the human genome contains five SUMO variants, *SUMO1-SUMO5*. *SUMO2* and *SUMO3* are 97% similar each other but only share 50% of sequence similarity with *SUMO1*. Accordingly, *SUMO1* and *SUMO2/3* seem to be functionally distinct with different substrate set [4, 5, 6] role has been yet identified for SUMO4 and even its

aptness to be conjugated *in vivo* remains unclear [7].

A consensus motif for SUMOylation was proposed soon after the mapping of the first SUMO-modified lysine residues. Studies of the first conjugation sites suggested that the acceptor lysines were contained within the consensus ψKxE (where ψ is a large hydrophobic amino acid and x any amino acid) [8]. This motif together with a particular 3D structure in the substrate was proposed to allow the binding of the E2 enzyme, Ubc9, and the consequent transfer of SUMO [9]. In addition to the simple 4 amino acid consensus motif, two more extended versions have also been identified. The first one was the phosphorylation-dependent SUMOylation motif (PDSM), which consists of the core motif succeeded by a phosphorylated serine and a proline (ψKxExxpSP) [10]. The second extended motif is the negatively charged amino-acid dependent SUMOylation motif (NDSM), consisting of the core motif succeeded by two or more acidic amino acids in the C-terminal tail [11]. Although previously described motifs are found in many substrates, some exceptions have been identified, e.g. the K14 in E2-25K (*H. sapiens*) and the K164 in PCNA (*S. cerevisiae*). It is still unclear how Ubc9 recognizes these sites. Whether these unorthodox motifs mimic the 3D structure present at the SUMO consensus motif, or whether they are no more than rare exceptions remains to be answered. However, they make predictions of whether and where SUMOylation might occur in a given protein challenging.

The identification and quantification of SUMOylation by mass spectrometry (MS) is specially challenging. Most of the *in vivo* modified proteins have low steady-state SUMOylation and conjugated SUMO is very likely to be lost during the protein extraction and purification. Hence, input protein sample for MS are likely to contain low amounts of SUMOylated peptides. In addition, SUMO-modified lysines keep an amino acid (aa) side chain (5 residues in case of Smt3) after trypsin digestion which belongs to the SUMO modifier. During tandem MS, this aa side chain generates overlapping fragment ions with the ones from the target protein peptide. Standard database matching logarithms find it challenging to assign correct sequences to such a complex ion spectrum. Therefore, our knowledge of site-specific SUMOylation of proteomes is poor particularly when compared to other PTMs, like phosphorylation.

Several studies in mammalian systems have used clever strategies to improve the identification of SUMOylation sites. One of these strategies involves the mutation of all internal lysines in SUMO to arginines to make the mutant SUMO immune to digestion when Lys-C protease is used [12]. This allows digestion of the entire lysate and enrichment of SUMOylated peptides, greatly diminishing the sample complexity. In this study, we have used a similar proteomic approach to identify SUMO-modified proteins and their conjugation sites in the budding yeast proteome. We report over 200 potential SUMO sites. We have chosen a handful of newly identified SUMOylated substrates and demonstrate that mutation of the identified

A

B

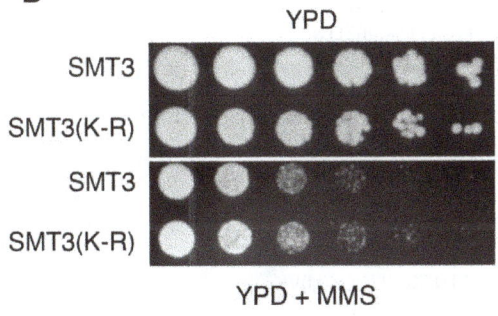

C

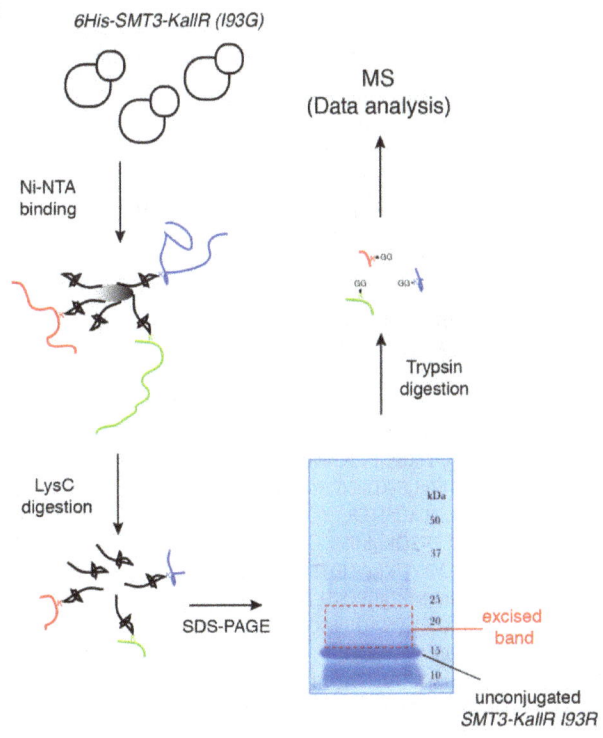

FIGURE 1: Proteomic screen to identify SUMO sites in budding yeast proteins. (A) Sequences of wildtype SUMO (*SMT3*) and the lysine-deficient His6 tagged mutant (*SMT3-KallR I93R*) used in the method. **(B)** Wild-type and *SMT3-KallR I93R* strains plated on full media (YPD) and media containing methyl methanesulphonate (MMS). **(C)** Diagrammatic representation of the purification strategy employed to enrich for *SMT3*-conjugated peptides. Cell lysates from yeast expressing *SMT3-KallR I93R* were digested with the endoprotease LysC, cleaving after lysines. The digested lysates were run on SDS-PAGE. Unconjugated *SMT3-KallR I93R* (indicated) and *SMT3-KallR I93R* conjugated to target protein fragments were identified. Gel area above the unconjugated *SMT3-KallR I93R* band were excised, digested with trypsin, and analyzed by nano LC-MS/MS. And database searched to identify SUMO acceptor lysines in the yeast proteome.

SUMO-conjugation sites prevents their modification *in vivo*. We present this resource to aid future efforts in the functional characterisation of SUMOylation in these yeast proteins.

RESULTS

Strategy to enrich SUMO-bearing peptides

To identify SUMOylation sites in the yeast proteome we adapted a strategy recently used to discover novel acceptor lysines for SUMO2 in Hela cells [12]. We used a His6-tagged *SMT3* (yeast SUMO) allele where all lysines had been replaced by arginines, *SMT3-KallR* (Fig. 1A). Similar alleles have been employed previously to identify sites of poly-SUMO chain formation [3]. In addition a mutation at the C-terminus of Smt3 was introduced, isoleucine at position 96 was substituted by arginine (*SMT3-KallR-I96R*; Fig. 1A) [13]. Importantly, we confirmed that the SMT3-*KallR-I96R* allele is conjugated *in vivo* (Fig. S1) and able to support growth similar to its wild-type Smt3 counterpart (Fig. 1B). Smt3-KallR-I96R protein is unsensitive to digestion by endoprotease LysC, an enzyme that specifically cleaves after lysine residues. Therefore unconjugated Smt3-KallR and Smt3-KallR covalently attached to peptides from target proteins can be easily separated by SDS-PAGE from the rest of the proteome fragments after digestion of protein ex-

tracts with LysC (Fig. 1C). Excision of the gel area above unconjugated Smt3-KallR selectively isolates Smt3-modified peptides (Fig. 1C). The excised gel fragments containing the peptides modified were digested with trypsin, which cleaves after arginine and lysine, and therefore removes most of Smt3-KallR-I96R from the substrate peptides. This strategy generates diglycine-modified isopeptides that are more compatible with mass spectrometry identification compared to wildtype conjugates.

Identification of SUMO-acceptor lysines by mass spectrometry

Smt3 purifications for mass spectrometry analysis were performed using 9 liters of cells (harvested at O.D. 0.9) grown in YPD. The Smt3 purifications were performed as described in Material and Methods. Lys-C digested peptides were separated by size in a SDS-PAGE gel. The area that corresponds to peptides between 15 and 25 kDa (containing peptides conjugated to Smt3) was cut from the gel (Fig. 1C), fragmented in smaller horizontal bands and digested with trypsin. These fractions were digested and loaded on the mass spectrometer separately, so that the complexity of the peptide mixture could be reduced. Trypsin digested peptides were analyzed by nanoscale

TABLE 1. Summary of all sumo-acceptor lysines detected by mass spectrometry.

Gene	Modified Sequence	Gene	Modified Sequence
ABD1	NISPIIK(gl)IR	BRX1	AEAAVERK(gl)IK
ABF1	MDK(gl)IVVNYYEYK	BSP1	NIK(gl)KEEEDSIPEAIK ·
ABF1	QQGVTIK(gl)NDTEDDSINK	BUD27	LEDFK(gl)EYNK
ABF1	VSNDSK(gl)LDFVTDDLEYHLANTHPDDTNDK	BUD3	FFEIEEELK(gl)EELK
ADE1	SITK(gl)TELDGILPLVAR	BUD3	NK(gl)QENINSSSNLFPEGK
AIM44	LNMEK(gl)DIK	BUD3	TGNEDVGNNNPSNSIPK(gl)IEKPPAFK
ALY2	FAPLDK(gl)VTLHR	BUD4	AGNK(gl)QENNEINIKAEEEIEPMTQQETDGLK
AOS1	MDMK(gl)VEKISEDEIAIYDRQIR	BUD4	QENNEINIK(gl)AEEEIEPMTQQETDGIK
AOS1	SIIEVTTRKDEEDEK(gl)K	CBF2	DNQPIK(gl)KEENIVNEDGPNTSR
AOS1	VEK(gl)LSEDEIALYDR	CBF5	EDFVIK(gl)PEAAGASTDTSEWPLLLK
APA1	ALTFFQDWLNENPELK(gl)K	CBF5	VNENTPEQWK(gl)K
APC5	K(gl)K(gl)TDELLESLSVEEDR	CDC11	EAKIK(gl)QEE
ARP7	LAPLIK(gl)EENDMENMADEQK	CDC12	IRLNGDLEEIQGK(gl)VK
ARP8	LTK(gl)EIKDLEGHYVNAPDK	CDC12	K(gl)YFTDQVK
ASG1	LLSNIK(gl)TER	CDC12	YK(gl)EEENALK
ATG2	DEPVSQK(gl)ISK	CDC19	IIVK(gl)IENQQGVNNFDEILK
BAF1	QQGVTIK(gl)NDTEDDSINK	CDC3	FEAAESDVK(gl)VEPGLGMGITSSQSEK
BDP1	ARQEFK(gl)PLHSLTKEEQEEEEEK	CDC3	KLQK(gl)SETELFAR
BDP1	DK(gl)LLNADIPESDRK	CDC3	SLK(gl)EEQVSIK
BDP1	K(gl)AHTAIQLK	CDC3	SLKEEQVSIK(gl)QDPEQEER
BDP1	K(gl)TEVVLGTIDDLK	CDC48	EVK(gl)VEGEDVEMTDEGAK
BDP1	KGSGGIMTNDLK(gl)VYR	CET1	KIAGNAVGSVVK(gl)KEEEANAAVDNIFEEK
BDP1	NTAK(gl)EEDQTAQR	CIN5	MTDTAFVPSPPVGFIK(gl)EENK
BFR2	SIADQISDIAIK(gl)PVNK	CMD1	SSNITEEQIAEFK(gl)EAFAIFDK
BIR1	EISGIK(gl)KETDDGK	CMR1	IFIFTDDSGTIK(gl)QEE
BIR1	ILEDVSVK(gl)NETPNNEMLLFETGTPIASQENK	CMR1	LSDLIK(gl)DEDESALLEK
BIR1	VIK(gl)PEFEPVPSVAR	CRZ1	IESGIVNIK(gl)NELDDTSK
BMH2	IVSSIEQK(gl)EESKEK	CRZ1	PK(gl)IESGIVNIK
BNI1	K(gl)LDEINR	CWC15	NK(gl)VEDK
BOP3	IGASAVAALNDNISIK(gl)EEDVAR	CYC8	QPTHAIPTQAPATGITNAEPQVK(gl)K
BRE1	KIK(gl)LELSDPSEPLTQSDVIAFQK	DAD3	MEHNISPIQQEVIDK(gl)YK

Gene	Modified Sequence	Gene	Modified Sequence
DEP1	LSSLVK(gl)QETLTESLK	MPP10	VK(gl)LDLFADEEDEPNAEGVGEASDK
DUN1	IVFGK(gl)SCSFIFK	MRP8	EFK(gl)DIPDLK
EAF7	EVK(gl)FEDEEK	MRPL11	IK(gl)QTGGKLTK
EBP2	SQELK(gl)KEEPTIVTASNLK	MRPL22	SSMK(gl)KATLLLR
ENO2	IEEELGDK(gl)AVYAGENFHHGDKL	MZM1	K(gl)VDGSSTKEPR
ERG10	AGAK(gl)FGQTVLVDGVER	NCB2	LHHNSVSDPVK(gl)SEDSS
ESC1	VNEGEEPEHQAVDIPVKVEVK(gl)EEQEEMPSK	NET1	DIDNSK(gl)PDPR
FBA1	DYIMSPVGNPEGPEK(gl)PNK	NET1	EKEDTNDK(gl)LLEK
FHL1	HPQNTTTDIENEVENPVTDDNGNLK(gl)LELPDNLDNADFSK	NET1	IK(gl)SSIVEEDIVSR
FLP1	EMIALK(gl)DETNPIEEWQHIEQLK	NET1	ISEIEK(gl)ELKEGPSSPASILPAK
GCN2	LMIDSPHLK(gl)K	NET1	K(gl)IKSSIVEEDIVSR
GCN4	FIK(gl)TEEDPIIK	NET1	K(gl)SQAEPSGIVEPK
GCN4	TEEDPIIK(gl)QDTPSNLDFDFALPQTATAPDAK	NET1	K(gl)VRPSLSSLSDLVSR
GCN5	VK(gl)LENNVEEIQPEQAETNKQEGTDK	NET1	NEIDLDDSAPVSLYK(gl)SVK
GPD1	PFK(gl)VTVIGSGNWGTTIAK	NET1	NESAQIDRQQK(gl)ETTSR
GPM1	GLVK(gl)HLEGISDADIAK	NET1	SDLFK(gl)MIEGDDTDLPQWFK
GSH1	ASGEIPTTAK(gl)FFR	NET1	SQAEPSGIVEPK(gl)R
GZF3	AISNVK(gl)TETTPPHFIPFLQSSK	NET1	VADLK(gl)SANIGGEDLNK
HAA1	IGSQENSVK(gl)QENYSK	NFI1	NENQGTVK(gl)QEQDYDSR
HAP1	VK(gl)QESSDELKK	NHP10	KISNIDADDDKEENEQK(gl)IK
HDA1	MDSVMVK(gl)K	NHP10	VADSK(gl)GGEDGSIVSSN
HHF1	K(gl)ILRDNIQGITKPAIR	NOP12	LLNEEAEAEDDK(gl)PTVTK
HHT1	RFQK(gl)STELLIR	NOP12	SSAIDNIFGNIDEK(gl)KIESSVDK
HHT1	STGGK(gl)APRK	NOP56	PTLK(gl)NELAIQEAMELYNK
HHT1	YK(gl)PGTVAIR	NOP7	LDPTEIEEDVK(gl)VESLDASTLK
HMO1	DAIIAAPVK(gl)AVR	NSR1	LSWSIDDEWLK(gl)K
HMO1	TTDPSVK(gl)IK	NTG1	IK(gl)QEEVVPQPVDIDWVK
HMS1	DSSLLSAASIVK(gl)KEQLSGFENFLPLSK	NTG1	LENDISVK(gl)VED
HPC2	MQTQTDTNAEVLNTDNSIK(gl)K	NTG1	RPLVK(gl)TETGPESELLPEK
HSC82	KPK(gl)IEEVDEEEEEK	NUM1	ESLSDK(gl)IEELTNQKK
HTA1/2	ATK(gl)ASQEL	PAA1	ELIK(gl)EEYDN
HTB1/2	AVTK(gl)YSSSTQA	PDC1	LTQDK(gl)SFNDNSK
HTB1/2	KPASK(gl)APAEK	PDR1	TSLEGTTVQVK(gl)EETDSSSTSFSNPQR
HTB2	SSAAEK(gl)KPASK	PGI1	TLSVK(gl)QEFQK
IES4	EPADEDPEVK(gl)QLEK	PGK1	VK(gl)ASKEDVQK
IES4	GSEFTASDVK(gl)GSDDK	PGK1	VLENTEIGDSIFDK(gl)AGAEIVPK
IES4	K(gl)KEPADEDPEVK	POB3	KEESSNEVVPK(gl)KEDGAEGEDVQMAVEEK
IES4	SQESSVLSESQEQLANNPK(gl)IEDTSPPSANSR	POL30	DLSQLSDSINIMITK(gl)ETIK
IKI1	DIK(gl)DENR	POL30	LMDIDADFLK(gl)IEELQYDSTLSLPSSEFSK
INO80	SIAVIINKEDK(gl)DISDFSK	PRE2	VK(gl)EEEGSFNNVIG
IRC20	K(gl)LEEADDK	PRP45	DVSEK(gl)IILGAAK
ISW1	AK(gl)IEDTSNVGTEQLVAEK	PRP45	K(gl)QTSTVAR
ISW1	DIISPLLLNPTK(gl)R	PRP45	LDEAVNVK(gl)SEGASGSHGPIQFTK
ISW1	LK(gl)EEGSR	PTA1	KIK(gl)METEPLAEEPEEPEDDDRMQK
KAP123	TSLLQTAFSEPK(gl)ENVR	PUS1	K(gl)ADFDDEK(gl)DKK
KRR1	DFIAPEEEAYK(gl)PNQN	RAD16	NDNDEIIEIK(gl)EER
LIF1	ISNQSVIK(gl)MEDDDFDDFQFFGISK	RAD52	K(gl)PVFGNHSEDIQTKLDK
MAG1	IK(gl)REYDEIIK	RAD52	NLVK(gl)IENTVSR
MCD1	ELSEEK(gl)EVIFTDVLK	RAD59	NEANTNYNLLSATNSKPTFIK(gl)LEDAK
MCM1	QQPQQQQPQQQQQVINAHANSIGHINQDQVPA-GAIK(gl)QEVK	RAP1	DSIRPK(gl)TEIISTNTNGATEDSTSEK
MET12	MEMLRNTGLEK(gl)	RBA50	DVHFIK(gl)EESQNEINIEKIDINDPNFNDK
MET28	VAATTAVVVK(gl)EEEAPVSTSNEIDK	REB1	AIIDADSITQHPDFQQYINTAADTDDNEK(gl)IK
MET4	MK(gl)QEQSHEGDSYSTEFINIFGK	REB1	ELVDYFSSNISMK(gl)TEN
MGA2	ALK(gl)EEEEDEHENK	REH1	KGMK(gl)KMQQIEK
MLP1	KIK(gl)TEDEEEKETDK	REP2	GAYK(gl)LQNTITEGPK
MLP2	RVK(gl)EEEYDIWQSR	REP2	MDDIETAK(gl)NITVK

Gene	Modified Sequence	Gene	Modified Sequence
REP2	NLTVK(gl)AR	RPS9B	K(gl)AEASGEAAEEAEDEE
RHR2	TYDAIAK(gl)FAPDFADEEYVNKLEGEIPEK	RPT6	K(gl)IEFPPPSVAAR
RLP7	SSTQDSK(gl)AQTINSNPEIIIRK	RPT6	YGEPQK(gl)VVLK
RNR2	DSK(gl)SNLNK	RRG9	RIIK(gl)SNWKR
RNR2	STK(gl)QEAGAFTFNEDF	RRP15	IFNAIIATQVK(gl)TEK
RPA34	VEGLK(gl)LEHFATGYDAEDFHVAEEVK	RRP9	TIDEYNNFDAGDLDK(gl)DIIASR
RPB4	HLK(gl)HENANDETTAVEDEDDDLDEDDVNADDDDFMH-SETREK	RSC2	TSVK(gl)RESEPGTDTNNDEDYEATDMDIDNPK
RPC37	SEEVK(gl)AEDDTGEEEEDDPVIEEFPLK	RSC4	LIAKPETVQSEVK(gl)NER
RPC37	SIDNK(gl)LFVTEEDEEDRTQDR	RSC58	VK(gl)QEELLNTNEEGINR
RPC53	EPTPSVK(gl)TEPVGTGLQSYLEER	RSC8	LENNGNSVK(gl)K
RPC53	GFIK(gl)SEGSGSSIVQK	RSC8	PFLPENVIK(gl)QEVEGGDGAEPQVKK
RPC53	LPAFERPAVKEEK(gl)EDMETQASDPSK	RTF1	NAEHVK(gl)KEDSNNFDSK
RPC53	MAK(gl)YLNNTHVISSGPLAAGNFVSEK	RVB1	K(gl)EIVVNDVNEAK
RPC53	PAVK(gl)EEKEDMETQASDPSK	SAT4	DLK(gl)PENLLLTHDGVLK
RPC53	VK(gl)LEEESK	SCC2	K(gl)SEIVSRPEAK
RPC82	LK(gl)TEDGFVIPALPAAVSK	SCM4	TLK(gl)PESER
RPD3	DAEDLGDVEEDSAEAK(gl)DTK	SDC1	SVTNQNVK(gl)IEESSSTNSVIEESSEPK
RPL13A/B	GFTLAEVK(gl)AAGLTAAYAR	SEF1	DSK(gl)VSVQTYLSR
RPL18A	APK(gl)GQNTLILR	SGS1	QLENDIK(gl)LEVIR
RPL18A/B	ALK(gl)QEGAANK	SHS1	EIK(gl)QENENLIR
RPL18A/B	VVLK(gl)ALFLSK	SHS1	FLNSPDLPERTK(gl)LR
RPL20A	DIK(gl)FPIPHR	SHS1	SIK(gl)TESSPK
RPL20A	TADVK(gl)R	SIC1	LTDEEK(gl)R
RPL25	APK(gl)YASK	SIR2	IK(gl)VAQPDSLR
RPL25	AVK(gl)ELYEVDVLK	SIR3	K(gl)IK(gl)IEPSADDDVNNGNIPSQR
RPL25	LDSYK(gl)VIEQPITSETAMK	SIR4	APFIK(gl)SESKPFSSDALSK
RPL28	FVSK(gl)LAEEK	SIZ1	NFLQNALVVGK(gl)SDPYR
RPL28	IPNVPVIVK(gl)AR	SIZ1	STNTDILTEK(gl)GSSAPSR
RPL34A/B	AFLIEEQK(gl)IVK	SIZ1	TLDPK(gl)SYNIVASETTTPVTNR
RPL35B	EQIASQIVDIK(gl)K	SIZ1	VIPEYLGNSSSYIGK(gl)QLPNILGK
RPL4A/B	LNPYAK(gl)VFAAEK	SKO1	DTNVVK(gl)SENAGYPSVNSRPIILDK
RPL5	VAAK(gl)IAALAGQQ	SLI15	EVK(gl)NYYQSPVR
RPL7A	VTK(gl)ATLELLK	SLI15	NNVYMNTLK(gl)YEDK
RPL8A/B	NFGIGQAVQPK(gl)R	SMC5	LDDIVSK(gl)ISAR
RPL8B	LVSTIDANFADK(gl)YDEVKK	SNF2	DIGAELK(gl)R
RPO21	VDLLNTDHTLDPSLLESGSEILGDLK(gl)LQVLLDEEYK	SOD1	GDAGVSGVVK(gl)FEQASESEPTTVSYEIAGNSPNAER
RPP1B	ALEGK(gl)DLK	SOD1	K(gl)THGAPTDEVR
RPP2A	VSSVLSALEGK(gl)SVDELITEGNEK	SPA2	TIK(gl)REEEDEDFDRVNHNIQITGAYTK
RPS0A/B	TWEK(gl)LVLAAR	SPC24	LLK(gl)DLDGLER
RPS1	VSGFK(gl)DEVLETV	SPP41	GVTTPIK(gl)IEDSDANVPPVSIAVSTIEPSQDK
RPS10A	HEEIDTK(gl)NLYVIK	SPP41	IPEIK(gl)NESVDLGSNITDILSSTITNILPEITATDVK
RPS13	K(gl)GLTPSQIGVLLR	SPP41	PK(gl)SEDHEWPLSDSSASQNYDAHLK
RPS17A/B	GISFK(gl)LQEEER	SPP41	RPQIK(gl)PEVSVINLVQNLVNTK
RPS17A/B	YYPK(gl)ITIDFQTNK	SPP41	VK(gl)QQLDK
RPS1A	VTGFK(gl)DEVLETV	SPT15	DGTKPATTFQSEEDIK(gl)R
RPS20	RYIDLEAPVQIVK(gl)R	SPT7	NGFGTVIK(gl)QEDDDQIQFHNDHSINGNEAFEK
RPS20	SDFQK(gl)EKVEEQEQQQQQIIK	SSE1	GK(gl)LEEEYAPFASDAEK
RPS20	SDFQKEK(gl)VEEQEQQQQQIIK	SSM4	LSPK(gl)DLK
RPS21A	ADDHASVQINVAK(gl)VDEEGR	STB3	EVSPPQAISVK(gl)SEASSSIFSK
RPS24A/B	TQFGGGK(gl)SVGFGIVYNSVAEAK	STH1	LIQLDELPK(gl)VFR
RPS28A/B	MDSK(gl)TPVTIAK	STH1	VFREDIEEHFK(gl)KEDSEPIGR
RPS3	ALPDAVTIIEPK(gl)EEEPILAPSVK	STP1	IK(gl)SEVNAK
RPS31	LIFAGK(gl)QLEDGR	SUM1	IITIK(gl)SSSENSGNNTTNNNNTDNVIK
RPS31	TLSDYNIQK(gl)ESTLHLVLR	SUM1	IK(gl)NEIPINSLLPSSK
RPS6A	K(gl)GEQELEGLTDTTVPK	SUM1	K(gl)TPGDEETTTFVPLENSQPSDTIRK
RPS8A/B	NVK(gl)EEETVAK	SUM1	LPSGPK(gl)DDVDTLALTSAQNQANSLR

Gene	Modified Sequence	Gene	Modified Sequence
SUM1	VNVEENK(gl)TEK	TUP1	INDTGSATTATTTTATETEIK(gl)PK(gl)EEDATPASLHQDHY LVPYNQR
SUM1	YFVEPSTK(gl)QESLLLSAPSSSR	TUP1	IWNIQNANNK(gl)SDSK
SWA2	YLEILK(gl)SK	TUP1	LQNQK(gl)DYDFK
SWC3	TTAESTQVDVK(gl)K	TYE7	K(gl)QDEDGAETAATTPIPSAAATSTK
SWI3	IQKEEEPENNTVIEGVK(gl)EESQPDENTK	TYE7	LQQIIPWVASEQTAFEVGDSVK(gl)K
SWR1	AGGEQDLADLK(gl)FR	TYE7	SSETTLIK(gl)PESEFDNWLSDENDGASHINVNK
SWR1	LLAQAEDEDDVK(gl)AANLAMR	TYE7	TNIDAK(gl)ETK
SWR1	YDHIAK(gl)VEEPSEAFTIK	UBA2	IK(gl)QETNELYELQK
TAF12	SAIFK(gl)QTEPAIPISENISTK	UBA2	LLAIENLWK(gl)TR
TAF14	TGSASTVK(gl)GSVDLEK	UBA2	SHIFNIPMK(gl)SVFDIK
TAH11	EK(gl)MPDSQANLMDRLR	UBC9	EGTNWAGGVYPITVEYPNEYPSK(gl)PPKVK
TDH1/2/3	TASGNIIPSSTGAAK(gl)AVGK	UBC9	VLLQAK(gl)QYSK
TDH2/3	VVDLVEHVAK(gl)A	UBI1	LIFAGK(gl)QLEDGR
TEC1	K(gl)IENFIK	UBI1	TLSDYNIQK(gl)ESTLHLVLR
TEF1	LPLQDVYK(gl)IGGIGTVPVGR	UBI4	LIFAGK(gl)QLEDGR
TFG1	AVDSSNNASNTVPSPIK(gl)QEEGLNSTVAER	UBI4	TLSDYNIQK(gl)ESTLHLVLR
TFG1	ENESPVK(gl)KEEDSDTLSK	UME1	STIDIAEDNKIK(gl)NEEFK
TFG1	GSLVK(gl)KDDPEYAEEREK	UME6	DREITDPNVK(gl)LDENESK
TFG1	VK(gl)DEDPNEYNEFPLR	UPC2	ADGSVESDSSVDLPPTIK(gl)K
TFP1	AIK(gl)EESQSIYIPR	UTP7	TNSDIPDVK(gl)PDVK
TIF4631	SAEPEVK(gl)QETPAEEGEQGEK	VBA5	AENK(gl)GIIQQIK
TOA1	IEVK(gl)PEIELTINNANITTVENIDDESEK	VHR1	NLFNIIINK(gl)NK
TOF2	FKPTGETK(gl)VQK	VHR2	LQK(gl)FDIEDQPLESEQEYDFIAK
TOF2	LHQSQGK(gl)EALFR	VIP1	SGIK(gl)KEPIESDEVPQQETK
TOF2	LVEKEFPDK(gl)SLGAASSTSHAK	VMA1	AIK(gl)EESQSIYIPR
TOP1	IK(gl)TEPVQSSSLPSPPAK	VPS3	K(gl)TEDDSLR
TOP1	KIK(gl)KEDGDVK	VPS72	SDIK(gl)RDETTNEDSDDQVR
TOP2	KIK(gl)IEDK	VPS72	VNSDELK(gl)PTALPDVTLDAIANK
TOP2	TEEEENAPSSTSSSSIFDIK(gl)KEDK	YAP5	QK(gl)LETLTLK
TOP2	TPSVSETK(gl)TEEEENAPSSTSSSSIFDIKKEDK	YJR129C	IK(gl)IEETPNLISAASTTGFR
TRI1	EIK(gl)LENESLPNLSG	YLR455W	NSISIK(gl)EDPEDNQK
TRI1	HLFNPDEIVK(gl)HEEEQKQTPEK	YMR111C	IK(gl)PEPGLSDFENGEYDGNESDENATTR
TRI1	VIIPK(gl)NDIISRDQEISIR	YRR1	YLK(gl)LTR
TRI1	VLLSAPLQK(gl)FLGSEELPR	YSH1	IEPIK(gl)EENEDNLDSQAEK
TRX1	FSEQYPQADFYK(gl)LDVDELGDVAQK	YTA7	VGYETQIK(gl)DENGIIHTTTR
TUP1	APESTLK(gl)ETEPENNNTSK	ZEO1	AETAAQDVQQK(gl)LEETK
TUP1	DAYEEEIK(gl)HLK	ZEO1	GQEVK(gl)EQAEAESIDNIK
TUP1	DYDFK(gl)MNQQLAEMQQIR	ZEO1	NEATPEAEQVK(gl)K
TUP1	ETTTLPSVK(gl)APESTLK		

liquid chromatography coupled to high-resolution mass spectrometers (LTQ-Orbitrap Velos) and identified using MaxQuant/Andromeda software [14]. A total of 257 SUMO-acceptor lysines were identified on asynchronous growing yeast cultures (Table 1 & Suppl. Table S2). We were able to detect many SUMO-acceptor lysines previously described in several studies for yeast proteins (Fig. 2A). Interestingly, our SUMO dataset had a significant overlap (over 50 proteins/sites) with that reported in a recent study were a different strategy was followed to identify site-specific SUMOylation [15]. Ubiquitination or neddylation also produce diglycine-modified lysines after trypsin digestion, therefore to ensure that our strategy was able to enrich for diglycine-modified lysines that represented SUMO conjugation of Smt3-KallR-I96R we used an Smt3-KallR where I96 was not mutated to arginine. We identified just over 20 diglycine-modified lysines in this sample (Fig. 2B). None of these diglycine-modified lysines were placed within a SUMO consensus sequence. This result suggests that the large majority of diglycine-modified lysines in the

strains carrying Smt3-KallR-I96R (Table 1) are indeed the product of SUMO-conjugation.

Verification of novel SUMO substrates

To confirm that some of the yeast proteins newly identified as potential SUMO substrates are indeed modified by SUMO we tagged the proteins of interest with either Myc- or HA-epitopes in strains where Smt3 contains a HIS/Flag tag (HF-Smt3) at its N-terminus. Cells were grown and lysed under denaturing conditions and Smt3 was purified using Ni-nitrilotriacetic acid (Ni-NTA) beads. The immunoprecipitates were separated by SDS-PAGE, and the different forms of the tagged proteins were resolved by immunoblotting against the epitope tags. It is important to note that immunoprecipitation of SMT3 using histidine tags allows us to work under denaturing conditions, which significantly reduces loss of SUMO conjugation in the lysates due to endogenous SUMO proteases. We confirmed the SUMOylation of several newly identified substrates, including the transcription factor Tfg1, the nucleotide excision repair

(NER) factor Rad16, the replication fork barrier protein Fob1 (Fig. 3A), the RNA Polymerase III subunit Rpc53 (Fig. 3B) and the base excision repair protein Ntg1 (Fig. 3C). Multiple SUMOylated species were observed for Tfg1, Rad16 and Fob1 (Fig. 3A).

Mutational analysis of SUMO-acceptor sites identified by mass spectrometry

Following the confirmation that we had identified novel SUMO-substrates during our proteomic analysis, we proceeded to test whether mutation of the lysine residues identified as potential conjugation sites prevented SUMOylation of these substrates. We identified four potential conjugation sites on Rpc53 (K51, 115, 237, 322) and two on Ntg1 (K38, 396). Strains carrying a tagged wild-type copy or a mutant allele where all acceptor lysines identified by mass spectrometry had been substituted for arginine were analyzed for Rpc53 and Ntg1. SUMOylated forms were detected in the immunoprecipitates for all the proteins when Smt3 was tagged (Fig. 3B-C), however SUMO forms were absent when the SUMO-acceptor lysines identified by mass spectrometry were mutated to arginine (Fig. 3B-C), thus these lysines represent bona fide SUMO-acceptor lysines in the target proteins and validate the success of our approach.

DISCUSSION

Identifying lysine residues where SUMO is conjugated to

A

Denison *et al.* 2005 [16]- Cdc3, Rpl25, Reb1.
Zhou *et al.* 2004 [17]- Pol30, Sod1, Spp41, Mrp8.
Silver *et al.* 2011 [18]- Rad16.
Chen *et al.* 2007 [19]- Top1, Tri1.
Nathan *et al.* 2006 [20]- H2B.
Johnson *et al.* 1999 [8]- Gcn5, Cdc11, Shs1, Cdc3.
Sacher *et al.* 2006 [21]- Rad52.
Albuquerque *et al.* 2015 [15]- Cdc3, Shs1, Bud3, Mlp1, Mlp2, Reb1, Rpo21, Rpo26, Tfg1, Gcn5, Spt7, Rpc37, Rpc53,Bdp1, Cin5, Tup1, Cyc8, Sko1, Crz1, Hap1, Mot1, Ngg1, Sum1, Tye7, Vhr1, Cet1, Prp45, Stb3, Hir2, Hpc2, H2B, Isw1, Pob3, Rsc8, Sth1, Snf2, Vps72, Net1, Rap1, Sir3, Sir4, Tof2, Mcd1, Smc5, Bir1, Cbf2, Abf1, Pol30, Top2, Uba2, Ubc9, Siz1, Bop3, Cdc48.

B

GENE	Modified Sequence
ATG14	RNK(gl)MK(gl)CR
BNI1	K(gl)LDEINR
BUD27	LEDFK(gl)EYNK
CCT8	GLMK(gl)PSGGK
HSP104	YFAIPDIK(gl)K
IFH1	LYK(gl)KTQK(gl)PSTR
INP2	K(gl)K(gl)NSPGLK(gl)R
IOC4	TK(gl)AK(gl)K(gl)IVLK
MLP2	SLKNVTEK(gl)NR
MOD5	MLKGPLKGCLNMSK(gl)K
NET1	EK(gl)TSKSNEK
NMD2	NIK(gl)K(gl)IVLK
NUM1	RGLQIALTTK(gl)EDKK
RIF1	NLK(gl)GPLK
RIT1	K(gl)ENK(gl)SVR
RPC53	DTK(gl)DALSTR
SKI2	K(gl)HKEILNGESAKGAPSK
SNI2	QGFIPSTVIHAK(gl)K
SNU56	YSEGNK(gl)PGFMTQDEIK(gl)QHCIGTIK
SWA2	YLEILK(gl)SK

FIGURE 2: SUMO-acceptor lysines detected in our mass spectrometry analysis were compared with SUMO-acceptor lysines already described in previous studies done in *S. cerevisiae*. (A) Previous studies were based on either MS or site-directed mutagenesis/immunoblotting, or a combination of both. Most of the SUMO-acceptor lysines previously found were also detected in this study. SUMO substrates (total of 257) identified in the yeast proteome of cells grown asynchronously. Previously published SUMO-substrates were obtained [8, 15-21]. **(B)** To ensure that diglycine-modified lysines detected by mass spectrometry after Smt3 purification are not due to modification by ubiquitin or Nedd8, a centromeric plasmid 8His-SMT3-KallR-REQIGG-pRS415 expressing the Smt3 variant used previously for the Smt3 purification protocol with the difference that the RGG conjugating terminus was replace for the native RIEQGG C-terminus was employed. SUMO-acceptor lysines modified by the 8His-Smt3-KallR-REQIGG keep a side chain of 5 aa after trypsin digestion (EQIGG). Therefore, any diglycine-modified lysines detected by mass spectrometry under these conditions can only be due to either false positive hits, or to ubiquitinated or neddylated contaminants. A large culture of 9 l of the strain expressing the 8His-SMT3-KallR-REQIGG variant was grown in YPD and harvested at O.D. 0.9. Smt3 purification and mass spectrometry analysis was performed as described in Material and Methods. We detected 23 diglycine-modified lysines. None of these corresponded with previously detected diglycine-modified lysines in our Smt3-RGG pulldowns. In addition none of these diglycine-modified lysines are within SUMO consensus sequences. This strongly indicates that sumo-acceptor lysines identified after purification of Smt3-RGG pulldowns represent bona fide SUMOylation sites.

substrates is a challenging process that in many cases is resolved by site-directed mutagenesis of potential acceptor lysines. This approach is not only labour intensive but often unable to discriminate between bona fide conjugation sites and lysines that alter the protein biology in a manner that leads to reduced SUMOylation levels (at other domains of the protein). Recent studies in mammalian cells have by-passed the challenges of spectrometric analysis of protein SUMOylation using differential protease cleavage of a modified SUMO2 lacking lysine residues [12]. Here we have employed a similar approach to map acceptor lysines for SUMO in endogenous substrate proteins purified from yeast cells under several conditions. Our analysis revealed a total of 257 SUMO-acceptor lysines. Over 70% of the sites identified adhere to previously described SUMO motifs, including the core consensus motif ([VILP]KxE) [22], the phosphorylation-dependent motif ([VILP]Kx[ED]x[S or xS]) [10], the negatively charged/acidic consensus motif ([VILP]Kx[ED][ED][ED]) [11] and the inverse consensus motif ([ED]xK[VILP]) [12].

SUMOylation substrates were found in all major cellular compartments. However, the presence of SUMOylated proteins was lower in the ER, Golgi's apparatus, mitochondrion, membrane and cytoplasm when proportionally compared with the whole proteome. On the other hand, the amount of SUMOylated proteins is proportionally higher in ribosomes and much higher in the nucleus. A closer look at nuclear proteins shows that a high proportion of SUMOylated proteins is found in association with chromatin (i.e. transcription factors) (Fig. S2). Functionally,

SUMOylated proteins were involved in a wide range of biological processes. There were a large number of SUMOylated proteins involved in nuclear processes such as chromatin organization, transcription and DNA damage. A few SUMO-acceptor lysines were found on proteins from the ubiquitination and protein degradation pathway, including ubiquitin itself. This raises the possibility that SUMOylation of ubiquitin might affect the ubiquitination pathway directly.

SUMOylation is known to play an important role in the metabolism of the chromatin and in gene expression. We identified a large number of SUMO-acceptor lysines on nuclear proteins. Most of these proteins are related to transcription and chromatin organization. In accordance with previous studies we found SUMO-acceptor lysines on components of the general transcription machinery as well as on components of gene-specific transcription pathways. SUMO-acceptor lysines were identified in six subunits of RNA polymerases and on components of the TFIID, TFIIF and TFIIIB, including the TATA-binding protein (TBP). Other components of the general transcription found SUMOylated included the transcription repressors Ncb2, Tup1 and Cyc8. SUMOylated proteins related to gene-specific transcription include the transcription factors Crz1 and Sko1 (stress response genes), Gzf3 (of nitrogen catabolic genes), and Cin5 (drug response genes).

Control of chromatin function is likely to involve not only SUMOylation of core histones but also proteins involved in the modification and exchange of histones. We found SUMO-acceptor lysines on histone acetylases (i.e.

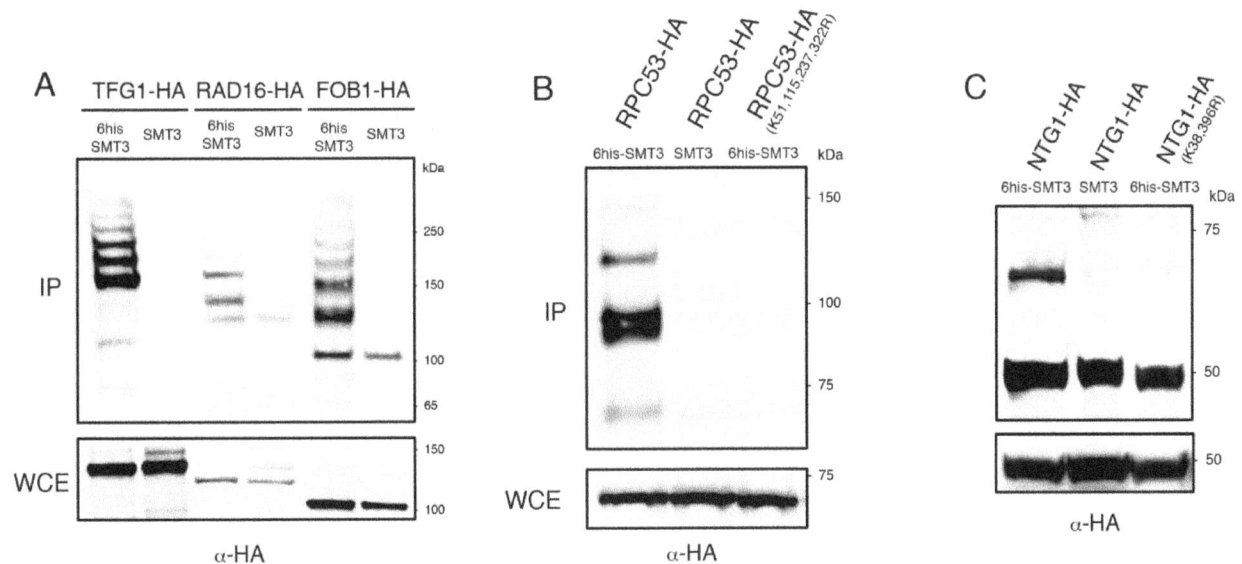

FIGURE 3: Mutational analysis of SUMO sites identified. (A) Histidine pulldowns from cells carrying TFG1, RAD16 and FOB1 tagged with 3 HA epitopes in strains expressing SMT3 tagged with 6 histidines (6his-SMT3) or wild-type SMT3 (untagged control). Western blot analysis using α-HA antibodies in the pulldown demonstrates the presence of SUMO conjugates for TFG1, RAD16 and FOB1. (B) Histidine pulldowns from cells carrying RPC53 or RPC53 (K38,115,237,322R) tagged with 3 HA epitopes. The analysis demonstrates that lysines 38, 115, 237 and 322 (identified in our screen) are sites for the conjugation of SMT3 in RPC53. (C) Histidine pulldowns from cells carrying NTG1 or NTG1 (K38, 396R) tagged with 3HA epitopes. The analysis demonstrates that lysines 38 and 396 (identified in our screen) are sites for the conjugation of SMT3 in NTG1.

Gcn5), and on components of the RPD3 complex (involved in histone deacetylation), on the Swr1 complex (involved in histone exchange) and of the COMPASS complex (involved in histone methylation). In addition, we found SUMO ylation on various chromatin remodelling complexes.

SUMOylation has also been linked to centromeric function. One of the key SUMO-substrate in this regulatory function is Top2, which controls local chromatin structures in the centromeric region. We mapped SUMO-acceptor lysines not only on Top2 but also on other kinetochore proteins like Cbf2, and on components of other centromeric complexes (CPC complex, Ndc80 or Dam1 complexes). These complexes are important players in the regulation of chromosome segregation and kinetochore clustering. The detailed functional significance of Top2 SUMOylation and of SUMOylation of other components with centromeric functions and in chromosome segregation pathways remains unclear. We hope that the SUMO-acceptor dataset presented here will be useful for future functional studies of SUMOylation in yeast.

MATERIALS AND METHODS
Yeast strains
Strains used in this study are isogenic to BY4741 (*MATa his3Δ1 leu2Δ0 met15Δ0 ura3Δ0*), W303 (*Mata can1-100 leu2-3 his3-11 trp1-1 ura3-1 ade2-1*) or DF5a (*MATα his3Δ200 leu2-3 lys2-801 trp1-1 ura3-52*). Strains used are listed in Supplemental Table S1.

Yeast media and cell cycle synchronizations
Cells were grown on complete media YPED in broth or solid form (3% yeast extract, 1% peptone, 1% glucose/dextrose, 2% agar for solid media). For plates containing MMS (in DMSO), the genotoxin was added to warm YPD.

Drops analysis by growth tests
10-fold dilutions of fresh cells were made in PBS. Cells were spotted (as 2 - 5 μl drops) onto solid media, incubated at the appropriate temperature for 3 - 5 days and then photographed.

Purification of His-tagged proteins
For His-tag purifications 100 OD of cells were harvested (4000 rpm, 2 minutes), washed once with water, and the cell pellet frozen at -80°C. The cell pellet was resuspended in 500 μl Buffer A (8 M Urea, 100 mM NaH$_2$PO$_4$, 10 mM Tris HCl, 0.05% Tween pH 8), an equal volume of glass beads were added, and the cells lysed by 1 45s cycle, power 6 in a FastPrep FP120 (BIO 101) machine. Tubes were pierced with a hot needle and placed onto fresh eppendorfs and spun (1000 rpm, 2 minutes) to collect lysate minus glass beads. Cell lysate was clarified by centrifugation (14000 rpm, 15 minutes, 4°C) Protein concentration was determined using a Bradford Assay, and 15 mg of protein in 1ml was added to 50 μl of a 50:50 slurry of Ni-NTA beads (Qiagen) (prewashed in Buffer A). Imidazole was added to a final concentration of 20 mM, to reduce non-specific binding. Proteins were bound for 2 - 3 hours at 4°C on a rotating platform, before the beads were washed 3 times in Buffer A containing 2 mM imidazole, followed by 5 washes in Buffer B (8 M Urea, 100 mM NaH$_2$PO$_4$, 10 mM Tris

HCl, 0.05% Tween pH 6.3). Bound proteins were eluted off the beads using 30 μl x2 NuPAGE loading buffer (Invitrogen) supplemented with 4% β-mercaptoethanol and 200 mM EDTA. Eluates were loaded onto 3-8% Tris-Acetate or 4-12% Bis-Tris pre-cast NuPAGE gels (Invitrogen) and analysed by western blotting.

Sample preparation for western blot analysis
Samples were prepared by TCA extraction. Extracts were prepared as follows, cells were collected by centrifugation (4000 rpm, 2 minutes) and washed with 20% TCA. The TCA was aspirated and the pellets frozen at -80°C. All of the following purification steps were performed on ice with pre-chilled solutions. Cells were resuspended in 250 μl 20% TCA, glass beads were added and the cells broken by 1 40s cycle, power 5.5 in a FastPrep™ FP120 (BIO 101) machine. Tubes were pierced with a hot needle and placed onto fresh eppendorfs and spun (1000 rpm, 2 minutes) to collect lysate minus glass beads. The glass beads were washed with 1 ml 5% TCA and this was added to the lysate, and mixed by pipetting. The precipitated proteins were collected by centrifugation (14000 rpm, 10 minutes 4°C) and then pellets were washed with 750 μl 100% ethanol. Proteins were solubilized in 50 μl 1M Tris pH 8 and 100 μl x2 SDS-PAGE loading buffer (60 mM Tris pH 6.8, 2% SDS, 10% glycerol, 0.2% bromophenol blue) and boiled for 5 minutes at 95°C. Insoluble material was removed by centrifugation (14000 rpm, 5 minutes, room temperature) and the supernatant either stored at -20°C or loaded immediately onto a SDS-PAGE mini-gel. Samples were either run on an 8% acrylamide gel in Tris-Glycine SDS running buffer using the Bio Rad Mini-PROTEAN 3 system or on Pre-Cast 4-12% Bis-Tris gels in NuPAGE MOPS (all Invitrogen). SDS-PAGE gels were transferred to polyvinylidene fluoride transfer membrane (Hybond-P, Amersham Biosciences) in either the Bio-Rad Mini Trans-Blot Electrophoretic Transfer Cell or by using the XCell SureLock Mini Cell Transfer module (Invitrogen). The Bio-Rad system was used in conjunction with Tris-Glycine blotting buffer (National Diagnostics) containing 20% methanol, and run for 1h at 200V or overnight at 30V. The NuPAGE system was used with NuPAGE transfer buffer containing 20% methanol and run for 1h at 30V.

Immunological detection
Membranes were blocked in 5% skimmed milk powder in PBS with 0.1% Tween 20 (PBS-T) for 1h or overnight at 4°C, then incubated with either mouse monoclonal anti-c-myc IgG1κ antibody 9E10 (Roche) or anti-HA IgG1 antibody 12CA5 (Roche) in blocking solution for between 1 hour at room temperature to overnight at 4°C. Following several washes in PBS-T, membranes were incubated with the sheep anti-mouse IgG Horseradish Peroxidase-linked antibody (GE Healthcare) at a 1/10000 dilution in blocking solution. After several further washes in PBS-T, membranes were incubated with the ECL Plus Western Blotting Detection System (GE Healthcare) followed by exposure to ECL Hyperfilm (GE Healthcare), to detect the secondary antibody.

Purification of SUMOylated proteins for mass spectrometry
Nine liters of culture harvested at O.D.0.9 were used in each experiment. Cells were pelleted and washed in chilled PBS with 20 mM NEM. Cells were resuspended in 250 ml of lysis buffer (1.85 N NaOH, 1.85% β-mercaptoethanol) and incubat-

ed in ice for 30 minutes. TCA (trichloroacetic acid) was added to a final concentration of 25% and kept in ice for a further 30 minutes. Sample were spun at 15.000 rpm for 10 minutes in acetone-resistant tubes. The pellet washed with 250 ml acetone and spun again. The pellet was air-dried for 10 minutes and resuspended in 100 ml of binding buffer (6 M guanidine hydrochloride, 100 mM NaH_2PO4, 10 mM Tris-HCl, 20 mM NEM, pH 8) by vortexing for 10 minutes at RT. Resuspended proteins were recovered after a spun at 15.000 rpm. for 10 minutes. All spins were performed at 4°C. Protein extraction after elution from agarose beads (see Purification of His-tagged proteins) was adjusted to pH 7 and filtered through a 50 kDa MWCO protein filter (Vivaspin 20, GE Healthcare) up to 1 ml (one filter was used for every 12 ml of elution sample). Buffer A (w/o Tween®-20) was added in a proportion 10:1 and the sample filter again up to 1 ml. Sample was placed in a 10 kDa MWCO protein filter (Vivaspin 2, GE Healthcare), and filtered up to a volume 1/100000 of the original volume of the culture (i.e. 90 µl from a 9 l culture). Protein sample was then digested with Lys-C endonuclease (Roche) at final concentration of 0.02 µg/µl. The reaction was incubated at 37°C for 12 hours. A standard 4% β-mercaptoethanol/2% SDS loading buffer was added to the sample, left on the bench for 15 minutes and loaded into a 12% Tris-Bis NuPAGE® gel (Invitrogen). Gel was run using an Invitrogen system with NuPAGE® MOPS SDS running buffer (Invitrogen) for 2 h at 140 V. After the running, the gel was stained using SimplyBlue™ SafeStain (Invitrogen) following manufacturer's protocol. The band of interest was cut from the gel. Peptides were in-gel digested using trypsin. First, gel bands were destained using 40% acetonitrile, 200 mM ammonium bicarbonate. Gel pieces were shrunk with acetonitrile and enough trypsin solution (10 ng/µl trypsin in 10% acetonitrile in 50 mM ammonium bicarbonate) was added to cover the gel pieces. Gel pieces were incubated overnight at 37°C. Peptides were extracted with 5% formic acid in acetonitrile. The extracts were dried to completeness in a vacuum concentrator and resuspended in 10 µl of 0.1% trifluoric acetic acid (TFA). Peptide mixtures were loaded onto

a 75 µm i.d. c18 column (pepmap, ThermoFisher) and eluted over a 60 minute gradient stretching from 3% acetonitrile to 50% acetonitrile (both in 0.1% formic acid) at 300 nl/minute using an RSLC Ultimate 3000 onto a LTQ-Orbitrap-Velos (both ThermoFisher). Profile spectra were acquired in the Orbitrap (60,000 resolution) and the 6 most intense ions excluding +1 charge state were selected for fragmentation in the linear ion trap (LTQ). Raw data was processed using MaxQuant and searched using the embedded Andromeda routine [14]. Data was searched against a *Saccharomyces cerevisiae* database (to which the recombinant Smt3 was manually added) using default settings. Mass tolerances for Ms and MS/MS data were 10 part per million (ppm) and 0.5 Da respectively. Methionine oxidation and glygly modification of lysine were allowed as variable modification. The false discovery rates at the peptide, protein and site level were 0.005, 0.01 and 0.01 respectively. Statistical analysis and charts were created using the SPSS software environment. Graphic representation of the local context of the SUMO-acceptor lysines was created using ice-Logo [23]. SUMO motifs were assigned to sumo-acceptor lysines using the 3of5 web application for pattern matching [24]. Subcellular distribution and function of SUMOylated proteins was obtained from the Gene Ontology (GO) slim mapper part of the *Saccharomyces* Genome Database (SGD, http://www.yeastgenome.org).

ACKNOWLEDGMENTS

We would like to thank members of the Aragon laboratory for discussions and critical reading of the manuscript. This work was funded by the Intramural research programme of the Medical Research Council UK and ERC grant (Chrm Stability-202337).

CONFLICT OF INTEREST

The authors declared that there is no conflict of interest arising from this work.

REFERENCES

1. Melchior F (**2000**). SUMO - nonclassical ubiquitin. **Annu Rev Cell Dev Biol** 16:591-626.

2. Geiss-Friedlander R, Melchior F (**2007**). Concepts in sumoylation: a decade on. **Nat Rev Mol Cell Biol** 8(12): 947-956.

3. Bylebyl GR, Belichenko I, Johnson ES (**2003**). The SUMO isopeptidase Ulp2 prevents accumulation of SUMO chains in yeast. **J Biol Chem** 278(45): 44113-44120.

4. Saitoh H, Hinchey J (**2000**). Functional heterogeneity of small ubiquitin-related protein modifiers SUMO-1 versus SUMO-2/3. **J Biol Chem** 275(9): 6252-6258.

5. Vertegaal AC, Andersen JS, Ogg SC, Hay RT, Mann M, Lamond AI (**2006**). Distinct and overlapping sets of SUMO-1 and SUMO-2 target proteins revealed by quantitative proteomics. **Mol Cell Proteomics** 5(12): 2298-2310.

6. Liang YC, Lee CC, Yao YL, Lai CC, Schmitz ML, Yang WM (**2016**). SUMO5, a Novel Poly-SUMO Isoform, Regulates PML Nuclear Bodies. **Sci Rep** 6:26509.

7. Owerbach D, McKay EM, Yeh ET, Gabbay KH, Bohren KM (**2005**). A proline-90 residue unique to SUMO-4 prevents maturation and sumoylation. **Biochem Biophys Res Commun** 337(2): 517-520.

8. Johnson ES, Blobel G (**1999**). Cell cycle-regulated attachment of the ubiquitin-related protein SUMO to the yeast septins. **J Cell Biol** 147(5): 981-994.

9. Bernier-Villamor V, Sampson DA, Matunis MJ, Lima CD (**2002**). Structural basis for E2-mediated SUMO conjugation revealed by a complex between ubiquitin-conjugating enzyme Ubc9 and RanGAP1. **Cell** 108(3): 345-356.

10. Yang XJ, Gregoire S (**2006**). A recurrent phospho-sumoyl switch in transcriptional repression and beyond. **Mol Cell** 23(6): 779-786.

11. Yang SH, Galanis A, Witty J, Sharrocks AD (**2006**). An extended consensus motif enhances the specificity of substrate modification by SUMO. **EMBO J** 25(21): 5083-5093.

12. Matic I, Schimmel J, Hendriks IA, van Santen MA, van de Rijke F, van Dam H, Gnad F, Mann M, Vertegaal AC (**2010**). Site-specific identification of SUMO-2 targets in cells reveals an inverted SUMOylation motif and a hydrophobic cluster SUMOylation motif. **Mol Cell** 39(4): 641-652.

13. Wohlschlegel JA, Johnson ES, Reed SI, Yates JR, 3rd (**2006**). Improved identification of SUMO attachment sites using C-terminal SUMO mutants and tailored protease digestion strategies. **J Proteome Res** 5(4): 761-770.

14. Cox J, Neuhauser N, Michalski A, Scheltema RA, Olsen JV, Mann M (**2011**). Andromeda: a peptide search engine integrated into the MaxQuant environment. **J Proteome Res** 10(4): 1794-1805.

15. Albuquerque CP, Yeung E, Ma S, Fu T, Corbett KD, Zhou H (**2015**). A Chemical and Enzymatic Approach to Study Site-Specific Sumoylation. **PLoS One** 10(12): e0143810.

16. Denison C, Rudner AD, Gerber SA, Bakalarski CE, Moazed D, Gygi SP (**2005**). A proteomic strategy for gaining insights into protein sumoylation in yeast. **Mol Cell Proteomics** 4(3): 246-254.

17. Zhou W, Ryan JJ, Zhou H (**2004**). Global analyses of sumoylated proteins in Saccharomyces cerevisiae. Induction of protein sumoylation by cellular stresses. **J Biol Chem** 279(31): 32262-32268.

18. Silver HR, Nissley JA, Reed SH, Hou YM, Johnson ES (**2011**). A role for SUMO in nucleotide excision repair. **DNA Repair (Amst)** 10(12): 1243-1251.

19. Chen XL, Silver HR, Xiong L, Belichenko I, Adegite C, Johnson ES (**2007**). Topoisomerase I-dependent viability loss in Saccharomyces cerevisiae mutants defective in both SUMO conjugation and DNA repair. **Genetics** 177(1): 17-30.

20. Nathan D, Ingvarsdottir K, Sterner DE, Bylebyl GR, Dokmanovic M, Dorsey JA, Whelan KA, Krsmanovic M, Lane WS, Meluh PB, Johnson ES, Berger SL (**2006**). Histone sumoylation is a negative regulator in Saccharomyces cerevisiae and shows dynamic interplay with positive-acting histone modifications. **Genes Dev** 20(8): 966-976.

21. Sacher M, Pfander B, Hoege C, Jentsch S (**2006**). Control of Rad52 recombination activity by double-strand break-induced SUMO modification. **Nat Cell Biol** 8(11): 1284-1290.

22. Rodriguez MS, Dargemont C, Hay RT (**2001**). SUMO-1 conjugation in vivo requires both a consensus modification motif and nuclear targeting. **J Biol Chem** 276(16): 12654-12659.

23. Colaert N, Helsens K, Martens L, Vandekerckhove J, Gevaert K (**2009**). Improved visualization of protein consensus sequences by iceLogo. **Nat Methods** 6(11): 786-787.

24. Mehrle A, Rosenfelder H, Schupp I, del Val C, Arlt D, Hahne F, Bechtel S, Simpson J, Hofmann O, Hide W, Glatting KH, Huber W, Pepperkok R, Poustka A, Wiemann S (**2006**). The LIFEdb database in 2006. **Nucleic Acids Res** 34(90001): D415-418.

14

Balanced CoQ$_6$ biosynthesis is required for lifespan and mitophagy in yeast

Isabel González-Mariscal, Aléjandro Martín-Montalvo, Cristina Ojeda-González, Adolfo Rodríguez-Eguren, Purificación Gutiérrez-Ríos, Plácido Navas, and Carlos Santos-Ocaña*

Centro Andaluz de Biología del Desarrollo, Universidad Pablo de Olavide-CSIC, CIBERER Instituto de Salud Carlos III, Sevilla, 41013, Spain.
* Corresponding Author:
C. Santos-Ocaña, E-mail: csanoca@upo.es

ABSTRACT Coenzyme Q is an essential lipid with redox capacity that is present in all organisms. In yeast its biosynthesis depends on a multiprotein complex in which Coq7 protein has both catalytic and regulatory functions. Coq7 modulates CoQ$_6$ levels through a phosphorylation cycle, where dephosphorylation of three amino acids (Ser/Thr) by the mitochondrial phosphatase Ptc7 increases the levels of CoQ$_6$. Here we analyzed the role of Ptc7 and the phosphorylation state of Coq7 in yeast mitochondrial function. The conversion of the three Ser/Thr to alanine led to a permanently active form of Coq7 that caused a 2.5-fold increase of CoQ$_6$ levels, albeit decreased mitochondrial respiratory chain activity and oxidative stress resistance capacity. This resulted in an increase in endogenous ROS production and shortened the chronological life span (CLS) compared to wild type. The null *PTC7* mutant (*ptc7Δ*) strain showed a lower biosynthesis rate of CoQ$_6$ and a significant shortening of the CLS. The reduced CLS observed in *ptc7Δ* was restored by the overexpression of *PTC7* but not by the addition of exogenous CoQ$_6$. Overexpression of *PTC7* increased mitophagy in a wild type strain. This finding suggests an additional Ptc7 function beyond the regulation of CoQ biosynthesis. Genetic disruption of *PTC7* prevented mitophagy activation in conditions of nitrogen deprivation. In brief, we show that, in yeast, Ptc7 modulates the adaptation to respiratory metabolism by dephosphorylating Coq7 to supply newly synthesized CoQ$_6$, and by activating mitophagy to remove defective mitochondria at stationary phase, guaranteeing a proper CLS in yeast.

Keywords: coenzyme Q$_6$, regulation, mitochondria, yeast, mitophagy, chronological life span.

Abbreviations:
CLS - chronological life span.

INTRODUCTION

Coenzyme Q (CoQ) deficiency is a syndrome that belongs to the family of mitochondrial diseases [1]. CoQ, a lipid embedded in cell membranes, main function is to act as an electron carrier and as an antioxidant. CoQ deficiency is classified by the observed phenotype as it is a multiple-caused syndrome [2]. There are primary and secondary CoQ deficiencies; the deficiency can be a consequence of mutations in genes involved directly in the CoQ biosynthesis (primary deficiency) [3, 4] or a consequence of defects or mutations in genes not directly related to CoQ biosynthesis (secondary deficiency) [5, 6]. The existence of secondary deficiencies introduces the idea of regulatory mechanisms of CoQ biosynthesis that could also be related to a general regulation of mitochondrial metabolism. The main defect caused by CoQ deficiency is a depletion of ATP

in tissues. The variety of symptoms and the diversity of CoQ functions introduce a source of complexity in the analysis of CoQ deficiency syndrome that requires the analysis of the mechanisms regulating the biosynthesis of CoQ [7].

In yeast, CoQ$_6$ is synthesized in the mitochondria after the condensation of an activated polyisoprenoid tail with a hydroxybenzoic ring; the ring is further modified by several reactions [8]. The enzymes that catalyze these reactions are encoded by nuclear *COQ* genes [9]. Most of the *COQ* genes encode proteins responsible for enzymatic activity; however, three proteins without enzymatic activity, Coq4, Coq8 and Coq9, play a structural role. Coq4 has been reported to support the assembly of the CoQ$_6$ biosynthetic complex in yeast [10]. Coq8 has been included in a family

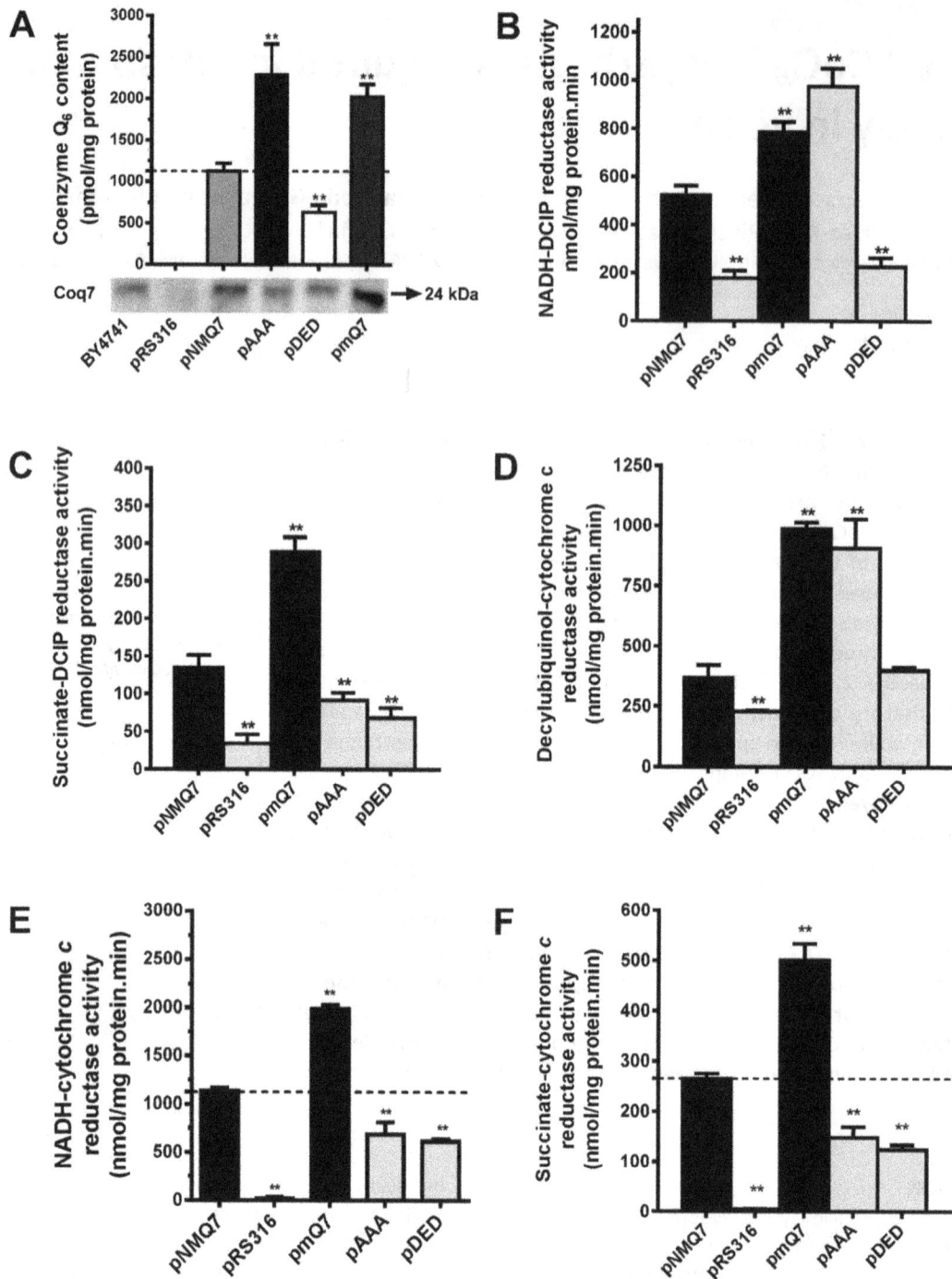

FIGURE 1: CoQ$_6$-dependent mitochondrial respiratory chain activities are negatively affected in yeast expressing Coq7 phosphomutant versions. Mitochondrial activities were measured with mitochondria purified from the $coq7\Delta$ strain harboring the indicated plasmids. pNMQ7 corresponds to the wild type *COQ7* gene, pRS316 corresponds to the empty vector, pAAA and pDED correspond to the Coq7 phosphomutant versions and pmQ7 corresponds to a multicopy expression of wild type *COQ7* gene. **(A)** Top panel: CoQ$_6$ quantification, bottom panel Western blots of mitochondrial samples probed with antibodies against Coq7. Mitochondrial activities: **(B)** NADH-DCIP reductase, **(C)** Complex II: succinate-DCIP reductase, **(D)** Complex III: Decylubiquinol-cytochrome *c* reductase, **(E)** NADH-cytochrome *c* reductase and **(F)** Complex II+III: Succinate-cytochrome *c* reductase. Results are expressed as nmol/mg mitochondrial protein·min. Data are mean ± SD, N ≥ 3 independent assays. ** P ≤ 0.001 compared to positive control samples.

of unusual kinases [11], which also includes other proteins involved in CoQ biosynthesis [12]. Recent analysis of the function of human ADCK3 protein, homologous of yeast Coq8, showed that the inhibition of ADCK3 kinase activity is required for the activation of CoQ_6 biosynthesis [13, 14]. Coq9 is a membrane protein located at the matrix side of the mitochondrial inner membrane and it belongs to the CoQ_6 biosynthetic complex, where it co-migrates with Coq3 and Coq4 at a molecular mass of approximately 1 MDa [15] and it binds to Coq7 to promote CoQ_6 biosynthesis [16].

Yeast biosynthesis of CoQ_6 occurs in a multi-protein complex (Q-synthome). The assembly of the Q-synthome requires the post-translational modification of Coq proteins. Several studies in the last years have demonstrated the existence of the Q-synthome [17–19] and several models for the assembly of the complex have been proposed [9, 20]. The complex assembly starts with a nucleation around the quinone-like lipid polyprenyl benzoate bound to a nucleating Coq protein such as Coq4. The nucleation step is ended with the assembly of a pre-complex that accumulates a CoQ_6 intermediate, the demethoxy quinone (DMQ_6) [19, 21]. DMQ_6 is converted to CoQ_6 after the activation of Coq7 by dephosphorylation [22]. Coq7 catalyzes the next to last reaction of the pathway [23], the DMQ_6 hydroxylation. Several studies have reported the existence of phosphoproteins in the family of Coq proteins: Coq3, Coq5 and Coq7 [24, 25], but only phosphorylation of Coq7 is known to have a physiological relevance [22]. Coq7 phosphorylation leads to a low activity state, therefore accumulating DMQ_6, while its dephosphorylation activates Coq7 and increases CoQ_6 levels. Both activation states of Coq7 can be achieved by changing the carbon source in the culture media [22]. These results were confirmed in COQ7 null mutants yeast strains (coq7Δ) expressing either a Coq7 version that is permanently dephosphorylated (Coq7-AAA), associated with a sharp increase of CoQ_6 concentration, or a permanently phosphorylated version (Coq7-DED) that is associated with a significant decrease of CoQ_6 levels [22]. The Coq7-AAA version was obtained by site-directed mutagenesis of residues S20, S28 and T32 to alanine, while in the Coq7-DED version these residues were mutated to glutamic or aspartic acid. Recent studies have demonstrated the presence of another phosphorylatable residue in Coq7, the S114 [26], whose modification affects the catalytic function of Coq7. The activation (i.e. dephosphorylation) of Coq7 is carried out by Ptc7, a phosphatase that belongs to the type 2C Mg^{2+} or Mn^{2+} dependent protein phosphatases, PPM [27, 28]. Other mitochondrial members of this family (Ptc5) have been related to the activation of pyruvate dehydrogenase (PDH) [29, 30] and with the activation of mitophagy (Ptc6) [31, 32]. Null mutants of PTC7 gene (ptc7Δ) have decreased levels of CoQ_6, decreased mitochondrial respiratory chain activities and decreased resistance to oxidative stress [28]. The expression of Coq7-AAA in ptc7Δ did not disrupt the high amount of CoQ_6 produced by this Coq7 version, demonstrating a relationship between Coq7 and Ptc7 [28].

Here we investigate the effect of both Coq7 mutants, Coq7-AAA and Coq7-DED on mitochondrial physiology and

its relationship with Ptc7 function. Our results indicate that although Coq7 mutants modify the mitochondrial physiology in a similar fashion than ptc7Δ relative to CoQ_6 levels, they have different effects on chronological life span, respiratory complexes interactions and on mitophagy activation.

RESULTS

Coq7-phosphorylation mutants show defects in mitochondrial respiratory chain activities

We measured CoQ_6 content and the activities of mitochondrial respiratory chain (MRC) in single or coupled complexes (Figure 1). The strain coq7Δ did not contain CoQ_6, which was rescued in the control strain (coq7Δ/pNMQ7). The strain expressing permanently dephosphorylated Coq7 (coq7Δ/pAAA) showed a dramatic increase of CoQ_6, while the strain expressing permanently phosphorylated Coq7 (coq7Δ/pDED) shows a significant decrease of CoQ_6 compared to control. Multicopy COQ7 transformed yeast (coq7Δ/pmQ7) also significantly increased CoQ_6 (Figure 1A). Under the experimental conditions and in all the strains analyzed, Coq7 was detected by western blotting at the expected size of 24 kDa. NADH-Q reductase activity, measured as NADH-DCIP reductase, was decreased in the coq7Δ/pDED strain in a similar manner than negative control (coq7Δ/pRS316), but in the coq7Δ/pAAA strain, the activity was increased over the control, equivalent to the activity measured in the coq7Δ/pmQ7 strain (Figure 1B). Complex II activity measured as succinate-DCIP reductase showed a moderated decrease in both coq7Δ/pDED and coq7Δ/pAAA strains (Figure 1C). However, complex II activity was increased significantly in the coq7Δ/pmQ7 strain. Complex III activity (decylubiquinol-cytochrome c reductase) showed changes comparable to those in complex I (Figure 1D). Coupled MRC activities require CoQ_6 as electron carrier, which is not added exogenously in the assay. Complexes activities such as NADH-cytochrome c reductase and succinate-cytochrome c reductase (Figures 1E and 1F) were decreased in both coq7Δ/pAAA and coq7Δ/pDED strains compared to control. Interestingly, the decrease in NADH-cytochrome c reductase activity in the coq7Δ/pAAA strain does not correlate with the increased activity observed in single complexes (Figure 1B and 1D) and with the high amount of CoQ_6 found in mitochondria of the coq7Δ/pAAA strain. The activities of these complexes in the coq7Δ/pmQ7 strain were significantly higher than in control.

Oxidative stress conditions in Coq7 phosphomutants

Due to the changes observed in MRC, we analyzed the endogenous oxidative stress, measured as H_2O_2 generation in mitochondria, from coq7Δ yeast expressing the different versions of Coq7 (Figure 2A). Expression of both Coq7-pAAA and Coq7-pDED showed an increased generation of H_2O_2 in mitochondria compared to control, while the coq7Δ/pmQ7 strain showed a decreased amount, even lower than control. To determine the oxidative stress pro-

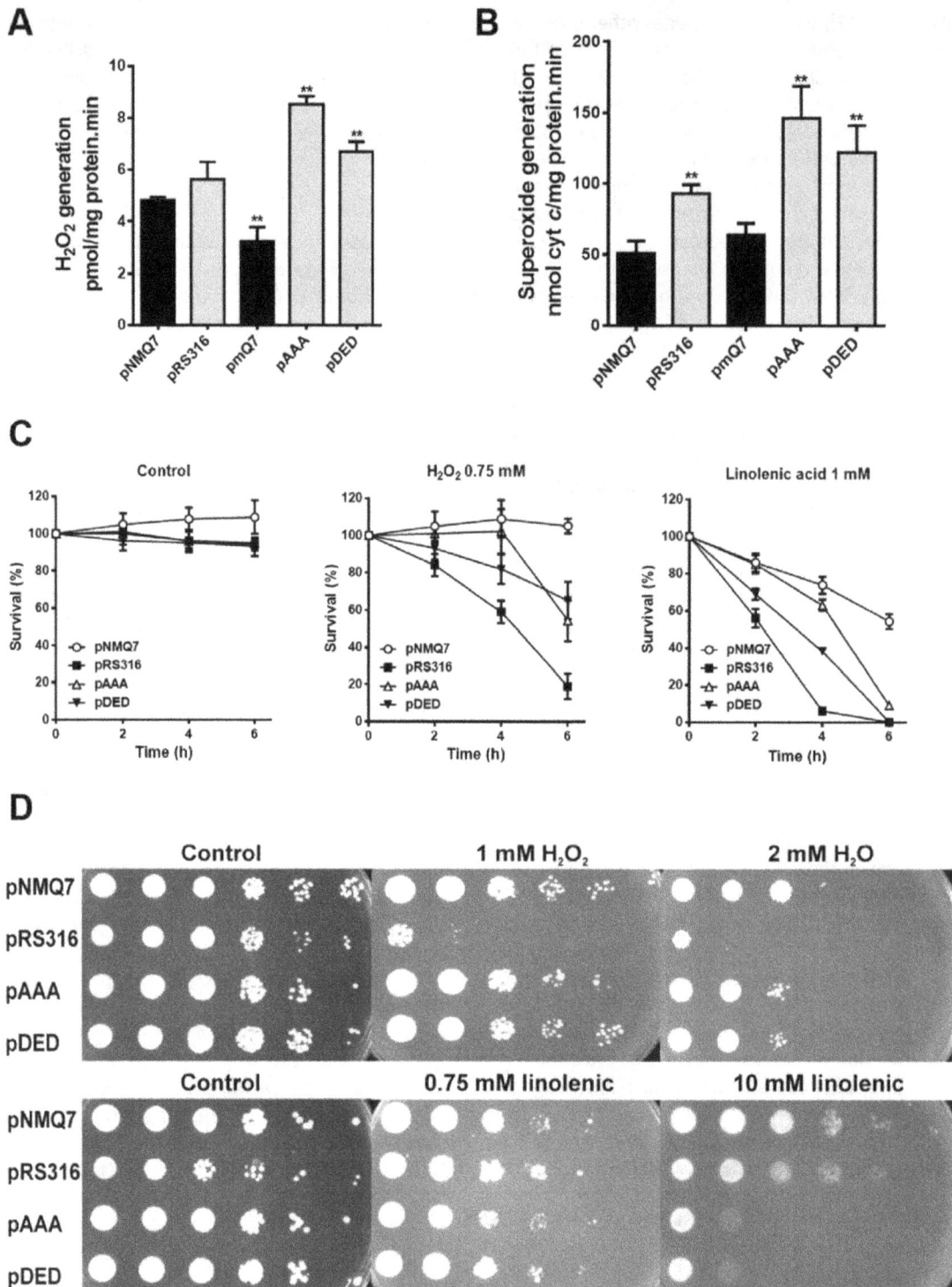

FIGURE 2: Expression of Coq7 phosphomutant versions increases both the endogenous generation and the sensitivity to oxidative stress. **(A)** Quantification of hydrogen peroxide generation was performed in presence of NADH and succinate as electron donors and 50 μg of purified mitochondria from the indicated strains. Data are mean ± SD, N ≥ 3 independent assays. **(B)** Superoxide anion generation was measured using acetylated cytochrome c as electron acceptor. Quantification was produced in presence of reduced decylubiquinone as electron donor and 50 μg of purified mitochondria from the indicated strains. Data are mean ± SD, N ≥ 3 independent assays. ** P ≤ 0.001 compared to positive control ($coq7\Delta$/pNMQ7). **(C)** Oxidative stress sensitivity was measured in the indicated strains. Cells were subjected to oxidative stress treatment by H_2O_2 (0.75 mM) or linolenic acid (1 mM) in 0.1 M sodium phosphate buffer plus 0.2 % glucose buffer at 150 x 10^6 cells/ml. At the indicated times cells were harvested, diluted in sterile water and seeded in triplicate in YPD plates to calculate the number of CFU. The experiment is representative of a set of three. **(D)** Cell viability was monitored under induced oxidative stress. Yeast cells were grown in SDc –ura glucose. After two days, cells were spotted by serial dilutions (1/10) from 0.5 OD_{660nm}/ml onto YPD plates containing the indicated concentrations of oxidative stress agents. Plates were incubated at 30°C for 3 days and imaged.

duced specifically by the complex III, we analyzed the superoxide anion generation using reduced decylubiquinone and acetylated cytochrome c (Figure 2B) [33]. Strains expressing both mutated versions of Coq7 produced significantly higher amounts of superoxide, from 200 to 400%, compared to wild type. Also, superoxide was higher in the $coq7\Delta$/pRS316 strain. On the contrary, the $coq7\Delta$/pmQ7 strain showed a superoxide production comparable to control.

In yeast, CoQ_6 acts as a powerful antioxidant and it is required to protect against lipophilic oxidants such as linolenic acid [34–36]. The sensitivity to either H_2O_2 or linolenic acid was analyzed in a $coq7\Delta$ strain expressing both Coq7 versions harboring phosphosite modifications (Figures 2C-D). Survival was not compromised in all strains without oxidative stress insult (Figure 2C) but after incubation with H_2O_2 or linolenic acid the wild type strain showed the higher survival rate at 2, 4 and 6 hours, while the negative control ($coq7\Delta$/pRS316) showed a lower survival rate. Interestingly, both treatments in $coq7\Delta$/pAAA and $coq7\Delta$/pDED strains produced a similar effect on survival, being higher than negative control but lower than control. A similar effect was found when the assay was performed in agar plates with H_2O_2 (Figure 2D); the lack of CoQ_6 in the $coq7\Delta$/pRS316 strain compromised the growth but $coq7\Delta$/pAAA and $coq7\Delta$/pDED growth was only slightly lower compared to control. When the assay was performed with linolenic acid at higher concentration, both coq7 mutants, $coq7\Delta$/pAAA and $coq7\Delta$/pDED, were more sensitive to the treatment, although both retain CoQ_6 production. That result is surprising since null Coq mutants do not produce CoQ_6 and show a high sensitivity to linolenic acid [35, 36]. In contrast, $coq7\Delta$/pAAA and $coq7\Delta$/pDED strains synthesize CoQ_6, with higher production than control for $coq7\Delta$/pAAA strain.

Chronological life span is differentially affected in Coq7 phosphomutants

Chronological lifespan (CLS) measurement refers to yeast cells longevity in an exhausted culture media after the onset of stationary phase [37]. The extension of CLS depends on the existence of an intact respiratory metabolism [38–41] and also by the expression of intact antioxidant defenses [42–44]. Given that these requirements are affected in cells expressing both Coq7-pAAA and Coq7-pDED, CLS have been measured in these cells (Figure 3A and Table S1). The $coq7\Delta$/pRS316 strain showed shorter mean CLS (2.8 ± 0.2 days) compared to both $coq7\Delta$/pNMQ7 (12.2 ± 0.7 days) and $coq7\Delta$/pmQ7 strains (14 ± 0.8 days). The $coq7\Delta$/pDED strain showed a slightly shorter mean CLS (11.4 ± 0.8 days) while the $coq7\Delta$/pAAA strain had a clearly shorter mean CLS (9.1 ± 0.7 days). These results compared to CoQ_6 content (Figure 1A) support a model where the phosphorylation state of Coq7 affects CLS in yeast. Comparing the levels of CoQ_6 and CLS in the Coq7 versions we found an assortment of results. Thus, CoQ_6 level alone is not a factor that can explain the changes observed in CLS. Ptc7 regulates CoQ_6 biosynthesis by dephosphorylation of Coq7 [28]. We studied whether the lack of Ptc7 would re-

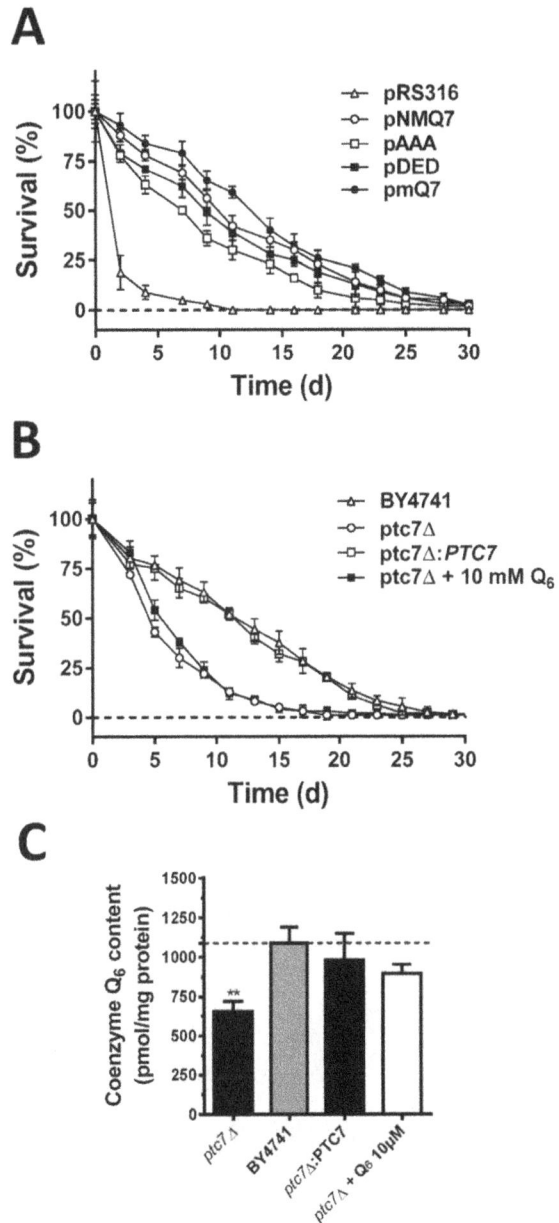

FIGURE 3: Chronological life span (CLS) is modified according to the phosphorylation state of Coq7. SDc-ura liquid cultures with 2% glucose as carbon source were inoculated at 0.1 OD_{660nm}/ml with the indicated strains. After 5 days of growth, samples of each culture were harvested to measure the cell viability in YPD plates. The third day was used as control (100%). Plot shows a representative experiment repeated three times with similar results. See full dataset in Supplementary Material. **(A)** CLS in *COQ7* strains. Average life was coq7Δ/pNMQ7 (12.1 ± 0.8), coq7Δ/pRS316 (2.8 ± 0.2), coq7Δ/pAAA (9.1 ± 0.7), coq7Δ/pDED (11.4 ± 0.8) and coq7Δ/pmQ7 (14 ± 0.8). **(B)** CLS in *PTC7* strains. Average life was BY4741 (12.7 ± 0.7), ptc7Δ (6.8 ± 0.4), ptc7Δ + 10 μM CoQ6 (6.9 ± 0.8) and ptc7Δ/PTC7 (12.7 ± 0.5). **(C)** CoQ6 of yeast strains analyzed in CLS experiment of *PTC7* gene strains. Similar data for *COQ7* gene CLS experiment was showed in Figure 1A. Cells from the indicated strains were subjected to mitochondrial purification, lipid extraction and quinone quantification. Samples were injected three times and the experiment was repeated at least three times.

sult in a compromised CLS. The *ptc7Δ* strain displayed a shortened mean CLS compared to wild type (6.8 ± 0.4 days versus 12.7 ± 0.7 days) (Figure 3B and Table S2). These results indicated that this phosphatase might regulate the normal function of longevity-associated pathways in yeast. CoQ$_6$ content of the *ptc7Δ* strain was significantly lower than both wild type and *ptc7Δ* strains rescued with the wild type allele (Figure 3C). As a result, CoQ$_6$ supplementation in CoQ$_6$ deficient *coq7Δ* yeast, rescues respiratory growth and oxidative stress resistance [45]. This was demonstrated by measuring the CLS of our strains in CoQ$_6$ supplemented media (Figure 3B). Interestingly, the addition of exogenous CoQ$_6$ to the *ptc7Δ* strain increased mitochondrial CoQ$_6$ to wild type levels (Figure 3C) but did not rescue CLS of the strain (7 ± 0.8 days), suggesting that the decreased CoQ$_6$ content is not responsible for the shortened CLS of the *ptc7Δ* strain. These results indicate that CoQ$_6$ levels cannot explain alterations of CLS in both *coq7Δ*

and *ptc7Δ* mutants. Moreover, the reduction in CLS in the *ptc7Δ* strain indicates that there might be additional functions of Ptc7 besides Coq7 activation.

Respiratory supercomplexes are altered in Coq7 phosphomutants

Mutant versions of Coq7 showed a clear defect on MRC activities focused in coupled reactions (NADH-cytochrome *c* reductase and succinate-cytochrome *c* reductase) that cannot be explained by the levels of CoQ$_6$. In the *coq7Δ/pAAA* strain the NADH-cytochrome *c* reductase activity is 40% less than wild type while the simple activity of both separated complexes is even higher than wild type. However, *coq7Δ/pAAA* mitochondria contain 250% of CoQ$_6$ compared to wild type (Figure 1). One possibility is that the expression of Coq7 versions can modify the stability or assembly of respiratory complexes. To this end, we analyzed the assembly of respiratory complexes by BN-PAGE in

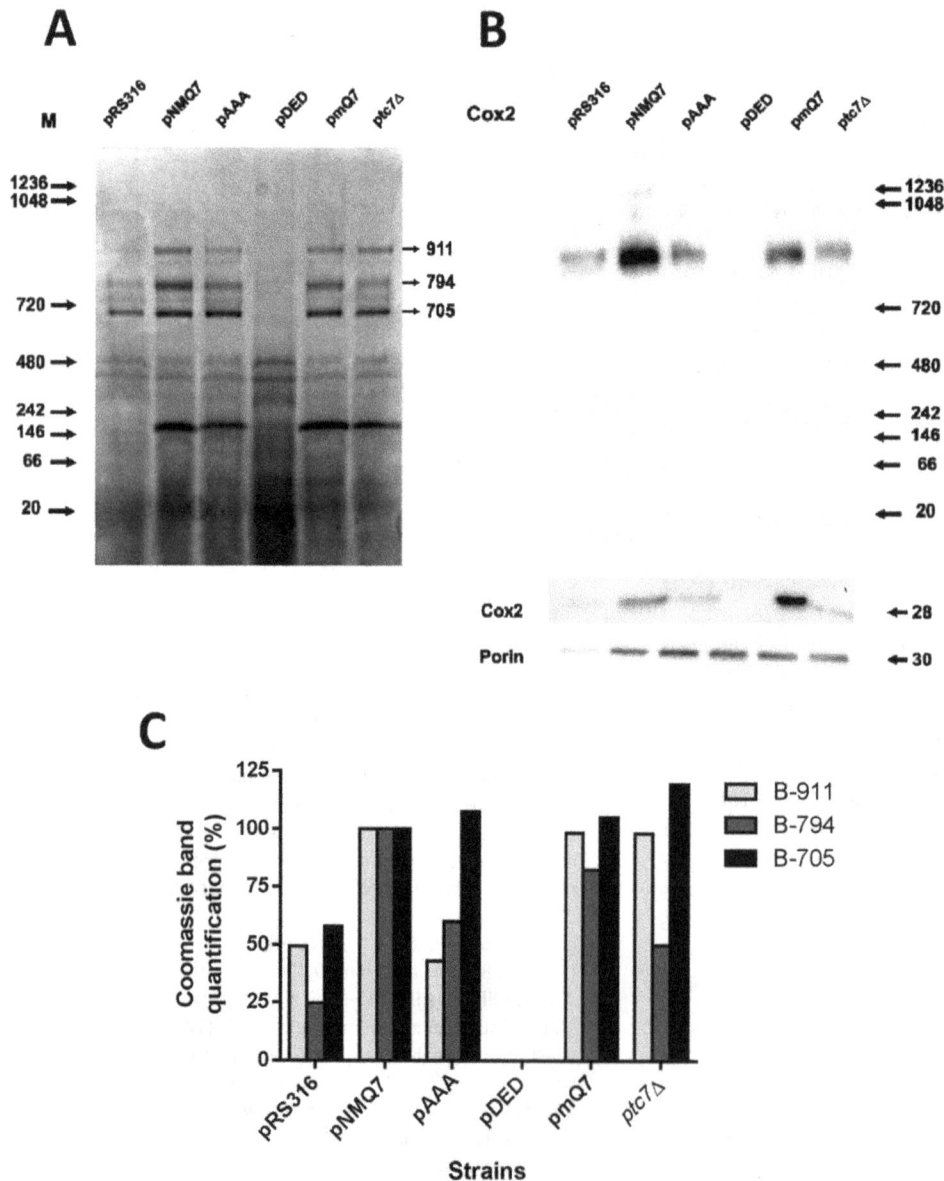

FIGURE 4: Mitochondrial supercomplexes stability is affected in yeast expressing Coq7 phosphomutant versions. Pure mitochondrial samples obtained from the indicated strains were subjected to digitonin solubilization and 200 μg of each sample were analyzed by 1D BN-PAGE. **(A)** Coomassie staining of 1D BN-PAGE gels. Molecular weight markers are indicated on the left. On the right side, the three arrows indicate the calculated molecular weight of bands analyzed by densitometry in Figure 4C. **(B)** Upper panel, Western blot analysis blotted with antibody against the subunit Cox2 of Complex IV of a 1D BN-PAGE gel obtained in parallel to the Coomassie stained gel shown in Figure 4A using the same original mitochondrial digitonin solubilized samples. Lower panel, analysis by SDS-PAGE and Western blot of mitochondrial samples blotted with antibodies against the subunit Cox2 of Complex IV and porin. **(C)** Densitometry analysis of bands indicated with arrows in 4A. The analysis was performed with the software Image Lab 4.0 of Biorad. Band optical density was normalized with the full lane density and compared with the optical density of bands corresponding to the positive control, *coq7Δ/pNMQ7*. Results are representative of a set of two.

permeabilized mitochondria isolated from *coq7Δ* yeast expressing several versions of Coq7 (Figure 4). Cells were cultured first in YPD to increase cell mass and then the culture was transferred to YPG to activate mitochondrial metabolism. The *coq7Δ*/pNMQ7 strain showed a typical profile of mitochondrial supercomplexes that mostly agrees with previously published data in wild type yeast [46], with three prominent bands at 911, 794 and 705 kDa, a double band around 480 kDa and a band located at 242 kDa (Figure 4A). The mass-spectrometry analysis of the three larger bands of *coq7Δ*/pNMQ7 yeast indicated that the 911 kDa band corresponds mainly to complex V, the band of 794 kDa to complex III + complex IV and 705 KDa correspond to complex V. The detailed list of detected and identified proteins can be consulted at the Table S3 of Supplementary Material. Full MASCOT analysis is included in Supplementary Material. Immunoblot of this BN-PAGE gel with anti-Cox2 antibody (Figure 4B) showed differential intensities of supercomplex at 794 kDa but not in other bands. No reaction was observed in mitochondria from *coq7Δ*/pDED yeast, and very low in mitochondria from *coq7Δ*/pRS316 yeast. Lower intensities were also found in both *coq7Δ*/pAAA and *ptc7Δ* strains. Higher intensity was observed in control *coq7Δ*/pNMQ7 and *coq7Δ*/pmQ7 strains. The quantitative analysis of Coomassie staining showed that the 911 kDa and 794 kDa bands were significantly affected in both Coq7 mutants (Figure 4C). The expression of modified versions of Coq7 induced alterations in the assembly profile of respiratory complexes, being

more dramatic in the *coq7Δ*/pDED strain.

Mitophagy but not autophagy is affected in *ptc7Δ* mutant strains

Mitophagy is a general process that degrades damaged or non-useful mitochondria, which is required to maintain CLS in eukaryotic cells [31, 47, 48]. The *ptc7Δ* strain combines a low amount of CoQ_6 with a low CLS while the *coq7Δ*/pDED strain shows a low amount of CoQ_6 but a CLS comparable to wild type (Figures 1A and 3A). Both strains are equivalent in terms of Coq7 phosphorylation state and therefore Ptc7 must have another function independent of CoQ_6 biosynthesis that is responsible for the low CLS measured. We have previously demonstrated that the survival of human fibroblasts deficient in CoQ_{10} production depends on a proper recycling of dysfunctional mitochondria by mitophagy [49]. We speculated that Ptc7 might regulate this process. Therefore, we analyzed whether the general process of autophagy and/or mitophagy could be compromised in *ptc7Δ*. Macroautophagy was analyzed by monitoring the Atg8p proteolysis using a plasmid expressing Atg8-GFP tag at the C-terminal under endogenous promoter (ATG8-GFP) (Figure 5A) [50]. Yeast cells were grown in YPD for 16 hours and subsequently were resuspended in nitrogen deprived medium (SDc–N) to induce macroautophagy. The positive control strain (BY4741; wild type) resulted in a marked increase of Atg8-GFP degradation, visible as free-GFP starting at 2 hours, which was maintained until 72 hours. According to previous reports, *pep4Δ* and *atg5Δ* strains,

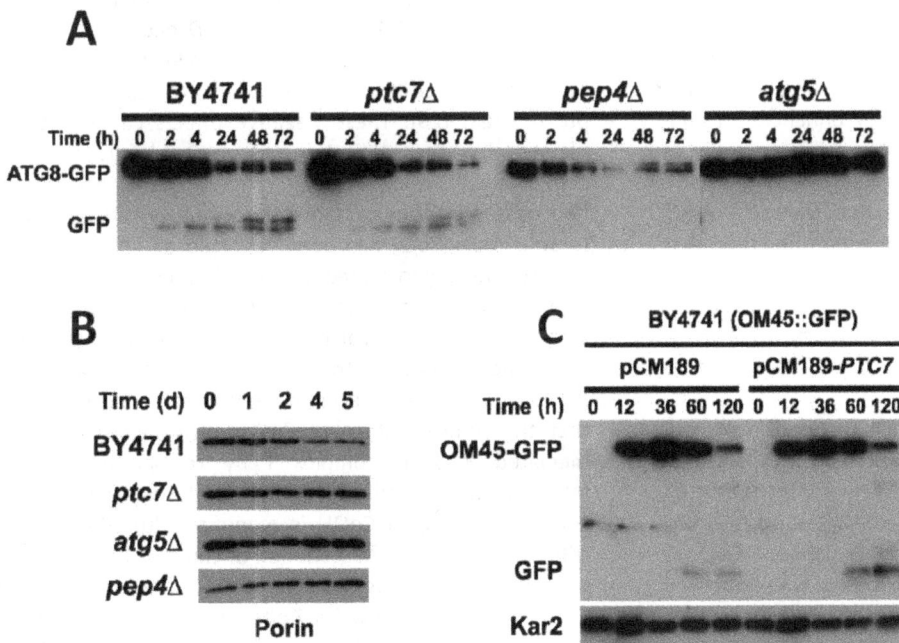

FIGURE 5: Mitophagy induction but no autophagy is compromised in Ptc7 yeast. (A) Yeast strains harboring an expression plasmid expressing the ATG8-GFP gene were grown in YPD 16 hours. Then, yeasts were washed and media was replaced by SDc-N with 2% glucose to activate autophagy. 0.5 $OD_{660\,nm}$ of Yeast cells were harvested at indicated time points. Protein extracts were analyzed by SDS-PAGE and Western blot using anti-GFP antibody. (B) Mitophagy analysis by porin degradation. Yeast strains were grown in YPD for 16 hours. Then, yeasts were washed and media was replaced by SDc-N with 2% glucose to activate mitophagy. Protein extracts were analyzed using anti-porin antibody. (C) Ptc7 overexpression increases the mitophagy induction. Yeast strains expressing the OM45-

GFP gene were grown in YPD for 16 hours. Then, yeasts were washed and media was replaced by YP Lactate. Protein extracts were analyzed using anti-GFP antibody for GFP-Om45, an outer membrane mitochondrial protein, and anti-Kar2 antibody for Kar2, an endoplasmic reticulum marker used as control for non-mitochondrial protein degradation. Original film plates used to prepare the panels are shown in Figure S1. Analysis of densitometry of panels B and C are shown in the Figure S2, S3 and S4. Data are mean ± SD, N ≥ 3 independent assays in all the experiments.

which are deficient in protein degradation in the vacuole or autophagy respectively, showed impaired autophagy induction [51] (Figure 5A). In accordance with induced autophagy, a decrease of intact Atg8-GFP and a subsequent increase on free GFP was also observed in ptc7Δ strain, indicating that macroautophagy induction is not compromised in this strain.

To further study the potential role of Ptc7 in mitochondrial recycling, mitophagy was analyzed by monitoring the degradation of porin. Yeasts were grown in YPL for 16 hours and media was replaced with a nitrogen deprived media (SDc−N) to induce mitophagy (Figure 5B). The wild type strain showed a decreased amount of porin starting at day 2 and was gradually reduced until day 5. Porin degradation was not observed in the ptc7Δ strain under similar conditions, suggesting that Ptc7 is involved in mitochondrial recycling. Porin degradation was not observed in the autophagy-deficient yeast strains atg5Δ and pep4Δ. A protein expression analysis normalized with total protein loaded corroborated previous data (Figure 2S). To further determine the involvement of Ptc7 in mitophagy induction we determined the effect of Ptc7 overexpression on mitophagy (Figure 5C). Wild type yeast harboring the GFP protein fused in frame with the mitochondrial outer membrane protein Om45 were transformed with the empty yeast expression plasmid (pCM189) or containing the yeast PTC7 coding sequence (pCM189-PTC7). Yeast cells were cultured in YPD for 16 hours and media was replaced with YPL, a non-fermentable carbon source, to induce mitochondrial biosynthesis. Yeast cells were grown through prolonged stationary phase to induce the selective recycling of mitochondria by mitophagy [31]. PTC7 overexpression produced increased GFP free levels starting at 60 hours (270%) and at 120 hours (470%) of growth, indicating that the over-expression of PTC7 enhances mitophagy induction (Figure S3). Remarkably, Kar2, a marker of endoplasmic reticulum, was not affected (Figure S4), indicating that macroautophagy was not activated by PTC7 overexpression. Taken together, these data indicate that Ptc7 regulates or participates specifically in mitophagy but not in general macroautophagy, suggesting that this specific process may compromise CLS in ptc7Δ yeast.

DISCUSSION

Our understanding of the role of mitochondria on cell physiology and metabolism has evolved in the past decades from only a bioenergetics role to an interconnected organelle whose functions exceed energy supply. A key factor for part of these mitochondrial functions is CoQ, as proper regulation of its biosynthesis pathway exceeds cell bioenergetics and is tightly linked to cellular homeostasis. This idea supports the pleiotropic effect observed in patients with primary CoQ deficiency [7, 52]. Although most Coq proteins are involved in enzymatic steps required for CoQ_6 biosynthesis [8], some Coq proteins play a structural function and are required to stabilize the biosynthetic complex [8, 53]. In yeast, the shift from fermentation to respiration activates the expression of COQ genes to accommodate CoQ_6 biosynthesis to respiratory metabolism

[19, 54], which is associated to regulatory mechanisms such as biosynthesis complex assembly and post-translational modifications [9, 22, 25, 55].

The protein encoded by the COQ7 gene has its catalytic activity controlled by complex regulatory mechanisms [16, 22, 25, 28, 56, 57]. Coq7 is a di-iron carboxylate protein with hydroxylase activity [58] that converts DMQ_6 in demethyl ubiquinone. The lack of Coq7 in yeast fully abolishes CoQ_6 biosynthesis and other intermediates although the expression of some point mutants such as e2519 (E223K) accumulates DMQ_6 [34], which would indicate that this step is a regulatory step in this pathway. DMQ_6 is also accumulated in wild type cells after the post-diauxic shift supporting this hypothesis [19, 34]. A model of CoQ_6 biosynthesis complex assembly, based on BN-PAGE, size exclusion chromatography and immunoprecipitation data, show that a precomplex of about 700 kDa is formed, which accumulates DMQ_6 [10] that, ultimately, will be converted into CoQ_6 after the recruitment of Coq7 to the complex [9, 55]. Coq7 is a phosphoprotein that, in the dephosphorylated state, activates CoQ_6 biosynthesis probably by interacting with the 700 kDa precomplex [22]. Here we have shown that the phosphomimetic Coq7 (Coq7-DED), which mimics Coq7 phosphorylation status in ptc7Δ strain [28], and the fully dephosphorylated Coq7 (Coq7-AAA) induced a decreased and an increased levels of CoQ_6 respectively, indicating that the regulatory step on CoQ_6 biosynthesis is at least partially controlled by Coq7 phosphorylation.

The physiological analysis of these strains confirms the catalytic function of Coq7 in CoQ_6 biosynthesis and its regulatory function in mitochondrial metabolism. Expression of both Coq7-AAA and Coq7-DED decreases antioxidant protection and increases the production of endogenous oxidative stress. A similar result was reported for the ptc7Δ strain [28]. These negative effects cannot be explained by the alteration of CoQ_6 levels, as the coq7Δ/pmQ7 strain, which also shows high levels of CoQ_6, had an endogenous oxidative stress and sensitivity similar to wild type. CoQ_6 has been reported as an antioxidant molecule mainly to protect cell membranes [35, 36, 59] but also as a pro-oxidant agent under physio-pathological conditions [60]. CoQ is involved in ROS production in the MRC mainly because of the transfer of electrons to complex III [61, 62]. It has been recently demonstrated in Drosophila that the reduced stage of CoQ ($CoQH_2$) causes $CoQH_2$-mediated ROS in complex I by retrograde electron transport and contributes to extend lifespan [63]. Similar conditions of higher $CoQH_2$ in mammal-cultured cells contribute to the partial degradation of complex I by the same mechanism [64]. Thus, we propose that the production of CoQ_6 in strains with low Coq7 phosphorylation are generating unbalanced CoQ_6 levels and redox stages out of the MRC complexes, which does not occur in the coq7Δ/pmQ7 strain. The unbalanced levels of CoQ_6 in these strains increase their sensitivity to external oxidative stress, decreasing their longevity. It has been reported previously that linolenic acid can induce mitochondrial oxidative stress [65]. The effect of linolenic acid may be enhanced by the mitochondrial dysfunction generated by the expression of pAAA and pDED versions. In fact, the expression of pAAA or pDED induces a higher level of endogenous oxidative stress and also affects the stability of respiratory complexes, which possibly make those strains more susceptible to linolenic acid-induced oxidative stress.

CoQ is a component of the respirasome and it is proposed that pools of CoQ are bound to specific super assembly stages of respiratory complexes [66], which depends on the carbon source [64]. Here our data show that super-complexes are dissociated in yeast strains with unbalanced CoQ_6 concentrations, which also showed mitochondrial dysfunction, but not in the COQ7 multicopy transformed yeast strain (coq7Δ/pmQ7) that exhibit a phenotype similar to wild type in both respiratory activities and assembly profile. Previous analyses of CoQ_6 biosynthesis complex showed that Coq7 is partially located in a large size complex but it can also be detected in smaller ones even as a monomer [10]. Coq7 interacts with Coq9, which shows lipid-binding activity and Coq7 interacting domains [16]. Several steps are required to integrate the CoQ_6 biosynthesis complex in yeast. First, there is an initial nucleation of Coq proteins around Coq4 [10, 67] to build up the Q-synthome, which requires stabilization by Coq8 [68], an unusual protein kinase that makes the pre-complex formation [20] and demethoxy-Q_6 accumulation possible [19]. At this step Coq7 is mostly phosphorylated and it is not a component of the pre-complex [10]; Coq7 must then be dephosphorylated by Ptc7 to get activated and to bind to the fully active Q-synthome [9, 28, 55]. We speculate that CoQ_6 biosynthesis must be balanced with the components of the respiratory complexes. In light of our data, we hypothesize that the interaction of Coq7 with other proteins of the biosynthesis complex is a requirement for its integration in the MRC, and that extreme phosphorylation stages might prevent this possibility. This is supported by the data obtained from the ptc7Δ strain that mimics Coq7-DED expressing cells without eliminating all regulation of Coq7.

In fact, yeast contains two other mitochondrial phosphatases that belong to the same family of Ptc7, Ptc5 that participates in pyruvate dehydrogenase complex (PDH) regulation [30, 69] and Ptc6/Aup1 that participates in PDH regulation and is required for mitophagy activation [31, 70]. It is possible that these mitochondrial phosphatases may dephosphorylate Coq7 in the absence of Ptc7, which would explain the less severe effect on mitochondrial functions. Interestingly, the ptc7Δ strain shows shorten CLS compared to the coq7Δ/pDED strain and it is not rescued by CoQ_6 supplementation, although this strain shows a significant decrease of CoQ_6 content. CLS is only rescued when it is transformed with the homologous gene, which is also able to rescue the CoQ_6 content; overall these data indicate that Ptc7 has at least a dual function in yeast. As indicated above, Ptc6 is required for mitophagy activation [31, 70] and we have shown here that Ptc7 induces mitophagy as a mechanism to recycle defective mitochondria caused by CoQ_6 deficient MRC.

The lack of mitophagy in ptc7Δ yeast, which is a requisite to extend CLS [71, 72], can explain its shorten CLS and the negative recovery after CoQ_6 supplementation. We propose a dual and temporary differential function of Ptc7 on mitochondrial physiology and homeostasis (Figure 6). The entry of yeast on the post-diauxic shift (PDS) increases CoQ_6 biosynthesis and therefore respiratory metabolism by Coq7 activation. However, cell homeostasis during PDS requires the recycling of the excess and defective mitochondria. Ptc7, which would trigger mitophagy by dephosphorylating an unknown target involved in this process, would be a key regulator of mitochondria homeostasis in yeast by coordinating mitochondria recycling with CoQ_6 biosynthesis.

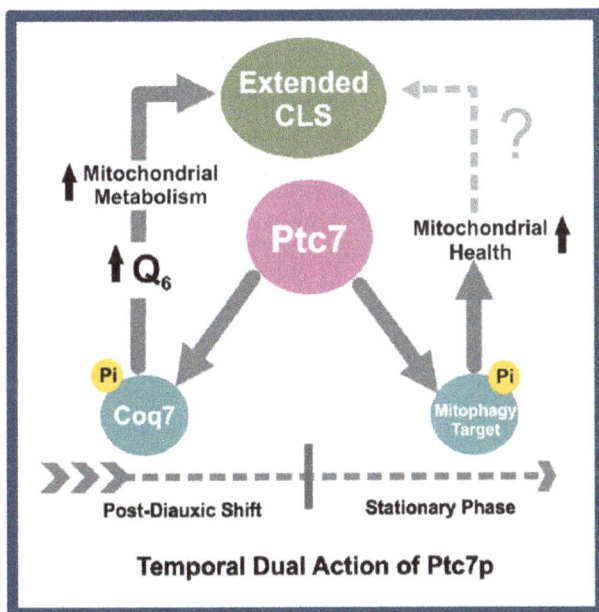

FIGURE 6: Model of Ptc7/Coq7 action to promote CLS extension. Ptc7 can regulate mitochondrial metabolisms in two different ways that are not directly related. After post diauxic shift, Coq7 expression and its activation by Ptc7 dephosphorylation leads to high levels of CoQ_6 concomitant with an increase of mitochondrial biogenesis. CLS extension requires a respiratory growth phase previous to the entrance in the stationary phase. The excess of mitochondria after the start of the stationary phase can be recycled by mitophagy to maintain an appropriated mitochondrial homeostasis. Ptc7 participates on mitophagy most likely by dephosphorylating of unknown target(s). Both functions, at different time frames, can help to promote a CLS extension in yeast.

MATERIAL AND METHODS

Yeast strains and growth media

Yeast strains used in this study are listed in Table 1. Growth media for yeast and bacteria were prepared as described previously [22]. Yeast cells were grown at 30°C with shaking (200 rpm).

Mitochondrial purification and BN-PAGE

Yeast cultures were grown in the appropriate culture media and mitochondria were purified according to the described method [73]. To solubilize mitochondria, 2 mg of pure mito-

TABLE 1. Strains used in this study.

Strains	Genotype	Source
coq7Δ	BY4741; MATa; his3Δ1; leu2Δ0; met15Δ0; ura3Δ0; YOR125c::KanMX4	Euroscarf
JM8	MAT α; Δade1;ρ⁰	[74]
JM6	MAT a; Δhis4; ρ⁰	[74]
ptc7Δ	BY4742; MATα; his3Δ1; leu2Δ0; lys2Δ0; ura3Δ0; YHR076w::KanMX4	Euroscarf
BY4742	MATα; his3Δ1; leu2Δ0; lys2Δ0; ura3Δ0.	Euroscarf
atg5Δ	BY4742, MATα; his3Δ1; leu2Δ0; lys2Δ0; ura3Δ0; YPL149w::kanMX4	Euroscarf
pep4Δ	BY4742; MATα; his3Δ1; leu2Δ0; lys2Δ0; ura3Δ0, YPL154c::kanMX4	Euroscarf
OM45::GFP	BY4741; MATα; his3Δ1; leu2Δ0; met15Δ0; ura3Δ0; YIL136W::GFP(S65T)-HIS3MX6	ATCC

chondria was incubated in 240 µl of solubilization buffer containing digitonin in a ratio 4:1 with protein, 1 mM PMSF, 10% glycerol, 150 mM potassium acetate and 30 mM HEPES, pH 7.4 for 30 min at 4°C. Solubilized samples were subjected to two rounds of centrifugation in a Beckman Coulter Microfuge 22R (15,000 × g, 15 min, at 4°C) and the supernatant was collected to BN-PAGE. Proteins quantification was performed by Bradford method (Biorad). BN-PAGE was performed with precast 3–12% gradient gels (NativePAGE™ Novex® Bis-Tris Gels) using the Xcell Sure Lock™ Mini-Cell electrophoresis system, including NativePAGE™ anode and cathode buffers according to the company instructions at 4°C. Lanes were loaded with 200 µg of mitochondrial solubilized supernatant and as MW marker NativeMark™ Unstained Protein Ladder was used. Gels were stained with Coomassie solution or were blotted onto PVDF, blocked in 5% Blocking Reagent (Biorad)

and phosphate-buffered saline. Proteins were detected by ECL using Luminata Crescendo (Millipore) and luminescence detected by Gel Doc XR+ image processing software (Bio-Rad). Cox2p (Novex Life Technologies), porin (Invitrogen) and Coq7 (Gift of Dr. C. F. Clarke, UCLA, USA) primary antibodies were used at 1: 2,000, 1: 1,000 and 1: 2,000 respectively. Goat anti-mouse and anti-rabbit secondary antibodies conjugated to horseradish peroxidase (Calbiochem) were used at a 1:5,000 dilutions.

Mitochondrial respiratory chain (MRC) activities
Fresh mitochondria were used to measure NADH-cytochrome *c* reductase, succinate-cytochrome *c* reductase activities and superoxide generation. Other MRC activities (NADH-DCIP reductase, succinate-DCIP reductase and decylubiquinol-cytochrome *c* reductase), were performed with fresh samples

TABLE 2. Vectors used in this study.

Name	Description	Specification	Source
pRS316	Yeast expression centromeric vector with *URA3* auxotrophy	Empty control	[34]
pRS426	Yeast expression episomic vector with *URA3* auxotrophy	Empty control	[34]
pNMQ7	pRS316 harboring the full yeast *COQ7* with promotor and terminator sequences	Positive control	[34]
pAAA	pRS316 with loss of function mutations in *COQ7* (S20A, S28A and S32A)	Non-phosphorylatable *COQ7* version	[22]
pDED	pRS316 with gain of function mutations in *COQ7* (S20D, S28E and S32D)	Mimicking a permanent phosphorylated *COQ7* version	[22]
pmQ7	pRS426 harboring yeast *COQ7*	*COQ7* multicopy complementation	[34]
ATG8-GFP	pRS316 harboring *ATG8*-GFP tag	*ATG8* tagged with GFP	[75]
pCM189	Yeast centromeric expression vector with *URA3* auxotrophy	Empty control	[76]
pCM189-PTC7	pCM189 harboring yeast *PTC7*	*PTC7* null mutant complementation	[28]

subjected to one freeze-thaw cycle. All MRC activities were determined according to previously published methods [34]. Superoxide generation was measured using the same method than for measuring complex III activity (decylubiquinol-cytochrome c reductase) but using acetylated cytochrome c instead cytochrome c [33]. H_2O_2 generation was performed using Amplex Red kit (Invitrogen) according to the manufacturer instructions.

Chronological Life Span (CLS)

Analysis was performed as previously described [37]. Briefly, cells were incubated in YPD and CLS was monitored starting at day 3 by quantification of colony forming units (CFUs) every 48 hours using the software OpenCFU 3.9 beta. The number of CFUs at day three was considered as 100% survival. Survival log-rank analyses Sigmastat 3.0 (SPSS) were calculated for each pair of lifespan analyses and average lifespan were shown in the corresponding dataset in Supplementary Material.

Protein mass spectrometry identification

Acrylamide gel bands were distained in NH_4HCO_3 25 mM water/ACN 50:50. For reduction of cysteines, samples were incubated at 56°C for 60 min in 10 mM DTT (NH_4HCO_3 25 mM). Cysteine carbamidomethylation was performed embedding bands in 55 mM IAA solution and incubating at room temperature for 30 min. Samples were in gel digested by trypsin (0.2 µg/µl in 1 mM HCl) diluted in NH_4HCO_3. Gel bands were covered with enzyme solution and incubated at 30°C overnight. Reaction was stopped with acetonitrile, and peptide were extracted adding 0.2% TFA. Prior to protein identification by MALDI-MS, we used nano-liquid chromatography for reversed phase peptide separation. Peptide fractioning was performed in a Proxeon EASY-nLC II apparatus with a C18 column (EASY-column, 75 µm x 100 mm) and mobile phases: Buffer A: 0.1% TFA (H_2O) and Buffer B: 0.1% TFA (ACN). Elution process was divided in the following flow steps: 0-48 min gradient 2%-45% B; 48-50 min gradient 45-100% B; 50-60 min isocratic 100% B. Fractions were collected every 15" using a Bruker Proteineer fc fraction collector and spotted onto a MALDI target plate. 192 samples were spotted and overlaid with 0.5 µl drops of HCCA matrix solution and left air dry. MALDI measurements were performed in a Bruker Ultraflextreme MALDI-TOF/TOF system, using Bruker Peptide Calibration standards as mass standards. A MALDI fingerprint spectrum was obtained for each fraction and peaks with higher intensity in each spot were selected as mass precursors for MS/MS peptide frag-

mentation experiments. Bruker WARP-LC software was used to process spectra and to integrate chromatography fractions data. Protein identification was carried out using MASCOT server. Ammonium bicarbonate, DL-dithiothreitol (DTT), Iodo-acetamide (IAA), trypsin from porcine pancreas, trifluoroacetic acid (TFA) and α-cyano-hydroxycinnamc acid (HCCA) were purchased from Sigma-Aldrich. Water (HPLC grade) and acetonitrile (HPLC grade) were purchased from Fluka. Peptide calibration standards were purchased from Bruker.

Total yeast protein extraction

Cells (10.10^6 in 500 µl of water) were disrupted with 100 µl 2 M NaOH and 35% β-mercaptoethanol for 15 min on ice. Proteins were precipitated after the addition of 100 µl 3 M TCA for 15 min on ice. The pellet obtained after 15 min centrifugation on a microcentrifuge at full speed was washed with acetone, dried and resuspended in 30 µl of SDS-PAGE 1 x LB.

Other methods

CoQ_6 quantification was performed using mitochondrial samples according to previously published methods [28]. Densitometry analysis was carried out with a Gel Doc XR+ (Bio-Rad) with Image Lab 4.0 as software analysis. Statistical (t-Student) analyses were carried out using the Sigmastat 3.0 (SPSS) statistical package. Mitochondrial DNA integrity was checked in all strains using two ρ^0 strains, JM6 and JM8 strains. All results are expressed as the average ± SD. Statistical analyses were carried out using the Sigmastat 3.0 (SPSS) statistical package.

ACKNOWLEDGMENTS

The research group is funded by the Andalusian Government as the BIO177 group through FEDER funds (European Commission), by the Ministerio de Economía y Competitividad, Instituto de Salud Carlos III (FIS PI14/01962) and by the International Q_{10} Association.

AMMS received a predoctoral fellowship from the Consejería de Innovación Ciencia y Empresa, Junta de Andalucía (Spain). IGM received a predoctoral fellowship from the Plan Propio of the Universidad Pablo de Olavide de Sevilla.

The authors thank the group components for the critical reading of the manuscript and also to Ana Sanchez Cuesta for her technical help.

CONFLICT OF INTEREST

The authors declare no conflict of interest.

REFERENCES

1. Doimo M, Desbats M a., Cerqua C, et al (**2014**) Genetics of Coenzyme Q10 Deficiency. **Mol Syndromol** 156–162.

2. Emmanuele V, López LC, López L, et al (**2012**) Heterogeneity of coenzyme Q10 deficiency: patient study and literature review. **Arch Neurol** 69:978–83.

3. DiMauro S, Quinzii CM, Hirano M (**2007**) Mutations in coenzyme Q10 biosynthetic genes. **J Clin Invest** 117:587–9.

4. Peng M, Falk MJ, Haase VH, et al (**2008**) Primary Coenzyme Q Deficiency in Pdss2 Mutant Mice Causes Isolated Renal Disease. **PLoS Genet** 4:e1000061.

5. Quinzii CM, Hirano M (**2012**) Primary and secondary CoQ(10) deficiencies in humans. **Biofactors** 37:361–5.

6. Yubero D, Montero R, Martín MA, et al (**2016**) Secondary coenzyme Q10 deficiencies in oxidative phosphorylation (OXPHOS) and non-OXPHOS disorders. **Mitochondrion** 30:51–58.

7. Santos-Ocaña C, Salviati L, Navas P (**2015**) The genes of CoQ10. In: Iain P. Hargreaves, Ph.D. (Neurometabolic Unit, National Hospital, Queen Square, London, UK) Assistant Editor: April K. Hargreaves, Ph.D. (Trinity University, Dublin I (ed) Coenzyme Q10 From fact to Fict., 1st ed. Nova Science Publisher, New York, pp 205–226

8. Tran UC, Clarke CF (**2007**) Endogenous synthesis of coenzyme Q in eukaryotes. **Mitochondrion** 7 Suppl:S62-71.

9. González-Mariscal I, García-Testón E, Padilla S, et al (**2014**) Regulation of coenzyme Q biosynthesis in yeast: A new complex in the block. **IUBMB Life** 66:63–70.

10. Marbois B, Gin P, Gulmezian M, Clarke CF (**2009**) The yeast Coq4 polypeptide organizes a mitochondrial protein complex essential for coenzyme Q biosynthesis. **Biochim Biophys Acta - Mol Cell Biol Lipids** 1791:69–75.

11. Leonard CJ, Aravind L, Koonin E V (**1998**) Novel families of putative protein kinases in bacteria and archaea: evolution of the "eukaryotic" protein kinase superfamily. **Genome Res** 8:1038–1047.

12. Macinga DR, Cook GM, Poole RK, Rather PN (**1998**) Identification and characterization of aarF, a locus required for production of ubiquinone in Providencia stuartii and Escherichia coli and for expression of 2'-N-acetyltransferase in P. stuartii. **J Bacteriol** 180:128–135.

13. Stefely J a, Reidenbach AG, Ulbrich A, et al (**2015**) Mitochondrial ADCK3 Employs an Atypical Protein Kinase-like Fold to Enable Coenzyme Q Biosynthesis. **Mol Cell** 57:83–94.

14. Stefely JA, Licitra F, Laredj L, et al (**2016**) Cerebellar Ataxia and Coenzyme Q Deficiency through Loss of Unorthodox Kinase Activity. **Mol Cell** 63:608–620.

15. Hsieh EJ, Gin P, Gulmezian M, et al (**2007**) Saccharomyces cerevisiae Coq9 polypeptide is a subunit of the mitochondrial coenzyme Q biosynthetic complex. **Arch Biochem Biophys** 463:19–26.

16. Lohman DC, Forouhar F, Beebe ET, et al (**2014**) Mitochondrial COQ9 is a lipid-binding protein that associates with COQ7 to enable coenzyme Q biosynthesis. Proc. Natl. Acad. Sci. U. S. A.

17. Gin P, Clarke CF (**2005**) Genetic evidence for a multi-subunit complex in coenzyme Q biosynthesis in yeast and the role of the Coq1 hexaprenyl diphosphate synthase. **J Biol Chem** 280:2676–81.

18. Marbois B, Gin P, Faull KF, et al (**2005**) Coq3 and Coq4 define a polypeptide complex in yeast mitochondria for the biosynthesis of coenzyme Q. **J Biol Chem** 280:20231–20238.

19. Padilla S, Tran UC, Jiménez-Hidalgo M, et al (**2009**) Hydroxylation of demethoxy-Q6 constitutes a control point in yeast coenzyme Q6 biosynthesis. **Cell Mol Life Sci** 66:173–86.

20. He CH, Xie LX, Allan CM, et al (**2014**) Coenzyme Q supplementation or over-expression of the yeast Coq8 putative kinase stabilizes multi-subunit Coq polypeptide complexes in yeast coq null mutants. **Biochim Biophys Acta** 1841:630–644.

21. Xie LX, Ozeir M, Tang JY, et al (**2012**) Over-expression of the Coq8 kinase in Saccharomyces cerevisiae coq null mutants allows for accumulation of diagnostic intermediates of the Coenzyme Q6 biosynthetic pathway. **J Biol Chem** 287:23571–81.

22. Martin-Montalvo A, Gonzalez-Mariscal I, Padilla S, et al (**2011**) Respiratory-induced coenzyme Q biosynthesis is regulated by a phosphorylation cycle of Cat5p/Coq7p. **Biochem J** 440:107–114.

23. Clarke CF (**1996**) The COQ7 Gene Encodes a Protein in Saccharomyces cerevisiae Necessary for Ubiquinone Biosynthesis. **J Biol Chem** 271:2995–3004.

24. Tauche A, Krause-Buchholz U, Rödel G, Rodel G (**2008**) Ubiquinone biosynthesis in Saccharomyces cerevisiae: the molecular organization of O-methylase Coq3p depends on Abc1p/Coq8p. **FEMS Yeast Res** 8:1263–75.

25. Xie LX, Hsieh EJ, Watanabe S, et al (**2011**) Expression of the human atypical kinase ADCK3 rescues coenzyme Q biosynthesis and phosphorylation of Coq polypeptides in yeast coq8 mutants. **Biochim Biophys Acta** 1811:348–360.

26. Busso C, Ferreira-Júnior JR, Paulela JA, et al (**2015**) Coq7p relevant residues for protein activity and stability. **Biochimie** 119:92–102.

27. Barford D, Das AK, Egloff MP (**1998**) The structure and mechanism of protein phosphatases: insights into catalysis and regulation. **Annu Rev Biophys Biomol Struct** 27:133–164.

28. Martin-Montalvo A, Gonzalez-Mariscal I, Pomares-Viciana T, et al (**2013**) The Phosphatase Ptc7 Induces Coenzyme Q Biosynthesis by Activating the Hydroxylase Coq7 in Yeast. **J Biol Chem** 288:28126–28137.

29. Krause-Buchholz U, Gey U (**2006**) YIL042c and YOR090c encode the kinase and phosphatase of the Saccharomyces cerevisiae pyruvate dehydrogenase complex. **FEBS Lett** 580:2553–2560.

30. Gey U, Czupalla C, Hoflack B, et al (**2008**) Yeast pyruvate dehydrogenase complex is regulated by a concerted activity of two kinases and two phosphatases. **J Biol Chem** 283:9759–9767.

31. Tal R, Winter G, Ecker N, et al (**2007**) Aup1p, a yeast mitochondrial protein phosphatase homolog, is required for efficient stationary phase mitophagy and cell survival. **J Biol Chem** 282:5617–5624.

32. Ruan H, Yan Z, Sun H, Jiang L (**2007**) The YCR079w gene confers a rapamycin-resistant function and encodes the sixth type 2C protein phosphatase in Saccharomyces cerevisiae. **FEMS Yeast Res** 7:209–215.

33. Santos-Ocaña C, Villalba JM, Córdoba F, et al (**1998**) Genetic evidence for coenzyme Q requirement in plasma membrane electron transport. **J Bioenerg Biomembr** 30:465–75.

34. Padilla S, Jonassen T, Jiménez-Hidalgo MA, et al (**2004**) Demethoxy-Q, an intermediate of coenzyme Q biosynthesis, fails to support respiration in Saccharomyces cerevisiae and lacks antioxidant activity. **J Biol Chem** 279:25995–6004.

35. Poon WW, Do TQ, Noelle Marbois B, Clarke CF (**1997**) Sensitivity to treatment with polyunsaturated fatty acids is a general characteristic of the ubiquinone-deficient yeast coq mutants. **Mol Aspects Med** 18:121–127.

36. Do TQ, Schultz JR, Clarke CF (**1996**) Enhanced sensitivity of ubiquinone-deficient mutants of Saccharomyces cerevisiae to products of autoxidized polyunsaturated fatty acids. **Proc Natl Acad Sci U S A** 93:7534–9.

37. Fabrizio P, Longo VD (**2003**) The chronological life span of Saccharomyces cerevisiae. **Aging Cell** 2:73–81.

38. Barros MH, Bandy B, Tahara EB, Kowaltowski AJ (**2004**) Higher respiratory activity decreases mitochondrial reactive oxygen release and increases life span in Saccharomyces cerevisiae. **J Biol Chem** 279:49883–8.

39. Aerts AM, Zabrocki P, Govaert G, et al (**2009**) Mitochondrial dysfunction leads to reduced chronological lifespan and increased apoptosis in yeast. **FEBS Lett** 583:113–117.

40. Breitenbach M, Laun P, Dickinson JR, et al (**2012**) The role of mitochondria in the aging processes of yeast. **Subcell Biochem** 57:55–78.

41. Ocampo A, Liu J, Schroeder EA a, et al (**2012**) Mitochondrial respiratory thresholds regulate yeast chronological life span and its extension by caloric restriction. **Cell Metab** 16:55–67.

42. Longo VD, Liou L-L, Valentine JS, Gralla EB (**1999**) Mitochondrial Superoxide Decreases Yeast Survival in Stationary Phase. **Arch Biochem Biophys** 365:131–142.

43. Fabrizio P, Liou L-L, Moy VN, et al (**2003**) SOD2 functions downstream of Sch9 to extend longevity in yeast. **Genetics** 163:35–46.

44. Gonidakis S, Longo VD (**2008**) Oxidative Stress and Aging in the

Budding Yeast Saccharomyces cerevisiae. In: Miwa S, Beckman KB, Muller FL (eds) Oxidative Stress Aging. Humana Press, Totowa, NJ, pp 67–80

45. Santos-Ocaña C, Do TQ, Padilla S, et al (**2002**) Uptake of exogenous coenzyme Q and transport to mitochondria is required for bc1 complex stability in yeast coq mutants. J. Biol. Chem. 277:

46. Schägger H, Pfeiffer K, Scha H, et al (**2000**) Supercomplexes in the respiratory chains of yeast and mammalian mitochondria. **EMBO J** 19:1777–83.

47. Youle RJ, Narendra DP (**2011**) Mechanisms of mitophagy. **Nat Rev Mol Cell Biol** 12:9–14.

48. Richard VR, Leonov A, Beach A, et al (**2013**) Macromitophagy is a longevity assurance process that in chronologically aging yeast limited in calorie supply sustains functional mitochondria and maintains cellular lipid homeostasis. **Aging (Albany NY)** 5:234–69.

49. Rodriguez-Hernandez A, Cordero MD, Salviati L, et al (**2009**) Coenzyme Q deficiency triggers mitochondria degradation by mitophagy. **Autophagy** 5:19–32.

50. Shintani T, Klionsky DJ (**2004**) Autophagy in health and disease: a double-edged sword. **Science (80-)** 306:990–995.

51. Kissova I, Deffieu M, Manon S, et al (**2004**) Uth1p is involved in the autophagic degradation of mitochondria. **J Biol Chem** 279:39068–39074.

52. Laredj LN, Licitra F, Puccio HM (**2014**) The molecular genetics of coenzyme Q biosynthesis in health and disease. **Biochimie** 100:78–87.

53. Kawamukai M (**2015**) Biosynthesis of coenzyme Q in eukaryotes. **Biosci Biotechnol Biochem** 8451:1–11.

54. Hagerman RA, Willis RA (**2002**) The yeast gene COQ5 is differentially regulated by Mig1p, Rtg3p and Hap2p. **Biochim Biophys Acta** 1578:51–58.

55. Allan CM, Awad AM, Johnson JS, et al (**2015**) Identification of Coq11, a new coenzyme Q biosynthetic protein in the CoQ-synthome in Saccharomyces cerevisiae. **J Biol Chem** jbc.M114.633131.

56. Brea-Calvo G, Siendones E, Sanchez-Alcazar JA, et al (**2009**) Cell survival from chemotherapy depends on NF-kappaB transcriptional up-regulation of coenzyme Q biosynthesis. **PLoS One** 4:e5301.

57. Cascajo M V, Abdelmohsen K, Noh JH, et al (**2015**) RNA-binding proteins regulate cell respiration and coenzyme Q biosynthesis by post-transcriptional regulation of COQ7. **RNA Biol** 0.

58. Stenmark P, Grünler J, Mattsson J, et al (**2001**) A new member of the family of di-iron carboxylate proteins. Coq7 (clk-1), a membrane-bound hydroxylase involved in ubiquinone biosynthesis. **J Biol Chem** 276:33297–300.

59. Maroz A, Anderson RF, Smith RA, Murphy MP (**2009**) Reactivity of ubiquinone and ubiquinol with superoxide and the hydroperoxyl radical: implications for in vivo antioxidant activity. **Free Radic Biol Med** 46:105–109.

60. Nohl H, Gille L, Schönheit K, Liu Y (**1996**) Conditions allowing redox-cycling ubisemiquinone in mitochondria to establish a direct redox couple with molecular oxygen. **Free Radic Biol Med** 20:207–213.

61. Boveris A, Cadenas E, Stoppani AOCQ sintesi. y. funcio. (**1976**) Role of ubiquinone in the mitochondrial generation of hydrogen peroxide. **J Biol Chem** 156:435–444.

62. Turrens JF, Alexandre A, Lehninger AL (**1985**) Ubisemiquinone is the electron donor for superoxide formation by complex III of heart mitochondria. **Arch Biochem Biophys** 237:408–414.

63. Scialo F, Mallikarjun V, Stefanatos R, Sanz A (**2013**) Regulation of Lifespan by the Mitochondrial Electron Transport Chain: Reactive Oxygen Species-Dependent and Reactive Oxygen Species-Independent Mechanisms. **Antioxid Redox Signal** 19:1953–1969.

64. Guarás A, Perales-Clemente E, Calvo E, et al (**2016**) The CoQH2/CoQ Ratio Serves as a Sensor of Respiratory Chain Efficiency. **Cell Rep** 15:197–209.

65. Colquhoun A, Schumacher RI (**2001**) gamma-Linolenic acid and eicosapentaenoic acid induce modifications in mitochondrial metabolism, reactive oxygen species generation, lipid peroxidation and apoptosis in Walker 256 rat carcinosarcoma cells. **Biochim Biophys Acta** 1533:207–219.

66. Lapuente-Brun E, Moreno-Loshuertos R, Acin-Perez R, et al (**2013**) Supercomplex Assembly Determines Electron Flux in the Mitochondrial Electron Transport Chain. **Science (80-)** 340:1567–1570.

67. González-Mariscal I, García-Testón E, Padilla S, et al (**2014**) The regulation of coenzyme q biosynthesis in eukaryotic cells: all that yeast can tell us. **Mol Syndromol** 5:107–18.

68. Do TQ, Hsu a Y, Jonassen T, et al (**2001**) A defect in coenzyme Q biosynthesis is responsible for the respiratory deficiency in Saccharomyces cerevisiae abc1 mutants. **J Biol Chem** 276:18161–8.

69. Ariño J, Casamayor A, González A, et al (**2011**) Type 2C protein phosphatases in fungi. **Eukaryot Cell** 10:21–33.

70. Journo D, Mor A, Abeliovich H (**2009**) Aup1-mediated regulation of Rtg3 during mitophagy. **J Biol Chem** 284:35885–35895.

71. Abeliovich H (**2011**) Stationary-Phase Mitophagy in Respiring Saccharomyces cerevisiae. **Antioxid Redox Signal** 14:2033–2011.

72. Sampaio-marques B, Burhans WC, Ludovico P (**2014**) Longevity pathways and maintenance of the proteome : the role of autophagy and mitophagy during yeast ageing. **Microb Cell** 1:118–127.

73. Glick BS, Pon LA (**1995**) Isolation of highly purified mitochondria from Saccharomyces cerevisiae. **Methods Enzymol** 260:213–223.

74. Proft M, Kötter P, Hedges D, et al (**1995**) CAT5, a new gene necessary for derepression of gluconeogenic enzymes in Saccharomyces cerevisiae. **EMBO J** 14:6116–26.

75. Shintani T, Klionsky DJ (**2004**) Cargo Proteins Facilitate the Formation of Transport Vesicles in the Cytoplasm to Vacuole Targeting Pathway. **J Biol Chem** 279:29889–29894.

76. Belli G, Gari E, Piedrafita L, et al (**1998**) An activator/repressor dual system allows tight tetracycline-regulated gene expression in budding yeast [published erratum appears in Nucleic Acids Res 1998 Apr 1;26(7):following 1855]. **Nucleic Acids Res** 26:942–947.

15

The frequency of yeast [*PSI⁺*] prion formation is increased during chronological ageing

Shaun H. Speldewinde[1] and Chris M. Grant[1],*

[1] University of Manchester, Faculty of Biology, Medicine and Health, The Michael Smith Building, Oxford Road, Manchester, M13 9PT, UK.
* Corresponding Author:
Chris M. Grant, The University of Manchester, Faculty of Life Sciences, The Michael Smith Building, Oxford Road; Manchester, M13 9PT, UK; E-mail: chris.grant@manchester.ac.uk

ABSTRACT Ageing involves a time-dependent decline in a variety of intracellular mechanisms and is associated with cellular senescence. This can be exacerbated by prion diseases which can occur in a sporadic manner, predominantly during the later stages of life. Prions are infectious, self-templating proteins responsible for several neurodegenerative diseases in mammals and several prion-forming proteins have been found in yeast. We show here that the frequency of formation of the yeast [*PSI⁺*] prion, which is the altered form of the Sup35 translation termination factor, is increased during chronological ageing. This increase is exacerbated in an *atg1* mutant suggesting that autophagy normally acts to suppress age-related prion formation. We further show that cells which have switched to [*PSI⁺*] have improved viability during chronological ageing which requires active autophagy. [*PSI⁺*] stains show increased autophagic flux which correlates with increased viability and decreased levels of cellular protein aggregation. Taken together, our data indicate that the frequency of [*PSI⁺*] prion formation increases during yeast chronological ageing, and switching to the [*PSI⁺*] form can exert beneficial effects via the promotion of autophagic flux.

Keywords: prion, yeast, chronological ageing, autophagy.

INTRODUCTION

Biological ageing involves a progressive decline in the ability of an organism to survive stress and disease. It is a complex process which is influenced by both genetic and environmental factors [1]. Common features of ageing include decreased resistance to stress, increased rates of apoptosis, a decline in autophagy and an elevated accumulation of protein aggregates [2]. In humans, ageing correlates with an increased frequency of age-related diseases including heart disease, metabolic syndromes and neurodegenerative diseases such as Alzheimer's, Parkinson's and dementia [3].

Prions are aberrant, infectious protein conformations which can self-replicate [4]. They are causally responsible for transmissible spongiform encephalopathies (TSEs) that cause several incurable neurodegenerative diseases in mammals [5]. The underlying cause of TSEs is the structural conversion of a soluble prion protein (PrP^C) into a prion form (PrP^sc) that is amyloidogenic. The amyloid form of the protein can subsequently convert other soluble molecules into the prion form thus resulting in the accumulation of the aberrant proteins in neuronal cells [6, 7]. Human prion diseases are predominantly sporadic constituting approximately 70% of all cases with higher frequencies occurring during advanced age [8]. There are several prion-forming proteins in yeast with the best-characterized being [*PSI⁺*] and [*PIN⁺*], which are formed from the Sup35 and Rnq1 proteins, respectively [9, 10]. [*PSI⁺*] is the altered conformation of the Sup35 protein, which normally functions as a translation termination factor during protein synthesis. The *de novo* formation of [*PSI⁺*] is enhanced by the presence of the [*PIN⁺*] prion, which is the altered form of the Rnq1 protein whose native protein function is unknown [11].

We have previously shown that autophagy protects against Sup35 aggregation and *de novo* [*PSI⁺*] prion formation [12]. Autophagy is an intracellular quality control pathway that degrades damaged organelles and protein aggregates via vacuolar/lysosomal degradation [13]. It proceeds in a highly sequential manner leading to the formation of a double-membrane-bound vesicle called the autophagosome. Fusion of the autophagosome with vacuoles/lysosomes introduces acidic hydrolases that degrade the contained proteins and organelles. Autophagy has been implicated in the ageing process and, for example, pharmacological interventions which induce autophagy result in lifespan extension during yeast chronological ageing [14, 15]. Autophagy appears to play a protective role in the ageing process since dysregulated autophagy is implicated in the accumulation of abnormal proteins associated with several age-related diseases including Alzheimer's, Parkinson's and Huntington's diseases [3, 16, 17].

Yeast cells can survive for prolonged periods of time in culture and have been used as a model of the chronological life span (CLS) of mammals, particularly for tissues composed of non-dividing populations [18]. Additionally, ageing is followed by replicative lifespan, which is defined as the number of budding daughter cells that originates from a particular mother yeast cell before it reaches senescence [19, 20]. Given that amyloidoses are typically diseases of old-age, yeast prions might be expected to form at a higher frequency in aging yeast cells. However, a study using the yeast replicative ageing model found that ageing does not increase the frequency of prion formation [21]. In this current study we have examined [PSI⁺] prion formation using the yeast CLS model and found that the frequency of prion formation is increased during ageing. Furthermore, this frequency is elevated in an autophagy mutant suggesting that autophagy normally acts to suppress age-dependent prion formation. We show that the prion-status of cells influences CLS in an autophagy-dependent manner suggesting that prions can be beneficial in aged populations of yeast cells.

RESULTS

The frequency of *de novo* [*PSI*⁺] formation increases during chronological ageing

We examined yeast CLS to determine whether there is an increased frequency of [*PSI*⁺] appearance during ageing. Cultures were grown to stationary phase in liquid SCD media and prion formation measured over time. [*PSI*⁺] prion formation was quantified using the *ade1-14* mutant allele which confers adenine auxotrophy and prions differentiated from nuclear gene mutations by their irreversible elimination in guanidine hydrochloride (GdnHCl). The frequency

of *de novo* [*PSI*⁺] prion formation in a control [*PIN*⁺][*psi*⁻] strain grown to stationary phase was estimated to be 1.1×10^{-5} comparable to previously reported frequencies [12, 22]. This frequency increased during CLS and a 39-fold increase was observed by day 12 (Fig. 1). We next examined the frequency of [*PSI*⁺] prion formation in an *atg1* autophagy mutant. Atg1 is a serine/threonine kinase which is responsible for the initiation of autophagy [13, 23]. The frequency of [*PSI*⁺] prion formation was further elevated in the *atg1* mutant compared with a wild-type strain, suggesting that autophagy normally acts to suppress [*PSI*⁺] prion formation during ageing (Fig. 1).

[*PSI*⁺] increase longevity in a yeast CLS model

We next examined cell survival to determine whether the [*PSI*⁺] prion status of cells influences longevity. For this analysis, flow cytometry was used to monitor propidium iodide uptake to assess yeast cell death [24]. The [*PIN*⁺][*PSI*⁺] strain showed a modest increase in maximal lifespan compared with a [*PIN*⁺][*psi*⁻] strain (Fig. 2A). Additionally, cell death was lower in the [*PIN*⁺][*PSI*⁺] strain over the entire lifespan compared with the [*PIN*⁺][*psi*⁻] strain suggesting that the [*PSI*⁺] prion improves viability during ageing. To verify this difference in ageing between [*PIN*⁺][*PSI*⁺] and [*PIN*⁺][*psi*⁻] strains, viability was monitored using colony formation assays. Whilst lifespan measured using the colony forming assay was shorter compared with the propidium iodide uptake assay, it confirmed that the [*PSI*⁺] strain maintained viability longer and had an increased lifespan compared with the [*psi*⁻] strain (Fig. 2B).

Treating cells with GdnHCl blocks the propagation of most yeast prions by inhibiting the key ATPase activity of Hsp104, a molecular chaperone that is absolutely required for yeast prion propagation [25, 26]. GdnHCl cures yeast cells of [*PSI*⁺] and all known prions. Curing the [*PIN*⁺][*PSI*⁺] strain with GdnHCl dramatically decreased maximal lifespan to 10 days suggesting that prions are beneficial during CLS (Fig. 2A). It should be noted however, that GdnHCl treatment can also potentially affect Hsp10-related

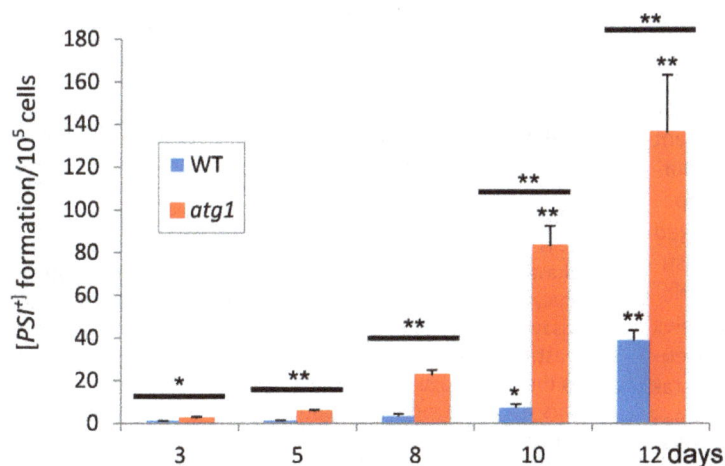

FIGURE 1: Increased frequency of [*PSI*⁺] prion formation during chronological lifespan. [*PIN*⁺][*psi*⁻] versions of the wild-type and *atg1* mutant stains were grown to stationary phase in SCD media and the frequency of [*PSI*⁺] formation measured over time. [*PSI*⁺] formation was quantified using the *ade1-14* mutant allele by growth on media lacking adenine and differentiated from nuclear mutations by their irreversible elimination in GdnHCl. Data shown are the means of three independent biological repeat experiments expressed as the number of colonies per 10^5 viable cells. Error bars denote standard deviation. Significance is shown comparing the wild-type and *atg1* mutant strains over time (above bars) as well as between the wild-type and *atg1* mutant at each time-point (between bars). Statistical analysis was performed by one-way ANOVA: *$p < 0.05$, **$p < 0.01$.

processes which are unrelated to prions. Autophagy is known to be required for chronological longevity and for example loss of *ATG1* reduces CLS [27]. Similarly, we found that loss of *ATG1* reduced CLS in the 74D-694 yeast strain used for our studies (Fig. 2A). Interestingly, longevity was comparable in the [*PIN*$^+$][*PSI*$^+$], [*PIN*$^+$][*psi*$^-$] and cured strains indicating that active autophagy is required for prion-dependent effects on longevity.

[*PSI*$^+$] cells have an increased rate of autophagy and decreased concentrations of amorphous protein aggregates

Given that the [*PSI*$^+$]-prion status of cells affects CLS in an autophagy-dependent manner, we examined whether autophagy is altered in [*PSI*$^+$] cells. We utilized a GFP-Atg8 fusion construct to follow the autophagy-dependent proteolytic liberation of GFP from GFP-Atg8, which is indicative of autophagic flux [28]. In late exponential phase cells (day 1), more free GFP was detected in the [*PIN*$^+$][*PSI*$^+$] strain compared with the [*PIN*$^+$][*psi*$^-$] and cured [*pin*$^-$][*psi*$^-$] strains indicative of increased autophagic flux (Fig. 3A). By day three, there was also an increase in autophagic flux in the [*PIN*$^+$][*psi*$^-$] and cured [*pin*$^-$][*psi*$^-$] strains detected as an increase in free GFP. As expected, no autophagic activity was observed for the *atg1* knockout mutants. This suggests that cells carrying the [*PSI*$^+$] prion have increased autophagic activity which may be beneficial during CLS.

We next examined whether the increased autophagic activity in the [*PIN*$^+$][*PSI*$^+$] strain affects amorphous protein aggregation, which seemed likely given the previous studies which have suggested that autophagy plays a role in the clearance of misfolded and aggregated proteins [16, 17]. For this analysis, we used a biochemical approach where we grew cells to stationary phase (day 3) and isolated aggregated proteins by differential centrifugation, and removed any contaminating membrane proteins using detergent washes [29]. The levels of protein aggregation were decreased in the [*PIN*$^+$][*PSI*$^+$] strain compared with the [*PIN*$^+$][*psi*$^-$] and cured [*pin*$^-$][*psi*$^-$] strains (Fig. 3B). This reduction in protein aggregation required autophagy since the levels of protein aggregation were comparable in the *atg1*-mutant version strains.

DISCUSSION

The majority of prion disease cases in humans occur in a sporadic manner, predominantly manifesting during later stages of life [8]. Similarly, we found an age-dependent increase in the frequency of *de novo* [*PSI*$^+$] formation during yeast chronological ageing, with the frequency of spontaneous formation increased approximately 40-fold in aged cells. Yeast has emerged as a powerful model to investigate the stochasticity of the ageing process and its contributing factors. The yeast CLS model more closely resembles the ageing of non-dividing cells such as neurons [18]. Neuronal cells are post-mitotic in nature and rely on proteostasis mechanisms such as autophagy to facilitate the elimination of superfluous and damaged material. The age-dependent increase in the *de novo* formation of [*PSI*$^+$] was exacerbated in an *atg1* mutant suggesting that autophagy

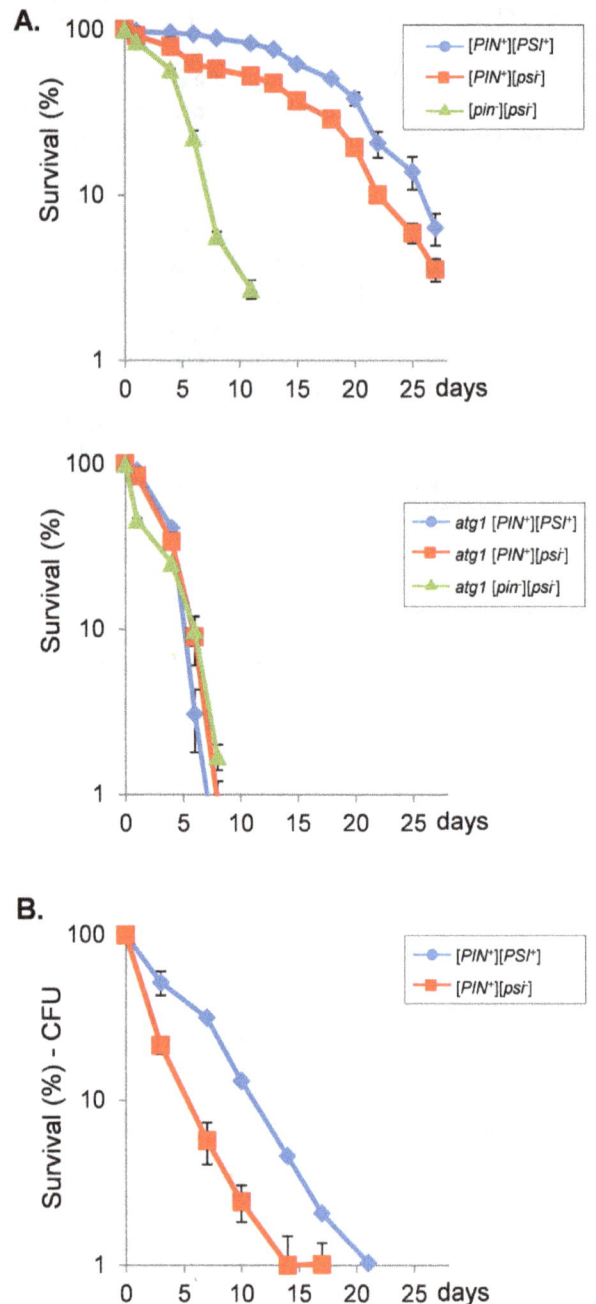

FIGURE 2: Prions improve chronological lifespan. (A) Chronological lifespan analysis, as determined by propidium iodide uptake to assess yeast cell death, is shown for [*PIN*$^+$][*PSI*$^+$], [*PIN*$^+$][*psi*$^-$] and cured [*pin*$^-$][*psi*$^-$] versions of wild-type and *atg1* mutant strains. Cells were grown in SCD media for 3 days to reach stationary phase and then aliquots taken every 2-3 days for flow cytometry analysis based on propidium iodide uptake by non-viable cells as assayed through flow cytometry. Data shown are the means of at least three independent biological repeat experiments expressed as the percentage of viable cells out of 10000 cells analyzed. Error bars denote standard deviation. **(B)** Viability measurements are shown for [*PIN*$^+$][*PSI*$^+$] and [*PIN*$^+$][*psi*$^-$] strains grown under the same conditions as for (A) above. Viability is expressed as a percentage of day zero.

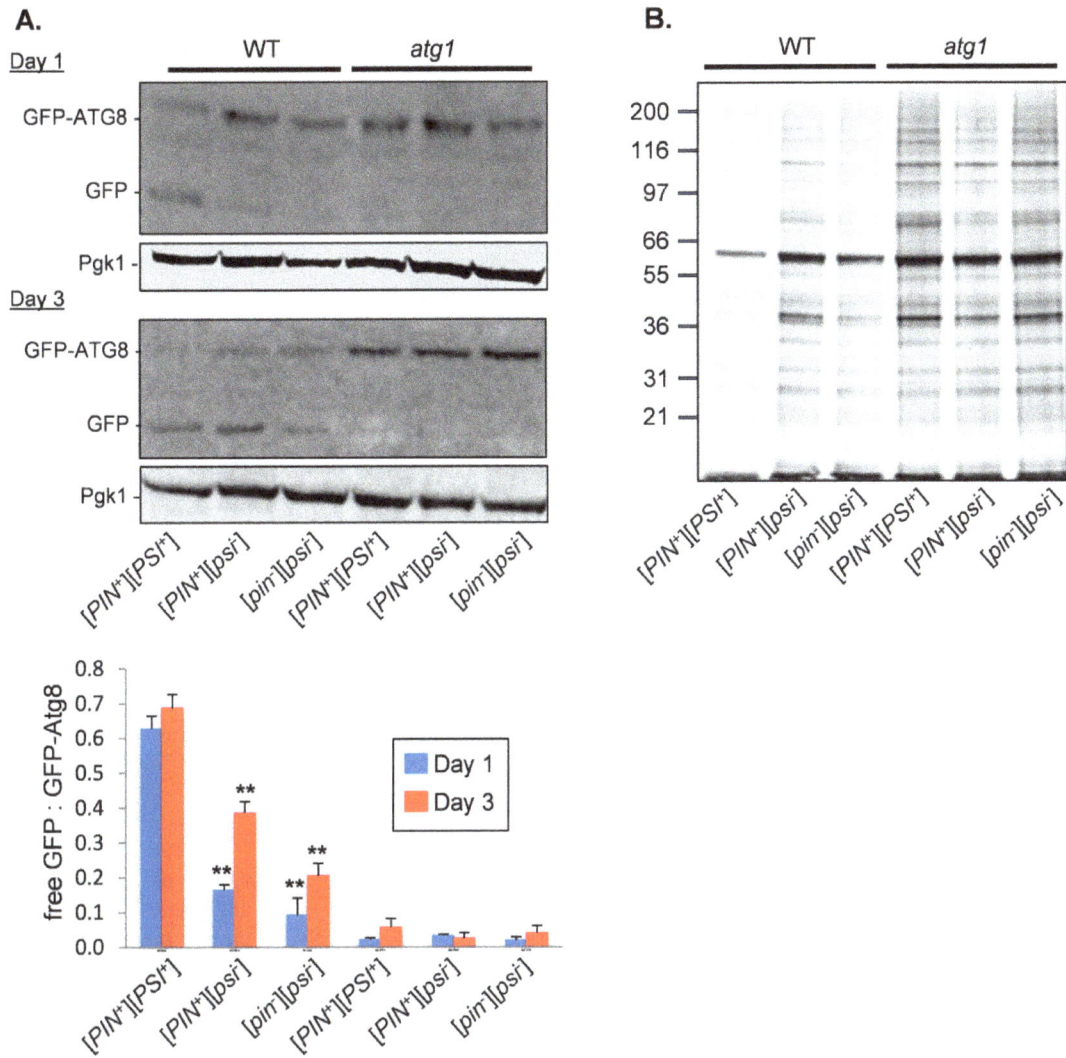

FIGURE 3: Increased autophagic flux in [*PSI⁺*] strains. **(A)** Autophagic flux was monitored in in the indicated strains expressing GFP-Atg8 during late exponential (day 1) and stationary phase growth (day 3). Increased free GFP is detected in the [*PIN⁺*][*PSI⁺*] strain indicative of autophagic flux. No free GFP is detected in the *atg1* mutant strains which cannot initiate autophagy. Representative images are shown from triplicate experiments with quantification comparing the ratios of free GFP with GFP-Atg8 below. Significance is shown comparing the [*PIN⁺*][*PSI⁺*] versions of the wild-type and *atg1* mutant strains with their corresponding [*PIN⁺*][*psi⁻*] and cured [*pin⁻*][*psi⁻*] versions. Statistical analysis was performed by one-way ANOVA: *$p < 0.05$, **$p < 0.01$. **(B)** Protein aggregates were isolated from the same strains as shown in panel A at day three of chronological growth and analyzed by SDS-PAGE and silver staining.

normally acts to suppress spontaneous prion formation during chronological ageing. Similarly, loss of autophagy has been found to cause neurodegenerative diseases in mice [16, 17], supporting a protective role for autophagy in defending against age-associated abnormal protein accumulation and aggregation.

We found that that the presence of the [*PSI⁺*] prion confers a beneficial advantage during yeast chronological ageing, which correlates with enhanced autophagic flux. Given that the [*PSI⁺*]-status of cells improves viability during ageing, this may result in selection for [*PSI⁺*] in aged cells. Our data indicate that the presence of the [*PSI⁺*] prion acts to simulate autophagy which results in improved via-

bility during ageing. There is previous evidence to suggest that the increased formation and accumulation of protein aggregates may exert a stimulatory effect on the autophagy pathway. For example, there is a correlation between the accumulation of PrP^SC and the enhanced activity of quality control pathways including endoplasmic reticulum chaperones, the unfolded protein response and autophagy [30]. In agreement with the idea that enhanced autophagy aids protein homeostasis during ageing, we found reduced levels of amorphous protein aggregation in a [*PSI⁺*] strain, suggesting that autophagy provides a beneficial effect during chronological ageing by removing potentially harmful protein aggregates, including both amorphous and amyloid forms. Increasing autophagic flux would also presumably act to prevent further amyloid aggregation, potentially protecting against any negative impact of [PSI⁺] aggregation altering translation termination efficiency.

The [PSI⁺] prion causes a loss of function phenotype where translation termination activity is reduced due to the aggregation of the normally soluble Sup35 protein [10]. The shift to the [PSI⁺] prion is thought to allow cells to re-program gene expression such that new genetic traits become uncovered which may aid survival during altered conditions [31-33]. However, as well as providing a selective advantage through altered gene expression, our data indicate that the [PSI⁺] prion can improve viability during ageing via modulation of autophagic flux. It is unclear what triggers the increased frequency of [PSI⁺] prion formation during ageing. One possibility is oxidative stress, since ROS-induced protein aggregation and mitochondrial dysfunction is a common feature in age-related diseases [34, 35]. In addition, ROS and oxidative stress are known to induce yeast and mammalian prion formation [36] which may account for increased [PSI⁺] formation observed during chronological ageing. Further research will be required to elucidate the exact signaling pathways and the range of quality control mechanisms that may be modulated through the direct and indirect action of the [PSI⁺] prion during yeast ageing.

MATERIALS AND METHODS
Yeast Strains
[PIN⁺][PSI⁺], [PIN⁺][psi⁻] and [pin⁻][psi⁻] derivatives of the wild-type yeast strain 74D-694 (MATa ade1-14 ura3-52 leu2-3,112 trp1-289 his3-200) were used for all experiments. The strain deleted for ATG1 (atg1::HIS3) has been described previously [12].

Growth conditions
Yeast strains were grown at 30°C, 180 rpm in minimal SCD medium (2% w/v glucose, 0.17% yeast nitrogen base without amino acids, supplemented with Kaiser amino acid mixes, Formedium, Hunstanton, England). Chronological life span experiments were performed in liquid SCD media supplemented with a four-fold excess of uracil, leucine, tryptophan, adenine and histidine to avoid any possible artefacts arising from the auxotrophic deficiencies of the strains. Strains were cured by five rounds of growth on YEPD agar plates containing 4 mM GdnHCl.

De novo [PSI⁺] formation
[PSI⁺] prion formation was scored by growth in the absence of adenine as described previously [12]. [PSI⁺] formation was calculated based on the mean of at least three independent biological repeat experiments.

Yeast Chronological Life Span Determination
CLS experiments were performed according to [37]. Briefly, cells were cultured in liquid SCD media for 3 days to reach stationary phase and then aliquots taken every 2-3 days for flow cytometry analysis. 50 µl of 4 mM of propidium iodide (P.I.) was added to 950 µl of culture and cell viability was measured based on propidium iodide uptake by non-viable cells as assayed through flow cytometry. Flow cytometry readings were performed using a Becton Dickinson (BD) LSRFortessa™ cell analyser, BD FACSDiva 8.0.1 software) after staining with propidium iodide. For the colony forming assay, cultures were serially diluted and plated onto YEPD plates. Viable counts were recorded following three days growth and were expressed as a percentage of the starting viability.

Protein analysis
Protein extracts were electrophoresed under reducing conditions on SDS-PAGE minigels and electroblotted onto PVDF membrane (Amersham Pharmacia Biotech). Bound antibody was visualised using WesternSure® Chemiluminescent Reagents (LI-COR) and a C-DiGit® Blot Scanner (LI-COR). Insoluble protein aggregates were isolated as previously described [38, 39], with the following minor adjustments [29]. Cell breakage was achieved by sonication (Sonifier 150, Branson; 8 x 5 s, Level 4) and samples were adjusted to equal protein concentrations before isolation of protein aggregates. Insoluble fractions were resuspended in detergent washes through sonication (4 x 5 s, Level 4). Insoluble fractions were resuspended in reduced protein loading buffer, separated by reducing SDS/PAGE (12% gels) and visualized by silver staining with the Bio-Rad silver stain plus kit. The induction of autophagy was confirmed by examining the release of free GFP due to the proteolytic cleavage of GFP-Atg8 [28].

ACKNOWLEDGEMENTS
S.H.S. was supported by a Wellcome Trust (grant number 099733/Z/12/Z) funded studentship.

CONFLICT OF INTEREST
The authors declare no conflict of interest.

REFERENCES
1. Kenyon CJ (2010). The genetics of ageing. Nature 464(7288): 504-512.

2. Levine B, Kroemer G (2008). Autophagy in the pathogenesis of disease. Cell 132(1): 27-42.

3. Rubinsztein DC, Marino G, Kroemer G (2011). Autophagy and aging. Cell 146(5): 682-695. doi: 10.1016/j.cell.2011.07.030.

4. Prusiner SB (1998). Prions. Proc Natl Acad Sci U S A 95(23): 13363-13383.

5. Aguzzi A, O'Connor T (2010). Protein aggregation diseases: patho-genicity and therapeutic perspectives. Nat Rev Drug Discov 9(3): 237-248.

6. Collinge J, Clarke AR (2007). A general model of prion strains and their pathogenicity. Science 318(5852): 930-936.

7. Prusiner SB (2013). Biology and genetics of prions causing neuro-degeneration. Annu Rev Genet 47(601-623). DOI: 10.1146/annurev-genet-110711-155524.

8. Appleby BS, Lyketsos CG (2011). Rapidly progressive dementias and the treatment of human prion diseases. Expert Opin Pharmacother 12(1): 1-12.

9. Derkatch IL, Bradley ME, Zhou P, Chernoff YO, Liebman SW (1997). Genetic and environmental factors affecting the de novo appearance

of the [PSI+] prion in Saccharomyces cerevisiae. **Genetics** 147(2): 507-519.

10. Wickner RB (**1994**). [URE3] as an altered URE2 protein: evidence for a prion analog in Saccharomyces cerevisiae. **Science** 264:566-5699.

11. Treusch S, Lindquist S (**2012**). An intrinsically disordered yeast prion arrests the cell cycle by sequestering a spindle pole body component. **J Cell Biol** 197(3): 369-379.

12. Speldewinde SH, Doronina VA, Grant CM (**2015**). Autophagy protects against de novo formation of the [PSI+] prion in yeast. **Mol Biol Cell** 26(25): 4541-4551.

13. Parzych KR, Klionsky DJ (**2014**). An overview of autophagy: morphology, mechanism, and regulation. **Antioxid Redox Signal** 20(3): 460-473.

14. Eisenberg T, Knauer H, Schauer A, Buttner S, Ruckenstuhl C, Carmona-Gutierrez D, Ring J, Schroeder S, Magnes C, Antonacci L, Fussi H, Deszcz L, Hartl R, Schraml E, Criollo A, Megalou E, Weiskopf D, Laun P, Heeren G, Breitenbach M, Grubeck-Loebenstein B, Herker E, Fahrenkrog B, Frohlich KU, Sinner F, Tavernarakis N, Minois N, Kroemer G, Madeo F (**2009**). Induction of autophagy by spermidine promotes longevity. **Nat Cell Biol** 11(11): 1305-1314.

15. Alvers AL, Wood MS, Hu D, Kaywell AC, Dunn WA, Jr., Aris JP (**2009**). Autophagy is required for extension of yeast chronological life span by rapamycin. **Autophagy** 5(6): 847-849.

16. Komatsu M, Waguri S, Chiba T, Murata S, Iwata J, Tanida I, Ueno T, Koike M, Uchiyama Y, Kominami E, Tanaka K (**2006**). Loss of autophagy in the central nervous system causes neurodegeneration in mice. **Nature** 441(7095): 880-884.

17. Hara T, Nakamura K, Matsui M, Yamamoto A, Nakahara Y, Suzuki-Migishima R, Yokoyama M, Mishima K, Saito I, Okano H, Mizushima N (**2006**). Suppression of basal autophagy in neural cells causes neurodegenerative disease in mice. **Nature** 441(7095): 885-889.

18. Longo VD, Shadel GS, Kaeberlein M, Kennedy B (**2012**). Replicative and chronological aging in Saccharomyces cerevisiae. **Cell Metab** 16(1): 18-31.

19. Wasko BM, Kaeberlein M (**2014**). Yeast replicative aging: a paradigm for defining conserved longevity interventions. **FEMS Yeast Res** 14(1): 148-159.

20. Mortimer RK, Johnston JR (**1959**). Life span of individual yeast cells. **Nature** 183(4677): 1751-1752.

21. Shewmaker F, Wickner RB (**2006**). Ageing in yeast does not enhance prion generation. **Yeast** 23(16): 1123-1128.

22. Lancaster AK, Bardill JP, True HL, Masel J (**2010**). The spontaneous appearance rate of the yeast prion [PSI$^+$] and its implications for the evolution of the evolvability properties of the [PSI$^+$] system. **Genetics** 184(2): 393-400.

23. Matsuura A, Tsukada M, Wada Y, Ohsumi Y (**1997**). Apg1p, a novel protein kinase required for the autophagic process in Saccharomyces cerevisiae. **Gene** 192(2): 245-250.

24. Pan Y, Schroeder EA, Ocampo A, Barrientos A, Shadel GS (**2011**). Regulation of yeast chronological life span by TORC1 via adaptive mitochondrial ROS signaling. **Cell Metab** 13(6): 668-678.

25. Jung G, Masison DC (**2001**). Guanidine hydrochloride inhibits Hsp104 activity in vivo: a possible explanation for its effect in curing yeast prions. **Curr Microbiol** 43(1): 7-10.

26. Ferreira PC, Ness F, Edwards SR, Cox BS, Tuite MF (**2001**). The elimination of the yeast [PSI+] prion by guanidine hydrochloride is the result of Hsp104 inactivation. **Mol Microbiol** 40(6): 1357-1369.

27. Alvers AL, Fishwick LK, Wood MS, Hu D, Chung HS, Dunn WA, Jr., Aris JP (**2009**). Autophagy and amino acid homeostasis are required for chronological longevity in Saccharomyces cerevisiae. **Aging cell** 8(4): 353-369.

28. Noda T, Matsuura A, Wada Y, Ohsumi Y (**1995**). Novel system for monitoring autophagy in the yeast Saccharomyces cerevisiae. **Biochem Biophys Res Commun** 210(1): 126-132.

29. Weids AJ, Grant CM (**2014**). The yeast peroxiredoxin Tsa1 protects against protein-aggregate-induced oxidative stress. **J Cell Sci** 127(Pt 6): 1327-1335.

30. Joshi-Barr S, Bett C, Chiang WC, Trejo M, Goebel HH, Sikorska B, Liberski P, Raeber A, Lin JH, Masliah E, Sigurdson CJ (**2014**). De novo prion aggregates trigger autophagy in skeletal muscle. **J Virol** 88(4): 2071-2082.

31. Tyedmers J, Madariaga ML, Lindquist S (**2008**). Prion switching in response to environmental stress. **PLoS Biol** 6:e294.

32. True HL, Lindquist SL (**2000**). A yeast prion provides a mechanism for genetic variation and phenotypic diversity. **Nature** 407:477-483.

33. True HL, Berlin I, Lindquist SL (**2004**). Epigenetic regulation of translation reveals hidden genetic variation to produce complex traits. **Nature** 431:184-187.

34. Shacka JJ, Roth KA, Zhang J (**2008**). The autophagy-lysosomal degradation pathway: role in neurodegenerative disease and therapy. **Front Biosci** 13:718-736.

35. Lee J, Giordano S, Zhang J (**2012**). Autophagy, mitochondria and oxidative stress: cross-talk and redox signalling. **Biochem J** 441(2): 523-540.

36. Grant CM (**2015**). Sup35 methionine oxidation is a trigger for de novo [PSI+] prion formation. **Prion** 9:257-265.

37. Ocampo A, Barrientos A (**2011**). Quick and reliable assessment of chronological life span in yeast cell populations by flow cytometry. **Mech Ageing Dev** 132(6-7): 315-323.

Rand JD, Grant CM (**2006**). The Thioredoxin System Protects Ribosomes against Stress-induced Aggregation. **Mol Biol Cell** 17:387-401.

39. Jacobson T, Navarrete C, Sharma SK, Sideri TC, Ibstedt S, Priya S, Grant CM, Christen P, Goloubinoff P, Tamas MJ (**2012**). Arsenite interferes with protein folding and triggers formation of protein aggregates in yeast. **J Cell Sci** 125:5073-5083.

16

The neuroprotective steroid progesterone promotes mitochondrial uncoupling, reduces cytosolic calcium and augments stress resistance in yeast cells

Slaven Stekovic[1,*], Christoph Ruckenstuhl[1,+,*], Philipp Royer[1], Christof Winkler-Hermaden[1], Didac Carmona-Gutierrez[1], Kai-Uwe Fröhlich[1], Guido Kroemer[3-8], and Frank Madeo[1,2,+]

[1] Institute of Molecular Biosciences, NAWI Graz, University of Graz, 8010 Graz, Austria.
[2] BioTechMed Graz, Austria.
[3] Equipe 11 labellisée par la Ligue contre le Cancer, Centre de Recherche des Cordeliers, Paris, France.
[4] INSERM, U1138, Paris, France.
[5] Université Paris Descartes, Sorbonne Paris Cité, Paris, France.
[6] Cell Biology & Metabolomics Platforms, Gustave Roussy Comprehensive Cancer Center, Villejuif, France.
[7] Pôle de Biologie, Hôpital Européen Georges Pompidou, AP‑HP, Paris, France.
[8] Karolinska Institute, Department of Women's and Children's Health, Karolinska University Hospital, 17176 Stockholm, Sweden,
* These authors contributed equally.
+ Corresponding Author:
Frank Madeo, Institute of Molecular Biosciences, University of Graz, Humboldtstrasse 50; 8010 Graz, Austria; E-mail: frank.madeo@uni-graz.at
Christoph Ruckenstuhl, Institute of Molecular Biosciences, University of Graz, Humboldtstrasse 50; 8010 Graz, Austria; E-mail: ru.ruckenstuhl@uni-graz.at

ABSTRACT The steroid hormone progesterone is not only a crucial sex hormone, but also serves as a neurosteroid, thus playing an important role in brain function. Epidemiological data suggest that progesterone improves the recovery of patients after traumatic brain injury. Brain injuries are often connected to elevated calcium spikes, reactive oxygen species (ROS) and programmed cell death affecting neurons. Here, we establish a yeast model to study progesterone-mediated cytoprotection. External supply of progesterone protected yeast cells from apoptosis-inducing stress stimuli and resulted in elevated mitochondrial oxygen uptake accompanied by a drop in ROS generation and ATP levels during chronological aging. In addition, cellular Ca^{2+} concentrations were reduced upon progesterone treatment, and this effect occurred independently of known Ca^{2+} transporters and mitochondrial respiration. All effects were also independent of Dap1, the yeast orthologue of the progesterone receptor. Altogether, our observations provide new insights into the cytoprotective effects of progesterone.

Keywords: TBI; traumatic brain injury, cell protection, cell stress, cell death, neuroprotection, progesterone, mitochondrial uncoupling.

Abbreviations:
DHE – dihydroethidium,
PCD – programmed cell death,
ROS – reactive oxygen species,
TBI – traumatic brain injury.

INTRODUCTION

Progesterone is a sterol-derived hormone that is crucial for female reproductive capacity and plays major regulatory roles in the monthly menstrual cycle and upon conception as well as during pregnancy and embryogenesis. In addition, it also serves as a neurosteroid, thus playing an important role in brain function in both sexes [1]. For instance, progesterone inhibits the neuronal nicotinic acetylcholine receptor and stimulates the synthesis of myelin proteins [1]. Of note, progesterone has been linked

to the gender-specific risk and outcome of brain injuries that is more favorable for females [2]. Interestingly, preclinical data strongly suggest that (high doses of) progesterone may positively affect recovery from traumatic brain injury (TBI) in model organisms [3–7], if administered before or shortly after TBI. Two clinical studies could confirm a neuroprotective effect of progesterone when administered shortly after TBI [8,9], while some more recent clinical data seem to disprove this hypothesis [10–12]. Therefore, it remains an open question

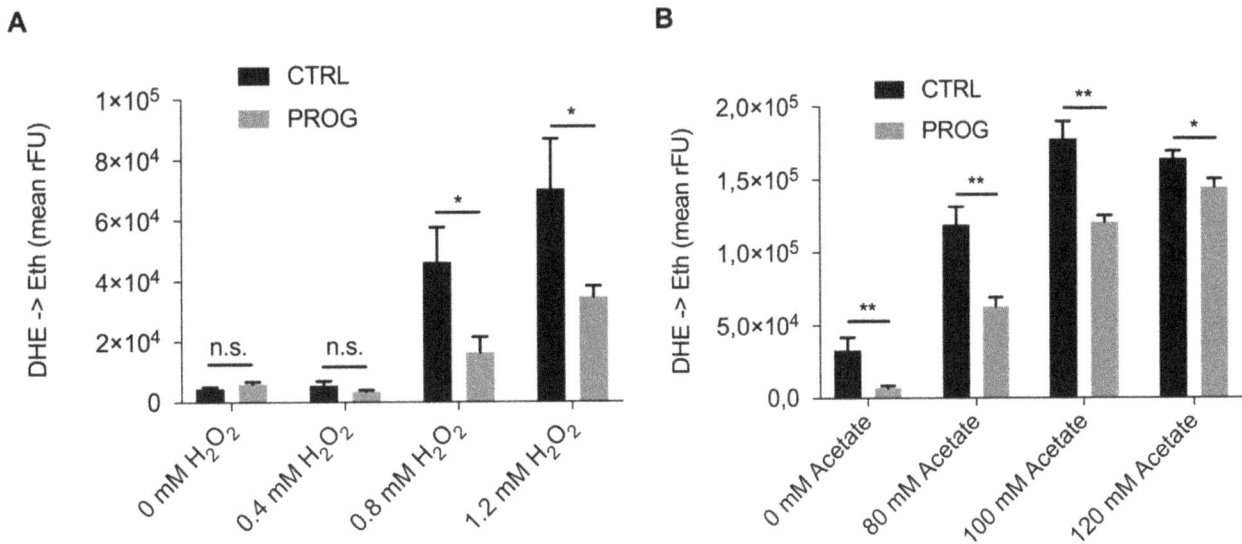

FIGURE 1: Progesterone treatment increases resistance of wildtype yeast to external stressors. ROS accumulation in yeast cells treated with progesterone (10 µg/ml) or left untreated as shown by the DHE to ethidium turnover rate upon hydrogen peroxide **(A)** or acetate **(B)** challenge during logarithmic phase. All data represent mean values (n = 3 ± SEM). Statistical analysis was conducted using non-paired Student's t-test. * = $p < 0.05$; ** = $p < 0.01$; *** = $p < 0.001$; n.s. = non-significant, Prog = progesterone, ctrl = control.

if progesterone affects the recovery and survival after TBI in humans and to which extent it promotes cellular restauration.

In order to investigate the cytoprotective potential of progesterone, we took advantage of *Saccharomyces cerevisiae*, knowing that this organism has repeatedly been shown to be suitable for mechanistic studies of programmed cell death (PCD) [13–19]. Yeast is especially useful as a model to study neuroprotection at the cellular level [20–27]. Here, we describe the positive impact of progesterone on several parameters of cellular physiology. Importantly, our results also suggest a possible receptor-independent mechanism for these effects, since deletion of *DAP1* – a heme-binding protein related to the mammalian membrane progesterone receptor – did not alter susceptibility towards progesterone treatment. Altogether, we reveal that progesterone exerts potent cytoprotective effects in yeast.

RESULTS

Progesterone increases stress tolerance

Traumatic brain injury is connected to elevated PCD and ROS accumulation in the brain tissue [28,29]. Therefore, we tested if progesterone would render yeast cells less susceptible towards different stressors that are connected to an increase in ROS production. Upon addition of progesterone, wild type yeast cultures treated with H_2O_2 or acetate, which are both well-known PCD inducers in yeast [14,30–34], showed reduced ROS accumulation as measured by the ROS-driven conversion of dihydroethidium (DHE) to fluorescent ethidium (Figure 1A and B). Furthermore, under physiological culture conditions, in the absence of PCD inducers, progesterone significantly reduced ROS levels as compared to the

untreated control (Figure 2A). Altogether, progesterone dampens ROS production in yeast, both in normal culture conditions and in the presence of external stress factors.

Progesterone impacts mitochondria by acting as a mild respiration-uncoupler

To further explore the mechanisms underlying progesterone cytoprotection, we next examined the physiology of mitochondria, since these organelles constitute one of the main sources of ROS [35–38]. Interestingly, while O_2 consumption was significantly enhanced during progesterone treatment, ATP levels were reduced (Figure 2B and C). Altogether, this indicates an uncoupling phenotype with diminished oxidative phosphorylation. Accordingly, we observed reduced growth of wild type yeast upon progesterone treatment on a non-fermentable carbon source (glycerol), while no changes were detected on a fermentable carbon source (glucose) (Figure 2D and E). Importantly, this effect was also observed in a mutant strain lacking the heme-binding protein Dap1, which is the sole yeast orthologue of the human progesterone receptor (Figure 2D and E) [39]. Furthermore, we could demonstrate that stress protection by progesterone is respiration-dependent, since progesterone treatment did not confer stress resistance in respiration-deficient rho[0] cells (Figure 2F). Altogether, it appears that progesterone impacts yeast mitochondrial respiration in a receptor-independent fashion.

Progesterone administration diminishes cytosolic Ca^{2+} concentrations both under physiological as well as under high calcium conditions

Next we investigated progesterone effects on Ca^{2+} homeostasis, knowing that mitochondria are one of the

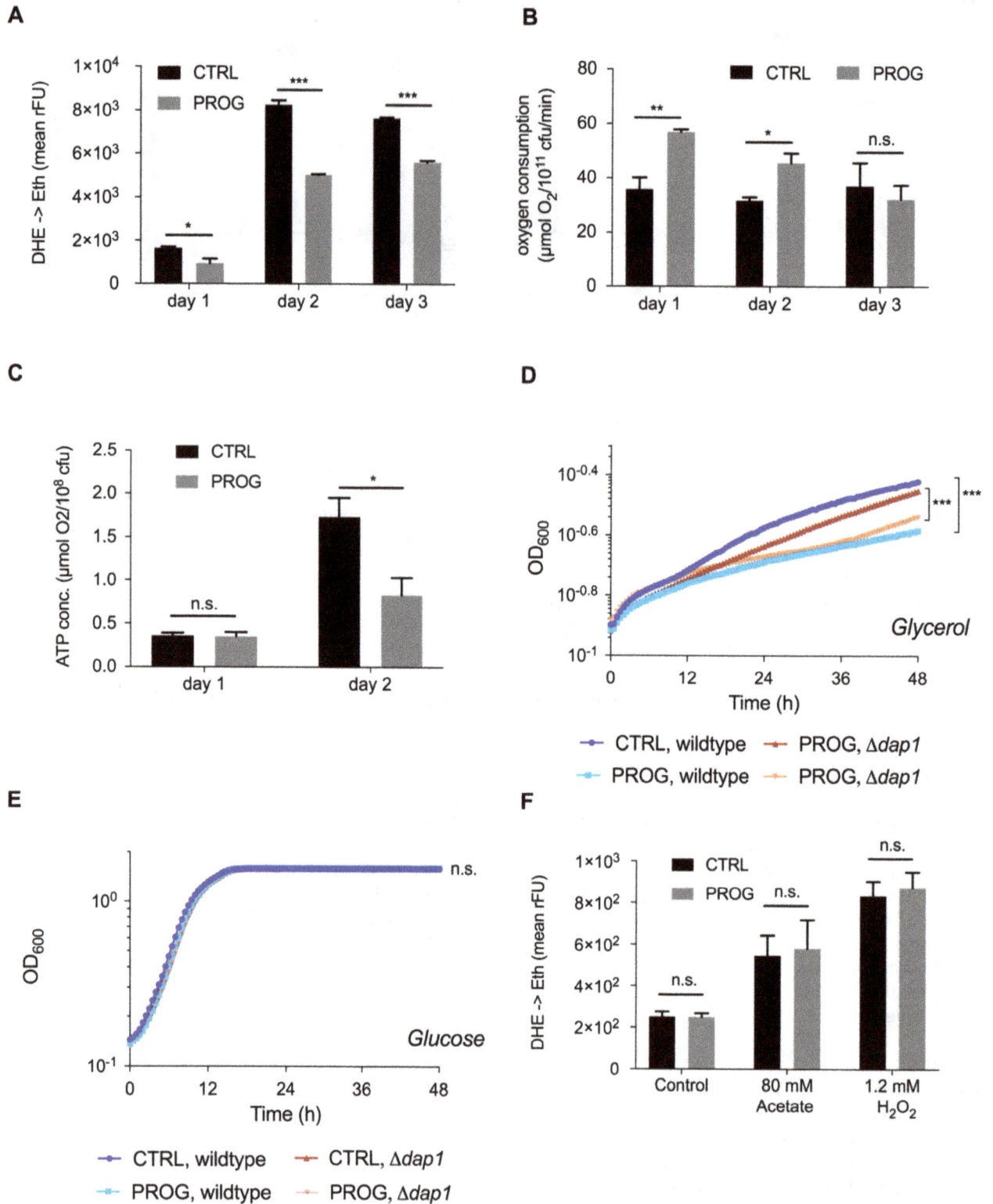

FIGURE 2: Progesterone impacts energy metabolism and reduces oxygen stress accumulation in wildtype yeast. Wildtype yeast were treated with 10 µg/ml progesterone and assayed for **(A)** ROS accumulation via DHE to ethidium turnover, **(B)** oxygen consumption via respirometry, and **(C)** ATP production. Growth curves of wildtype as well as Δ*dap1* strains, with or without progesterone treatment, on glycerol (respiratory carbon source) **(D)** and glucose (fermentative carbon source) media **(E)**. ROS accumulation in rho[0] yeast cells +/- progesterone (10 µg/ml) treated or untreated with H_2O_2 or acetate during logarithmic phase **(F)**. All data represent mean values (n = 3-5 ± SEM). Statistical analysis was conducted using non-paired Student t-test (A-C, F) or using a two-way repeated measurement ANOVA and multiple comparison post-hoc Tukey's test (D, E). * = p<0.05; ** = p<0.01; *** = p<0.001; n.s. = non-significant. ROS = reactive oxygen species, rFU = relative fluorescence units, Prog = progesterone, ctrl = control.

organelles

responsible for buffering cytosolic Ca^{2+} under normal conditions [40]. Importantly, TBI, stroke, and even some forms of dementia cause Ca^{2+} accumulation in the cytosol of neurons followed by cell death and neurodegeneration [41]. Thus, we examined the capacity of yeast cells to process Ca^{2+} uptake under the influence of progesterone. Specifically, wild type yeast cell cultures were challenged with 150 mM $CaCl_2$ and transient concentrations of cytoplasmic Ca^{2+} levels ([Ca^{2+}]cyt) / responses were monitored. Progesterone caused a significantly reduced Ca^{2+} uptake capacity alongside with a faster reduction of cytoplasmic Ca^{2+} levels (Figure 3A and B). Of note, basal Ca^{2+} levels before and after the Ca^{2+} pulse were already lowered when cells were treated with progesterone (Figure 3B). However, mitochondrial respiration was not involved in this phenotype, since progesterone treatment continued to affect basic cytosolic Ca^{2+} levels in respiration-deficient rho^0 cells (Figure 3C and D).

To further investigate the observed phenotypes, we tested single-gene deletion mutants of all currently known Ca^{2+} channels/transporters in yeast, including the cytoplasmic membrane transporters Cch1 and Mid 1, the organelle transporters Vcx1, Pmr1, Cod1, Yvc1, and Pmc1 as well as Emc7, an ER protein associated to Ca^{2+} homeostasis. Although Ca^{2+} uptake and clearance was influenced by some of these gene deletion, all mutants continued to exhibit significantly reduced Ca^{2+} uptake when treated with progesterone (compare Supplemental Figure 1A-G to H). Thus, the effects observed in wild type cells could not be reversed by single gene deletions in any of these transporters. Similarly, the effects of progesterone treatment on Ca^{2+} homeostasis/uptake were independent of the mammalian membrane progesterone receptor homolog Dap1 (Figure 3E and F). Taken together, progesterone seems to influence Ca^{2+} homeostasis/uptake in a general manner, independently from known Ca^{2+} transporters and respiration capacity.

DISCUSSION

Here, we establish *S. cerevisiae* as a model to investigate cytoprotection by progesterone. We observed that progesterone increased stress tolerance of yeast to the well-known PCD inducers H_2O_2 and acetate [14,30–34] as well as under physiological (control) conditions. Interestingly, progesterone treatment led to a mild uncoupling phenotype with higher O_2 consumption (+50%) but lower ATP levels (-50%), arguing for a mitochondrial uncoupling effect. Indeed, growth on the non-fermentable carbon source glycerol was diminished in the presence of progesterone. Notably, mild uncoupling induced by chemical substances (such as dinitrophenol), caloric restriction or ectopic expression of mammalian uncoupling proteins in yeast - *S. cerevisiae* does not possess any known uncoupling proteins [42] - is known to increase lifespan [43–45]. Similarly, in mammalian aging cells, changes in mitochondrial energy metabolism caused by mitochondrial uncoupling seem to improve cellular fitness [46]. Progesterone treatment of human cells has been demonstrated to strongly increase the levels of mRNAs coding for uncoupling proteins [47]. Increased O_2 consumption with decreased ^{32}P uptake (as a parameter for ATP production) has been reported for isolated rat mitochondria treated with progesterone [48]. Collectively, our data combined with those reported in the literature highlight the possibility to investigate progesterone-mediated effects in the yeast model. The uncoupling aspect of progesterone, in fact, could represent one of the mechanisms of neuroprotection conferred by this steroid. In fact, the stress tolerance of a respiration-deficient rho^0 strain was not influenced by progesterone treatment.

Progesterone had major effects on Ca^{2+} homeostasis and, in particular, on Ca^{2+} susceptibility/uptake. However, we could not identify any single Ca^{2+} channel in yeast that would influence these effects. However, we cannot exclude that yet unidentified Ca^{2+} channels or a combinations of known Ca^{2+} channels mediate these effects [49]. Another possible mode of action of progesterone on Ca^{2+} homeostasis could reside in its direct interaction with biological membranes. Since the chemical structure of progesterone shows four-ring as well as hydrophobic backbone and polar groups at both ends of the molecule, it could directly interact with cellular and mitochondrial membranes [50] and possibly influence their permeability towards inorganic cations (e.g. Ca^{2+}, H^+). This mode of action could connect our observations of mitochondrial uncoupling and modulation of Ca^{2+} homeostasis. Of note, a progesterone-treated rho^0 strain still showed Ca^{2+} effects but no enhanced stress tolerance, suggesting that altered Ca^{2+} homeostasis may lie upstream of mitochondrial uncoupling. However, these mechanistic hypotheses remain to be empirically tested.

Certainly, the putative relevance of the herein described progesterone effects for TBI pathology remains to be explored. In some mammalian cell types, progesterone leads to a significant increase of intracellular Ca^{2+} [51,52], partly by activating protein kinase C [53] and depleting endocannabinoids by activating α/β hydrolase domain-containing protein 2 (ABHD2) [54]. However, in other cell types, progesterone withdrawal leads to an increased level of cytosolic Ca^{2+} [55]. While progesterone was not able to reduce estrogen-induced Ca^{2+} uptake in the rabbit myometrial smooth muscle cells, it increased the accumulation of Ca^{2+} in mitochondria [55]. This suggests that progesterone withdrawal reduces both myometrial cytosolic Ca^{2+} levels as well as the capacity of these cells to accumulate Ca^{2+} in different cellular compartments. Similar effects were reported for other types of smooth muscles [56,57] and are believed to be caused by regulation of the inward current through L-type Ca^{2+} channels [56,58]. In neurons, the influence on Ca^{2+} signaling and the following inhibition of excitotoxic neuron death seem to be the neuroprotective mechanism induced by acute administration of progesterone after various neuronal injuries [59–61]. Indeed, progesterone might mediate broad neuroprotective effects, not only in the context of TBI but also in other pathologies [62,63].

The role of progesterone in the pathological develop-

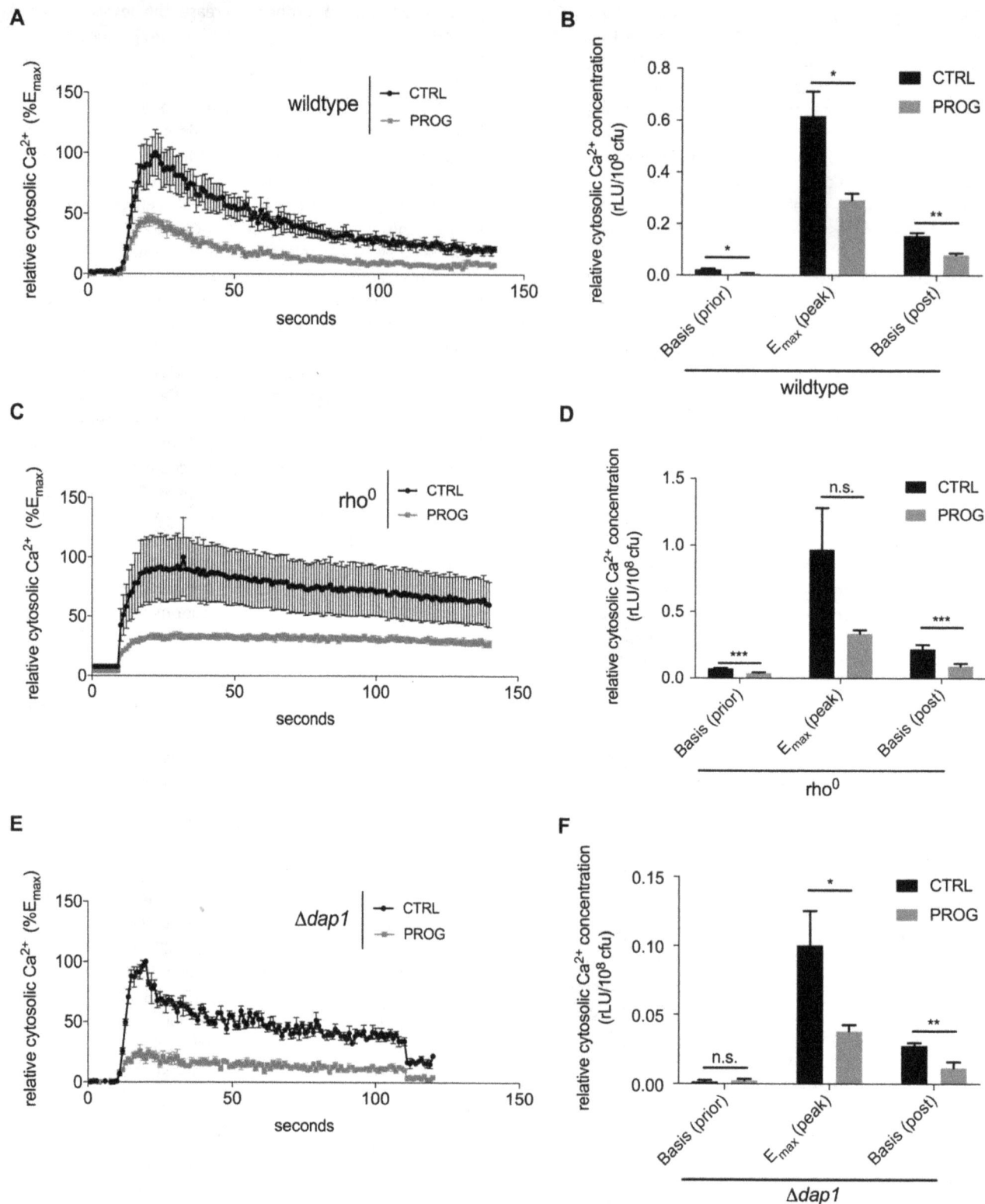

FIGURE 3: Cellular Ca²⁺ homeostasis is modulated by progesterone treatment in yeast. Cells were treated with progesterone (10 µg/ml) and challenged with high doses of Ca²⁺ (150 mM). Intake of Ca²⁺ as well as Ca²⁺-clearance in the cytosol to its basal level were measured in wild type **(A,B)**, a *DAP1*-deletion strain **(C,D)**, as well as in a rho⁰ strain, incapable of mitochondrial respiration **(E,F)**. Data are shown as mean values of at least three replicates including the standard error of the mean. Statistical analysis was conducted using non-paired Student's t-test. * = p<0.05; ** = p<0.01; *** = p<0.001; n.s. = non-significant. E$_{max}$ = global maximum of the respective ethanol-treated control; rLU = relative luminiscence units, Prog = progesterone, ctrl = control.

ment of TBI has been well described in recent years. It has been shown that progesterone improves the function of the blood-brain-barrier after TBI [64]. Progesterone also increases the level of circulating endothelial cells, which in turn improves neovascularization and vascular remodeling in the brain [65]. Furthermore, progesterone treatment reduces neuroinflammation and oxidative stress [66] as it improves remyelination and functional recovery [63].

Interestingly, the intracellular effects exerted by progesterone in our model - reduced intracellular Ca^{2+} levels, uncoupled mitochondria and ROS reduction – were not lost when the sole possible yeast orthologue of the human progesterone receptor was removed from the system. This suggests that progesterone mediates its broad cytoprotective effects through other proteins than steroid receptors or perhaps with cellular membrane lipids. We surmise that yeast constitutes an ideal platform for exploring these effects in further detail.

MATERIALS AND METHODS
Growth conditions
S. cerevisiae strains (Table 1) were inoculated to $5*10^5$ (for growth curve OD_{600} of 0.05) cells in SC medium containing 0.17% yeast nitrogen base (BD Diagnostics; without ammonium sulfate and amino acids), 0.5% $(NH_4)_2SO_4$, 30 mg/L of all amino acids (except 80 mg/L histidine and 200 mg/L leucine), 30 mg/L adenine, and 320 mg/L uracil with 2% glucose (SCD) or alternatively with 3% glycerol (SCGly), w/o treatment with progesterone (10 µg/ml; Sigma Aldrich, Catalogue Nr. P0130). Controls were treated with respective solvent (EtOH). Where

indicated, stress (acetate or H_2O_2) was inflicted as described previously in mid-log phase (~ 6h of growth, culture density 2-$4*10^6$ cells/ml). Due to the inherent reduced respiration-rate of BY4741 strains, TB50a strains were used for respiration-related experiments. *DAP1* deletion was carried out by classical homologous recombination [67,68].

Growth curve
Cells from ONC in SCD media were inoculated to an OD_{600} of 0.05 in SCD media and SCGly media with or without 10 µg/ml progesterone addition. Untreated cultures were supplemented with 0.1% EtOH for solvent control. To obtain growth curves, 300 µl of respective cultures per well were transferred into Honeycomb® plates, and measured with Bioscreen C MBR system (Oy Growth Curves Ab Ltd.) for a period of 48 hours at 28°C, using continuous shaking and OD_{600} measurements every 30 minutes.

Oxygen consumption measurement
Oxygen consumption was measured using a FireSting oxygen electrode (Pyro-Science) under constant stirring at a temperature of 28.0 ± 0.2°C in sealed 2 ml bottles. The corresponding cell counts were measured using a CASY Cell Counter, whereas percentage of living cells in the sample were established by flow cytometry with propidium iodide (PI: 100 ng/ml) stained samples. The slope of the oxygen concentration as the function of time in its linear region was calculated and normalized to the number of living cells in the sample.

ROS accumulation (DHE) assay
Oxidation of non-fluorescent di-hydroethidium (DHE) to fluo-

Table 1. Strains used in this study.

Strain	Genotype	Reference
TB50a wildtype	MATα; leu2-3,112 ura3-52 trp1 his3 rme1 HMLα	[69]
TB50a Δdap1	MATα; leu2-3,112 ura3-52 trp1 his3 rme1 HMLα dap1::kanMX	This study
BY4741 wildtype	MATa his3Δ1 leu2Δ0 met15Δ0 ura3Δ0	Euroscarf
BY4741 Δdap1	MATa his3Δ1 leu2Δ0 met15Δ0 ura3Δ0 dap1::kanMX	Euroscarf
BY4741 Δcch1	MATa his3Δ1 leu2Δ0 met15Δ0 ura3Δ0 cch1::kanMX	Euroscarf
BY4741 Δmid1	MATa his3Δ1 leu2Δ0 met15Δ0 ura3Δ0 mid1::kanMX	Euroscarf
BY4741 Δvcx1	MATa his3Δ1 leu2Δ0 met15Δ0 ura3Δ0 vcx1::kanMX	Euroscarf
BY4741 Δpmr1	MATa his3Δ1 leu2Δ0 met15Δ0 ura3Δ0 pmr1::kanMX	Euroscarf
BY4741 Δcod1	MATa his3Δ1 leu2Δ0 met15Δ0 ura3Δ0 cod1::kanMX	Euroscarf
BY4741 Δyvc1	MATa his3Δ1 leu2Δ0 met15Δ0 ura3Δ0 yvc1::kanMX	Euroscarf
BY4741 Δpmc1	MATa his3Δ1 leu2Δ0 met15Δ0 ura3Δ0 pmc1::kanMX	Euroscarf
BY4741 Δemc7	MATa his3Δ1 leu2Δ0 met15Δ0 ura3Δ0 emc7::kanMX	Euroscarf

rescent ethidium was used to measure ROS accumulation in yeast cells [38]. Approximately $5*10^6$ cells from each sample were collected, washed and incubated with DHE solution (2.5 µg/ml in PBS) for 10 min in the dark. After washing samples were re-suspended in PBS buffer and measured using flow cytometry. The relative mean fluorescence measured for the cell population was used for analysis [70].

Boiling ethanol extraction of ATP and ATP measurement

ATP extraction was done with flash-frozen cells by adding 0.5 ml preheated (90°C) BES buffer and incubation at 90°C for 3 minutes. After centrifugation, supernatants were stored at -80°C until the measurement. ATP levels were determined by using the ATP detection kit from Invitrogen in a Luminoskan (Thermo Scientific).

Cytosolic Ca^{2+} measurements

$[Ca^{2+}]$cyt were measured using yeast strains carrying the vector pYX212 encoding the bioluminescent protein aequorin under the control of a TPI promoter. For analysis of the cellular response to high doses of external Ca^{2+}, an equivalent of $6*10^6$ cells was harvested, resuspended in 200 µl SCD containing 4 µM coelenterazine and incubated for 1 h in the dark. After washing cells were measured in a Luminoskan for 10 s and then challenged with high dose of Ca^{2+} (pump injection of 150 mM Ca^{2+}). Kinetics were recorded over 120 s. The luminescence signal was normalized to the OD_{600} of each well and reported in relative luminescence units, normalized to the global maximum value of the ethanol treated control of the respective run for better comparability.

ACKNOWLEDGEMENTS

We thank Silvia Dichtinger for technical assistance. FM is grateful to the Austrian Science Fund FWF (Austria) for grants P23490-B20, P29262, P24381, P29203 P27893, I1000 and 'SFB Lipotox' (F3012), as well as to BMWFW and the Karl-Franzens University for grant 'Unkonventionelle Forschung' and grant DKplus Metabolic and Cardiovascular Diseases (W1226). We acknowledge support from NAWI Graz and the BioTechMed-Graz flagship project "EPIAge". GK is supported by the Ligue contre le Cancer Comité de

Charente-Maritime (équipe labelisée); Agence National de la Recherche (ANR) – Projets blancs; ANR under the frame of E-Rare-2, the ERA-Net for Research on Rare Diseases; Association pour la recherche sur le cancer (ARC); Cancéropôle Ile-de-France; Institut National du Cancer (INCa); Inserm (HTE); Institut Universitaire de France; Fondation pour la Recherche Médicale (FRM); the European Commission (ArtForce); the European Research Council (ERC); Fondation Carrefour; the LeDucq Foundation; the LabEx Immuno-Oncology; the RHU Torino Lumière, the SIRIC Stratified Oncology Cell DNA Repair and Tumor Immune Elimination (SOCRATE); the SIRIC Cancer Research and Personalized Medicine (CARPEM); and the Paris Alliance of Cancer Research Institutes (PACRI).

CONFLICT OF INTEREST

The authors declare no conflict of interest.

REFERENCES

1. Baulieu E and Schumacher M (2000). Progesterone as a neuroactive neurosteroid, with special reference to the effect of progesterone on myelination. Steroids 65(10–11): 605–612.

2. Vagnerova K, Koerner IP, and Hurn PD (2008). Gender and the Injured Brain. Anesth Analg 107(1): 201–214.

3. Meffre D, Labombarda F, Delespierre B, Chastre A, De Nicola AF, Stein DG, Schumacher M, and Guennoun R (2013). Distribution of membrane progesterone receptor alpha in the male mouse and rat brain and its regulation after traumatic brain injury. Neuroscience 231: 111–124.

4. Si D, Wang H, Wang Q, Zhang C, Sun J, Wang Z, Zhang Z, and Zhang Y (2013). Progesterone treatment improves cognitive outcome following experimental traumatic brain injury in rats. Neurosci Lett 553: 18–23.

5. Soltani Z, Khaksari M, Shahrokhi N, Mohammadi G, Mofid B, Vaziri A, and Amiresmaili S (2016). Effect of estrogen and/or progesterone administration on traumatic brain injury-caused brain edema: the changes of aquaporin-4 and interleukin-6. J Physiol Biochem 72(1): 33–44.

6. Carswell HV, Anderson NH, Clark JS, Graham D, Jeffs B, Dominiczak AF, and Macrae IM (1999). Genetic and gender influences on sensitivity to focal cerebral ischemia in the stroke-prone spontaneously hypertensive rat. Hypertens Dallas Tex 1979 33(2): 681–685.

7. Alkayed NJ, Murphy SJ, Traystman RJ, Hurn PD, and Miller VM (2000). Neuroprotective effects of female gonadal steroids in reproductively senescent female rats. Stroke 31(1): 161–168.

8. Xiao G, Wei J, Yan W, Wang W, and Lu Z (2008). Improved outcomes from the administration of progesterone for patients with acute severe traumatic brain injury: a randomized controlled trial. Crit Care Lond Engl 12(2): R61.

9. Wright DW, Kellermann AL, Hertzberg VS, Clark PL, Frankel M, Goldstein FC, Salomone JP, Dent LL, Harris OA, Ander DS, Lowery DW, Patel MM, Denson DD, Gordon AB, Wald MM, Gupta S, Hoffman SW, and Stein DG (2007). ProTECT: A Randomized Clinical Trial of Progesterone for Acute Traumatic Brain Injury. Ann Emerg Med 49(4): 391–402.e2.

10. Skolnick BE, Maas AI, Narayan RK, van der Hoop RG, MacAllister T, Ward JD, Nelson NR, Stocchetti N, and SYNAPSE Trial Investigators (2014). A clinical trial of progesterone for severe traumatic brain injury. N Engl J Med 371(26): 2467–2476.

11. Lin C, He H, Li Z, Liu Y, Chao H, Ji J, and Liu N (2015). Efficacy of progesterone for moderate to severe traumatic brain injury: a meta-analysis of randomized clinical trials. Sci Rep 5: 13442.

12. Zeng Y, Zhang Y, Ma J, and Xu J (2015). Progesterone for Acute Traumatic Brain Injury: A Systematic Review of Randomized Controlled Trials. PloS One 10(10): e0140624.

13. Madeo F, Fröhlich E, and Fröhlich KU (1997). A yeast mutant showing diagnostic markers of early and late apoptosis. J Cell Biol 139(3): 729–734.

14. Madeo F, Fröhlich E, Ligr M, Grey M, Sigrist SJ, Wolf DH, and Fröhlich KU (1999). Oxygen stress: a regulator of apoptosis in yeast. J Cell Biol 145(4): 757–767.

15. Madeo F, Herker E, Maldener C, Wissing S, Lächelt S, Herlan M, Fehr M, Lauber K, Sigrist SJ, Wesselborg S, and Fröhlich KU (2002). A caspase-related protease regulates apoptosis in yeast. Mol Cell 9(4): 911–917.

16. Herker E, Jungwirth H, Lehmann KA, Maldener C, Fröhlich K-U, Wissing S, Büttner S, Fehr M, Sigrist S, and Madeo F (2004). Chronological aging leads to apoptosis in yeast. J Cell Biol 164(4): 501–507.

17. Wissing S, Ludovico P, Herker E, Büttner S, Engelhardt SM, Decker

T, Link A, Proksch A, Rodrigues F, Corte-Real M, Fröhlich K-U, Manns J, Candé C, Sigrist SJ, Kroemer G, and Madeo F (**2004**). An AIF orthologue regulates apoptosis in yeast. **J Cell Biol** 166(7): 969–974.

18. Büttner S, Ruli D, Vögtle F-N, Galluzzi L, Moitzi B, Eisenberg T, Kepp O, Habernig L, Carmona-Gutierrez D, Rockenfeller P, Laun P, Breitenbach M, Khoury C, Fröhlich K-U, Rechberger G, Meisinger C, Kroemer G, and Madeo F (**2011**). A yeast BH3-only protein mediates the mitochondrial pathway of apoptosis. **EMBO J** 30(14): 2779–2792.

19. Galluzzi L, Kepp O, and Kroemer G (**2016**). Mitochondrial regulation of cell death: a phylogenetically conserved control. **Microb Cell** 3(3): 101–108.

20. Büttner S, Habernig L, Broeskamp F, Ruli D, Vögtle FN, Vlachos M, Macchi F, Küttner V, Carmona-Gutierrez D, Eisenberg T, Ring J, Markaki M, Taskin AA, Benke S, Ruckenstuhl C, Braun R, Van den Haute C, Bammens T, van der Perren A, Fröhlich K-U, Winderickx J, Kroemer G, Baekelandt V, Tavernarakis N, Kovacs GG, Dengjel J, Meisinger C, Sigrist SJ, and Madeo F (**2013**). Endonuclease G mediates α-synuclein cytotoxicity during Parkinson's disease. **EMBO J** 32(23): 3041–3054.

21. Büttner S, Broeskamp F, Sommer C, Markaki M, Habernig L, Alavian-Ghavanini A, Carmona-Gutierrez D, Eisenberg T, Michael E, Kroemer G, Tavernarakis N, Sigrist SJ, and Madeo F (**2014**). Spermidine protects against α-synuclein neurotoxicity. **Cell Cycle Georget Tex** 13(24): 3903–3908.

22. Heinisch JJ and Brandt R (**2016**). Signaling pathways and posttranslational modifications of tau in Alzheimer's disease: the humanization of yeast cells. **Microb Cell** 3(4): 135–146.

23. Menezes R, Tenreiro S, Macedo D, Santos C, and Outeiro T (**2015**). From the baker to the bedside: yeast models of Parkinson's disease. **Microb Cell** 2(8): 262–279.

24. Amen T and Kaganovich D (**2016**). Yeast screening platform identifies FDA-approved drugs that reduce Aβ oligomerization. **Microb Cell** 3(3): 97–100.

25. Shrestha A and Megeney L (**2015**). Yeast proteinopathy models: a robust tool for deciphering the basis of neurodegeneration. **Microb Cell** 2(12): 458–465.

26. Bond M, Brown R, Rallis C, Bahler J, and Mole S (**2015**). A central role for TOR signalling in a yeast model for juvenile CLN3 disease. **Microb Cell** 2(12): 466–480.

27. Carmona-Gutierrez D, Hughes AL, Madeo F, and Ruckenstuhl C (**2016**). The crucial impact of lysosomes in aging and longevity. **Ageing Res Rev** 32: 2–12. doi: 10.1016/j.arr.2016.04.009.

28. Raghupathi R (**2004**). Cell death mechanisms following traumatic brain injury. **Brain Pathol Zurich Switz** 14(2): 215–222.

29. Cheng G, Kong R, Zhang L, and Zhang J (**2012**). Mitochondria in traumatic brain injury and mitochondrial-targeted multipotential therapeutic strategies: Mitochondria in traumatic brain injury. **Br J Pharmacol** 167(4): 699–719.

30. Ludovico P, Sousa MJ, Silva MT, Leão C, and Côrte-Real M (**2001**). Saccharomyces cerevisiae commits to a programmed cell death process in response to acetic acid. **Microbiol Read Engl** 147(Pt 9): 2409–2415.

31. Rockenfeller P, Ring J, Muschett V, Beranek A, Buettner S, Carmona-Gutierrez D, Eisenberg T, Khoury C, Rechberger G, Kohlwein SD, Kroemer G, and Madeo F (**2010**). Fatty acids trigger mitochondrion-dependent necrosis. **Cell Cycle Georget Tex** 9(14): 2836–2842.

32. Eisenberg T, Carmona-Gutierrez D, Büttner S, Tavernarakis N, and Madeo F (**2010**). Necrosis in yeast. **Apoptosis Int J Program Cell Death** 15(3): 257–268.

33. Carmona-Gutiérrez D, Bauer MA, Ring J, Knauer H, Eisenberg T, Büttner S, Ruckenstuhl C, Reisenbichler A, Magnes C, Rechberger GN, Birner-Gruenberger R, Jungwirth H, Fröhlich K-U, Sinner F, Kroemer G, and Madeo F (**2011**). The propeptide of yeast cathepsin D inhibits programmed necrosis. **Cell Death Dis** 2: e161.

34. Braun RJ, Sommer C, Carmona-Gutierrez D, Khoury CM, Ring J, Büttner S, and Madeo F (**2011**). Neurotoxic 43-kDa TAR DNA-binding protein (TDP-43) triggers mitochondrion-dependent programmed cell death in yeast. **J Biol Chem** 286(22): 19958–19972.

35. Holmström KM and Finkel T (**2014**). Cellular mechanisms and physiological consequences of redox-dependent signalling. **Nat Rev Mol Cell Biol** 15(6): 411–421.

36. Braun RJ, Sommer C, Leibiger C, Gentier RJG, Dumit VI, Paduch K, Eisenberg T, Habernig L, Trausinger G, Magnes C, Pieber T, Sinner F, Dengjel J, van Leeuwen FW, Kroemer G, and Madeo F (**2015**). Accumulation of Basic Amino Acids at Mitochondria Dictates the Cytotoxicity of Aberrant Ubiquitin. **Cell Rep** 10(9): 1557–1571.

37. Braun R, Sommer C, Leibiger C, Gentier R, Dumit V, Paduch K, Eisenberg T, Habernig L, Trausinger G, Magnes C, Pieber T, Sinner F, Dengjel J, van Leeuwen F, Kroemer G, and Madeo F (**2015**). Modeling non-hereditary mechanisms of Alzheimer disease during apoptosis in yeast. **Microb Cell** 2(4): 136–138.

38. Ruckenstuhl C, Büttner S, Carmona-Gutierrez D, Eisenberg T, Kroemer G, Sigrist SJ, Fröhlich K-U, and Madeo F (**2009**). The Warburg Effect Suppresses Oxidative Stress Induced Apoptosis in a Yeast Model for Cancer. **PLoS ONE** 4(2): e4592.

39. Hand RA, Jia N, Bard M, and Craven RJ (**2003**). Saccharomyces cerevisiae Dap1p, a novel DNA damage response protein related to the mammalian membrane-associated progesterone receptor. **Eukaryot Cell** 2(2): 306–317.

40. Gregor A, Kocyłowski M, and Kostrzewska E (**1986**). Evaluation of the diagnostic usefulness of determining porphobilinogen deaminase activity in the erythrocytes in patients with acute intermittent porphyria and in carriers of the gene of this type of porphyria. **Przegl Lek** 43(11): 703–705.

41. Tubiana N, Mishal Z, le Caer F, Seigneurin JM, Berthoix Y, Martin PM, and Carcassonne Y (**1986**). Quantification of oestradiol binding at the surface of human lymphocytes by flow cytofluorimetry. **Br J Cancer** 54(3): 501–504.

42. Roussel D, Harding M, Runswick MJ, Walker JE, and Brand MD (**2002**). Does any yeast mitochondrial carrier have a native uncoupling protein function? **J Bioenerg Biomembr** 34(3): 165–176.

43. Barros MH, Bandy B, Tahara EB, and Kowaltowski AJ (**2004**). Higher respiratory activity decreases mitochondrial reactive oxygen release and increases life span in Saccharomyces cerevisiae. **J Biol Chem** 279(48): 49883–49888.

44. Skulachev VP (**1998**). Uncoupling: new approaches to an old problem of bioenergetics. **Biochim Biophys Acta BBA - Bioenerg** 1363(2): 100–124.

45. Mookerjee SA, Divakaruni AS, Jastroch M, and Brand MD (**2010**). Mitochondrial uncoupling and lifespan. **Mech Ageing Dev** 131(7–8): 463–472.

46. Amara CE, Shankland EG, Jubrias SA, Marcinek DJ, Kushmerick MJ, and Conley KE (**2007**). Mild mitochondrial uncoupling impacts cellular aging in human muscles in vivo. **Proc Natl Acad Sci** 104(3): 1057–1062.

47. Rodriguez AM, Monjo M, Roca P, and Palou A (**2002**). Opposite actions of testosterone and progesterone on UCP1 mRNA expression in cultured brown adipocytes. **Cell Mol Life Sci CMLS** 59(10): 1714–1723.

48. Wade Ruth and Jones Howard W. Jr. (**1956**). Effect of progesterone on mitochondrial adenosinetriphospatase. **JBC** 220: 547–551.

49. Liu W (**2012**). Control of Calcium in Yeast Cells. Introduction to Modeling Biological Cellular Control Systems. Springer Milan, Milano; pp. 95–122.

50. Ren ZW (**1992**). Radiofrequency ablation of left-sided atrioventricular accessory tract to treat supraventricular tachycardia. **Zhonghua Xin Xue Guan Bing Za Zhi** 20(4): 212–214.

51. Romarowski A, Sánchez-Cárdenas C, Ramírez-Gómez HV, Puga Molina L del C, Treviño CL, Hernández-Cruz A, Darszon A, and Buffone MG (**2016**). A Specific Transitory Increase in Intracellular Calcium Induced by Progesterone Promotes Acrosomal Exocytosis in Mouse Sperm. **Biol Reprod** 94(3): 63.

52. Li L-F, Xiang C, Zhu Y-B, and Qin K-R (**2014**). Modeling of progesterone-induced intracellular calcium signaling in human spermatozoa. **J Theor Biol** 351: 58–66.

53. Bonaccorsi L (**1998**). Progesterone-stimulated intracellular calcium increase in human spermatozoa is protein kinase C-independent. **Mol Hum Reprod** 4(3): 259–268.

54. Miller MR, Mannowetz N, Iavarone AT, Safavi R, Gracheva EO, Smith JF, Hill RZ, Bautista DM, Kirichok Y, and Lishko PV (**2016**). Unconventional endocannabinoid signaling governs sperm activation via the sex hormone progesterone. **Science** 352(6285): 555–559.

55. Batra S (**1986**). Effect of estrogen and progesterone treatment on calcium uptake by the myometrium and smooth muscle of the lower urinary tract. **Eur J Pharmacol** 127(1–2): 37–42.

56. Barbagallo M, Dominguez LJ, Licata G, Shan J, Bing L, Karpinski E, Pang PKT, and Resnick LM (**2001**). Vascular Effects of Progesterone : Role of Cellular Calcium Regulation. **Hypertens Dallas Tex 1979** 37(1): 142–147.

57. He Y, Gao Q, Han B, Zhu X, Zhu D, Tao J, Chen J, and Xu Z (**2016**). Progesterone suppressed vasoconstriction in human umbilical vein via reducing calcium entry. **Steroids** 108: 118–125.

58. Wu Z and Shen W (**2010**). Progesterone inhibits L-type calcium currents in gallbladder smooth muscle cells. **J Gastroenterol Hepatol** 25(12): 1838–1843.

59. Luoma JI, Kelley BG, and Mermelstein PG (**2011**). Progesterone inhibition of voltage-gated calcium channels is a potential neuroprotective mechanism against excitotoxicity. **Steroids** 76(9): 845–855.

60. Luoma JI, Stern CM, and Mermelstein PG (**2012**). Progesterone inhibition of neuronal calcium signaling underlies aspects of progesterone-mediated neuroprotection. **J Steroid Biochem Mol Biol** 131(1–2): 30–36.

61. Cai W, Zhu Y, Furuya K, Li Z, Sokabe M, and Chen L (**2008**). Two different molecular mechanisms underlying progesterone neuroprotection against ischemic brain damage. **Neuropharmacology** 55(2): 127–138.

62. Brotfain E, Gruenbaum SE, Boyko M, Kutz R, Zlotnik A, and Klein M (**2016**). Neuroprotection by Estrogen and Progesterone in Traumatic Brain Injury and Spinal Cord Injury. **Curr Neuropharmacol** 14(6): 641–653.

63. Wei J and Xiao G (**2013**). The neuroprotective effects of progesterone on traumatic brain injury: current status and future prospects. **Acta Pharmacol Sin** 34(12): 1485–1490.

64. Pascual JL, Murcy MA, Li S, Gong W, Eisenstadt R, Kumasaka K, Sims C, Smith DH, Browne K, Allen S, and Baren J (**2013**). Neuroprotective effects of progesterone in traumatic brain injury: blunted in vivo neutrophil activation at the blood-brain barrier. **Am J Surg** 206(6): 840-845; discussion 845-846.

65. Li Z, Wang B, Kan Z, Zhang B, Yang Z, Chen J, Wang D, Wei H, Zhang J, and Jiang R (**2012**). Progesterone increases circulating endothelial progenitor cells and induces neural regeneration after traumatic brain injury in aged rats. **J Neurotrauma** 29(2): 343–353.

66. Webster KM, Wright DK, Sun M, Semple BD, Ozturk E, Stein DG, O'Brien TJ, and Shultz SR (**2015**). Progesterone treatment reduces neuroinflammation, oxidative stress and brain damage and improves long-term outcomes in a rat model of repeated mild traumatic brain injury. **J Neuroinflammation** 12: 238.

67. Gueldener U, Heinisch J, Koehler GJ, Voss D, and Hegemann JH (**2002**). A second set of loxP marker cassettes for Cre-mediated multiple gene knockouts in budding yeast. **Nucleic Acids Res** 30(6): e23.

68. Güldener U, Heck S, Fielder T, Beinhauer J, and Hegemann JH (**1996**). A new efficient gene disruption cassette for repeated use in budding yeast. **Nucleic Acids Res** 24(13): 2519–2524.

69. Loewith R, Jacinto E, Wullschleger S, Lorberg A, Crespo JL, Bonenfant D, Oppliger W, Jenoe P, and Hall MN (**2002**). Two TOR complexes, only one of which is rapamycin sensitive, have distinct roles in cell growth control. **Mol Cell** 10(3): 457–468.

70. Kainz K, Tadic J, Zimmermann A, Pendl T, Carmona-Gutierrez D, Ruckenstuhl C, Eisenberg T, and Madeo F (**2017**). Methods to Assess Autophagy and Chronological Aging in Yeast. **Methods in Enzymology**. 588: 367–394.

17

Salt stress causes cell wall damage in yeast cells lacking mitochondrial DNA

Qiuqiang Gao[1], Liang-Chun Liou[2], Qun Ren[2], Xiaoming Bao[3] and Zhaojie Zhang[2]*

[1] State Key Laboratory of Bioreactor Engineering, East China University of Science and Technology, 130 Meilong Road, Shanghai 200237, China.
[2] Department of Zoology and Physiology, University of Wyoming, Laramie, WY 82071, USA.
[3] State Key Laboratory of Microbial Technology, Shandong University, Jinan, China.
* Corresponding Author: Dr. Zhaojie Zhang, Department of Zoology and Physiology, University of Wyoming; Laramie, WY, 82071, USA; Email: zzhang@uwyo.edu

ABSTRACT The yeast cell wall plays an important role in maintaining cell morphology, cell integrity and response to environmental stresses. Here, we report that salt stress causes cell wall damage in yeast cells lacking mitochondrial DNA (ρ^0). Upon salt treatment, the cell wall is thickened, broken and becomes more sensitive to the cell wall-perturbing agent sodium dodecyl sulfate (SDS). Also, *SCW11* mRNA levels are elevated in ρ^0 cells. Deletion of *SCW11* significantly decreases the sensitivity of ρ^0 cells to SDS after salt treatment, while overexpression of *SCW11* results in higher sensitivity. In addition, salt stress in ρ^0 cells induces high levels of reactive oxygen species (ROS), which further damages the cell wall, causing cells to become more sensitive towards the cell wall-perturbing agent.

Keywords: cell wall damage; reactive oxygen species; SCW11; salt stress; Saccharomyces cerevisiae.

INTRODUCTION

The yeast *Saccharomyces cerevisiae* cell wall occupies about 15% of the total cell volume and plays important roles in maintaining cell shape, cell integrity and protection against environmental stresses [1]. The yeast cell wall forms a microfibrillar network complex composed of glucan, mannoprotein and chitin. Glucan is mainly (80 to 90%) composed of β-1,3-glucan chains with some β-1,6-linked glucan branches. Glucan represents 50-60% of the cell wall mass and its main function is to maintain cell elasticity that can adapt to different physiological states (sporulation or budding) and various stress conditions, including exposure to cell wall-perturbing agents or hypo-osmotic shock [2,3].

Scw11p is a cell wall protein similar to glucanases, whose main function is to break down glucans [4,5]. Whole-genome transcriptional analysis suggests that Scw11p is required for efficient cell separation following cytokinesis [6], and is also associated with cell wall metabolism [7]. Mutation of *scw11* suppresses the phenotype of Δscw4/Δscw10 double mutants, or Δscw4/Δscw10/Δbgl2 triple mutants, suggesting that Scw11p has an activity antagonistic to that of Scw4p, Scw10p and Bgl2p [8].

Reactive oxygen species (ROS) play a crucial role in the signal transduction pathways that regulate yeast cell death [9,10]. Several studies have shown that ROS function as inducers of yeast apoptosis [11], while others have demonstrated that ROS play a crucial role as mediators of apopto-

sis [12,13]. It has been shown that ROS accumulate in yeast cells after induction of apoptotic death by various stimuli, and are necessary and sufficient to induce an apoptotic phenotype in yeast [11,14-16].

Previously, we showed that yeast lacking mitochondrial DNA (ρ^0) undergoes apoptosis upon salt stress and that cell death is mediated by cytochrome *c* [17]. In this study, we further report that salt stress causes cell wall damage (CWD) in ρ^0 cells and that this damage is related to elevated levels of *SCW11* and salt stress-induced ROS.

RESULTS AND DISCUSSION

Salt stress causes cell wall damage (CWD) in ρ^0 cells

We previously reported that ρ^0 cells were sensitive to salt stress and consequently underwent apoptotic cell death [17]. In the current study, our TEM analysis revealed that salt stress caused an abnormal cell wall structure in ρ^0 cells. When treated with NaCl, the cell wall of more than 50% of the ρ^0 cells became thicker and uneven, compared to ρ^0 cells without NaCl treatment or to the wild type (WT). Certain areas of the cell wall also appeared abrupt, or damaged (Fig. 1A). To further examine possible salt stress effects on the cell wall, we tested the sensitivity of ρ^0 cells towards sodium dodecyl sulfate (SDS), a typical cell wall-perturbing agent, before and after salt stress. As shown in Fig. 1B, no apparent change was observed in WT cells when treated with NaCl, SDS or both. ρ^0 cells, however,

A

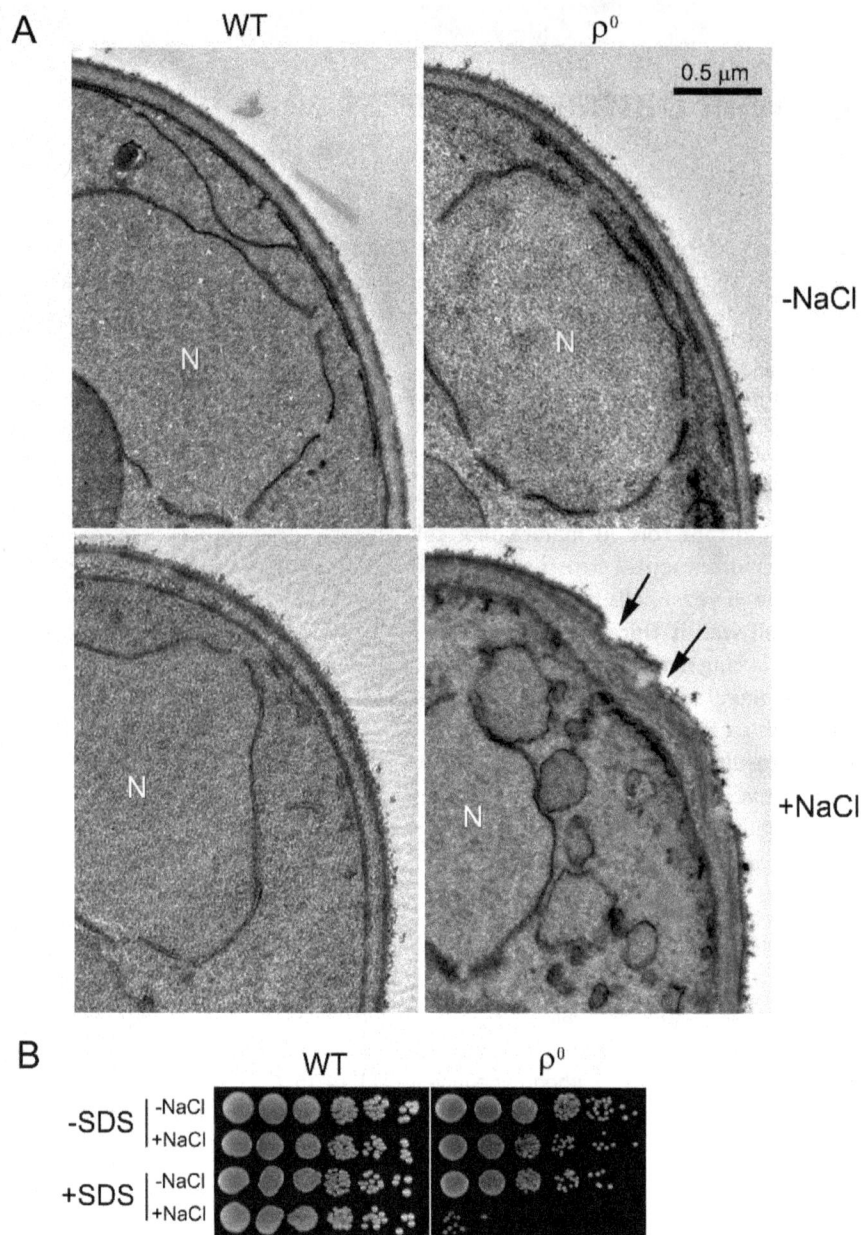

FIGURE 1: **(A)** TEM images of WT and ρ^0 cells growing in YPD medium with or without treatment of 0.6 M NaCl for 1 hr. A normal cell wall structure was observed in WT either with or without NaCl. The cell wall structure was damaged in ρ^0 cells upon NaCl treatment. Arrows indicate damaged cell wall. N = nucleus. **(B)** ρ^0 cells are hypersensitive to SDS after NaCl treatment. Cells were first grown in YPD liquid media containing 0.6 M NaCl for 1 hr, then washed, treated with 0.1% SDS for 0.5 hr, spotted on YPD plates using 4 µl of 1:5 serial dilutions and incubated for 2-3 days at 30°C.

were hypersensitive to SDS after salt stress compared to cells treated with salt or SDS alone, further suggesting that the cell wall was damaged by the salt stress.

SCW11 is involved in CWD of ρ^0 cells under salt stress

Scw11p is a cell wall protein with an endo-1,3-β-glucanase activity. It has been previously reported that the *SCW11* gene is up-regulated in ρ^0 cells [18]. However, it is still elusive how *SCW11* is involved in the salt stress response in ρ^0 cells. In this study, *SCW11* gene expression was first examined using RT-PCR with or without salt treatment. Consistent with microarray analyses [18], we showed that in the absence of NaCl, the *SCW11* expression level was significantly (P < 0.001) higher in ρ^0 cells than in the WT cells. Under salt stress conditions, *SCW11* expression in ρ^0 cells

was decreased but still significantly higher (P < 0.05) than in WT cells (Fig. 2A, 2B).

To further assess the role of *SCW11* during salt stress response, *SCW11* was deleted or overexpressed in both WT and ρ^0 cells. As shown in Fig. 2C, deletion or overexpression of *SCW11* had no apparent effect on the WT. However, deletion of *SCW11* greatly enhanced the survival rate of ρ^0 cells, especially when the cells were treated with both NaCl and SDS. *SCW11* overexpression caused ρ^0 cells to become more sensitive towards salt stress and SDS (Fig. 2D). These results suggest that, under salt stress conditions, *SCW11* is involved in cell wall organization and its down-regulation in ρ^0 cells (Fig. 2A) is part of a protection mechanism against salt stress.

The yeast cell wall is a dynamic structure that can adapt to different physiological states and various stress condi-

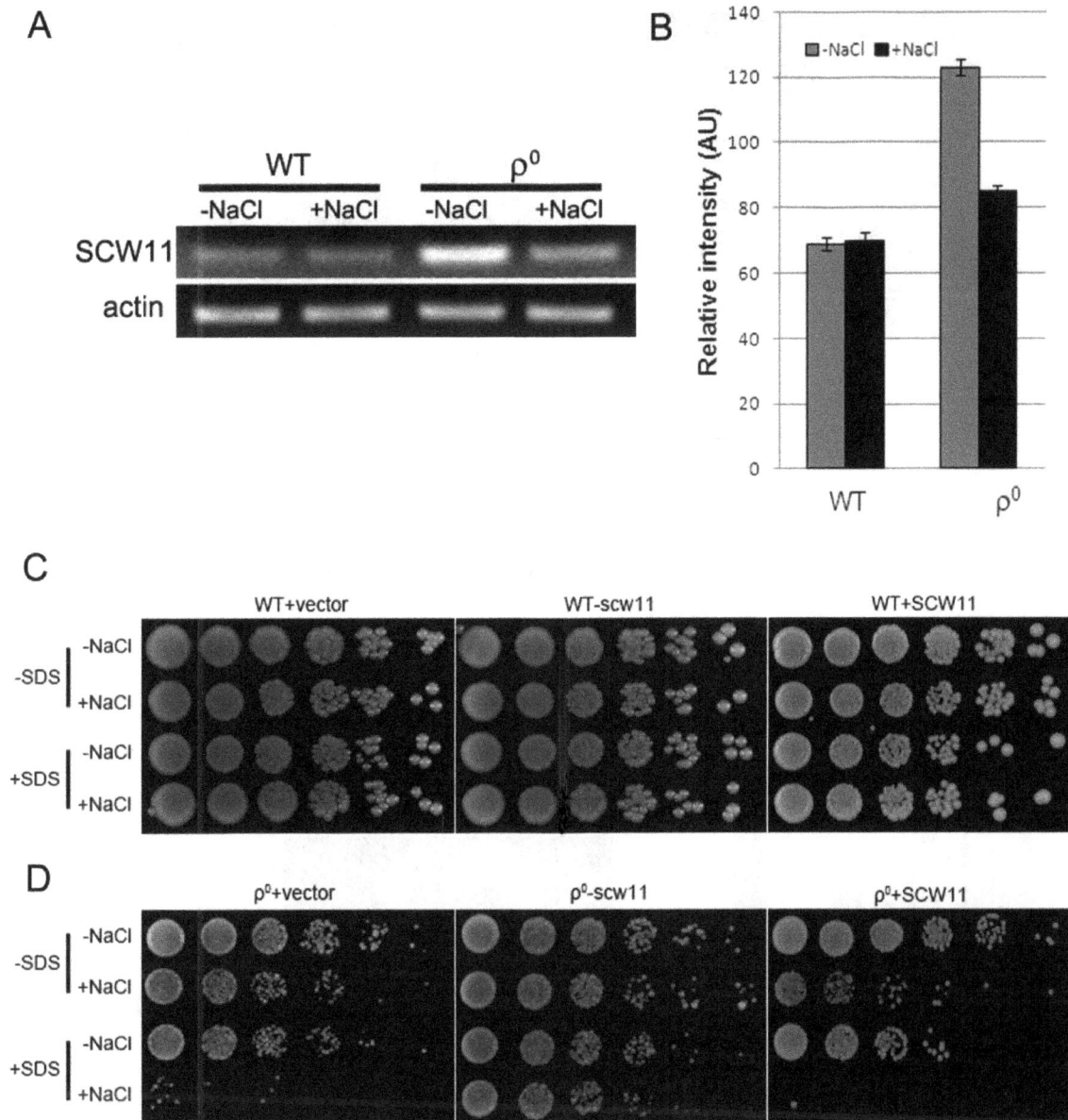

FIGURE 2: (A) RT-PCR analysis of ρ⁰ cells showing *SCW11* expression upregulated compared to WT. *SCW11* expression in ρ⁰ cells was decreased when treated with 0.6 M NaCl for 30 min, but still higher than in WT cells. **(B)** Quantitative analysis of the RT-PCR bands from three independent experiments. **(C) and (D)** The cell wall gene *SCW11* is involved in salt stress-induced cell death in ρ⁰ cells. **(C)** When treated with NaCl, SDS, or both, no apparent change is observed in the growth of WT cells when *SCW11* is either deleted or overexpressed. **(D)** Deletion of *SCW11* in ρ⁰ cells (ρ⁰-scw11) increases their resistance to salt; SCW11 overexpression (ρ⁰+SCW11) causes cells to become more sensitive to SDS after NaCl treatment (0.6 M NaCl for 1 hr). The BG1805 empty vector (WT/ρ⁰+vector) was used as a negative control. Cells were spotted on YPG using 4 µl of 1:5 serial dilutions and cultured at 30°C for 3-4 days.

tions [2,3]. The endo-β-1,3-glucanase is an abundant protein in the yeast cell wall making it rigid by inserting intrachain linkages into 1,3-β-glucan [19]. Overexpression of the endo-β-1,3-glucanase Bgl2p has been shown to lead to defects in the cell wall structure and to sensitivity towards osmotic stress [20]. Similar to this result, we show that the *SCW11* gene, coding for a glucanase like-protein, is upregulated in ρ⁰ cells, where it leads to a higher sensitivity towards SDS and salt stress.

ROS is involved in CWD of ρ⁰ cells under salt stress

Mitochondria are the main source for ROS, which play an important role in yeast apoptosis. With the lack of mitochondrial DNA, it is unclear if ρ⁰ cells produce any ROS, and if ROS are involved in salt stress response of the ρ⁰ cells. We first examined the ROS level using dihydroethidium (DHE), a fluorescence dye that has specificity toward superoxide for detecting ROS production [16,21,22]. Little ROS was detected in wild type in the absence or presence

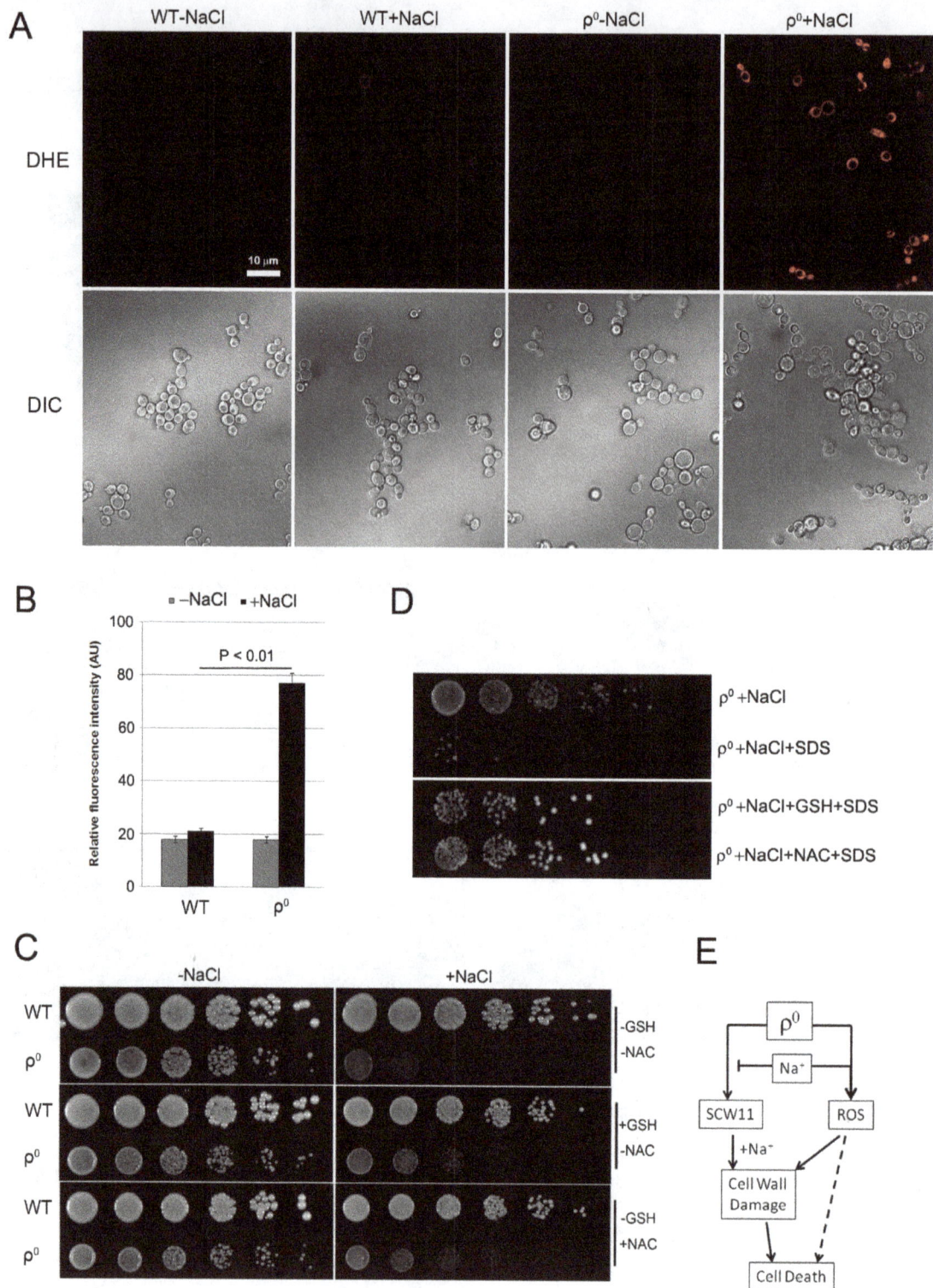

FIGURE 3: (A) and (B) Reactive oxygen species (ROS) production in the WT and ρ⁰ cells treated with 0.6 M NaCl for 15 min. **(A)** After salt stress, ρ⁰ cells displayed a higher fluorescence level than WT cells. Staining with 5 µM dihydroethidium (DHE) was used. **(B)** Quantification of ROS production. Relative fluorescence intensities were measured by the ImageJ software. Values presented are means of three independent experiments. About 300 cells were measured during each experiment. **(C)** Antioxidant GSH or NAC slightly decreases the salt sensitivity of ρ⁰ cells. Log phase cells were spotted directly onto YPD plates containing 0.6 M NaCl and GSH (10 mM) or NAC (20 mM). Cells were cultured at 30°C for 2 days. **(D)** Addition of antioxidant GSH or NAC reduced the sensitivity of ρ⁰ cells towards SDS after salt stress. Cells were treated with 0.6 M NaCl in presence of GSH (10 mM) or NAC (20 mM) for 1 hr. Cells were washed, treated with SDS for 0.5 hr and spotted on YPD plates. Cells were then cultured at 30°C for 2 days. **(E)** Diagram showing the possible involvement of *SCW11* and ROS in salt stress-induced cell wall damage and cell death of ρ⁰ cells.

of NaCl. In ρ^0 cells, however, NaCl treatment caused a significant increase of the ROS level (Fig. 3A, 3B). Addition of antioxidant GSH or NAC only slightly improved the cell survival of ρ^0 cells (Fig. 3C), suggesting that ROS may not directly be involved in salt stress response and cell death of the ρ^0 cells. We further tested if addition of the antioxidant could improve the sensitivity of ρ^0 cells to SDS. As shown in Fig. 3D, in the presence of GSH or NAC, ρ^0 cells were much less sensitive to SDS after NaCl treatment. This result suggests that ROS likely affect the salt sensitivity of ρ^0 cells, at least in part, via their damage to cell wall. Deletion or overexpression of *SCW11* in ρ^0 cells greatly affects the sensitivity of ρ^0 cells towards SDS after the salt stress (Fig. 2D). However, DHE staining showed that deletion or overexpression of *SCW11* did not affect the ROS level of the ρ^0 cells (Fig. S1, compared to Figs. 3A, 3B). This result further suggests that ROS are not directly involved in the cell survival or cell death of ρ^0 cells, but rather through their damage to cell wall. Deletion of *SCW11* makes ρ^0 cells more resistant, while *SCW11* overexpression makes them more sensitive, to ROS damage.

It has been reported that ROS cause damages to DNA, proteins and lipids [23], which in turn, induce apoptotic cell death [11,24], or aging [25]. To our best knowledge, this is the first report that ROS induce potential cell wall damage in ρ^0 cells. One possibility is that the increased level of Scw11 makes the cell wall more vulnerable to ROS.

In summary, we demonstrated that the *SCW11* level was elevated in ρ^0 cells and the high level of Scw11 caused ρ^0 cells prone to cell wall damage by salt stress. The salt stress-induced ROS further damaged the cell wall, causing decreased viability and cell death of the ρ^0 cells (Fig. 3E).

MATERIALS AND METHODS

Yeast strains and media

Yeast strains are derivatives of *S. cerevisiae* W303-1A (*MATa, leu2-3/112 ura3-1 trp1-1 his3-11/15 ade2-1 can1-100*). The *scw11* deletion was introduced by PCR-mediated gene replacement [26], replacing the complete sequence of YGL028C with G418 marker. *SCW11* overexpression was constructed using the BG1805 plasmid [27]. Yeast cells were routinely grown in YPD medium (1% yeast extract, 2% peptone, 2% glucose) at 30°C. For overexpression of *SCW11*, cells were preincubated in BG1805 selection medium (URA minus), then transferred to YPR (1% yeast extract, 2% peptone, 2% raffinose) overnight, and then to YPG (1% yeast extract, 2% peptone, 2% galactose).

Transmission electron microscopy (TEM)

Cells were cultured at 30°C in YPD medium to early log phase, treated with 0.6 M NaCl for 1 hr. Cells were then harvested and prepared for TEM according to [28].

REFERENCES

1. Feldmann H (**2012**). Yeast: molecular and cell biology. Wiley-Blackwell Publishers, Weinheim.

2. Levin DE (**2005**). Cell wall integrity signaling in Saccharomyces cerevisiae. **Microbiol Mol Biol Rev** 69(2):262-291.

Spot dilution growth assays

Cells were precultured in YPD liquid medium overnight. Cells were harvested by centrifugation, washed three times with ddH$_2$O, and then resuspended in ddH$_2$O. The cell density was normalized to 1×10^7 cells/ml. A five-fold serial dilution was made and 4 μl of each dilution was spotted onto appropriate YPD plates. For YPG medium growth, cells were precultured in YPR liquid overnight, and transferred to YPG liquid for 6 hr. A five-fold serial dilution was made as above, and spotted onto appropriate YPG plates. The antioxidants glutathione (GSH, 10 mM) or N-acetyl-L-cysteine (NAC, 20 mM) were added directly to the medium.

For the SDS sensitivity test, cells were grown in YPD liquid medium to early log phase, treated with 0.6 M NaCl (0.2 M NaCl for YPG mdium) for 1 hr, then washed with ddH$_2$O, resuspended in 2-[4-(2-Hydroxyethyl)-1-piperazinyl] ethanesulfonic acid (HEPES) buffer (10 mM pH 7.5), treated with 0.1% SDS for 0.5 hr, then washed one time with HEPES buffer. A five-fold serial dilution was made as above, and then spotted onto the YPD or YPG plates.

Semi-quantitative RT-PCR

Total RNA from yeast was extracted using RNeasy Protect Mini Kit (QIAGEN, CA). The RT-PCR and the amplification procedure were performed as in [29]. Yeast actin gene (*ACT1*) was used as control, and the primers are 5'-TGTCACCAACTGGGACGATA-3' and 5'-AACCAGCGTAAATTGGAACG-3'. Primers used for *SCW11* amplification are 5'-CCCCAACTGTCGAATTCCTA-3' and 5'-AAAGTAGAGGTTGGCTGCGA-3'. Quantitative analysis was conducted using ImageJ software (http://rsb.info.nih.gov/ij/).

Detection of reactive oxygen species (ROS)

The detection of ROS was performed according to [15]. Briefly, cells were grown to early log phase, treated with 0.8 M NaCl for 15 min, then washed with PBS for 3 times. Cells were stained with dihydroethidium (5 μM) (Sigma Chemical Co.) for 10 min in the dark, viewed with a Zeiss 710 laser scanning confocal microscope with excitation at 514 nm (Jena, Germany).

Statistical Analysis of Data

A two-tailed t-test was used for statistical analysis. $P \leq 0.05$ was considered as statistically significant.

ACKNOWLEDGMENTS

This study was supported in part by an Institutional Development Award (IDeA) from the National Institute of General Medical Sciences of the National Institutes of Health under grant number P30 GM103398, by the National Natural Science Foundation of China (No. 31300070), and the China Postdoctoral Science Foundation (No. 2012M520850).

CONFLICT OF INTEREST

The authors declare no conflict of interest.

3. Lesage G, and Bussey H (**2006**). Cell wall assembly in Saccharomyces cerevisiae. **Microbiol Mol Biol Rev** 70(2): 317-343.

4. Cappellaro C, Mrsa V, and Tanner W (**1998**). New potential cell wall glucanases of Saccharomyces cerevisiae and their involvement in mating. **J Bacteriol** 180(19): 5030-5037.

5. Zeitlinger J, Simon I, Harbison CT, Hannett NM, Volkert TL, Fink GR, and Young RA (**2003**). Program-specific distribution of a transcription factor dependent on partner transcription factor and MAPK signaling. **Cell** 113(3): 395-404.

6. Bidlingmaier S, Weiss EL, Seidel C, Drubin DG, and Snyder M (**2001**). The Cbk1p pathway is important for polarized cell growth and cell separation in Saccharomyces cerevisiae. **Mol Cell Biol** 21(7): 2449-2462.

7. Doolin MT, Johnson AL, Johnston LH, and Butler G (**2001**). Overlapping and distinct roles of the duplicated yeast transcription factors Ace2p and Swi5p. **Mol Microbiol** 40(2): 422-432.

8. Teparić R, Stuparević I, and Mrsa V (**2004**). Increased mortality of Saccharomyces cerevisiae cell wall protein mutants. **Microbiology** 150(Pt 10): 3145-3150.

9. Slater AF, Nobel CS, Maellaro E, Bustamante J, Kimland M, and Orrenius S (**1995**). Nitrone spin traps and a nitroxide antioxidant inhibit a common pathway of thymocyte apoptosis. **Biochem J** 306(Pt 3): 771-778.

10. Jiang S, Cai J, Wallace DC, and Jones DP (**1999**). Cytochrome c-mediated apoptosis in cells lacking mitochondrial DNA. Signaling pathway involving release and caspase 3 activation is conserved. **J Biol Chem** 274(42): 29905-29911.

11. Madeo F, Fröhlich E, Ligr M, Grey M, Sigrist SJ, Wolf DH, and Fröhlich KU (**1999**). Oxygen stress: a regulator of apoptosis in yeast. **J Cell Biol** 145(4): 757-767.

12. Fleury C, Mignotte B, and Vayssiere JL (**2002**). Mitochondrial reactive oxygen species in cell death signaling. **Biochimie** 84(2-3): 131-141.

13. Eisler H, **Fröhlich** KU, and Heidenreich E (**2004**). Starvation for an essential amino acid induces apoptosis and oxidative stress in yeast. **Exp Cell Res** 300(2): 345-353.

14. Del Carratore R, Della Croce C, Simili M, Taccini E, Scavuzzo M, and Sbrana S (**2002**). Cell cycle and morphological alterations as indicative of apoptosis promoted by UV irradiation in S. cerevisiae. **Mutat Res** 513(1-2): 183-191.

15. Ren Q, Yang H, Rosinski M, Conrad MN, Dresser ME, Guacci V, and Zhang Z (**2005**). Mutation of the cohesin related gene PDS5 causes apoptotic cell death in Saccharomyces cerevisiae during early meiosis. **Mutat Res** 570(2): 163-173.

16. Pérez-Gallardo RV, Briones LS, Díaz-Pérez AL, Gutiérrez S, Rodríguez-Zavala JS, and Campos-García J (**2013**). Reactive oxygen species production induced by ethanol in *Saccharomyces cerevisiae* increases because of a dysfunctional mitochondrial iron-sulfur cluster assembly system. **FEMS Yeast Res** 13(8): 804-819.

17. Gao Q, Ren Q, Liou LC, Bao X, and Zhang Z (**2011**). Mitochondrial DNA protects against salt stress-induced cytochrome c-mediated apoptosis in yeast. **FEBS Lett** 585(15): 2507-2512.

18. Woo DK, Phang TL, Trawick JD, and Poyton RO (**2009**). Multiple pathways of mitochondrial-nuclear communication in yeast: Intergenomic signaling involves ABF1 and affects a different set of genes than retrograde regulation. **Biochim Biophys Acta** 1789(2): 135-145.

19. Mrsa V, Klebl F, and Tanner W (**1993**). Purification and characterization of the *Saccharomyces cerevisiae* BGL2 gene product, a cell wall endo-beta-1,3-glucanase. **J Bacteriol** 175(7): 2102-2106.

20. Shimizu J, Yoda K, and Yamasaki M (**1994**). The hypo-osmolarity-sensitive phenotype of the *Saccharomyces cerevisiae* hpo2 mutant is due to a mutation in PKC1, which regulates expression of beta-glucanase. **Mol Gen Genet** 242(6): 641-648.

21. Benov L, Sztejnberg L, and Fridovich I (**1998**). Critical evaluation of the use of hydroethidine as a measure of superoxide anion radical. **Free Radic Biol Med** 25(7): 826-831.

22. Weinberger M, Mesquita A, Carroll T, Marks L, Yang H, Zhang Z, Ludovico P, and Burhans WC (**2010**). Growth signaling promotes chronological aging in budding yeast by inducing superoxide anions that inhibit quiescence. **Aging** 2(10): 709-726.

23. Farrugia G, and Balzan R (**2012**). Oxidative stress and programmed cell death in yeast. **Front Oncol** 2:64.

24. Carmona-Gutierrez D, Eisenberg T, Büttner S, Meisinger C, Kroemer G, and Madeo F (**2010**). Apoptosis in yeast: triggers, pathways, subroutines. **Cell Death Differ** 17(5): 763-773.

25. Fröhlich KU, and Madeo F (**2001**). Apoptosis in yeast: a new model for aging research. **Exp Gerontol** 37(1): 27-31.

26. Wach A, Brachat A, Pöhlmann R, and Philippsen P (**1994**). New heterologous modules for classical or PCR-based gene disruptions in Saccharomyces cerevisiae. **Yeast** 10(13): 1793-1808.

27. Gelperin DM, White MA, Wilkinson ML, Kon Y, Kung LA, Wise KJ, Lopez-Hoyo N, Jiang L, Piccirillo S, Yu H, Gerstein M, Dumont ME, Phizicky EM, Snyder M, and Grayhack EJ (**2005**). Biochemical and genetic analysis of the yeast proteome with a movable ORF collection. **Genes Dev** 19(23): 2816-2826.

28. Yang H, Ren Q, and Zhang Z (**2006**). Chromosome or chromatin condensation leads to meiosis or apoptosis in stationary yeast (Saccharomyces cerevisiae) cells. **FEMS Yeast Res** 6(8): 1254-1263.

29. Gao Q, Liu Y, Wang M, Zhang J, Gai Y, Zhu C, and Guo X (**2009**). Molecular cloning and characterization of an inducible RNA-dependent RNA polymerase gene, GhRdRP, from cotton (*Gossypium hirsutum L.*). **Mol Biol Rep** 36(1): 47-56.

18

Overexpression of the transcription factor Yap1 modifies intracellular redox conditions and enhances recombinant protein secretion

Marizela Delic[1,2], Alexandra B. Graf[2,3], Gunda Koellensperger[1,4], Christina Haberhauer-Troyer[1,4], Stephan Hann[1,4], Diethard Mattanovich[1,2], Brigitte Gasser[1,2,*]

[1] Department of Biotechnology, BOKU University of Natural Resources and Life Sciences Vienna, Vienna, Austria.
[2] Austrian Centre of Industrial Biotechnology (ACIB), Vienna, Austria.
[3] School of Bioengineering, University of Applied Sciences FH Campus Wien, Vienna, Austria.
[4] Department of Chemistry, BOKU University of Natural Resources and Life Sciences Vienna, Vienna, Austria.
* Corresponding Author: Brigitte Gasser, Department of Biotechnology, BOKU University of Natural Resources and Life Sciences Vienna, Muthgasse 18; 1190 Vienna, Austria; E-mail: brigitte.gasser@boku.ac.at

ABSTRACT Oxidative folding of secretory proteins in the endoplasmic reticulum (ER) is a redox active process, which also impacts the redox conditions in the cytosol. As the transcription factor Yap1 is involved in the transcriptional response to oxidative stress, we investigate its role upon the production of secretory proteins, using the yeast *Pichia pastoris* as model, and report a novel important role of Yap1 during oxidative protein folding. Yap1 is needed for the detoxification of reactive oxygen species (ROS) caused by increased oxidative protein folding. Constitutive co-overexpression of *PpYAP1* leads to increased levels of secreted recombinant protein, while a lowered Yap1 function leads to accumulation of ROS and strong flocculation. Transcriptional analysis revealed that more than 150 genes were affected by overexpression of *YAP1*, in particular genes coding for antioxidant enzymes or involved in oxidation-reduction processes. By monitoring intracellular redox conditions within the cytosol and the ER using redox-sensitive roGFP1 variants, we could show that overexpression of *YAP1* restores cellular redox conditions of protein-secreting *P. pastoris* by reoxidizing the cytosolic redox state to the levels of the wild type. These alterations are also reflected by increased levels of oxidized intracellular glutathione (GSSG) in the *YAP1* co-overexpressing strain. Taken together, these data indicate a strong impact of intracellular redox balance on the secretion of (recombinant) proteins without affecting protein folding per se. Re-establishing suitable redox conditions by tuning the antioxidant capacity of the cell reduces metabolic load and cell stress caused by high oxidative protein folding load, thereby increasing the secretion capacity.

Keywords: ER, cytosol, cellular redox regulation, oxidative protein folding, glutathione, redox sensitive roGFP, Pichia pastoris.

Abbreviations:
ER - endoplasmic reticulum,
PDI - protein disulfide isomerase,
ROS - reactive oxygen species,
roGFP - redox sensitive variants of green fluorescent protein,
SOD - superoxide dismutase,
TRP - trypsinogen,
UPR - unfolded protein response.

INTRODUCTION

Cells have evolved potent antioxidant mechanisms in order to circumvent oxidative stress and oxidative damage of cellular components when reactive oxygen species (ROS) occur during physiological conditions. Apart from the mitochondrial respiratory chain and the beta-oxidation of fatty acids occurring in the peroxisomes, also oxidative protein folding of secretory proteins within the endoplasmic reticulum (ER) was reported to contribute to the formation of ROS (reviewed by [1-4]).

Formation of disulfide bonds is attained through the oxidative protein folding machinery in the ER, using glutathione as redox buffer. In order to enable oxidative protein folding, the ER is specially equipped with dedicated enzymes such as oxidoreductases and chaperones and has a more oxidizing environment than the cytosol (for recent reviews see e.g. [5-7]). During *de novo* formation of disulfide bonds, protein disulfide isomerase (PDI) introduces disulfide bonds into nascent client proteins, which causes reduction of PDI. FAD-dependent ER oxidoreductin Ero1 is responsible for the re-oxidation of PDI [8]. Ero1 uses molecular oxygen as terminal electron acceptor, thereby generating stoichiometric amounts of hydrogen peroxide H_2O_2 [9, 10]. It has been suggested that glutathione acts as a redox buffer against ER-derived oxidative stress [11],[12].

Once generated, ROS can damage nucleic acids and lead to oxidation of proteins and peroxidation of lipids [13]. As a consequence, the expression of a set of proteins that eliminate ROS is induced [14], which is called "the oxidative stress response". Most organisms have evolved a combination of mechanisms for maintaining cellular redox balance, involving both ROS detoxifying enzymes with very high catalytic activity (such as superoxide dismutases (SODs), catalases and glutathione peroxidases), redox-regulating enzymes (e.g. thioredoxin, glutaredoxins, peroxiredoxins) as well as non-enzymatic compounds such as glutathione [3]. The oxidative stress response is a complex regulatory circuit controlled by the interplay of different transcription factors, which are often subject to redox induced structural changes [4, 15]. Among them, the yeast AP-1 transcription factor Yap1 plays a major role in the regulation of the transcriptional response to oxidative stress [16].

Upon treatment of the yeast *Saccharomyces cerevisiae* with hydrogen peroxide, at least 115 genes are overexpressed and 52 repressed as a consequence of oxidative stress in the cell [17], the majority of which is regulated by Yap1p [18]. Yap1p-mediated regulatory pathways are activated by redox-sensitive cysteine residues that serve as cellular sensors for changes in the intracellular redox balance (reviewed e.g. by [3] and [12]. Through the formation of different intramolecular disulfide bonds, Yap1 undergoes different conformational changes upon exposure to different oxidants (such as hydrogen peroxide or the superoxide generating oxidant menadione), which result in nuclear localization of Yap1 by masking the leucine-rich nuclear export signal [19]. Microarray data of a *YAP1* overexpressing *S. cerevisiae* strain revealed that oxidoreductases formed a remarkable fraction of the regulated genes. This group was supposed to have a protective function upon oxidative stress [20]. Another important function of Yap1 and its target genes is to balance cytosolic redox homeostasis [21].

Using the yeast *Pichia pastoris* as a model, we have recently shown that protein folding stress within the ER has a strong impact on the redox state of the cytosol, leading to a significant reduction of this compartment [22]. By applying redox sensitive variants of green fluorescent protein (roGFP) to measure *in vivo* glutathione redox conditions in ER and cytosol during oxidative protein folding, we observed a significant reduction of the redox state of the cytosol when ER resident proteins such as Ero1, Pdi1, or recombinant secretory proteins were overexpressed.

Cellular redox imbalance is stressful for the cells, and does not only lead to reduced productivity in biotechnological production of recombinant proteins, but is also associated with the development of many aging-related human diseases including diabetes mellitus, atherosclerosis, and neurodegenerative diseases such as Alzheimer's, amyotrophic lateral sclerosis and Parkinson's (reviewed e.g. by [23]). Therefore we aimed at restoring cytosolic redox rati-

os during conditions of increased oxidative protein folding in the ER.

Yano *et al.* recently identified the *P. pastoris* Yap1 homolog and reported the involvement of the transcription factor in the detoxification of formaldehyde and ROS in cells grown on methanol as carbon and energy source [24, 25]. Here we investigated the role of Yap1 during the production of recombinant secretory proteins in glucose based growth conditions in *P. pastoris*, and report a novel role of Yap1 during ER resident oxidative protein folding.

RESULTS AND DISCUSSION

Downregulation of the transcription factor Yap1 leads to accumulation of reactive oxygen species (ROS), whereas Yap1 overproduction has a positive influence on protein secretion

In our previous studies we reported that increased oxidative protein folding within the ER of *P. pastoris* led to changes in the redox state of the cytosol, independent of unfolded protein response (UPR) activation [22]. Oxidative protein folding has been implicated with the generation of stoichiometric amounts of H_2O_2 during regeneration of reduced Pdi1 by Ero1 [26], correspondingly, we detected ROS in strains with increased levels of Pdi1. However, we did not detect ROS accumulation upon increased folding load due to overexpression of secretory model proteins unless ER stress and UPR activation occurred [22]. Therefore we assumed that the cellular ROS scavenging system is effective to degrade ROS under physiological levels of oxidative folding, but is overwhelmed upon severe ER stress, thus leading to ROS accumulation. As Yap1 could be involved in the detoxification of ROS generated during oxidative folding of proteins in the ER, we examined the role of this transcription factor during the production of the secretory model protein trypsinogen (TRP) in more detail. In order to observe the specific effects of oxidative protein folding, we used a TRP secreting strain which is not induced for UPR [22].

In this strain background (trp), we investigated the influence of lowered (trpΔyap1) and increased constitutive (trpYAP1) expression levels of *YAP1*. Increased constitutive expression of *YAP1* was obtained by integrating an additional copy of *YAP1* under the control of the strong glycolytic glyceraldehyde-3-phosphate dehydrogenase P_{GAP} promoter into the *P. pastoris* genome [27], while downregulation of the *YAP1* expression level was achieved through the exchange of the native promoter of the respective gene with the serine repressible promoter P_{SER1} [28]. Repression of this promoter of the serine biosynthesis gene *SER1* was obtained through repeated addition of 10 mM serine to the synthetic M2 medium during the cultivation. We decided to go for a conditional *yap1* repression rather than a total gene knock out in order to be more flexible and prevent potential growth impairment as was reported for the *yap1* knock-out during growth on methanol [24, 25]. We did not analyse *YAP1* down-regulation in the wild type background, because so far no detectable phenotypic differences were reported for *yap1* mutants during normal non-stressed growth conditions ([25] for *P. pastoris*, [29] for *S. cerevisiae*).

Transcript level determination with quantitative real time PCR showed clearly that the trpΔyap1 strain expressed less than 10% of YAP1 mRNA compared to *P. pastoris* wild type X-33 or the parental trp strain (all grown in repressing conditions of P$_{SER1}$), whereas the trpYAP1 strain exhibited significantly higher YAP1 mRNA levels (approx. 8-fold higher levels as compared to X-33 and the parental trp strain) (Figure 1A).

Next, we determined the levels of accumulated ROS in the Yap1 deregulated strains compared to their parental strain. DHR and DHE were used for detection of H$_2$O$_2$ and superoxide anion as described previously [22]. As reported in [22], the non-UPR induced single copy trp strain did not accumulate significant levels of ROS compared to the wild type X-33 (data not shown). ROS analysis with flow cytometry revealed that trpΔyap1 indeed accumulated significantly higher levels of both ROS species compared to the non-engineered control strain (Figure 1B), indicating its essential function in detoxification of accumulated ROS. Accumulation of ROS in the YAP1 overexpressing strain trpYAP1 was indistinguishable from the non-engineered single copy trp strain. Moreover, we observed a strong phenotype of the Yap1-depleted strain trpΔyap1, which showed intensive flocculation in liquid cultures (Figure 1C). As no phenotype has been reported upon yap1 deletion previously, we hypothesize that flocculation is a response triggered by the enhanced level of oxidative protein folding.

Oxidative stress response has been implicated with re-

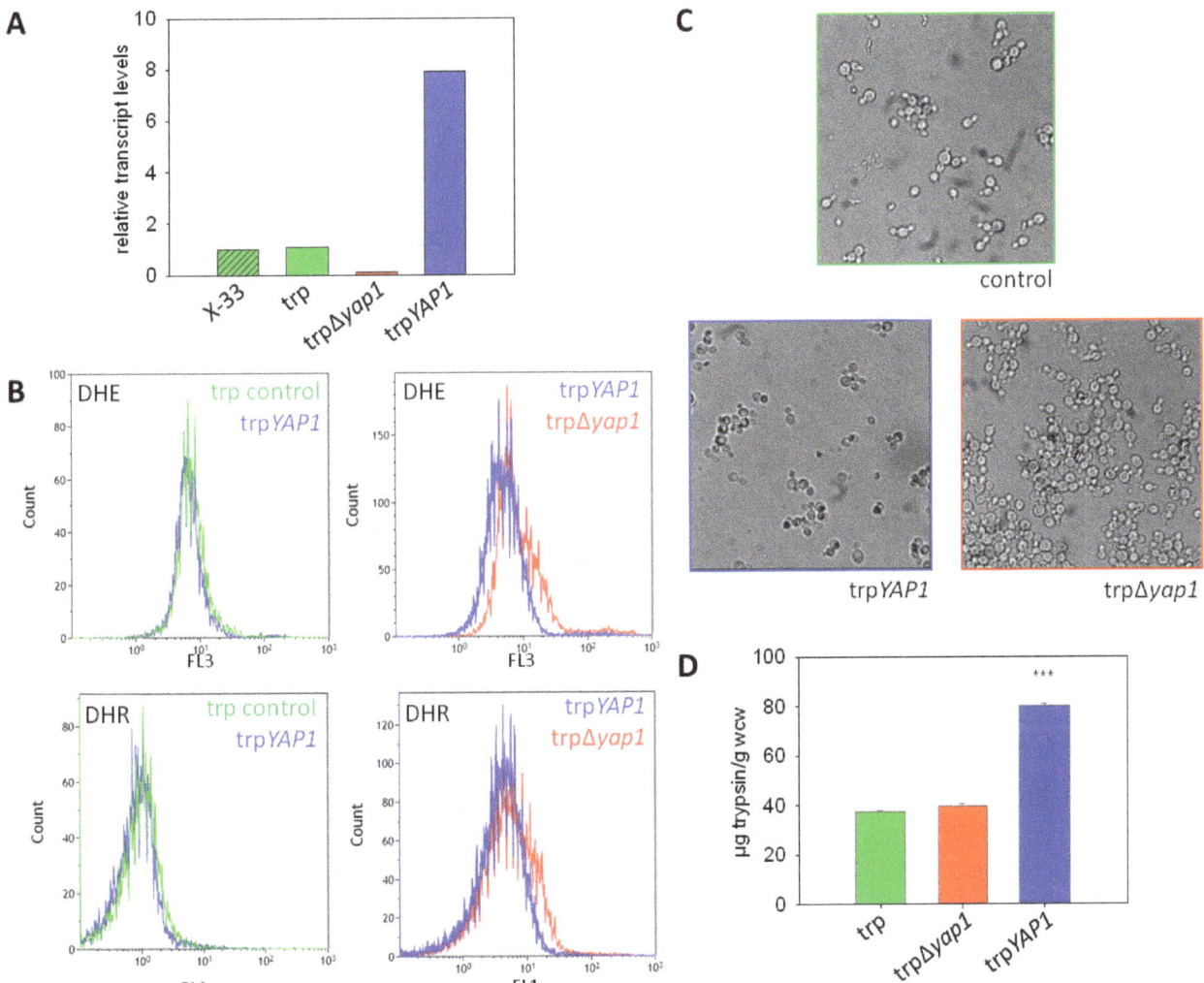

FIGURE 1: Downregulation of *YAP1* leads to accumulation of reactive oxygen species and flocculation, whereas *YAP1* co-overexpression enhances secretion of recombinant trypsinogen. (A) Transcript levels of *YAP1* measured by quantitative real time PCR in the strains X-33, trp, trpΔyap1 and trpYAP1. (B) ROS were measured with the fluorescent dyes DHE and DHR in the wild type (green), the trpYAP1 strain (blue) and the trpΔyap1 strain (red). (C) Microscopic images of X-33, trpYAP1 and trpΔyap1 after 20 h of cultivation in M2 medium. (D) Amounts of secreted trypsinogen in the strains trp, trpΔyap1 and trpYAP1 measured with TAME assay. The average of 10 clones per strain and the standard error of the mean (SEM) are shown, the significance of differences is indicated by *** (P < 0.01).

combinant protein secretion stress in *S. cerevisiae* quite recently, using a transcriptomics based approach. Although Yap1 itself did not show up within the differentially regulated genes, the Reporter Transcription Factors algorithm [30] identified Yap1 as significant based on downregulation of Yap1 target genes in a *S. cerevisiae* strain secreting α-amylase, a difficult to express recombinant protein [31]. On the contrary, during our transcriptomics analyses of different recombinant protein secreting strains of *P. pastoris*, we did not see any significant regulation of putative Yap1 target genes involved in oxidative stress de-

fense ([32, 33] and Stadlmayr et al. unpublished data). However, a significant downregulation of *YAP1*, and several of its target genes such as *AHP1*, *CTA1*, *GLR1*, *GRX3*, *GSH1*, *GSH2*, *SOD2*, and *TRX1* (among other genes) was observed in a *HAC1* overexpression *P. pastoris* strain (except for *SOD1* which was up-regulated; data from [34]) correlating with ROS accumulation in this strain with constitutive UPR induction [22].

Analysis of trypsinogen secretion levels in the strains deregulated for Yap1 transcript levels substantiate a role of Yap1 in the secretion process. *YAP1* overexpression en-

TABLE 1. Differentially regulated genes upon *YAP1* overexpression in *P. pastoris* secreting trypsinogen (trpYAP1/trp).

Functional group	total	up	down	Gene names
Amino acid metabolism, thereof glutamine family	10 6	2 0	8 6	⇧ *GLY1*, *HIS2* ⇩ *ARG1*, *ARG5,6*, *CPA1*, **GAD1**, *GCV2*, **GLT1**, *HIS7*, *MET13*
Cell wall & glycosylation	14	4	10	⇧ **PAS_chr048_0005**, *PAS_chr1-1_0135*, **PAS_chr3_0030**, *YJR061W* ⇩ **EXG1**, *FLO103*, *FLO104*, *FLO11*, **FLO5-2**, *GAS1-2*, *KTR1*, *PpBMT1*, *PpBMT2*, *UTR2*
Cofactor	4	1	3	⇧ *RIB3* ⇩ *SNZ3*, *THI4*, *THI7*
Lipid metabolism	4	2	2	⇧ *PDR16*, *PRY2* ⇩ **INO1**, *RSB1*
Metabolism, thereof oxido-reductase activity	9 4	5 3	4 1	⇧ *ACS1*, *ArbD*, *GOR1*, *GPD1*, **RGS2** ⇩ *ALD5*, *DAL7*, *DOG1*, **ICL1**
Metal ion homeostasis	20	11	9	⇧ *ATX1*, **FET3**, **FRE2**, **FRE3**, **FRE6**, *FTR1*, *MZM1*, **PAS_chr1-3_0013**, **PAS_chr2-2_0470**, *SIT1*, *YOR389W* ⇩ **CCC1**, **CTR1**, *FEP1*, *FRA1-2*, **FRE1**, *PAS_chr3_0141*, **PHO84**, *PIC2*, *SMF2*
Other	8	6	2	⇧ *CWC15*, *ESS1*, *FCY1*, *IMP2*, **PAS_chr1-4_0226**, *PAS_chr4_0991* ⇩ *CMK2*, *SLM1*
Response to oxidative stress	26	26	0	⇧ **AHP1**, **AIF1-1**, *CCS1*, *CYS3*, *ECM4*, *ETT1*, **GSH1**, *HAP1*, **HBN1**, *HYR1*, *MCR1*, **OYE2**, **OYE3**, **PAS_c034_0013**, *PCD1*, *PST2*, *PTC7*, **SNQ2**, **SOD1**, **SRX1**, *TRR1*, *TRX1*, **TRX2**, **TSA1**, *YAP1*, *YDL124*
Oxido-reductase activity	11	10	1	⇧ **ETR1**, **FDH1**, **PAS_c157_0001**, **PAS_chr3_0006**, **YDR541C-2**, **YDR541C-3**, **YDR541C-4**, **YDR541C-5**, **YNL134C-1**, **YNL134C-3** ⇩ **PAS_chr2-1_0307**
Proteolysis, thereof proteasome	5	4	1	⇧ *ADD66*, *PRE4*, *UBI4*, *YPS7* ⇩ *YPS1-4*
Transcriptional regulator, also metal-ion homeostasis	10	7	3	⇧ *ACA1*, **MPP1**, **PAS_chr1-3_0166**, *PAS_chr1-4_0652*, **TOS8**, *YPR013C*, *YPR022C-1* ⇩ *FEP1*, *FRA1-2*, **PAS_chr4_0324**, *RME1*, *YLR278C*
Transport, thereof polyamine transport	18	6	12	⇧ *ATO2*, **FLR1**, *PAS_chr1-4_0636*, *SNG1*, *TPO1*, *YOR1-1* ⇩ *AQR1*, *AQY1*, *GAP1*, *ITR1*, *ITR2*, *MEP1*, *PAS_chr4_0656*, **PDR5-2**, *PDR12*, **TPO3**, *TPO4*, *YHL008C*
Unknown	31	14	17	⇧ **BSC5**, **ECM13**, *PAS_chr1-1_0094*, *PAS_chr1-1_0209*, *PAS_chr3_0153*, *PAS_chr3_0187*, *PAS_chr3_0288*, **PAS_chr3_0837**, **PAS_chr4_0080**, *PAS_chr4_0328*, *PAS_chr4_0773*, **PAS_chr4_0820**, *PAS_chr4_0860*, **YPR127W** ⇩ *AIM17*, *APD1*, *DCG1*, **PAS_chr1-1_0257**, *PAS_chr1-3_0169*, **PAS_chr1-4_0689**, **PAS_chr2-1_0064**, *PAS_chr2-1_0240*, *PAS_chr2-1_0539*, **PAS_chr2-1_0642**, *PAS_chr3_0494*, *PAS_chr4_0602*, **PAS_chr4_0851**, *PAS_chr4_0947*, **TMA10**, *YBR056W*, *YMR244W*

Significantly up- and down-regulated genes were determined using an adjusted P-value < 0.05 as cut-off criterion. Genes that exceed an expression fold change of ± 1.5 fold are highlighted in bold.

hanced the level of secreted trypsinogen in the single copy trp strain more than 2-fold (Figure 1D). Notably, only the amount of correctly folded active protein is measured by the enzymatic TAME assay [35]. No influence on biomass specific trypsinogen yields were observed in the conditional yap1 knock-down strain (Figure 1D). Quantitative real time PCR analysis showed that transcript levels of genes involved in oxidative protein folding, such as ERO1 and PDI1, which have been shown to enhance trypsinogen secretion previously [22], remained unchanged in the YAP1 overexpressing strain (data not shown), ruling out one possible explanation for the enhanced secretion phenotype. Additionally, we confirmed the positive impact of YAP1 overexpression on protein secretion also in a high level secreting and UPR-induced P. pastoris strain (containing multiple copies of the pTRP expression cassette), leading on average to 1.58-fold higher secretion yields.

Genes involved in the antioxidant response are significantly up-regulated in the YAP1-overexpressing strain trpYAP1

In order to elucidate the cellular mechanisms behind the improved secretion of a recombinant protein through constitutive overexpression of YAP1, transcriptional analysis of this strain and its parental strain were performed during exponential growth of the cells using DNA microarrays [34]. The yap1 knock-down strain was excluded at this point, as its strong flocculation phenotype hampered quantitative sampling. This technical limitation was not an issue as our aim was to investigate the transcriptional regulation responsible for the increased secretion phenotype, and not to elucidate the Yap1 regulon upon oxidative stress in P. pastoris. Microarray analysis revealed that the constitutive overexpression of YAP1 (without any external oxidative stimulus) exerted significant regulation of 170 genes (cut-off criteria: adjusted P-value < 0.05), 98 thereof being induced, and 72 being downregulated (Table 1). Out of these genes, 146 (more than 85%) had at least one YAP binding site (ARE) in their promoter regions (determined using Matinspector, Genomatix, see Additional File 1). Basal regulation through YAP1 overexpression without stressors has also been reported for S. cerevisiae [20, 36]. Most probably, the transcriptional response without external stressor is due to overload of Crm1, the protein responsible for nuclear export of Yap1, and consequently increased levels of Yap1 localized to the nucleus (as suggested by [36]). Localization was analysed using Yap1, which was C-terminally fused to sGFP and found to be distributed all over the cell including the nucleus (data not shown).

GO Term Finder (P-value cutoff <0.02) was used to identify significantly regulated biological processes, molecular functions and cellular components (Table 1 and Additional File 1). Among the upregulated genes, GO processes "Response to oxidative stress", "Response to chemical stimulus", "Metal ion transport" and "Cellular homeostasis" were the most significantly influenced biological functions, while "glutamine family amino acid metabolism", "manganese ion transport" and "flocculation" are significantly down-regulated functions.

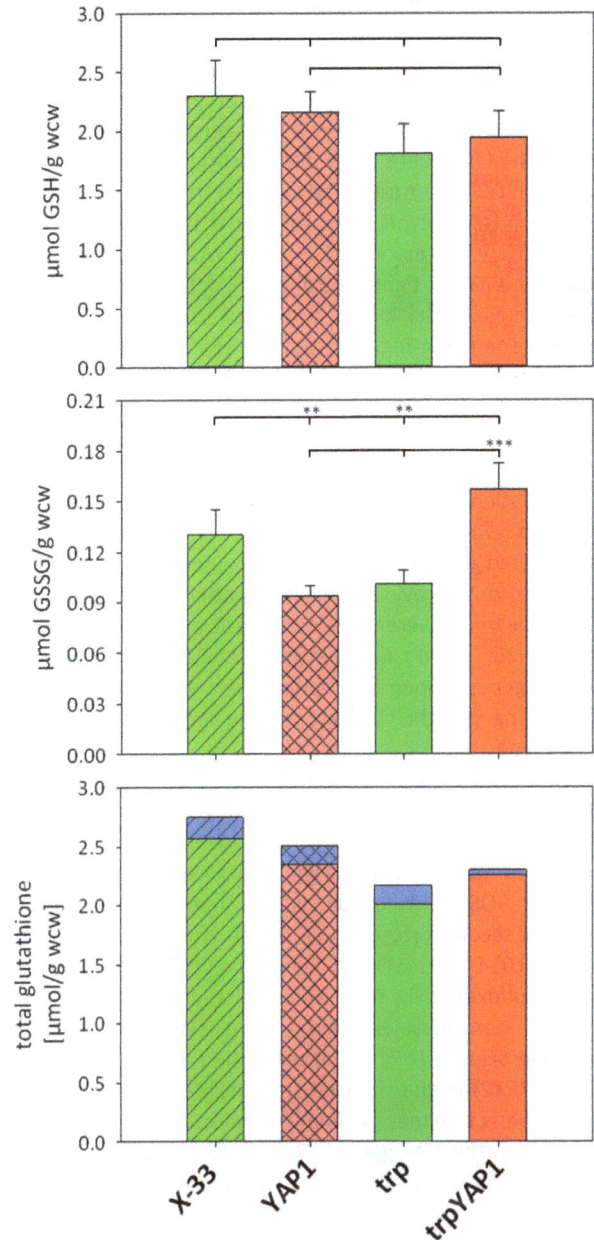

FIGURE 2: Comparison of intra- and extracellular glutathione (GSH) and glutathione disulfide (GSSG) levels in YAP1 overexpressing P. pastoris and controls. Upper panel: Intracellular GSH concentrations (µmol GSH per gram wet cell weight). Differences were not statistically significant (P-values > 0.1). **Middle panel:** Intracellular GSSG concentrations (µmol GSSG per gram wet cell weight). ** in the upper row indicate a significant difference to X-33 (P < 0.05). *** in the lower row indicates a significant difference to the trp strain (P < 0.01). **Lower panel:** Total glutathione (tGSH) concentration (intracellular: green (controls) or red (YAP1 overexpressing strains), excreted: blue) was determined using the equation $tGSH = [GSH] + 2*[GSSG]$.

Regulation of genes coding for multidrug transporter has been reported to be dependent on Yap1 [37-39], but has not been shown without externally applied stress previously. Transcriptional regulation of genes involved in cellular iron and copper homeostasis might look contradictory at first, as there is no difference in the external ion concentrations, but are clearly related to the lowered levels of the GATA type repressor of the iron regulon Fep1 [40] and the repressor Fra1-2 in the *YAP1* overexpressing strain. Seemingly, tight control of the uptake of these free redox active metal ions is a part of the antioxidant response. The down-regulation of the several cell wall associated genes including *FLO103*, *FLO104*, FLO11, *FLO5-2* upon *YAP1* overexpression may explain the strong flocculation phenotype of the *yap1* knock-down strain. Down-regulation of the flocculin genes might be due to the slight up-regulation of the gene encoding Nrg1 transcriptional repressor, which seems to be a key regulator of flocculation-related genes in *P. pastoris* (own unpublished results).

Contrary to previous regulation patterns observed in *S. cerevisiae* upon overexpression of Yap1 or oxidative stimulus [17, 20], we do not see chaperones or other folding related genes among the regulated genes in our analysis, thus ruling out their impact on the improved secretion phenotype.

Among the regulated genes, increased expression of genes coding for antioxidant enzymes, e.g., superoxide dismutase and peroxiredoxins, was predominant (Table 1). Strong up-regulation of the majority of genes involved in cytosolic ROS detoxification such as superoxide dismutase *SOD1*, a second cytosolic Cu/Zn SOD isoenzyme encoded by PAS_c034_0013, SOD co-chaperone *CCS1*, peroxiredoxin *AHP1*, sulfoxiredoxin *SRX1* which is needed to reduce thioredoxin peroxidase encoded by *TSA1*, cytoplasmic thioredoxin peroxidase *TRR1* which keeps the thioredoxin system (*TRX1*, *TRX2*) in the reduced state as well as glutathione biosynthesis enzymes gamma glutamylcysteine synthetase *GSH1*, cystathionine gamma-lyase *CYS3* and omega glutathione transferase *ECM4* was observed, indicating that they might be correlated to enhanced secretion.

Comparison of total intra- and extracellular glutathione (GSH) and glutathione disulfide (GSSG) levels show no significant differences, but an alteration in the GSH/GSSG ratio

As glutathione biosynthesis was among the up-regulated cellular processes in trpYAP1 (*GSH1*, *CYS3*), we investigated the levels of GSH and GSSG in the *YAP1* overexpressing strain and its parental control. No significant changes were determined for the concentration of reduced glutathione (Figure 2A). As shown previously [22], the strain secreting trypsinogen had a lower intracellular level of GSSG compared to the wild type strain X-33. Upon *YAP1* co-overexpression, GSSG levels in the trp strain are increased more than 50% and are even slightly higher than in X-33. This effect is specific for *YAP1* overexpression in the protein secreting strain, as GSSG levels were not increased in the wild type background (Figure 2B). On the contrary, GSSG levels are lower when overexpressing *YAP1* in the

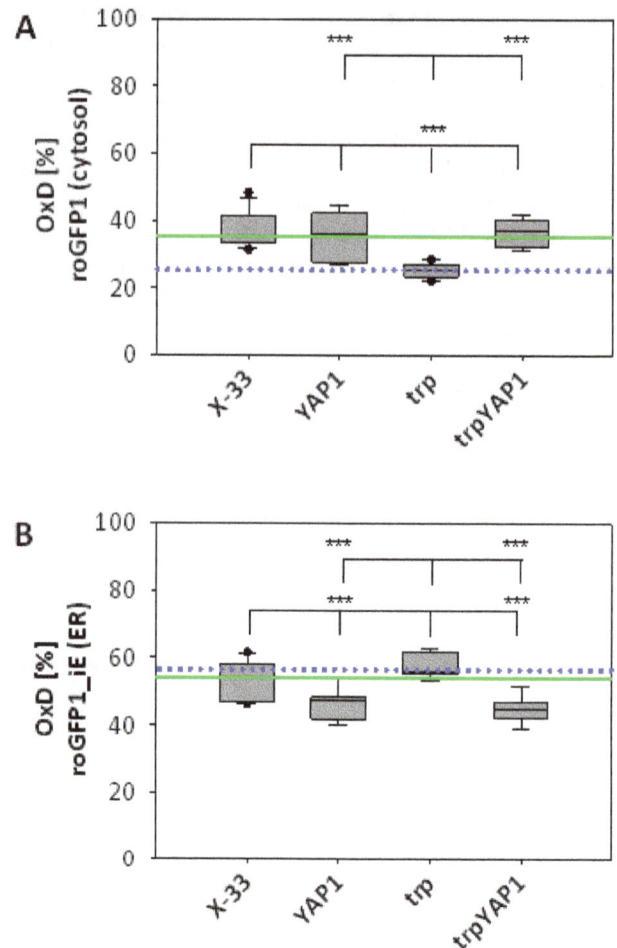

FIGURE 3: *YAP1* **co-overexpression with trypsinogen reoxidizes the cytosol redox state to the level of the wild type strain, while slightly reducing the ER redox ratio.** The OxD obtained with roGFP are represented as Box-and-Whisker plots. Each box is separated into two inter-quartiles by the statistical median as a horizontal line in the box. The extreme values extending from the inter-quartile at most 1.5 times from the upper or lower inter-quartile are the 'whiskers'. The green line indicates the median of X-33, the blue dotted line represents the median of the trp strain. **(A)** Cytosolic redox ratios of the respective strains represented as OxD of roGFP. Statistical significance is indicated as for (B). **(B)** ER redox ratios of the respective strains represented as OxD of roGFP-iE. *** in the lower row indicate a significant difference to X-33 (P < 0.01). *** in the upper row indicates a significant difference to the trp strain.

wild type background. Total glutathione content is slightly but not significantly lower in the engineered strains (Figure 2C). It seems, that enhanced Yap1 availability is able to restore the physiological redox conditions in the secreting strain to the wild type state. Again, GSH excretion does not seem to play a significant role, as extracellular GSH levels were almost identical in all strains (Figure 2C), while extracellular GSSG concentrations are below the limit of quantification (~0.007 µmol/g wet cell weight).

Overexpression of the transcription factor Yap1 causes a slight but significant reduction of the ER redox state, but restores cytosolic redox state in the trypsinogen secreting strain to the wild type level

As YAP1-overexpression seems to have a strong impact on intracellular redox conditions, we determined compartment-specific redox ratios in the cytosol and the ER using redox-sensitive GFP variants [27, 41]. We have already demonstrated that folding of proteins in the ER does not affect the redox environment of the ER, but leads to reduction of the redox state of the cytosol [22]. Overexpression of YAP1 in the strain producing the recombinant secretory protein reoxidizes the redox state of the cytosol to the level of the wild type strain (Figure 3A), which is also reflected by the increased levels of oxidized GSSG in the trpYAP1 strain. Again, the effect is specific for the recombinant protein secreting strain, which displayed the lower cytosolic redox state. Interestingly, YAP1 overexpression had also an effect on the ER redox state, and caused a slight but significant reduction of the ER through a yet unidentified mechanism (Figure 3B).

Conclusions

Oxidative protein folding in the ER generates stoichiometric amounts of ROS, in particular H_2O_2. At present, it is not clear whether these ROS stay within the ER or diffuse to the cytosol, but we could clearly show that Yap1 is involved in physiological detoxification of ROS formed upon oxidative folding in the ER. Yap1 is required to activate the antioxidant enzymes, which are quenching ROS. Cells with significantly lowered Yap1 levels react to increased secretory folding load with accumulation of ROS (e.g., hydrogen peroxide and superoxide anions) and strong flocculation. On the other hand, enhanced levels of Yap1 seem to create a more convenient environment for folding and secretion of (recombinant) proteins.

It should be noted that protein folding is not a stoichiometric process, but usually requires several rounds of folding per molecule. It has been reported previously, that per each disulfide bond formed one molecule of hydrogen peroxide is generated [9]. Additionally, folding also requires ATP, as correct folding of nascent peptides by chaperones occurs along multiple ATP-driven cycles of substrate binding and release (recently reviewed e.g. by [42]). Thus, increased mitochondrial activity to generate sufficient amounts of ATP might also contribute to ROS formation [43]. At present we cannot distinguish between ER and mitochondria derived ROS. Based on the amount of secreted trypsinogen after 48 h in our study, at least 10^6 molecules of trypsinogen were processed per cell in the trp strain, with each trypsinogen having six disulfide bonds to be correctly formed. Nevertheless, we did not observe ROS accumulation in this strain compared to the non-expressing wild type strain X-33, indicating that the cell's antioxidant response can cope with this amount of secretory proteins. However, we did see ROS accumulation in strains with higher oxidative protein folding and secretion levels [22], indicating that the additional amount of secretory load overwhelms the antioxidant capacity of the cells.

We have previously shown that increased oxidative folding of proteins in the ER has a strong effect on the redox environment of the cytosol, leading to more reducing conditions [22]. This cellular redox imbalance is stressful for the cells. Here we demonstrate that overexpression of YAP1 in such a strain reverses the effect on the redox status within these two compartments. It causes re-oxidation of the cytosol to the level of the wild type strain, while slightly reducing the redox state of the ER. Constitutive co-overexpression of YAP1 also leads to increased secretion levels of a recombinant protein. These alterations are also reflected by the levels of reduced and oxidized intracellular glutathione, whereby a significantly higher GSSG amount in the YAP1 co-overexpressing strain compared to its parental strain was measured. Total glutathione levels were thereby not remarkably affected among the analysed strains. Thereby, the degree of oxidation (OxD = 2*[GSSG] / [tGSH]) of the whole cell was shifted from 10% in the wild type and the trp strain to 14% in the trp_YAP1 co-overexpressing strain. Taken together, these data indicate a strong impact of intracellular redox balance on the secretion of (recombinant) proteins without affecting protein folding per se.

This hypothesis is further supported by microarray data of the respective P. pastoris strains, as genes involved in the antioxidant response were the predominant group of genes regulated by overexpressed YAP1, whereas no effect on the transcription of folding related genes was observed.

With these findings we conclude that the re-establishment of suitable redox conditions by tuning the antioxidant capacity of the cell reduces cell stress caused by recombinant protein production, and thereby increases the secretion capacity of P. pastoris. Moreover, these findings are thought to have implications not only for improving biotechnological production processes, but also impact our understanding of the development of many aging related diseases, as both cases share the underlying molecular processes, which connect cellular redox conditions with folding related stresses in the ER.

MATERIALS AND METHODS
Strains and vectors

The single copy trypsinogen expressing strain (trp) was described in [22]. Briefly, this strain contains a single expression cassette of porcine trypsinogen under control of the P_{GAP} (glyceraldehyde-3-phosphate-dehydrogenase) promoter using the S. cerevisiae alfa-mating factor pre-pro-leader for secretion and has been confirmed not to activate the UPR by qPCR previously. For overexpression of P. pastoris YAP1, the respective gene (PAS_chr4_0601) was amplified from X-33 genomic DNA, and cloned into a pPuzzle vector [44] under control of the P_{GAP} promoter. The Zeocin resistance marker was flanked by loxP sites. The vector was integrated into the native YAP1 locus of the P. pastoris genome after linearization in the respective sequence. After transformation by electroporation, positive transformants were selected on YPD plates with Zeocin.

For conditional downregulation of YAP1, the native P_{YAP1} promoter was exchanged for the serine repressible P_{SER} promoter using the strategy described in [28], resulting the trpΔyap1 strain. The used primers are summarized in Table 2.

Vectors containing redox-sensitive GFP variants roGFP1 and roGFP1_iE for targeting the cytosol and the ER were described in [27] and were used for the transformation of the strains mentioned above.

For the localization of overproduced Yap1, superfolder GFP (sGFP) was fused to the C-terminus of the *YAP1* gene, and expressed under control of P_{GAP}. These strains were then analysed by Western blot for expression and fluorescence microscopy for intracellular localization.

High level trypsinogen secreting strains were generated by overexpressing pTRP under control of P_{GAP} and selecting for the highest expressing *P. pastoris* strains containing multiple copies of the expression cassette on increased antibiotics (Zeocin) concentrations. For overexpression of *P. pastoris* *YAP1* in these strains, the Zeocin resistance marker in the pPuzzle vector was exchanged for the KanMX cassette. The vector was integrated into the native *YAP1* locus of the *P. pastoris* genome after linearization in the respective sequence. After transformation by electroporation, positive transformants were selected on YPD plates with G418.

Production of porcine trypsinogen

Screening media contained per liter: 10 g pea peptone, 10 g yeast extract, 10.2 g $(NH_4)_2HPO_4$, 1.24 g KCl, 910 µL 1M $CaCl_2$ solution and 1 mL biotin stock solution (0.2 g/L). Pre-culture was performed in 5 mL YPD for all strains. For the main culture, 10 clones per construct were inoculated into 10 mL of screening medium in 100 mL shake flasks (at an OD_{600} of 0.1) and incubated at 28°C with 170 rpm (rotations per minute) for 48 h.

The amount of secreted trypsinogen was determined in the supernatant by measuring the activity of enterokinase activated trypsin according to the protocol established by [45]. The sample buffer was exchanged to 1 mM HCl over PD-10 columns (Amersham Biosciences). Then, 50 µL of sample was incubated with 5 µg bovine enterokinase (Sigma) in 50 mM Tris/HCl buffer, pH 8.6, containing 50 mM $CaCl_2$, to activate trypsinogen, and finally trypsin activity was determined with the TAME assay. Therefore, aliquots of the samples were added to 40 mM Tris/HCl buffer pH 8.1 buffer containing 10 mM $CaCl_2$ and 1mM p-toluene sulfonyl-L-arginine ethylester-HCl (TAME), and the increase of absorbance at 247 nm was followed. Trypsinogen concentration is calculated using the extinction coefficient $\epsilon = 0.0101\ E_{247}\ cm^{-1}\ min^{-1}$.

Staining of reactive oxygen species (ROS)

The fluorescent probes DHE (dihydroethidium) and DHR (dihydrorhodamine 123) were used to determine levels of ROS as described in [22]. Briefly, cells of an OD_{600} of 0.4 were harvested by centrifugation and resuspended in 2 mL PBS. DHE and DHR were added to cells at a final concentration of 10 µg/mL and incubated at 30°C for 30 min. Flow cytometry analysis was performed using a FACS Canto with the settings: DHR green filter (525-550 nm), DHE red filter (600-650 nm).

Analysis of transcriptional regulation using DNA microarrays

The strains trp and trpYAP1 were cultivated in screening medium in 4 biological replicates each. 10 mL of main culture were inoculated from an overnight pre-culture with an OD_{600} of 4. After 6 h (exponential growth phase), samples were taken and fixed in phenol/ethanol, and stored at -80°C until total RNA extraction. Total RNA extraction, two-color labelling in a dye swap manner for each sample and cRNA hybridization to the *P. pastoris* specific microarrays were performed as in [34]. Samples were hybridized to the microarrays described in Graf et al. ([34]; 8x15K custom arrays, AMAD-ID 018045, Agilent) and to a novel in house designed *P. pastoris* specific oligonucleotide array, which was based on the improved annotation of *P. pastoris* strain GS115 (8x15K custom arrays, AMAD-ID 034821, Agilent, [46]). Raw data normalization was done using locally weighted MA-scatterplot smoothing (LOESS) followed by a between array 'Aquantile' normalization, both available within the limma package of R [47]. P-values were corrected for multiple testing using Benjamini & Yekutieli method [48]. Features were defined as differentially expressed if they had a P-value < 0.05. For the identification of stronger regulatory effects an additional cut-off for the fold change (FC) of ± 1.5 fold was applied. A high degree of correlation of significantly regulated genes between the two microarrays was deter-

TABLE 2. Primers used for generation of *YAP1* overexpression and knock down.

Primer name	Primer sequence
Yap1_fw (*Sbf*I)	TAGA<u>CCTGCAGG</u>ATGAGTGACGTGGTAAACAAG
Yap1_rv (*Sfi*I)	AATA<u>GGCCGAGGCGGCCC</u>TATTTAAACATGGAAAAATCG
P_{SER1}_fw (*Sbf*I)	TAGA<u>CCAAGGCCTTGG</u> CAGCAAATAATTAGCAGCC
P_{SER1}_rv (*Bst*XI)	AATA<u>CCTGCAGG</u> TGTATTATATGGTTAGTTCAAGATG
P_{YAP1}_fw (*Asc*I)	ATTA<u>GGCGCGCC</u>GCAAAAGCGAGTACTATTTCTCAAA
P_{YAP1}_rv (*Apa*I)	ATTA<u>GGGCCC</u>AGCGGAAGAACAACTTTAGTGAGTA

mined (R^2 = 0.902), thus the average FC of both microarrays was used.

Significantly regulated *P. pastoris* genes with a functional annotation were categorized using the "generic gene ontology (GO) term mapper" (http://go.princeton.edu/cgi-bin/GOTermMapper).

Determination of transcript levels by quantitative real time PCR

RNA isolation, cDNA synthesis and measurement of mRNA transcript levels using real time PCR was performed as described in [44] using the primers given in [22]. The reference gene for normalization was actin (*ACT1*), each gene transcript was correlated to *ACT1* as internal control. The wild type strain X-33 served as reference strain for relative transcript level determination using the delta-delta C_t method.

Cultivation conditions for redox and glutathione measurements

M2 minimal medium contained per liter: 20 g of glucose, 20 g of citric acid, 3.15 g of $(NH_4)_2HPO_4$, 0.03 g of $CaCl_2 \cdot 2H_2O$, 0.8 g of KCl, 0.5 g of $MgSO_4 \cdot 7H_2O$, 2 mL of biotin (0.2 g L^{-1}), 1.5 mL of trace salts stock solution. The pH was set to 5.0 with 5M KOH solution. Trace salts stock solution contained per liter: 6.0 g of $CuSO_4 \cdot 5H_2O$, 0.08 g of NaI, 3.0 g of $MnSO_4 \cdot H_2O$, 0.2 g of $Na_2MoO_4 \cdot 2H_2O$, 0.02 g of H_3BO_3, 0.5 g of $CoCl_2$, 20.0 g of $ZnCl_2$, 5.0 g of $FeSO_4 \cdot 7H_2O$, and 5.0 mL of H_2SO_4 (95-98% w/w).

Pre-culture was performed in 5 mL M2 medium. 10 mL M2 medium in 100 mL shake flasks were inoculated at an OD600 of 0.5, and were incubated at 28°C with 170 rpm (rotations per minute) for 24 h. For each strain, 12 individual clones were analysed.

Determination of redox ratios using roGFP

Redox states of all strains were measured during the exponential growth phase. Briefly, 840 µL culture were mixed with 60 µL 1.5 M K-MOPS buffer (pH 7.0). Total oxidation or reduction of the roGFPs was achieved by addition of 200 µL 6.3 mM 4,4'-dipyridyl disulfide (4-DPS) or 200 µL 1M dithiothreitol (DTT), respectively, and was compared to the untreated culture where 200 µL of PBS were added as control [49]. The fluorescence of the cells was detected in 96-well plates (FluoroNunc, Nunc) on a fluorescence photometer (Infinite M200 Tecan plate reader), where it was possible to measure the excitation of two wavelengths, 395 and 465 nm, corresponding to the oxidized and the reduced form of the protein. Each sample was measured four times [27]. The degree of oxidation (OxD) of the cytosol and the ER was calculated according to equation 1. The quotient of fluorescence intensities (equation 2) is the so called 'instrument factor IF' and corrects for variations in the signal strength from the instrument. R is the experimentally determined fluorescence ratio of the intensities for the two wavelengths 395 and 465 nm [50]. R_{ox} and R_{red} stand for the ratios of fully oxidized or fully reduced roGFP, respectively, while R is the ratio of the untreated sample.

Equation 1:
$$OxD = \left(\frac{R - R_{red}}{\frac{I465_{ox}}{I465_{red}}(R_{ox} - R) + (R - R_{red})} \right)$$

Equation 2:
$$IF = \frac{I465_{ox}}{I465_{red}}$$

Determination of glutathione concentrations by LC/MS-MS

Three pellets of 1 mL of exponential culture were collected by centrifugation at 3000 rpm for 5 min at 4°C. The cells were resuspended in 1 mL ice cold 0.1 M H_3PO_4, spiked with labeled glutathione-glycine-$^{13}C_2$,^{15}N (Sigma Aldrich) and GSSG- glutathione-glycine-$^{13}C_2$,^{15}N (synthesized in house) and heated to 75°C for 3 min to extract GSH and GSSG. The cell extracts were measured by LC/MS-MS (Agilent Ion Trap) using hydrophilic interaction chromatography (HILIC) and acetonitrile/water as eluent [51].

Statistical analyses

Secreted trypsinogen labels and glutathione concentrations are expressed as mean ± standard error of the mean. Statistically significant differences between the strains were determined using Student's t-test. All strains were compared to the wild type strain X-33 or the single copy trypsinogen secreting strain trp. The P-values are marked with *** for P < 0.01 and ** for P < 0.05.

ACKNOWLEDGEMENTS

The authors thank Dr. Martin Dragosits, Roland Prielhofer and Caroline Kotlowski for microarray hybridization, Corinna Rebnegger and Franziska Wanka for their help with qRT-PCR during the course of their diploma theses, Christine Sanystra and Martin Nagl for generation of high producing trp strains and their analysis, Stefan Pflügl for HPLC measurements, Verena Puxbaum for performing fluorescence microscopy and also for microsome preparation and glutathione assay together with Katharina Littringer. The authors declare that they have no competing interests.

Parts of this work have been supported by the Federal Ministry of Science, Research and Economy (BMWFW), the Federal Ministry of Traffic, Innovation and Technology (bmvit), the Styrian Business Promotion Agency SFG, the Standortagentur Tirol and ZIT – Technology Agency of the City of Vienna through the COMET-Funding Program managed by the Austrian Research Promotion Agency FFG. EQ BOKU VIBT GmbH is acknowledged for providing LC-MS/MS instrumentation. Finally, we thank the BOKU-VIBT Imaging Center for access and expertise with fluorescence microscopy equipment.

CONFLICT OF INTEREST

The authors declare no conflict of interest.

REFERENCES

1. Shimizu Y, Hendershot LM **(2009)**. Oxidative folding: cellular strategies for dealing with the resultant equimolar production of reactive oxygen species. **Antioxid Redox Signal** 11:2317-2331.

2. Malhotra JD, Kaufman RJ **(2007)**. Endoplasmic reticulum stress and oxidative stress: a vicious cycle or a double-edged sword? **Antioxid Redox Signal** 9:2277-2293.

3. Herrero E, Ros J, Bellí G, Cabiscol E **(2008)**. Redox control and oxidative stress in yeast cells. **Biochim Biophys Acta** 1780:1217-1235.

4. Morano KA, Grant CM, Moye-Rowley WS **(2012)**. The response to heat shock and oxidative stress in *Saccharomyces cerevisiae*. **Genetics** 190:1157-1195.

5. Margittai E, Sitia R **(2011)**. Oxidative protein folding in the secretory pathway and redox signaling across compartments and cells. **Traffic** 12:1-8.

6. Araki K, Nagata K **(2011)**. Protein folding and quality control in the ER. **Cold Spring Harb Perspect Biol** 3:a007526.

7. Gorlach A, Klappa P, Kietzmann T **(2006)**. The endoplasmic reticulum: folding, calcium homeostasis, signaling, and redox control. **Antioxid Redox Signal** 8:1391-1418.

8. Frand AR, Kaiser CA **(1999)**. Ero1p oxidizes protein disulfide isomerase in a pathway for disulfide bond formation in the endoplasmic reticulum. **Mol Cell** 1999, 4:469-477.

9. Gross E, Sevier CS, Heldman N, Vitu E, Bentzur M, Kaiser CA, Thorpe C, Fass D **(2006)**. Generating disulfides enzymatically: reaction products and electron acceptors of the endoplasmic reticulum thiol oxidase Ero1p. **Proc Natl Acad Sci U S A** 103:299-304.

10. Tu BP, Weissman JS **(2002)**.: The FAD- and O(2)-dependent reaction cycle of Ero1-mediated oxidative protein folding in the endoplasmic reticulum. **Mol Cell** 10:983-994.

11. Chakravarthi S, Jessop CE, Bulleid NJ **(2006)**. The role of glutathione in disulphide bond formation and endoplasmic-reticulum-generated oxidative stress. **EMBO Rep** 7:271-275.

12. Toledano MB, Delaunay-Moisan A, Outten CE, Igbaria A **(2013)**. Functions and cellular compartmentation of the thioredoxin and glutathione pathways in yeast. **Antioxid Redox Signal** 18:1699-1711.

13. Temple MD, Perrone GG, Dawes IW **(2005)**. Complex cellular responses to reactive oxygen species. **Trends Cell Biol** 15:319-326.

14. Linke K, Jakob U **(2003)**. Not every disulfide lasts forever: disulfide bond formation as a redox switch. **Antioxid Redox Signal** 5:425-434.

15. Brandes N, Schmitt S, Jakob U **(2009)**. Thiol-based redox switches in eukaryotic proteins. **Antioxid Redox Signal** 11:997-1014.

16. Harshman KD, Moye-Rowley WS, Parker CS **(1998)**. Transcriptional activation by the SV40 AP-1 recognition element in yeast is mediated by a factor similar to AP-1 that is distinct from GCN4. **Cell** 53:321-330.

17. Godon C, Lagniel G, Lee J, Buhler JM, Kieffer S, Perrot M, Boucherie H, Toledano MB, Labarre J **(1998)**. The H$_2$O$_2$ stimulon in *Saccharomyces cerevisiae*. **J Biol Chem** 273:22480-22489.

18. Lee J, Godon C, Lagniel G, Spector D, Garin J, Labarre J, Toledano MB **(1999)**. Yap1 and Skn7 control two specialized oxidative stress response regulons in yeast. **J Biol Chem** 274:16040-16046.

19. Kuge S, Toda T, Iizuka N, Nomoto A **(1998)**. Crm1 (Xpol) dependent nuclear export of the budding yeast transcription factor yAP-1 is sensitive to oxidative stress. **Genes Cells** 3:521-532.

20. DeRisi JL, Iyer VR, Brown PO **(1997)**. Exploring the metabolic and genetic control of gene expression on a genomic scale. **Science** 278:680-686.

21. Ayer A, Fellermeier S, Fife C, Li SS, Smits G, Meyer AJ, Dawes IW, Perrone GG **(2012)**. A genome-wide screen in yeast identifies specific oxidative stress genes required for the maintenance of sub-cellular redox homeostasis. **PLoS One** 7:e44278.

22. Delic M, Rebnegger C, Wanka F, Puxbaum V, Haberhauer-Troyer C, Hann S, Kollensperger G, Mattanovich D, Gasser B **(2012)**. Oxidative protein folding and unfolded protein response elicit differing redox regulation in endoplasmic reticulum and cytosol of yeast. **Free Radic Biol Med** 52:2000-2012.

23. Yoshida H: ER stress and diseases **(2007)**. **FEBS J** 274:630-658.

24. Yano T, Yurimoto H, Sakai Y **(2009)**. Activation of the oxidative stress regulator PpYap1 through conserved cysteine residues during methanol metabolism in the yeast *Pichia pastoris*. **Biosci Biotechnol Biochem** 73:1404-1411.

25. Yano T, Takigami E, Yurimoto H, Sakai Y **(2009)**. Yap1-regulated glutathione redox system curtails accumulation of formaldehyde and reactive oxygen species in methanol metabolism of *Pichia pastoris*. **Eukaryot Cell** 8:540-549.

26. Tu BP, Weissman JS **(2004)**. Oxidative protein folding in eukaryotes: mechanisms and consequences. **J Cell Biol** 164:341-346.

27. Delic M, Mattanovich D, Gasser B **(2010)**. Monitoring intracellular redox conditions in the endoplasmic reticulum of living yeasts. **FEMS Microbiol Lett** 306:61-66.

28. Delic M, Mattanovich D, Gasser B **(2013)**. Repressible promoters - A novel tool to generate conditional mutants in *Pichia pastoris*. **Microb Cell Fact** 12:6.

29. Wemmie JA, Wu AL, Harshman KD, Parker CS, Moye-Rowley WS (1994). Transcriptional activation mediated by the yeast AP-1 protein is required for normal cadmium tolerance. **J Biol Chem** 269:14690-14697.

30. Oliveira AP, Patil KR, Nielsen J **(2008)**. Architecture of transcriptional regulatory circuits is knitted over the topology of bio-molecular interaction networks. **BMC Syst Biol** 2:17.

31. Tyo KE, Liu Z, Petranovic D, Nielsen J **(2012)**. Imbalance of heterologous protein folding and disulfide bond formation rates yields runaway oxidative stress. **BMC Biol** 10:16.

32. Baumann K, Carnicer M, Dragosits M, Graf AB, Stadlmann J, Jouhten P, Maaheimo H, Gasser B, Albiol J, Mattanovich D, Ferrer P **(2010)**. A multi-level study of recombinant *Pichia pastoris* in different oxygen conditions. **BMC Syst Biol** 4:141.

33. Dragosits M, Stadlmann J, Graf A, Gasser B, Maurer M, Sauer M, Kreil D, Altmann F, Mattanovich D **(2010)**. The response to unfolded protein is involved in osmotolerance of *Pichia pastoris*. **BMC Genomics** 11:207.

34. Graf A, Gasser B, Dragosits M, Sauer M, Leparc G, Tuechler T, Kreil D, Mattanovich D **(2008)**. Novel insights into the unfolded protein response using *Pichia pastoris* specific DNA microarrays. **BMC Genomics** 9:390.

35. Walsh KA, Wilcox PE **(1970)**. Serine proteases. **Methods Enzymol** 19:31-41.

36. Nisamedtinov I, Kevvai K, Orumets K, Arike L, Sarand I, Korhola M, Paalme T **(2011)**. Metabolic changes underlying the higher accumulation of glutathione in *Saccharomyces cerevisiae* mutants. **Appl Microbiol Biotechnol** 89:1029-1037.

37. Oskouian B, Saba JD **(1999)**. YAP1 confers resistance to the fatty acid synthase inhibitor cerulenin through the transporter Flr1p in *Saccharomyces cerevisiae*. **Mol Gen Genet** 261:346-353.

38. Nguyen DT, Alarco AM, Raymond M **(2001)**. Multiple Yap1p-binding sites mediate induction of the yeast major facilitator FLR1

gene in response to drugs, oxidants, and alkylating agents. **J Biol Chem** 276:1138-1145.

39. Miyahara K, Hirata D, Miyakawa T **(1996)**. yAP-1- and yAP-2- mediated, heat shock-induced transcriptional activation of the multidrug resistance ABC transporter genes in *Saccharomyces cerevisiae*. **Curr Genet** 29:103-105.

40. Miele R, Barra D, Bonaccorsi di Patti MC **(2007)**. A GATA-type transcription factor regulates expression of the high-affinity iron uptake system in the methylotrophic yeast *Pichia pastoris*. **Arch Biochem Biophys** 465:172-179.

41. Lohman JR, Remington SJ **(2008)**. Development of a family of redox-sensitive green fluorescent protein indicators for use in relatively oxidizing subcellular environments. **Biochemistry** 47:8678-8688.

42. Simmen T, Lynes EM, Gesson K, Thomas G **(2010)**. Oxidative protein folding in the endoplasmic reticulum: tight links to the mitochondria-associated membrane (MAM). **Biochim Biophys Acta** 1798(8):1465-73.

43. Mattoo RU, Goloubinoff P **(2014)**. Molecular chaperones are nanomachines that catalytically unfold misfolded and alternatively folded proteins. **Cell Mol Life Sci** 71(17):3311-25.

44. Stadlmayr G, Mecklenbräuker A, Rothmüller M, Maurer M, Sauer M, Mattanovich D, Gasser B **(2010)**. Identification and characterisation of novel *Pichia pastoris* promoters for heterologous protein production. **J Biotechnol** 150:519-529.

45. Hohenblum H, Gasser B, Maurer M, Borth N, Mattanovich D **(2004)**. Effects of gene dosage, promoters, and substrates on unfolded protein stress of recombinant *Pichia pastoris*. **Biotechnol Bioeng** 85:367-375.

46. Prielhofer R, Maurer M, Klein J, Wenger J, Kiziak C, Gasser B, Mattanovich D **(2013)**. Induction without methanol: novel regulated promoters enable high-level expression in *Pichia pastoris*. **Microb Cell Fact** 12:5.

47. Smyth GK **(2005)**. Limma: linear models for microarray data. In **Bioinformatics and Computational Biology Solutions using R and Bioconductor** 397-420.

48. Reiner A, Yekutieli D, Benjamini Y **(2003)**. Identifying differentially expressed genes using false discovery rate controlling procedures. **Bioinformatics** 19:368-375.

49. Østergaard H, Tachibana C, Winther JR **(2004)**. Monitoring disulfide bond formation in the eukaryotic cytosol. **J Cell Biol** 166(3):337-45.

50. Meyer AJ, Dick TP **(2010)**. Fluorescent protein-based redox probes. **Antioxid Redox Signal** 13:621-650.

51. Haberhauer-Troyer C, Delic M, Gasser B, Mattanovich D, Hann S, Koellensperger G **(2013)**. Accurate quantification of the redox-sensitive GSH/GSSG ratios in the yeast *Pichia pastoris* by HILIC-MS/MS. **Anal Bioanal Chem** 405:2031-2039.

19

Arabidopsis Bax Inhibitor-1 inhibits cell death induced by pokeweed antiviral protein in *Saccharomyces cerevisiae*

Birsen Çakır[1,2],* and Nilgun E. Tumer[1]

[1] Biotechnology Center for Agriculture and the Environment and the Department of Plant Biology and Pathology, Rutgers University, New Brunswick, NJ 08901-8520, USA.
[2] Department of Horticulture, Faculty of Agriculture, Ege University, Izmir, Turkey.
* Corresponding Author: Birsen Çakir, Department of Horticulture, Faculty of Agriculture, Ege University; Izmir, Turkey;
E-mail: birsencakir@hotmail.com

ABSTRACT   Apoptosis is an active form of programmed cell death (PCD) that plays critical roles in the development, differentiation and resistance to pathogens in multicellular organisms. Ribosome inactivating proteins (RIPs) are able to induce apoptotic cell death in mammalian cells. In this study, using yeast as a model system, we showed that yeast cells expressing pokeweed antiviral protein (PAP), a single-chain ribosome-inactivating protein, exhibit apoptotic-like features, such as nuclear fragmentation and ROS production. We studied the interaction between PAP and AtBI-1 (*Arabidopsis thaliana* Bax Inhibitor-1), a plant anti-apoptotic protein, which inhibits Bax induced cell death. Cells expressing PAP and AtBI-1 were able to survive on galactose media compared to PAP alone, indicating a reduction in the cytotoxicity of PAP in yeast. However, PAP was able to depurinate the ribosomes and to inhibit total translation in the presence of AtBI-1. A C-terminally deleted AtBI-1 was able to reduce the cytotoxicity of PAP. Since anti-apoptotic proteins form heterodimers to inhibit the biological activity of their partners, we used a co-immunoprecipitation assay to examine the binding of AtBI-1 to PAP. Both full length and C-terminal deleted AtBI-1 were capable of binding to PAP. These findings indicate that PAP induces cell death in yeast and AtBI-1 inhibits cell death induced by PAP without affecting ribosome depurination and translation inhibition.

Keywords: ribosome inactivating proteins (RIPs), pokeweed antiviral protein (PAP), Arabidopsis thaliana Bax Inhibitor-1, apoptotic-like cell death.

Abbreviations:
PAP - pokeweed antiviral protein,
PCD - programmed cell death,
RIPs - ribosome inactivating proteins,
SRL - sarcin/ricin loop.

INTRODUCTION

Ribosome inactivating proteins (RIPs) that are toxins isolated from plants, fungus, or bacteria catalytically inactivate eukaryotic as well as prokaryotic ribosomes by removing single adenine residues from the universally conserved sarcin/ricin loop (SRL) of the large rRNA [1-4]. In addition to rRNA *N*-glycosidase activity, RIPs have broad spectrum antiviral activity against RNA and DNA from plant and animal viruses [5, 4].

Pokeweed antiviral protein (PAP), a single chain type I RIP, isolated from leaves of pokeweed plants (*Phytolacca americana*), removes specific adenine and guanine residues from the SRL [1, 6, 7]. This enzymatic activity interferes with the binding of eEF-2 (elongation factor 2) thereby inhibiting protein synthesis at the translocation step [8, 9]. PAP has antiviral activity against animal and plant viral pathogens including HIV, poliovirus, herpes simplex virus, influenza, potato virus X, and BMV [10-14]. Hudak *et al.* [7,

15] demonstrated that PAP could also inhibit translation of mRNAs and viral RNAs that are capped by binding to the cap structure and were depurinating the RNAs downstream of the cap. It has been reported that the antiviral activity of PAP can be separated from rRNA depurination [16]. These results suggested that PAP might interfere with virus replication by a mechanism other than host ribosome inactivation. One possible mechanism is that RIPs might target not only the SRL but also the nucleic acids of invading pathogens. Wang and Tumer [17] showed that PAP cleaved double stranded supercoiled DNA using the same active site required for ribosome depurination. Similar activity of other RIPs on supercoiled double-stranded DNA templates was observed with dianthin, gelonin, cinnamomin and saporin [18-20].

Besides inhibition of protein synthesis, RIPs are able to induce apoptosis in different cells [21-25]. Griffiths *et al.* [26] demonstrated that ricin and abrin induced apoptosis in mammalian cells. Many other bacterial as well as plant toxins were also found to induce apoptosis in mammalian cells [21-23, 25, 27-29]. Work from our laboratory has shown that the precursor form of the A chain of ricin (pre-RTA) in yeast cells induced the onset of apoptotic markers such as nuclear fragmentation, chromatin condensation, and accumulation of reactive oxygen species [30]. The ability to depurinate ribosomes and inhibit translation does not always correlate with ricin-mediated cell death [30]. The cell death induced by the RIPs, such as ricin, modeccin, diphtheria toxin and pseudomonas toxin involves caspases [31, 32]. In addition, trichosanthin, a type I RIP, has been shown to induce apoptosis by high levels of ROS production in human choriocarcinoma cells [33].

Apoptosis is co-regulated by the conserved family of Bcl-2 related proteins, which includes both antiapoptotic (e.g., Bcl-2 and Bcl-XL) and proapoptotic (e.g., bax and bak) members. A human Bax inhibitor-1 (BI-1) gene was isolated as a suppressor of bax induced cell death in yeast. BI-1 is evolutionary conserved ER protein that suppresses cell death in plants, yeast and animal cells [34]. Recently, a yeast BH3 domain containing protein (Ybh3p) was identified and regulates the mitochondrial pathway of apoptosis [33]. Interestingly, overexpression of Ybh3p sensitizes yeast cells to apoptotic stimuli, while its knockout reduces cell death [33]. Although no homologs of Bcl-2 family proteins have been identified in plants, BI-1 is widely conserved in organisms, including Caenorhabditis elegans and Xenopus leavis [34]. Subsequently, BI-1 homologs from plants have been isolated [34-39]. AtBI-1, a plant homologue of BI-1 from Arabidopsis thaliana suppresses Bax and H_2O_2 mediated cell death in yeast, animal, and plant cells [40-42]. AtBI-1 is mainly localized in the ER, contains 6 or 7 transmembrane domains with highly conserved C-terminal region that is required for the suppression of cell death [41]. Overexpression of AtBI-1 was shown to suppress cell death induced by biotic and abiotic stresses [36, 41-42]. Overexpression of Bax in plant cells causes ROS generation, organelle disruption, and ion leakage from cells [44, 45] and AtBI-1 prevents ion leakage, but not ROS generation, when overexpressed together with Bax in Arabidopsis [43]. Recently, Watanabe and Lam [46] demonstrated that AtBI-1 played an important role in attenuation of ER stress-induced cell death [46-47]. Another plant BI-1 homologue inCapsicum annuum has been shown to be induced under various abiotic stresses including high salinity, heavy metal stresses and ABA [48]. Moreover, it was demonstrated that AtBI-1 interacts with calmodulin (CAM) and the cell death suppression activities of AtBI-1 in plant cells are mediated by modulation of ion homeostasis. In addition, Oshimo et al. [49] reported that BI-1 requires a functional electron transport chain for cell death suppression in yeast [49].

Although these reports indicate that BI-1 regulates cell death mechanism in animals, yeast, plants, the molecular mechanism by which AtBI-1 inhibits cell death is still unclear. The cell-death induced by plant toxins that inhibit protein synthesis is also not well understood. Their mode of action on cell death needs to be studied further.

Here, we examined the ability of PAP to induce cell death in yeast cells. Yeast expressing PAP displayed apoptosis-like features such as nuclear fragmentation and ROS production. We then studied the interaction between PAP and AtBI-1 for a possible effect of AtBI-1 on PAP induced cell death in yeast. Our results showed that AtBI-1 inhibited cell death induced by PAP in yeast. PAP was able to depurinate the ribosomes and inhibit translation in the presence of AtBI-1. To our knowledge, this is the first report demonstrating that PAP induces cell death in yeast and AtBI-1 inhibits PAP-mediated cell death independent of ribosome depurination and translation inhibition.

RESULTS

PAP expression in yeast causes cell death and AtBI-1 expression attenuates PAP induced cell death

RIPs such as ricin and abrin are able to induce apoptosis in a wide variety of cells and cell lines [21]. Though many studies have been reported on toxin-induced apoptosis, we have very little knowledge of the mechanism of RIP-induced apoptosis. Because of the extreme toxicity of PAP, a type I RIP, in plant cells, the yeast, S. cerevisiae, has been used and demonstrated to be a powerful tool for genetic and biochemical characterization of PAP [50]. When PAP cDNA was expressed in yeast under the control ofGAL1 promoter, cell growth was inhibited [50]. Previous results indicated that ribosome depurination activity of PAP does not always correlate with its translation inhibition activity and is not sufficient for cytotoxicity [51]. In this study, we investigated the ability of PAP to induce cell death in yeast. PAP cDNA was transformed into yeast. Cells were grown in glucose containing medium, then switched to fresh medium containing galactose to induce expression. At different times after induction, cells were recovered from liquid medium by centrifugation and cell viability was determined on the basis of the ability to take up Evans blue dye. Fig. 1A presents results from a representative experiment, showing an increase in the number of cells taking up Evans blue dye in cultures of PAP transformants in galactose containing medium in a time dependent manner. By 24 h post-induction, very few cells survive. These results were confirmed using control cells harboring an empty plasmid which remained mostly dye negative indicating more viable cells (Fig. 1A).

Arabidopsis Bax Inhibitor-1 (AtBI-1), a plant antiapoptotic protein is able to suppress Bax mediated cell death in plants, as well as in yeast cells [52, 41]. To determine if AtBI-1 affects PAP-induced cell death, we cloned full length AtBI-1 cDNA upstream of the V5 tag in pYES 2.1 (Topo Cloning Kit, Invitrogen) vector, in which the expression is under the control of yeast GAL1 promoter. W303 yeast strain has been co-transformed with shuttle vectors harboring PAP and AtBI-1 cDNAs, grown in glucose containing medium, switched to galactose containing medium for induction before staining with Evans blue. As shown in Fig. 1A, yeast cells expressing PAP were stained with Evans blue dye, in contrast to cells expressing AtBI-1, which remained mostly dye negative. Yeast co-expressing AtBI-1

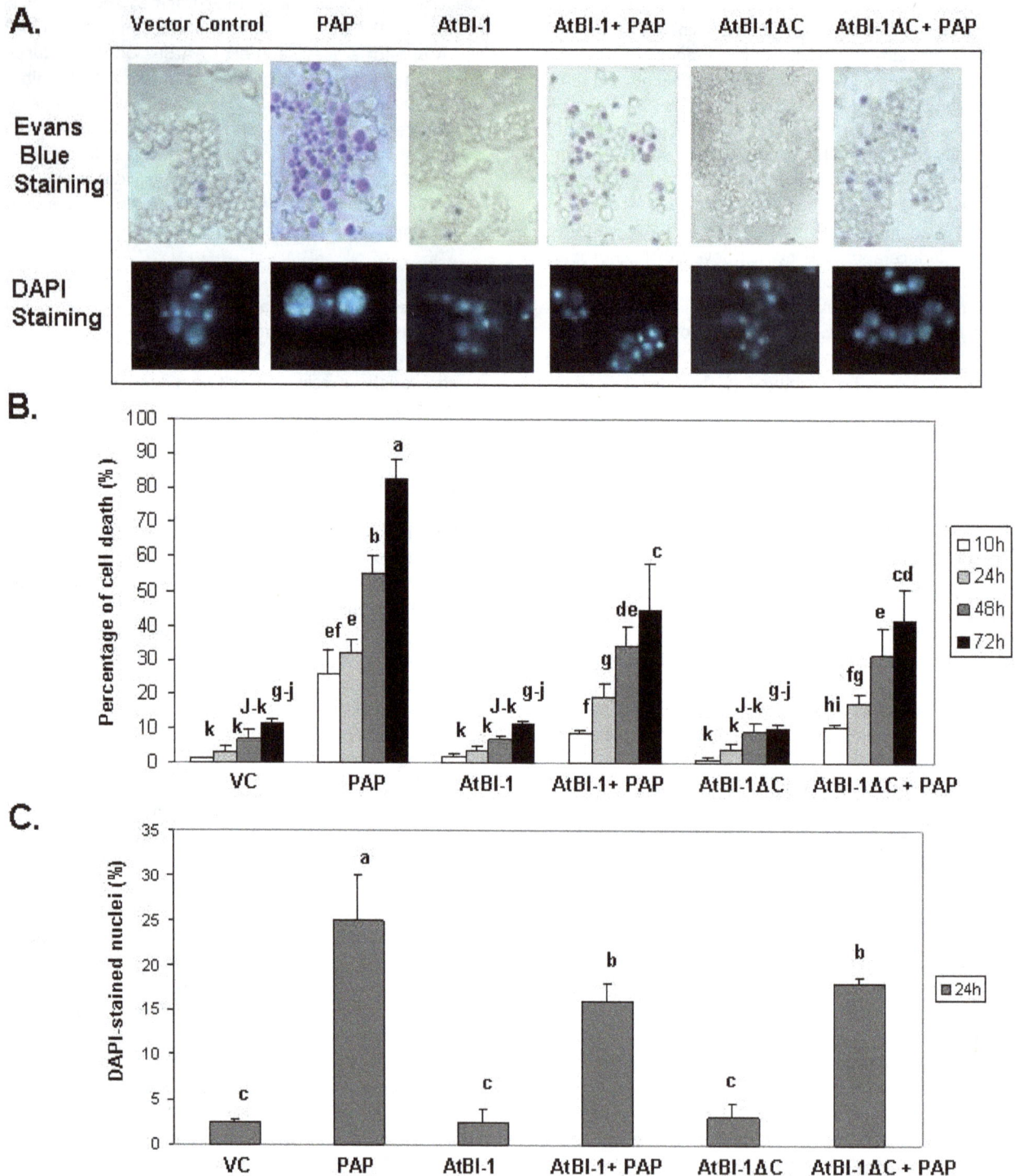

FIGURE 1: Analysis of cell death and nuclear fragmentation in yeast cells expressing PAP, AtBI-1 and AtBI-1ΔC. (A) Cells were stained with Evans Blue or DAPI at 24 h after induction and visualised using Zeiss Axiovert 200 inverted microscope (magnification, X 40) nuclei are shown enlarged 40 times relative to the yeast cells. (B) The percentage of the cell death at different hours after induction were quantified and are represented as the means ± standard deviation (n=3). (C) DAPI stained nuclei at 24 h post-induction were quantified and are represented as the means ± standard deviation (n=3). At least 100 cells were counted per experiment. The results represent three independent experiments. VC - vector control. Columns are statistically different according to ANOVA (P < 0.001) followed by a post-hoc Fisher's Least Significant Difference (LSD) test.

and PAP showed more Evans blue dye excluding cells, indicating an increase in cell viability (Fig. 1A).

Previous studies demonstrated that the C-terminal region of AtBI-1 is necessary for the inhibition of Bax induced cell death in yeast [43, 42]. The deletion of the last 14 amino acids completely abolished cell death suppression ability of AtBI-1 [43]. To determine the functional domain of AtBI-1 responsible for reduced cytotoxicity of PAP, we produced AtBI-1 C-terminal truncation mutant called AtBI-1ΔC (last 23 aa – 224 to 247 – were deleted) and subcloned it into pYES 2.1 vector upstream of V5 epitope. We next co-transformed W303 yeast strain with AtBI-1ΔC and PAP containing plasmids, grew in glucose containing medium then switched to galactose medium for induction. Cells were stained with Evans blue to test the possible effect of C-terminal deletion of AtBI-1 on cell viability in the presence of PAP. As shown in Fig. 1A, viability of cells expressing PAP and AtBI-1 was similar to cells expressing PAP and AtBI-1ΔC, suggesting that the deletion of C-terminal region did not affect the ability of AtBI-1 to suppress the cytotoxicity of PAP.

Apoptotic cell death is characterized by chromatin condensation, nuclear fragmentation and DNA fragmentation in mammalian and yeast cells [53-55]. We examined nuclear fragmentation in those cells to further characterize cell death process induced by PAP. Staining PAP expressing yeast cells with DAPI revealed nuclear fragmentation 24 h after induction, whereas PAP and AtBI-1 co-transformed yeast cells showed a significant decrease in the number of cells with nuclear fragmentation (Fig. 1A and 1C). Chromatin condensation and nuclear fragmentation had been already observed in yeast [56, 57]. After overexpression of PAP, cells showed accumulation of DAPI staining within the area of the nucleus to the appearance of multiple stained regions within a single cell. No nuclear fragmentation was observed in cells expressing AtBI-1 or vector control. These data further confirmed that PAP induces cell death in yeast and AtBI-1 expression attenuates PAP induced cell death. In addition the mutation at C-terminal domain of AtBI-1 does not affect the death suppression ability of AtBI-1 when expressed with PAP.

PAP induces ROS production in yeast

The accumulation of ROS is one of apoptotic cell death features involved in many forms of cell death [58]in animals, yeast and plants [55, 59, 60]. To determine whether ROS generation was involved in PAP induced cell death, we quantified intracellular ROS production by using DCDHF-DA oxidation as a marker to measure intracellular levels of H_2O_2. As shown in Fig. 2, H_2O_2 level was increased in cells expressing PAP up to 24 h post-induction, which correlated well with cell death. In contrast, we did not observe any increase in the level of H_2O_2 up to 24 h post-induction in cells expressing vector control. To determine whether cell death inhibitory activity of AtBI-1 was accompanied by the inhibition of ROS generation, cells expressing PAP and AtBI-1 were also examined. Measurement of ROS generation did not reveal any decrease in H_2O_2 level in cells expressing PAP and AtBI-1 or AtBI-1ΔC, indicating AtBI-1 function as a negative regulator of cell death independent of ROS accumulation, downstream of ROS or both. In addition, the C-terminal deletion of AtBI-1 did not affect H_2O_2 production in cells expressing PAP.

AtBI-1 reduces PAP toxicity in yeast

PAP expression is toxic to yeast cells. To investigate whether AtBI-1 overexpression inhibits PAP toxicity in yeast cells, we tested its ability to rescue against PAP toxicity by examining cell viability. We transformed yeast cells with AtBI-1 and PAP, then plated onto galactose selective

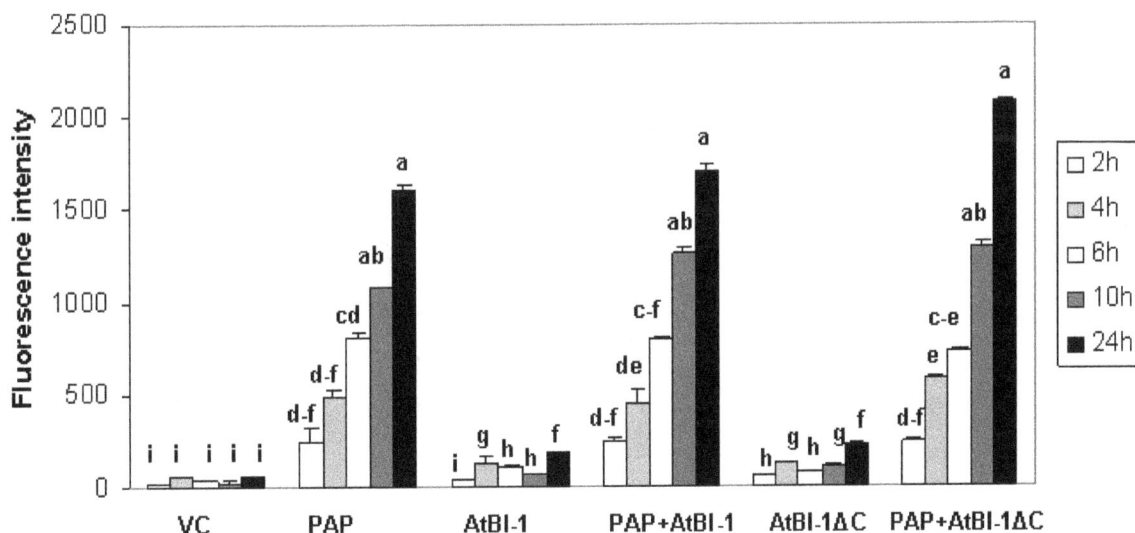

FIGURE 2: ROS generation in cells expressing PAP and AtBI-1. The amount of H_2O_2 production was quantified using DCDHF-DA. The results are represented as the means ± standard deviation (n=3). VC - vector control. The results represent three independent experiments. Columns are statistically different according to ANOVA (P < 0.001) followed by a post-hoc Fisher's Least Significant Difference (LSD) test.

media to induce PAP and AtBI-1 expression. The empty plasmid has been used as a negative control. We then investigated irreversible growth inhibition by carrying out cell viability assay. Yeast cells transformed with PAP and AtBI-1 have been induced in liquid selective medium containing galactose (different induction times were tried out), then they were plated on medium containing glucose. As shown in Fig. 3, PAP expression reduced cell viability at 4 h post-induction, whereas yeast cells expressing PAP in the presence of AtBI-1 slightly restored growth of colonies, indicating reduction of PAP toxicity. We conclude that AtBI-1 is capable of rescuing yeast cells from PAP toxicity.

Cells expressing PAP and AtBI-1ΔC restored growth of colonies at 4 h as compared to cells expressing only PAP. Interestingly, AtBI-1ΔC could protect yeast cells from cytotoxicity of PAP as well as the full length AtBI-1. These results indicate that C-terminus of AtBI-1 is not critical for reducing the cytotoxicity of PAP.

PAP depurinates ribosomes in the presence of AtBI-1

To determine if the reduction in PAP toxicity in the presence of AtBI-1 is due to reduced depurination of ribosomes, we examined ribosome depurination using a primer extension assay. After inducing PAP and AtBI-1 expression in galactose containing media in yeast, we isolated total RNA and examined depurination by using a previously described dual primer extension assay [61]. As shown in Fig. 4, ribosomes were depurinated in yeast cells expressing PAP and AtBI-1 at similar levels as in cells expressing PAP alone. Depurination in cells expressing PAP peaks by 4 h and de-

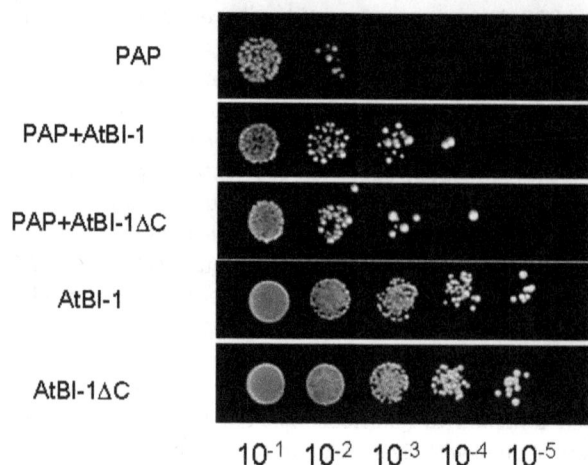

FIGURE 3: **Expression and functional characterization of AtBI-1.** Yeast cells containing AtBI-1 or PAP were grown in SD-U-L/raf overnight then diluted into SD-U-L/galactose for induction at 4 h for PAP and AtBI-1, then serial dilution were spotted on glucose containing plates and incubated for 48 h at 30°C.

creases gradually up to 10 h post-induction. Similarly, PAP depurinated ribosomes when it was co-expressed with AtBI-1 (Fig. 4A).

As shown in Fig. 4A and C, depurination decreased in cells expressing PAP by 10 h post-induction. Ribosomes were depurinated in cells co-expressing PAP and AtBI-1. However, depurination did not decrease in these cells, possibly because these cells did not die unlike cells ex-

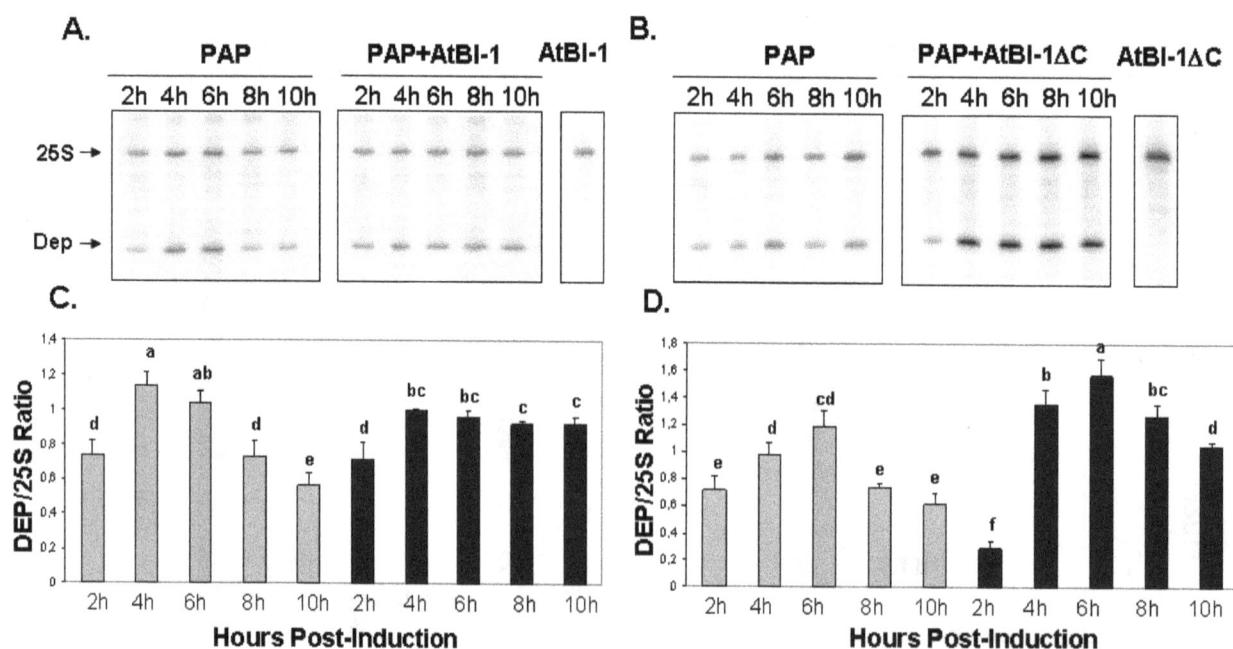

FIGURE 4: **Ribosome depurination in yeast cells.** Primer extension analysis in yeast cells expressing PAP and AtBI-1 **(A)** or PAP and AtBI-1ΔC **(B)** using two different end labeled primers, the depurination primer (Dep) was used to measure the extend of depurination and the 25S rRNA primer (25S) was used to measure the amount of 25S rRNA **(C)** and ratio of Dep/25S **(D)**. The results represent three independent experiments. Columns are statistically different according to ANOVA ($P < 0.001$) followed by a post-hoc Fisher's Least Significant Difference (LSD) test.

pressing PAP alone (Fig. 4C and D). The deletion of the C-terminal domain of AtBI-1 did not reduce the ribosome depurination activity of PAP as compared to that of the full length AtBI-1 in the presence of PAP. Moreover, cells expressing PAP and AtBI-1 survive better, even though their ribosomes are depurinated.

AtBI-1 does not affect translation inhibition by PAP

To determine whether reduction of PAP toxicity is related to reduction in translation inhibition activity of PAP, we examined total translation in yeast cells expressing PAP and AtBI-1 compared with cells expressing PAP alone and vector control. Total translation was examined by [^{35}S] methionine incorporation at 0, 4, 6, 10 h post-induction. As shown in Fig. 5A, in yeast cells expressing PAP and AtBI-1 at 4 h post-induction, translation was inhibited at the same level as with PAP alone, whereas in cells expressing AtBI-1, translation increased gradually over time. These results indicate that total translation is inhibited in cells co-expressing PAP and AtBI-1 at a similar level as in cells expressing PAP alone, indicating that AtBI-1 expression does not have any effect on the translation inhibition activity of PAP.

As shown in Fig. 5B, total translation was significantly inhibited in cells expressing PAP and AtBI-1ΔC at a similar level as PAP alone, whereas total translation was not inhibited in cells expressing AtBI-1 mutant or vector control. These results indicate that the reduction in the cytotoxicity of PAP in the presence of full length or C-terminally deleted AtBI-1 is not due to a decrease in the translation inhibitory activity of PAP.

AtBI-1 and PAP mRNAs are upregulated in yeast co-expressing AtBI-1 and PAP

To determine if the reduction in the cytotoxicity of PAP is due to a decrease in PAP expression we isolated total RNA and examined PAP mRNA expression pattern in yeast cells co-expressing PAP and AtBI-1 at various times post-induction (Fig. 6). PAP mRNA level in yeast expressing AtBI-1 was upregulated compared to yeast expressing PAP alone (Fig. 6A). PAP transcript level increased by 11-fold compared to that of PAP alone by 6 h post-induction, and then decreased up to 10 h post-induction. AtBI-1 mRNA level increased by 4 hours post-induction and decreased gradually up to 10 h post-induction in yeast expressing AtBI-1 alone, suggesting possible autoregulation. AtBI-1

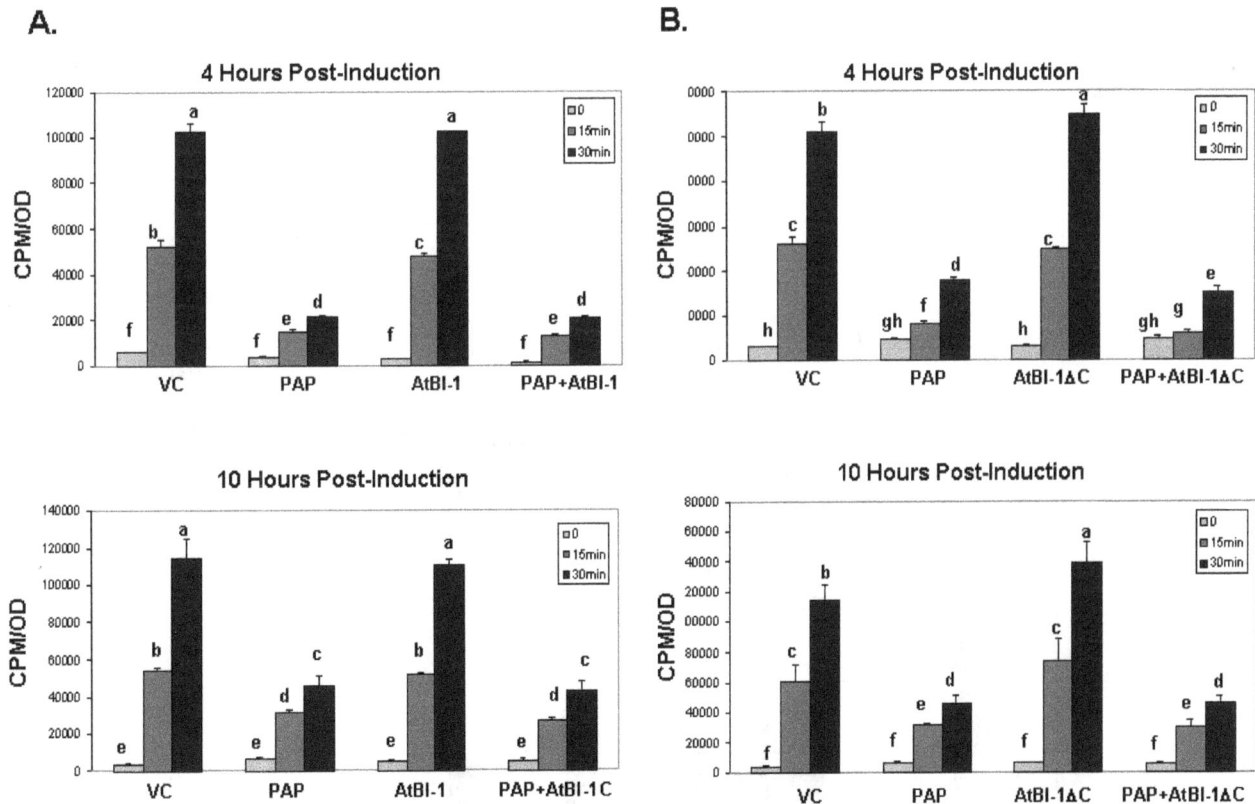

FIGURE 5: **Analysis of total translation in yeast cells. (A)** Total translation in yeast cells expressing PAP and AtBI-1. **(B)** Total translation in yeast cells expressing PAP and AtBI-1ΔC. Yeast cells were grown in SD-L-U-Met and 2% glucose overnight then switched to 2% galactose containing media to induce the expression. At time 0, [^{35}S] methionine was added to cells growing on galactose which expressed PAP or AtBI-1 and incorporation of [^{35}S] methionine was determined at the indicated times. Each point was repeated in duplicate. The results represent three independent experiments. VC - vector control. Columns are statistically different according to ANOVA (P < 0.001) followed by a post-hoc Fisher's Least Significant Difference (LSD) test.

mRNA accumulated at a higher level at 2h post-induction and stayed at a similar level up to 6 h post-induction in cells co-expressing PAP and either full-length or C-terminally deleted AtBI-1 (Fig. 6B). We conclude that AtBI-1 and PAP mRNA expression is upregulated in yeast cells expressing both proteins, consistent with the reduction in the cytotoxicity of PAP in these cells.

AtBI-1 and PAP proteins are expressed

We then investigated PAP expression level in yeast extracts to determine whether the reduced cytotoxicity of PAP is due to altered expression or subcellular localization. We fractionated yeast extracts at various times after induction into cytoplasmic and membrane fractions [62] and analyzed each fraction for the presence of PAP and AtBI-1 proteins. ER membrane protein, Dpm1p and the cytosolic protein Pgk1p have been used as controls for fractionation. It has been already reported that both the precursor form of PAP and the mature form are associated with the ER membrane in yeast [63]. As shown in Fig. 7A, PAP level was higher in both the membrane and the cytosol fraction at 4 and 6 h post-induction in yeast cells expressing PAP and AtBI-1 as compared to that of the cells expressing PAP alone. We analyzed AtBI-1 expression in both cytosolic and

membrane fractions (Fig. 7B). AtBI-1 was detected in the cytosol and membrane fraction in cells expressing AtBI-1 alone. However, AtBI-1 was detected only in the membrane fraction in cells expressing AtBI-1 together with PAP (Fig. 7B). AtBI-1 and PAP mRNA expression did not correlate to protein levels in yeast cells expressing both proteins. This was previously observed by Di et al. [64].

In contrast to full-length AtBI-1, we did not observe any AtBI-1ΔC in the cytosolic fraction (Fig. 7B). When co-transformed with AtBI-1ΔC and PAP, yeast cells slightly expressed AtBI-1ΔC in cytosolic fraction by 24 hours post-induction (Fig. 7B).

AtBI-1 binds to PAP in vitro

To examine the possibility that AtBI-1 may reduce the cytotoxicity of PAP by binding to it and forming a heterodimer, we used a co-immunoprecipitation assay in yeast cells expressing both PAP and AtBI-1. Total protein extracted from yeast co-expressing PAP and AtBI-1 were co-immunoprecipitated with the V5 monoclonal antibody. Total protein extracted from yeast expressing PAP alone and vector control was used as a negative control. PAP and AtBI-1 were co-immunoprecipitated with V5 antibody (Fig. 8A). Immunoblot analysis using total lysate was used to

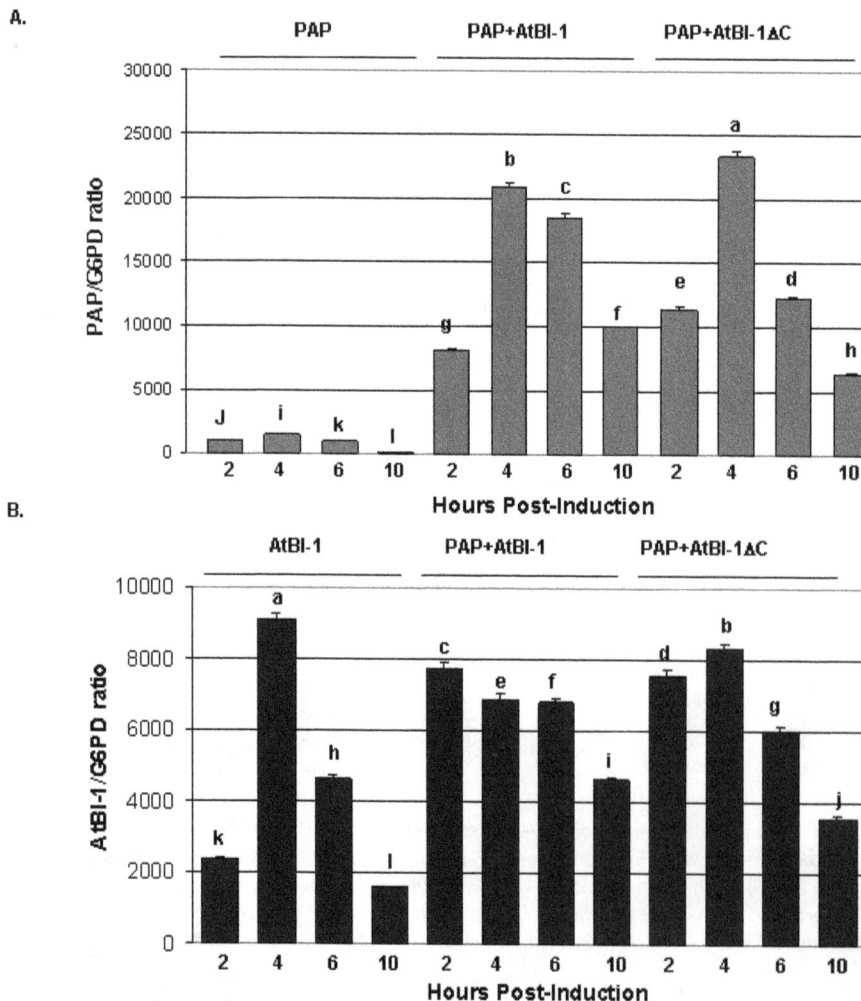

FIGURE 6: Real-time PCR analysis of mRNA levels in yeast cells. (A) Analysis of PAP mRNA in yeast cells expressing AtBI-1. (B) Analysis of AtBI-1 mRNA in yeast cells expressing PAP. Cells were grown on galactose for the hours indicated. The mRNA levels for the genes were normalized to G6PD mRNA using the ΔΔCT method from Applied Biosystems. The results represent three independent experiments. Columns are statistically different according to ANOVA (P < 0.001) followed by a post-hoc Fisher's Least Significant Difference (LSD) test.

show the level of expression of both proteins (Fig. 8B). We next performed co-immunoprecipitation assay using AtBI-1ΔC. These results indicated that AtBI-1 binds directly to PAP and C-terminal deletion of 23 aa did not change the binding capacity of AtBI-1 to PAP. These finding demonstrate that AtBI-1 can rescue yeast cells from cytotoxicity of PAP by binding it to form a heterodimer.

DISCUSSION

We present evidence here that cells expressing PAP, a type I ribosome inactivating protein, exhibit nuclear fragmentation characterized by DAPI stained multiple regions in the nucleus with extensive vacuolization. These alterations were absent in cells expressing vector control, AtBI-1 and AtBI-1ΔC (Fig. 1). The number of cells exhibiting nuclear fragmentation was decreased in cells expressing PAP and AtBI-1, indicating protective effect of AtBI-1 on PAP induced cell death. Besides inhibition of protein synthesis, PAP, a type I RIP, cleaves single-stranded [65] as well as double-stranded DNA [17] using the same active site required to depurinate rRNA. The cleavage of DNA in the nucleus and nuclear fragmentation are typical apoptotic features in yeast [66, 67]. It is tempting to think that the nuclear fragmentation in cell death induced by PAP is due to its nuclease activity. Similarly, some RIPs induce DNA damage by their nuclease activity [20, 68]. However, this

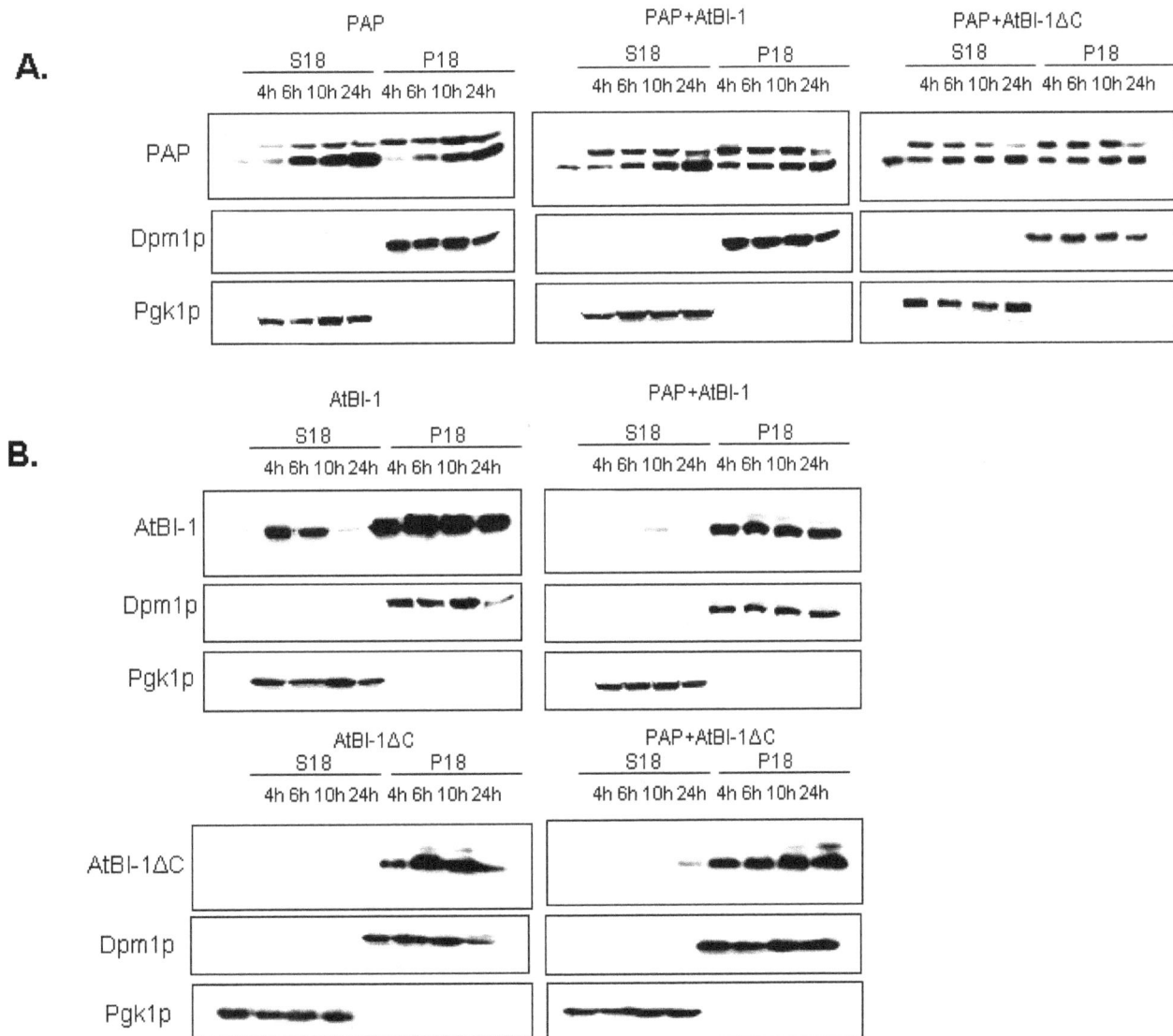

FIGURE 7: **Immunoblot analysis. (A)** PAP expression in yeast cells expressing AtBI-1 or AtBI-1ΔC. **(B)** AtBI-1 or AtBI-1ΔC expression in yeast cells expressing PAP. Yeast transformants were grown in SD-U-L glucose media. The expression of PAP and AtBI-1 was induced for indicated hours, lysed and fractionated into membrane (P18) and cytosolic (S18) components. The amount of 10 mg protein was separated on 15 % SDS-PAGE. Proteins were transferred into nitrocellulose membrane and probed with polyclonal PAP antiserum and monoclonal V5 antibody. The membrane marker Dpm1p and the cytosolic marker Pgk1p were used as controls to show equal amount of loading and lack of cross-contamination.

aspect needs to be studied in depth to determine whether RIPs can enter the nucleus to induce DNA damage by nuclease activity and trigger apoptosis.

ROS production is involved in many types of cell death processes in animals, yeast and plants [55, 59, 60]. Previous studies showed that ROS production is implicated in ricin-induced apoptotic cell death in mammalian cells as well as in yeast [28, 69]. Yeast cells exposed to oxidative stress or expressing mammalian Bax also induce ROS production [57]. To determine if ROS is accumulated in yeast expressing PAP, intracellular ROS levels were quantified. The ROS accumulation in cells expressing PAP at 24 hours post-induction was almost 2-fold higher than that of vector control suggesting that ROS may act as an effector of apoptosis and trigger cell death signaling pathways. ROS induction precedes cell death in PAP expressing cells, suggesting that ROS may act as an effector of apoptosis and trigger cell death signaling pathways in those cells. In this study, we demonstrated for the first time that PAP, a type I RIP induces ROS production. Since ribosome depurination and translation inhibition are not always correlated with the cytotoxicity of PAP, ROS production in cells expressing PAP may be an important step, leading to PAP induced cell death.

We present the first evidence that the cytotoxicity of PAP is not only due to the depurination and translation inhibition but also to cell death in yeast.

AtBI-1 suppresses Bax induced cell death in plants, mammalian and yeast cells [41, 44, 70]. Bax induced cell death in *Arabidopsis* protoplast system is inhibited by the overexpression of AtBI-1 through ROS independent processes [44]. We investigated the possibility that AtBI-1 can inhibit PAP toxicity through ROS dependent processes. We

did not observe any decrease in ROS production in cells expressing PAP and AtBI-1 or AtBI-1ΔC, indicating that AtBI-1 functions as a negative regulator of PAP induced cell death independent of ROS accumulation or downstream of ROS, or both. In addition, the deletion of C-terminal region of AtBI-1 did not affect ROS accumulation in cells expressing both proteins suggesting that both proteins inhibit PAP toxicity via ROS independent pathway. These results correlated well with the cell viability assay. Yeast expressing PAP and AtBI-1 were able to grow on glucose containing medium after induction for 4 h in galactose containing media (Fig. 3). To determine the role of the C-terminal region of AtBI-1 in reduced cytotoxicity of PAP, we co-transformed yeast cells with PAP and AtBI-1ΔC, and AtBI-1ΔC was able to rescue yeast cells from cytotoxicity of PAP at the same level as AtBI-1. It was recently reported that the deletion of C-terminal region of AtBI-1 abolishes the ability of AtBI-1 to suppress Bax-induced cell death in yeast [43]. The same authors demonstrated that the formation of coiled-coil structure in C terminus of AtBI-1 is essential for Bax-induced cell death inhibition in yeast. In our study, the deletion of 23 aa, which eliminated the predicted formation of coiled-coil structure, altered the level of AtBI-1 in the cytosol. The deletion of 23 aa at C-terminal region of AtBI-1 did not affect cell death suppression activity of AtBI-1 against PAP. Our results suggest that the C-terminal region of AtBI-1 may be critical for its transport to the cytosol. Since both AtBI-1 and AtBI-1ΔC are associated with the ER, they may inhibit PAP associated with the ER membrane fraction in yeast. We conclude that the C-terminal region of AtBI-1 is not critical for the interaction between AtBI-1 and PAP, and the cell death inhibition activity of AtBI-1 against PAP. The structural model of Human Bax inhibitor-1

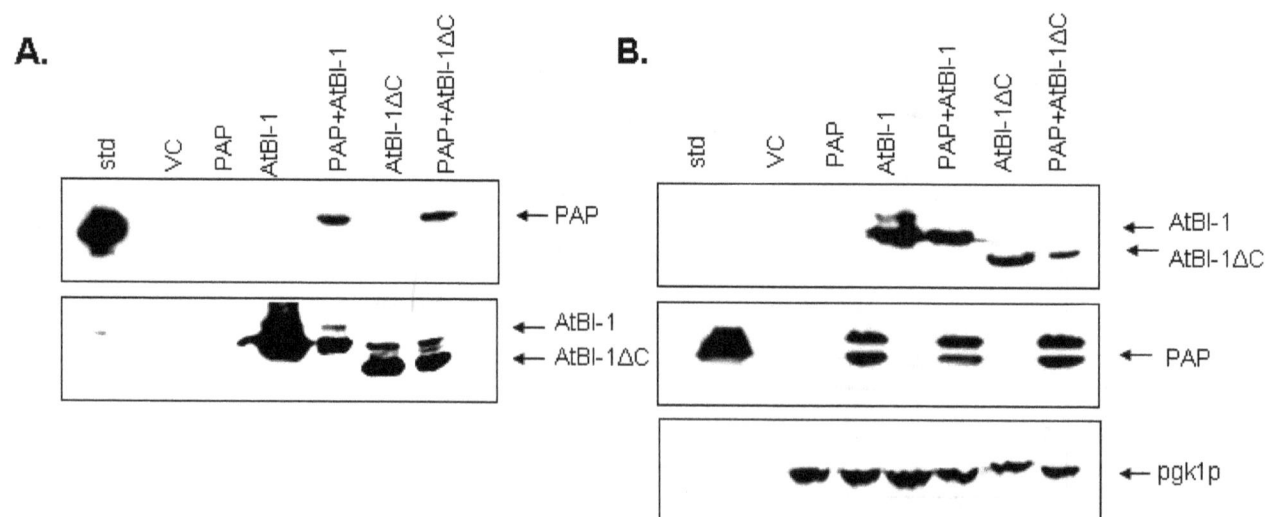

FIGURE 8: **Co-immunoprecipitation assay. (A)** Co-immunoprecipitation of PAP with either AtBI-1 or AtBI-1ΔC. Total proteins isolated from yeast cells induced to express PAP, AtBI-1, AtBI-1ΔC and vector control (VC) for 6 hours were incubated with V5-antibody and immunoprecipitated with protein A-Sepharose beads. Immunoprecipitated proteins were separated on 15 %SDS-PAGE, transferred to nitrocellulose and probed with affinity purified PAP antibody or with V5-antibody. **(B)** The total lysate from cells expressing PAP, AtBI-1 and AtBI-1ΔC subjected to SDS PAGE/immunoblot analysis using V5 and PAP-antibody to show the level of expression of both proteins.

(hBI-1) revealed a 6-TM topology with both N- and C-termini in the cytoplasm and exhibits PH-sensitive calcium leak activities, proposed to be mediated by the C-terminal region [71]. By homology, the C-terminal region of AtBI-1 may also have PH-sensitive calcium leak activity.

Hudak et al. [51] identified the PAP residues that are critical for ribosome depurination, inhibition of translation and cytotoxicity, and demonstrated that ribosome depurination is not sufficient for the inhibition of translation and cytotoxicity. Our results support this observation. Even though, rRNA depurination level was higher and translation inhibition was not affected in yeast expressing PAP and either AtBI-1 or AtBI-1ΔC, cells were able to survive on galactose containing medium. These results show that At-BI-1 inhibited cell death caused by PAP in yeast independent of ribosome depurination and translation inhibition.

BI-1 was shown to be Bcl-2 binding but not Bax-binding protein with antiapoptotic activity [70]. To further investigate a direct interaction between PAP and AtBI-1, we conducted co-immunoprecipitation assay with cells expressing both proteins. AtBI-1 as well as AtBI-1ΔC proteins were able to bind to PAP at 6 hours post-induction. At 6 hours post-induction, both AtBI-1 and AtBI-1ΔC were able to bind to the precursor form of PAP but not to mature PAP. The deletion of 23 aa at C-terminal region of AtBI-1 did neither abolish nor diminish the binding capacity of protein to PAP suggesting that the C-terminal region is not critical for this interaction.

Plant and animal BI-1 proteins are located mostly in the ER and the perinuclear region [38, 41, 70]. Although the precursor form of PAP is mostly associated with the ER membrane, it is not exclusively localized in the ER [72]. At 6 h post-induction we found the precursor form of PAP in cytosolic fraction as well as in the membrane fraction as described previously by Parikh et al. [72]. However, AtBI-1 was only associated with the membrane fraction in yeast co-expressing PAP and AtBI-1, suggesting that the binding may take place in the ER.

Ricin inhibits adaptation responses to ER stress by preventing HAC1 mRNA splicing and Ire1p signaling to downstream mediators of UPR [73]. The inability to activate UPR in response to ER stress contributes to ricin-mediated cell death. By analogy with ricin, we can speculate that PAP may interfere with UPR therefore causing ER stress induced cell death. We recently showed that Bax expression induced the UPR in yeast and this was associated with HAC1 mRNA splicing [74]. Yeast cells deficient for yeast bax inhibitor (Δbxi1) are not only more sensitive to ER stress-inducing drugs but also have a decreased UPR [75]. By homology with BXI1, AtBI-1 could also regulate PAP-induced cell death by UPR.

In summary, we show here that PAP induces cell death in yeast and AtBI-1 inhibits PAP induced cell death. We present evidence that the C-terminal region of AtBI-1 is not required to reduce PAP cytotoxicity. We demonstrate that AtBI-1 inhibits cell death induced by PAP independent of ribosome depurination and translation inhibition. Future experiments will characterize the mechanism by which AtBI-1 inhibits PAP cytotoxicity.

MATERIALS AND METHODS

Determination of cell viability by Evan`s blue staining, chromatin staining and ROS measurement

Cells were collected after induction at the times indicated, washed in PBS buffer and Evans blue was added to 1 ml of 0.6 OD cells at the concentration of 0.5% in PBS buffer and stained at room temperature for 30 min. After staining cells were washed several times with ddH$_2$O to remove unbound dye from cultures before observation. Cells were counted using Zeis Axiovert 200 inverted microscope. The percentage of cell death was calculated by counting ~800 total cells as described by Xu et al. [71].

To detect nuclear fragmentation, yeast cells were washed with PBS buffer, fixed in 100% ethanol at room temperature for 5 min. and washed again. For nuclear staining, samples were incubated for 5 min. with 0.5 μg*ml^{-1} diaminopheylindole (DAP) in PBS and analyzed after washing by Zeiss Axiovert 200 inverted microscope with the epifluorescence setting (Axiovision 3.0; Carl Zeiss Vision GmbH).

Intracellular production of H$_2$O$_2$ was measured using the antioxidant sensitive probe 2',7'-dichlorodihydrofluorescein diacetate (DCDHF-DA) (Invitrogen, Carlsbad, CA). 2 μl of fresh 5 mM DCDHF-DA was added to 1 ml of yeast culture (10^7 cells) and incubated at 28°C for 45 min. The cells were then washed twice in sterile distilled water and resuspended in 1 ml of 50 mM Tris-HCl, pH 7.5. After 20 μl of chloroform and 10 μl of 0.1% SDS were added, the cells were incubated for 15 min. and pelleted. The fluorescence of the supernatant was measured using an HTS700 Perkin Elmer bioassay reader (Wellesley, MA) with excitation at 485 nm and emission at 525 nm.

Plasmids

The cloning of PAP cDNA into NT198 under the control of GAL1 promoter used in this study was described previously [7, 73] . AtBI-1 (AB025927) was cloned into the yeast expression vector pYES2.1 (pYES2.1 TOPO TA expression kit, Invitrogen, USA) in upstream of V5 epitope by PCR using 5'GGATCCACGATGGATGCGTTCTCTTCCTTC3' and 5'GTTTCTCC-TTTTCTTCTTCTTCTC3' primers and into the pTKB175 without a tag. After the cloning, vectors were transformed into E. coli DH5α. The sequences were confirmed by sequencing two times using specific primers.

Yeast transformation and cell viability

The S. cerevisiae strain W303 (MATa ade2 trp-1 ura-3 leu2-3,112 his3-11,15 can1-100) (from B. Thomas, Columbia university, New York, NY) was used for all transformations.
Cells were transformed and co-transformed as described previously [51]. One-half of transformed yeast suspension was plated onto 2% glucose media, the other half was plated onto 2% galactose containing media. The toxicity of PAP was verified by re-plating the selected colonies onto both 2% glucose and 2% galactose media.

For cell viability, transformed and co-transformed yeast cells were grown on SD-Leu containing 2% glucose to an A_{600} of 0.3 and then transferred to selective medium containing 2% galactose to induce PAP and AtBI-1 expression. A serial dilution of cells was plated on selective media containing 2% glucose at 0, 4, 6, 10, and 12 h post-induction. Plates were incubated at 30°C for approximately 48 h.

Growth conditions

Yeast cells were grown in YPD rich medium or synthetic drop-out (SD) medium with appropriate amino acids at 30°C. Yeast cells transformed with PAP and AtBI-1 were grown initially at 30°C in a total volume of 100 ml of selective medium supplemented with 2% raffinose to a starting A_{600} of 0.6. Yeast cells were pelleted by centrifugation and washed with SD medium before replacing with 100 ml of selective medium containing 2% galactose to a starting A_{600} of 0.3. Then, 5 ml of culture were sampled for protein isolation, 10 ml of culture for RNA isolation and 1 ml for a growth reading (A_{600}) at different times post-induction.

Yeast protein expression analysis

Total protein extraction from frozen yeast cells collected at different times post-induction was extracted as described by Hudak et al. [76]. Samples were separated on 15% SDS-PAGE, transferred to nitrocellulose membrane (Roche) and probed with affinity purified anti-PAP polyclonal antibody (1:5000). The AtBI-1 and mutant forms of AtBI-1 proteins level were determined by using V5 monoclonal antibody (1:5000) that recognize V5 epitope at C-terminal of the protein. PAP, AtBI-1 and AtBI-1 mutants were visualized by chemiluminescence using the Renaissance kit (PerkinElmer Life Sciences). The blots were then stripped with 8 M guanidine hydrochloride for 30 min and reprobed with 3-phosphoglycerate kinase (Pgk1 p; Molecular probes) (1:10000) as an internal loading control.

For cell fractionations, protein from frozen yeast cells collected during various times post-induction was extracted as described by Frey et al. (2001). Briefly, after addition of low-salt (LS) buffer (20mM HEPES-KOH, pH 7.6, 100 mM potassium acetate, 5 mM magnesium acetate, 1 mM EDTA, 2mM DTT and 0.1 mM PMSF, yeast protease inhibitor cocktail (Sigma) and acid-washed glass beads (Sigma), cells were vortexed for 1 min and chilled for 1 min on ice for a total of 8 cycles. Crude lysates were spanned at 1200 g for 2 min. The same lysate was then centrifuged an additional 20 min at 18 000 g. The pellet was washed twice with ice cold water and resuspended in LS buffer. The supernatant and pellet fraction were stored at -80°C.

RNA analysis

Total RNA was extracted from yeast using hot phenol [17]. cDNA was synthesized from 1 μg of total RNA in a 20 μl reaction, containing 1 × first-strand buffer (Invitrogen), 40 U/μl RNA Guard RNase inhibitor (Promega, Madison, WI, USA), 0.5 μg poly d(T) oligonucleotide (Promega), 40 mM dNTPs and Superscript II (Invitrogen) reverse transcriptase. Quantification of transcript levels by real-time PCR analysis was performed using an ABI Prism 7000 Sequence Detection System using the manufacturers' protocols. For quantitative PCR, the primers used were as follows:

PAP, 5'-gggtaagatttcaacagcaattca-3' and 5'-caccactggcatccact-agct-3';

G6PD 5'-CAGCAATGACTTTCAACATC-GAA-3' and 5'-CCGGCAC-GCATCATGAT-3';

AtBI-1, 5'-GTTGTGCTCTTGTGGCGTCTGC-3' and 5'- TCAAGGG-GCCAACAGAAGCACCT-3'.

In vivo [35S] Methionine Incorporation

Yeast cells were grown to an A_{600} of 0.6 in SD selective medium supplemented with 2% raffinose. Cells were then resus-pended at an A_{600} of 0.3 in 2% galactose containing SD selective medium for 4-10 hours in order to induce either PAP, AtBI-1 or mutant forms of AtBI-1. At time zero, [35S] methionine was added to cells growing on galactose. At the various times post-induction, 600 ml of yeast cells were taken for growth measurements and an aliquot of 800 ml were assayed for methionine incorporation in triplicate as described by Parikh et al. 2002. Briefly, the yeast were added to 200 ml of 100% trichloroacetic acid and incubated for 10 min on ice followed by 20 min at 70°C. The precipitate then filtered through 24-mm glass microfiber filters (VWR), washed with ice-cold 5% trichloroacetic acid followed by 95% ethanol. Filters were dried overnight and incorporation was quantified in a scintillation counter. The Cpm was normalized to the A_{600} reading.

rRNA depurination Assay

Depurination of ribosomal RNA was performed by primer extension analysis in according to Hudak et al.[15]. 2 mg of total yeast RNA from transformants was incubated with (α-32P) ATP end labeled 5' reverse primer (5'- AGCGGATGGTGCTTCGCGG-CAATG-3') complementary to 73 nt 3' end of depurination site for depurination product and 5' reverse primer (5'-TTCACTCGCCGTTACTAAGG-3') specific to the 3' end of yeast 25S rRNA as an internal control. The presence of depurination was observed by synthesis of a 73 nt extension product corresponding to the depurination site. An aliquot of 4 ml of extension product was separated on a 6% polyacrylamide/7 M urea denaturing gel and visualized and quantified on a PhosphorImager (Amersham Biosciences).

Co-immunoprecipitation

PAP and AtBI-1 expressed in vivo yeast cells were co-immunoprecipitated with the monoclonal antibody against V5 epitope essentially as described by Otto and Lee [77]. Total protein extracts from cells induced to express PAP, AtBI-1 at 6 h post-induction were used as substrate for immunoprecipitation with Protein A-Sepharose beads. Proteins were eluted from the beads with SDS sample buffer and visualized by immunoblot analysis using the antibodies to PAP and AtBI-1.

Statistical analyses

The data were subjected to ANOVA test according to completely randomized factorial design. Differences between means were determined with Fisher's Least Significant (LSD) test. P value of ≤ 0.001 was considered statistically significant. All values are presented as the mean of three independent experiments with the corresponding Standard Deviation (SD).

ACKNOWLEDGMENTS

This work was supported by National Science Foundation grants MCB-0348299 and MCB-0130531 to Nilgun E. Tumer. We are grateful to Dr. Özlem Tuncay for her help in statistical analyses.

CONFLICT OF INTEREST
The authors declare no conflict of interest.

REFERENCES

1. Endo Y, Tsurugi K (**1988**) The RNA N-glycosidase activity of ricin A-chain. Thecharacteristics of the enzymatic activity of ricin A-chain with ribosomes and with rRNA. **J Biol Chem** 263: 8735-8739.

2. Endo Y, Mitsui K, Motizuki M, Tsurugi K (**1987**) The mechanism of action of ricin and related toxic lectins on eukaryotic ribosomes. The site and the characteristics of the modification in 28 S ribosomal RNA caused by the toxins. **J Biol Chem** 262: 5908-5912.

3. Wool IG, Gl̦ck A, Endo Y (**1992**) Ribotoxin recognition of ribosomal RNA and a proposal for the mechanism of translocation. **Trends Biochem Sci** 17: 266-269.

4. Barbieri L, Battelli M, Stirpe F (1993) Ribosome-inactivating proteins from plants. **Biochim Biophys Acta** 1154: 237-287.

5. Wang P, Turner NE (**2000**) Virus resistance mediated by ribosome inactivating proteins. **Adv Virus Res**: Academic Press. pp. 325-355.

6. Hartley MR, Legname G, Osborn R, Chen Z, Lord JM (**1991**) Single-chain ribosome inactivating proteins from plants depurinate Escherichia coli 23S ribosomal RNA. **FEBS Lett** 290: 65-68.

7. Hudak KA, Wang P, Tumer NE (**2000**) A novel mechanism for inhibition of translation by pokeweed antiviral protein: depurination of the capped RNA template. **RNA** 6: 369-380.

8. Montanaro L, Sperti S, Mattioli A, Testoni G, F S (**1975**) Inhibition by ricin of protein synthesis in vitro. Inhibition of the binding of elongation factor 2 and of adenosine diphosphate-ribosylated elongation factor 2 to ribosomes. **J Biochem** 146: 127-131.

9. Osborn RW, Hartley MR (**1990**) Dual effects of the ricin A chain on protein synthesis in rabbit reticulocyte lysate. **E J Biochem** 193: 401-407.

10. Aron GM, Irvin JD (**1980**) Inhibition of herpes simplex virus multiplication by the pokeweed antiviral protein. **Antimicrob Agents Chemother** 17: 1032-1033.

11. Lodge JK, Kaniewski WK, Tumer NE (**1993**) Broad-spectrum virus resistance in transgenic plants expressing pokeweed antiviral protein. **Proc Natl Acad Sci USA** 90: 7089-7093.

12. Tomlinson JA, Walker VM, Flewett TH, Barclay GR (**1974**) The Inhibition of Infection by Cucumber Mosaic Virus and Influenza Virus by Extracts from Phytolacca americana. **J Gen Virology** 22: 225-232.

13. Ussery MA, Irvin JD, Hardesty B (**1977**) Inhibition of Poliovirus Replication by a Plant Antiviral Paptide. **Ann New York Acad Sci** 284: 431-440.

14. Zarling JM, Moran PA, Haffar O, Sias J, Richman DD, Spina CA, Myers DE, Kuebelbeck V, Ledbetter JA, Uckun FM (**1990**) Inhibition of HIV replication by pokeweed antiviral protein targeted to CD4+ cells by monoclonal antibodies. **Nature** 347: 92-95.

15. Hudak KA, Bauman JD, Tumer NE (**2002**) Pokeweed antiviral protein binds to the cap structure of eukaryotic mRNA and depurinates the mRNA downstream of the cap. **RNA** 8: 1148-1159.

16. Tumer NE, Hwang D-J, Bonness M (**1997**) C-terminal deletion mutant of pokeweed antiviral protein inhibits viral infection but does not depurinate, host, ribosomes. **Proc Natl Acad Sci USA** 94: 3866-3871.

17. Wang P, Tumer NE (**1999**) Pokeweed antiviral protein cleaves double-stranded supercoiled DNA using the same active site required to depurinate rRNA. **Nuc Acids Res** 27: 1900-1905.

18. Huang PL, Chen HC, Kung HF, Huang P, Huang HI, Lee-Huang S (**1992**) Anti-HIV plant proteins catalyze topological changes of DNA into inactive forms. **Biofactors** 4(1):37-41.

19. Ling J, Liu W-y, Wang TP (**1994**) Cleavage of supercoiled double-stranded DNA by several ribosome-inactivating proteins in vitro. **FEBS Lett** 345: 143-146.

20. Roncuzzi L, Gasperi-Campani A (**1996**) DNA-nuclease activity of the single-chain ribosome-inactivating proteins dianthin 30, saporin 6 and gelonin. **FEBS Lett** 392: 16-20.

21. Narayanan S, Surolia A, Karande AA (**2004**) Ribosome-inactivating protein and apoptosis: abrin causes cell death via mitochondrial pathway in Jurkat cells. **The Biochem j** 377: 233-240.

22. Morimoto H, Bonavida B (**1992**) Diphtheria toxin- and Pseudomonas A toxin-mediated apoptosis. ADP ribosylation of elongation factor-2 is required for DNA fragmentation and cell lysis and synergy with tumor necrosis factor-alpha. **The J Immun** 149: 2089-2094.

23. Kochi SK, Collier RJ (**1993**) DNA Fragmentation and Cytolysis in U937 Cells Treated with Diphtheria Toxin or Other Inhibitors of Protein Synthesis. **Exp Cell Res** 208: 296-302.

24. Allam M, Bertrand R, Zhang-Sun G, Pappas J, Viallet J (**1997**) Cholera Toxin Triggers Apoptosis in Human Lung Cancer Cell Lines. **Cancer Res** 57: 2615-2618.

25. Brinkmann U, Mansfield E, Pastan I (**1997**) Effects of BCL-2 overexpression on the sensitivity of MCF-7 breast cancer cells to ricin, diphtheria and Pseudomonas toxin and immunotoxins. **Apoptosis** 2: 192-198

26. Griffiths GD, Leek MD, Gee DJ (**1987**) The toxic plant proteins ricin and abrin induce apoptotic changes in mammalian lymphoid tissues and intestine. **The J Pathol** 151: 221-229.

27. Chang MP, Bramhall J, Graves S, Bonavida B, Wisnieski BJ (**1989**) Internucleosomal DNA cleavage precedes diphtheria toxin-induced cytolysis. Evidence that cell lysis is not a simple consequence of translation inhibition. **J Biol Chem** 264: 15261-15267.

28. Jetzt AE, Cheng JS, Tumer NE, Cohick WS (**2009**) Ricin A-chain requires c-Jun N-terminal kinase to induce apoptosis in nontransformed epithelial cells. **Int J Biochem Cell Biol** 41:2503–10.

29. Jetzt AE, Cheng J-S, Li X-P, Tumer NE, Cohick WS (**2012**) A relatively low level of ribosome depurination by mutant forms of ricin toxin A chain can trigger protein synthesis inhibition, cell signaling and apoptosis in mammalian cells. **Int. J. Biochem Cell Biol** 44:2204–2211.

30. Li X-P, Baricevic M, Saidasan H, Tumer NE (**2006**) Ribosome Depurination Is Not Sufficient for Ricin-Mediated Cell Death in *Saccharomyces cerevisiae*. **Infect Immun** 75: 417-428.

31. Komatsu N, Oda T, Muramatsu T (**1998**) Involvement of Both Caspase-Like Proteases and Serine Proteases in Apoptotic Cell Death Induced by Ricin, Modeccin, Diphtheria Toxin, and Pseudomonas Toxin. **J Biochem** 124: 1038-1044.

32. Fujii J, Matsui T, Heatherly DP, Schlegel KH, Lobo PI, Yutsudo T, Ciraolo GM, Morris RE, Obrig T (**2003**) Rapid Apoptosis Induced by Shiga Toxin in HeLa Cells. **Infect Immun** 71: 2724-2735.

33. Zhang C, Gong Y, Ma H, An C, Chen D, Chen ZL (**2001**) Reactive oxygen species involved in trichosanthin-induced apoptosis of human choriocarcinoma cells. **The Biochem j** 355: 653-661.

34. Hückelhoven R (**2004**) BAX Inhibitor-1, an ancient cell death suppressor in animals and plants with prokaryotic relatives. **Apoptosis** 9: 299-307.

35. Kawai M, Pan L, Reed JC, Uchimiya H (**1999**) Evolutionarily conserved plant homologue of the Bax Inhibitor-1 (BI-1) gene capable of suppressing Bax-induced cell death in yeast. **FEBS Lett** 464: 143-147.

36. Matsumura H, Nirasawa S, Kiba A, Urasaki N, Saitoh H, Ito M, Kawai-Yamada M, Uchimiya H, Terauchi R (**2003**) Overexpression of Bax inhibitor suppresses the fungal elicitor-induced cell death in rice (*Oryza sativa* L.) cells. **The Plant J** 33: 425-434.

37. Sanchez P, De Torres Zabala M, Grant M (**2000**) AtBI-1, a plant homologue of Bax Inhibitor-1, suppresses Bax-induced cell death in

yeast and is rapidly upregulated during wounding and pathogen challenge. **The Plant J** 21: 393-399.

38. Bolduc N, Brisson LF (**2002**) Antisense down regulation of NtBI-1 in tobacco BY-2 cells induces accelerated cell death upon carbon starvation. **FEBS Lett** 532: 111-114.

39. Eichmann R, Schultheiss H, Kogel K-H, Hückelhoven R (**2004**) The Barley Apoptosis Suppressor Homologue Bax Inhibitor-1 Compromises Nonhost Penetration Resistance of Barley to the Inappropriate Pathogen *Blumeria graminis* f. sp. tritici. **Mol Plant-Microbe Interactions** 17: 484-490.

40. Chae H-J, Kim H-R, Xu C, Bailly-Maitre B, Krajewska M, Krajewski S, Banares S, Cui J, Digicaylioglu M, Ke N, Kitada S, Monosov E, Thomas M, Kress CL, Babendure JR, Tsien RY, Lipton SA, Reed JC (**2004**) BI-1 Regulates an Apoptosis Pathway Linked to Endoplasmic Reticulum Stress. **Mol Cell** 15: 355-366.

41. Kawai-Yamada M, Jin L, Yoshinaga K, Hirata A, Uchimiya H (**2001**) Mammalian Bax-induced plant cell death can be down-regulated by overexpression of Arabidopsis Bax Inhibitor-1 (AtBI-1). **Proc Natl Acad Sci USA** 98: 12295-12300.

42. Chae H-J, Ke N, Kim H-R, Chen S, Godzik A, Dickman M, Reed JC (2003) Evolutionarily conserved cytoprotection provided by Bax Inhibitor-1 homologs from animals, plants, and yeast. **Gene** 323: 101-113.

43. Kawai-Yamada M, Ohori Y, Uchimiya H (**2004**) Dissection of Arabidopsis Bax Inhibitor-1 Suppressing Bax-, Hydrogen Peroxide-, and Salicylic Acid-Induced Cell Death. **The Plant Cell** Online 16: 21-32.

44. Baek D, Nam J, koo YD, kim DH, Lee J, jeong JC, Kwak S-s, chung WS, lim CO, bahk JD, hong JC, lee SY, Kawai-yamada M, Uchimiya H, Yun D-j (2004) Bax-induced cell death of *Arabidopsis* is meditated through reactive oxygen-dependent and -independent processes. **Plant Mol Biol** 56: 15-27.

45. Yoshinaga K, Arimura S-i, Hirata A, Niwa Y, Yun D-J, Tsutsumi N, Uchimiya H, Kawai-Yamada M (**2005**) Mammalian Bax initiates plant cell death through organelle destruction. **Plant Cell Rep** 24: 408-417.

46. Watanabe N, Lam E (**2006**) Arabidopsis Bax inhibitor-1 functions as an attenuator of biotic and abiotic types of cell death. **The Plant J** 45: 884-894.

47. Watanabe N, Lam E (**2008**) BAX Inhibitor-1 Modulates Endoplasmic Reticulum Stress-mediated Programmed Cell Death in *Arabidopsis*. **J Biol Chem** 283: 3200-3210.

48. Isbat M, Zeba N, Kim SR, Hong CB (**2009**) A BAX inhibitor-1 gene in *Capsicum annuum* is induced under various abiotic stresses and endows multi-tolerance in transgenic tobacco. **J Plant Physiol** 166: 1685-1693.

49. Oshima R, Yoshinaga K, Ihara-Ohori Y, Fukuda R, Ohta A, Uchimiya H, Kawai-Yamada M (**2007**) The Bax Inhibitor-1 needs a functional electron transport chain for cell death suppression. **FEBS Lett** 581: 4627-4632.

50. Hur Y, Hwang DJ, Zoubenko O, Coetzer C, Uckun FM, Tumer NE (**1995**) Isolation and characterization of pokeweed antiviral protein mutations in *Saccharomyces cerevisiae*: identification of residues important for toxicity. **Proc Natl Acad Sci USA** 92: 8448-8452.

51. Hudak KA, Parikh BA, Di R, Baricevic M, Santana M, Seskar M, Tumer NE (**2004**) Generation of pokeweed antiviral protein mutations in Saccharomyces cerevisiae: evidence that ribosome depurination is not sufficient for cytotoxicity. **Nuc Acids Res** 32: 4244-4256.

52. Büttner S1, Ruli D, Vögtle FN, Galluzzi L, Moitzi B, Eisenberg T, Kepp O, Habernig L, Carmona-Gutierrez D, Rockenfeller P, Laun P,

Breitenbach M, Khoury C, Fröhlich KU, Rechberger G, Meisinger C, Kroemer G, Madeo F (2011) A yeast BH3-only protein mediates the mitochondrial pathway of apoptosis. **EMBO J.** 30(14):2779-92.

53. Oberhammer F, Wilson JW, Dive C, Morris ID, Hickman JA, E WA, Walker PR, Sikorska M (**1993**) Apoptotic death in epithelial cells: cleavage of DNA to 300 and/or 50 kb fragments prior to or in the absence of internucleosomal fragmentation. **EMBO J** 12: 3679-3684.

54. Ucker DS (**1991**) Death by suicide: one way to go in mammalian cellular development? **New Biol** 3: 103-109.

55. Madeo F, Herker E, Wissing S, Jungwirth H, Eisenberg T, Fröhlich K-U (2004) Apoptosis in yeast. Curr Opin Microbiol 7: 655-660.

56. Takayama S, Sato T, Krajewski S, Kochel K, Irie S, Milian JA, Reed JC (**1995**) Cloning and functional analysis of BAG-1: A novel Bcl-2-binding protein with anti-cell death activity. **Cell** 80: 279-284.

57. Ligr M, Madeo F, Frˆhlich E, Hilt W, Frˆhlich K-U, Wolf DH (**1998**) Mammalian Bax triggers apoptotic changes in yeast. **FEBS Lett** 438: 61-65.

58. Suzuki K, Nakamura M, Hatanaka Y, Kayanoki Y, Tatsumi H, Taniguchi N (**1997**) Induction of Apoptotic Cell Death in Human Endothelial Cells Treated with Snake Venom: Implication of Intracellular Reactive Oxygen Species and Protective Effects of Glutathione and Superoxide Dismutases. **J Biochem** 122: 1260-1264.

59. Jabs T (**1999**) Reactive oxygen intermediates as mediators of programmed cell death in plants and animals. **Biochem Pharm** 57: 231-245.

60. Madeo F, Fröhlich E, Ligr M, Grey M, Sigrist SJ, Wolf DH, Fröhlich K-U (**1999**) Oxygen Stress: A Regulator of Apoptosis in Yeast. **The J Cell Biol** 145: 757-767.

61. Parikh BA, Coetzer C, Tumer NE (**2002**) Pokeweed Antiviral Protein Regulates the Stability of Its Own mRNA by a Mechanism That Requires Depurination but Can Be Separated from Depurination of the alpha-Sarcin/Ricin Loop of rRNA. **J Biol Chem** 277: 41428-41437.

62. Frey S, Pool M, Seedorf M (**2001**) Scp160p, an RNA-binding, Polysome-associated Protein, Localizes to the Endoplasmic Reticulum of Saccharomyces cerevisiae in a Microtubule-dependent Manner. **J Biol Chem** 276: 15905-15912.

63. Parikh BA, Baykal U, Di R, Tumer NE (**2005**) Evidence for Retro-Translocation of Pokeweed Antiviral Protein from Endoplasmic Reticulum into Cytosol and Separation of Its Activity on Ribosomes from Its Activity on Capped RNA.. **Biochem** 44: 2478-2490.

64. Di R and Tumer NE (**2005**) Expression of a Truncated Form of Ribosomal Protein L3 Confers Resistance to Pokeweed Antiviral Protein and the Fusarium Mycotoxin Deoxynivalenol. **MPMI** 18 (8): 762–770.

65. Nicolas E, Beggs JM, Haltiwanger BM, Taraschi TF (**1998**) A New Class of DNA Glycosylase/Apurinic/Apyrimidinic Lyases That Act on Specific Adenines in Single-stranded DNA. **J Biol Chem** 273: 17216-17220.

66. Fahrenkrog B, Sauder U, Aebi U (**2004**) The *S. cerevisiae* HtrA-like proteinNma111p is a nuclear serine protease that mediates yeast apoptosis. **J Cell Science** 117: 115-126.

67. Atlas RM (**2002**) BIOTERRORISM: From Threat to Reality. **Ann Rev Microbiol** 56: 167-185.

68. Brigotti M, Alfieri R, Sestili P, Bonelli M, Petronini P, Guidarelli A, Barbieri L, Stirpe F, Sperti S (**2002**) Damage to nuclear DNA induced by Shiga toxin 1 and ricin in human endothelial cells. **The FASEB J** 16: 365-372.

69. Rao PVL, Jayaraj R, Bhaskar ASB, Kumar O, Bhattacharya R, Saxena P, Dash PK, Vijayaraghavan R (**2005**) Mechanism of ricin-induced apoptosis in human cervical cancer cells. **Biochem Pharm** 69: 855-865.

70. Xu Q, Reed JC (**1998**) Bax Inhibitor-1, a Mammalian Apoptosis Suppressor Identified by Functional Screening in Yeast. **Mol Cell** 1: 337-346.

71. Xu Q, Jürgensmeier JM, Reed JC (**1999**) Methods of Assaying Bcl-2 and Bax Family Proteins in Yeast. **Methods** 17: 292-304.

72. Parikh BA, Tumer NE (**2004**) Antiviral Activity of Ribosome Inactivating Proteins in Medicine. **Mini-Rev Medicin Chem** 4: 523-543.

73. Parikh B, Tortora A, Li X-P, Tumer N (**2008**) Ricin Inhibits Activation of the Unfolded Protein Response by Preventing Splicing of the *HAC1* mRNA. **J Biol Chem**: 6145-6153.

74. Cakir B (**2012**) Bax induces activation of the unfolded protein response by inducing *HAC1* mRNA splicing in *Saccharomyces cerevisiae*. **Yeast** 29(9):395-406.

75. Cebulski J, Malouin J, Pinches N, Cascio V, Austriaco N (**2011**) Yeast Bax Inhibitor, Bxi1p, is an ER-localized protein that links the unfolded protein response and programmed cell death in *Saccharomyces cerevisiae*. **PLoS One** 6(6):e20882.

76. Hudak KA, Dinman JD, Tumer NE (**1999**) Pokeweed Antiviral Protein Accesses Ribosomes by Binding to L3. **J Biol Chem** 274: 3859-3864.

77. Otto J, Lee S-W (**1993**) Immunoprecipitation methods. **Methods Cell Biol** 37119-37126.

Permissions

All chapters in this book were first published in MIC, by Shared Science Publishers OG; hereby published with permission under the Creative Commons Attribution License or equivalent. Every chapter published in this book has been scrutinized by our experts. Their significance has been extensively debated. The topics covered herein carry significant findings which will fuel the growth of the discipline. They may even be implemented as practical applications or may be referred to as a beginning point for another development.

The contributors of this book come from diverse backgrounds, making this book a truly international effort. This book will bring forth new frontiers with its revolutionizing research information and detailed analysis of the nascent developments around the world.

We would like to thank all the contributing authors for lending their expertise to make the book truly unique. They have played a crucial role in the development of this book. Without their invaluable contributions this book wouldn't have been possible. They have made vital efforts to compile up to date information on the varied aspects of this subject to make this book a valuable addition to the collection of many professionals and students.

This book was conceptualized with the vision of imparting up-to-date information and advanced data in this field. To ensure the same, a matchless editorial board was set up. Every individual on the board went through rigorous rounds of assessment to prove their worth. After which they invested a large part of their time researching and compiling the most relevant data for our readers.

The editorial board has been involved in producing this book since its inception. They have spent rigorous hours researching and exploring the diverse topics which have resulted in the successful publishing of this book. They have passed on their knowledge of decades through this book. To expedite this challenging task, the publisher supported the team at every step. A small team of assistant editors was also appointed to further simplify the editing procedure and attain best results for the readers.

Apart from the editorial board, the designing team has also invested a significant amount of their time in understanding the subject and creating the most relevant covers. They scrutinized every image to scout for the most suitable representation of the subject and create an appropriate cover for the book.

The publishing team has been an ardent support to the editorial, designing and production team. Their endless efforts to recruit the best for this project, has resulted in the accomplishment of this book. They are a veteran in the field of academics and their pool of knowledge is as vast as their experience in printing. Their expertise and guidance has proved useful at every step. Their uncompromising quality standards have made this book an exceptional effort. Their encouragement from time to time has been an inspiration for everyone.

The publisher and the editorial board hope that this book will prove to be a valuable piece of knowledge for researchers, students, practitioners and scholars across the globe.

List of Contributors

Harriet Allison, Amanda J. O'Reilly and Mark C. Field
Division of Biological Chemistry and Drug Discovery, University of Dundee, Dundee, Scotland, DD1 5EH

Jeremy Sternberg
School of Biological Sciences, University of Aberdeen, Aberdeen, AB24 2TZ, UK

Estéfani García-Ríos, Rosana Chiva and José Manuel Guillamon
Departamento de Biotecnología de los alimentos, Instituto de Agroquímica y Tecnología de los Alimentos (CSIC), Avda, Agustín Escardino, 7, E-46980-Paterna, Valencia, Spain

María López-Malo
Departamento de Biotecnología de los alimentos, Instituto de Agroquímica y Tecnología de los Alimentos (CSIC), Avda, Agustín Escardino, 7, E-46980-Paterna, Valencia, Spain
Biotecnologia Enològica. Departament de Bioquímica i Biotecnologia, Facultat d'Enologia, Universitat Rovira i Virgili, Marcelli Domingo s/n, 43007, Tarragona, Spain

António Rego, Ana Marta Duarte, Flávio Azevedo, Maria João Sousa, Manuela Côrte-Real and Susana R. Chaves
Centro de Biologia Molecular e Ambiental, Departamento de Biologia, Universidade do Minho, Braga, Portugal

Viktor Scheidt, André Jüdes, Roland Klassen and Raffael Schaffrath
Institut für Biologie, Abteilung Mikrobiologie, Universität Kassel, D-34132 Kassel, Germany

Christian Bär
Institut für Biologie, Abteilung Mikrobiologie, Universität Kassel, D-34132 Kassel, Germany
Molecular Oncology Program, Spanish National Cancer Centre (CNIO), Melchor Fernandez Almagro 3, Madrid, Spain

Fazal Shirazi and Dimitrios P. Kontoyiannis
Department of Infectious Diseases, Infection Control and Employee Health, The University of Texas M.D. Anderson Cancer Center, Houston, TX 77030, U.S.A

Shivatheja Soma, Kailu Yang, Maria I. Morales and Michael Polymenis
Department of Biochemistry and Biophysics, Texas A&M University, College Station, TX 77843, USA

Maksim I. Sorokin
Faculty of Bioengineering and Bioinformatics, Moscow State University, Vorobyevy Gory 1, Moscow, Russia
Institute of Mitoengineering, Moscow State University, Vorobyevy Gory 1, Moscow, Russia

Dmitry A. Knorre and Fedor F. Severin
Belozersky Institute of Physico-Chemical Biology, Moscow State University, Vorobyevy Gory 1, Moscow, Russia
Institute of Mitoengineering, Moscow State University, Vorobyevy Gory 1, Moscow, Russia

R. Roshini Beenukumar and R. Jürgen Dohmen
Institute for Genetics, University of Cologne, Biocenter, Zülpicher Str. 47a, D-50674 Cologne, Germany

Daniela Gödderz
Institute for Genetics, University of Cologne, Biocenter, Zülpicher Str. 47a, D-50674 Cologne, Germany
Karolinska Institute, Department for Cell- and Molecular Biology, Von Eulers väg 3, 171 77 Stockholm

R. Palanimurugan
Institute for Genetics, University of Cologne, Biocenter, Zülpicher Str. 47a, D-50674 Cologne, Germany
Center for Cellular and Molecular Biology (CCMB), Uppal Road, Hyderabad 500007, India

Donna Garvey Brickner, Robert Coukos and Jason H. Brickner
Department of Molecular Biosciences, Northwestern University, Evanston, IL USA 60201

Hazel Xinyu Koh
Department of Biological Sciences, National University of Singapore
Department of Microbiology, National University of Singapore

Htay Mon Aye and Cynthia Y. He
Department of Biological Sciences, National University of Singapore

Kevin S. W. Tan
Department of Microbiology, National University of Singapore

Sofiane Y. Mersaoui, Serge Gravel, Victor Karpov and Raymund J. Wellinger
Dept of Microbiology and Infectious Diseases, Faculty of Medicine and Health Sciences, Université de Sherbrooke, 3201, Rue Jean Mignault, Sherbrooke, J1E 4K8, Canada

Fazal Shirazi and Dimitrios P. Kontoyiannis
Department of Infectious Diseases, Infection Control
and Employee Health, The University of Texas M D
Anderson Cancer Center, Houston, TX, USA

Russel E. Lewis
Department of Infectious Diseases, Infection Control
and Employee Health, The University of Texas M D
Anderson Cancer Center, Houston, TX, USA
Department of Medical and Surgical Sciences,
University of Bologna, Bologna, Italy

**Miguel Esteras, Adam Jarmuz and Luis Aragon and
I-Chun Liu**
Cell Cycle Group, MRC London Institute of Medical
Sciences, Du Cane Road, London W12 0NN, UK

Ambrosius P. Snijders
Protein Analysis and Proteomics Platform, The Francis
Crick Institute, 1 Midland Road, London NW1 1AT, UK

**Isabel González-Mariscal, Aléjandro Martín-
Montalvo, Cristina Ojeda-González, Adolfo
Rodríguez-Eguren, Purificación Gutiérrez-Ríos,
Plácido Navas and Carlos Santos-Ocaña**
Centro Andaluz de Biología del Desarrollo, Universidad
Pablo de Olavide-CSIC, CIBERER Instituto de Salud
Carlos III, Sevilla, 41013, Spain

Shaun H. Speldewinde and Chris M. Grant
University of Manchester, Faculty of Biology, Medicine
and Health, The Michael Smith Building, Oxford Road,
Manchester, M13 9PT, UK

**Slaven Stekovic, Christoph Ruckenstuhl, Philipp
Royer, Christof Winkler-Hermaden, Didac Carmona-
Gutierrez and Kai-Uwe Fröhlich**
Institute of Molecular Biosciences, NAWI Graz,
University of Graz, 8010 Graz, Austria

Frank Madeo
Institute of Molecular Biosciences, NAWI Graz,
University of Graz, 8010 Graz, Austria
BioTechMed Graz, Austria

Guido Kroemer
Equipe 11 labellisée par la Ligue contre le Cancer,
Centre de Recherche des Cordeliers, Paris, France
INSERM, U1138, Paris, France
Université Paris Descartes, Sorbonne Paris Cité, Paris,
France
Cell Biology & Metabolomics Platforms, Gustave
Roussy Comprehensive Cancer Center, Villejuif, France
Pôle de Biologie, Hôpital Européen Georges Pompidou,
AP-HP, Paris, France

Karolinska Institute, Department of Women's and
Children's Health, Karolinska University Hospital,
17176 Stockholm, Sweden

Qiuqiang Gao
State Key Laboratory of Bioreactor Engineering, East
China University of Science and Technology, 130
Meilong Road, Shanghai 200237, China

Liang-Chun Liou and Qun Ren and Zhaojie Zhang
Department of Zoology and Physiology, University of
Wyoming, Laramie, WY 82071, USA

Xiaoming Bao
State Key Laboratory of Microbial Technology,
Shandong University, Jinan, China

**Marizela Delic, Diethard Mattanovich and Brigitte
Gasser**
Department of Biotechnology, BOKU University of
Natural Resources and Life Sciences Vienna, Vienna,
Austria
Austrian Centre of Industrial Biotechnology (ACIB),
Vienna, Austria

Alexandra B. Graf
Austrian Centre of Industrial Biotechnology (ACIB),
Vienna, Austria
School of Bioengineering, University of Applied
Sciences FH Campus Wien, Vienna, Austria

**Gunda Koellensperger, Christina Haberhauer-Troyer
and Stephan Hann**
Department of Biotechnology, BOKU University of
Natural Resources and Life Sciences Vienna, Vienna,
Austria
Department of Chemistry, BOKU University of Natural
Resources and Life Sciences Vienna, Vienna, Austria

Nilgun E. Tumer
Biotechnology Center for Agriculture and the
Environment and the Department of Plant Biology
and Pathology, Rutgers University, New Brunswick,
NJ 08901-8520, USA

Birsen Çakır
Biotechnology Center for Agriculture and the
Environment and the Department of Plant Biology
and Pathology, Rutgers University, New Brunswick,
NJ 08901-8520, USA
Department of Horticulture, Faculty of Agriculture,
Ege University, Izmir, Turkey

Index